1969

University of St. Francis Library
3 0301 00202435 0

D1547557

HANDBOOKS OF AMERICAN NATURAL HISTORY

ALBERT HAZEN WRIGHT, ADVISORY EDITOR

# Handbook of Salamanders

BY SHERMAN C. BISHOP

Texas Newt, *Triturus meridionalis* (Cope), Harlingen, Texas. Painted by Helen R. Bassett.

# Handbook of SALAMANDERS

The Salamanders of the United States, of Canada, and of Lower California

BY

SHERMAN C. BISHOP

Comstock Publishing Associates

A DIVISION OF

CORNELL UNIVERSITY PRESS

ITHACA, NEW YORK

FIRST PUBLISHED NOVEMBER 1943
SECOND PRINTING OCTOBER 1947
REISSUED 1967



PRINTED IN THE UNITED STATES OF AMERICA
BY THE VAIL-BALLOU PRESS, INC.

TO

ELMER J. BOND

AN EXCELLENT TEACHER AND FRIEND
WHOSE INTEREST AND SYMPATHETIC UNDERSTANDING
HELPED MANY A YOUTH TOWARD SOME LINE
OF USEFUL ENDEAVOR

# PREFACE

It is hoped that this Handbook may be of help to the growing number of persons interested in salamanders and perhaps lead some of them to more detailed studies of many of the forms than I have been able to make. In the preparation of the book I have drawn freely on published information but have been fortunate in being able to know most of the species and subspecies in the field. The majority of existing type specimens have been seen at one time or another, and collections have been made at a considerable number of type localities.

Many individuals and institutions have permitted the study of pertinent material, and I am glad to acknowledge my indebtedness. In this connection, thanks should go to Thomas Barbour and Arthur Loveridge, Museum of Comparative Zoology, Cambridge; Charles M. Bogert, American Museum of Natural History; C. S. Brimley and Roxie Collie Simpson, North Carolina State Museum; Archie F. Carr and Coleman J. Goin, University of Florida; E. B. Chamberlain, Charleston Museum; Doris M. Cochran, U. S. National Museum; E. R. Dunn, Haverford College; Helen T. Gaige, Museum of Zoology, University of Michigan; H. T. Gier, Ohio University; Francis Harper, Swarthmore, Pa.; L. M. Klauber, Zoological Society of San Diego; Charles E. Mohr, Philadelphia Academy; George A. Moore, Oklahoma A. and M. College; George S. Myers, V. C. Twitty, and Anita Daugherty, Stanford University; Leo T. Murray, Baylor University; M. G. Netting, Carnegie Museum; J. R. Slater, College of Puget Sound; Joseph R. Slevin, California Academy of Sciences; Edward S. Thomas, Ohio State Museum; Charles F. Walker, Ohio State University; and A. H. Wright and Harold Trapido, Cornell University.

For their efforts to secure particular species to complete the series of photographs I am particularly indebted to Dr. and Mrs. A. H. Wright, Mr. Joseph R. Slevin, Mr. M. B. Mittleman, Mr. N. B. Green, Mr. George Wilmott, Dr. V. C. Twitty and Catherine H. Brown. Students and associates have been most helpful. As companions in the field I have had the late Professor C. R. Crosby and Mrs. Crosby, Margaret R. Wright, Barbara H. Leonard, Hobart M. Smith, Arnold B. Grobman, and Joseph A. Tihen. Help in the preparation of the distribution maps was graciously given by Arnold B. Grobman, Hobart M. Smith, Barbara H. Leonard, Alice S. Bishop, and Beth S. Bishop.

The majority of the photographs are original; a few have appeared in earlier reports of the writer. A considerable number were made in the field under varying and often trying conditions. A number of fine photographs have been contributed and acknowledgment is made at appropriate places. Of particular value for their fine detail is the series by Arthur L. Smith of Cornell University. Especial thanks should go to Raymond Maas, technician of the Department of Zoology, University of Rochester, for his patience and skillful help. Many line drawings and several excellent water colors were made by Mr. Hugh P. Chrisp. Other line drawings were made by Marion L. Stalker and Helen M. Zorsch. The frontispiece in color is a drawing from life by Helen R. Bassett. Finally, I would probably not have been able to complete the work at this time except for the consideration shown me by Dr. Curt Stern, who generously supported my activities and helped greatly by easing a teaching schedule over a considerable period.

S. C. Bishop

*Rochester, New York*
*March, 1943*

# CONTENTS

PART I

## INTRODUCTION



PART II

## ACCOUNTS OF SPECIES

# PART I

# INTRODUCTION

PART I

# INTRODUCTION

The salamanders, the familiar frogs and toads, the less familiar cæcilians, and a number of fossil creatures comprise the Class Amphibia. The living Amphibia are cold-blooded animals which possess a glandular skin but lack a surface covering of scales, feathers, or fur. Some of the worm-like, burrowing cæcilians have minute scales imbedded in the skin, which may be found by dissection, but their presence does not invalidate the general rule that recent Amphibia have naked skins. Salamanders possess tails; cæcilians, as adults, have very short tails and no limbs, and the frogs and toads, as adults, are completely tailless.

In their development, amphibians pass through a larval, gill-bearing stage, and in this respect they differ from the reptiles, birds, and mammals. The eggs, which may be laid on land in moist situations, or in the water, lack a calcareous shell, and the young develop without the benefit of the protective embryonic membranes common to higher vertebrates. Some salamanders resemble certain of the reptiles in body form, the lizards, but the similarity is superficial, for lizards possess scales, claws on the toes, and—many of them—an external ear opening, structures lacking in the Amphibia.

The salamanders comprise the Order Caudata, the name having reference to the tailed condition. The group has had a long ancestry and the fossil types from which they may well have been derived constitute the order Phyllospondyli, the branchiosaurs and their relatives. These comparatively small, salamander-like creatures have been found in numbers in deposits of Carboniferous and Permian age and, recently, in the still older Devonian of Greenland. The recent salamanders retain many of the structural features of their ancient ancestors and the modifications that have occurred mainly involve the skull and elements of the pectoral girdle.

This Handbook includes all the species and subspecies from the United States, Canada, and Lower California which I regard as valid. Altogether seven families are represented, the number of species and subspecies in each indicated in tabular form below:

|  |  |
|---|---|
| Proteidae, the Mudpuppies | 7 species and subspecies |
| Amphiumidae, the Congo-eels | 2 subspecies |
| Cryptobranchidae, the Hellbenders | 2 species |
| Salamandridae, the Newts | 12 species and subspecies |
| Ambystomidae, the Blunt-mouthed salamanders | 19 species and subspecies |
| Plethodontidae, the Lungless salamanders | 79 species and subspecies |
| Sirenidae, the Sirens | 5 species and subspecies |

## RELATIONSHIPS

The relationships of the families are not in every instance clearly understood. The Proteidae and Sirenidae are somewhat aberrant groups, the adults of which have retained certain larval structures and in addition possess a number of distinctive characteristics not shared by other salamanders. The members of the Cryptobranchidae, Salamandridae, Amphiumidae, Ambystomidae, and Plethodontidae, while often differing markedly in appearance, present an ensemble of structural features which probably indicates common ancestry. On the basis of the character of the structures associated with reproduction, a somewhat different alignment would be made. The members of Cryptobranchidae, their Asiatic relatives the Hynobiidae, and, quite possibly, the Sirenidae, have external fertilization and the females lack the cloacal sperm receptacles which are present in all other families. External fertilization is probably the primitive method, but the Sirenidae, while retaining this method, are considerably advanced in some other respects. The members of Proteidae, otherwise differing from the remaining families in certain skeletal features, possess the rather highly specialized devices associated with internal fertilization, the males having cloacal glands, the females, sperm receptacles, the spermathecae. The principal characteristics of the several families are indicated at appropriate places in the text.

#### GENERAL HABITS

Salamanders may be terrestrial, aquatic, or at home in either environment. A few have become arboreal and many are efficient burrowers. Many plethodontids are terrestrial throughout life, a few entirely aquatic, and some seasonally adapted to one or the other type of habitat. Many newts have a terrestrial interlude in an otherwise essentially aquatic life. Most adult ambystomids are mainly terrestrial except during the breeding season, when they resort to quiet waters for mating and egg-laying. The cryptobranchids, proteids, sirenids, and amphiumids are aquatic types, although some, on occasion, may inhabit burrows and abandon the water temporarily.

Most salamanders are nocturnal or at least avoid direct light. Newts and some species of *Desmognathus* are commonly found abroad during the day, but these are exceptions to the general rule. The larvae of a number of salamanders, notably the ambystomids, are often active in the open while the adults of the same species may be in hiding. During the breeding season, species which are usually secretive may sometimes be found in exposed situations. Members of the family Sirenidae are sometimes taken while swimming freely in open water or among vegetation, even in bright sunlight. Among the species that perform a definite migration, either to the breeding places in the spring or to places of hibernation in the fall, are some that will expose themselves to light, although ordinarily they avoid it.

The food of most salamanders consists of living animals and a great many different types are taken. A few, such as *Cryptobranchus,* will sometimes gorge themselves on animal refuse, and some, *Gyrinophilus* and *Amphiuma,* are often cannibalistic. *Siren,* which usually feeds on crayfish, shrimp, and other animal types, will occasionally fill itself with filamentous algae.

Most terrestrial salamanders in the North are inactive during the winter and some may spend months in hibernation. Most *Ambystoma* and the terrestrial members of the family Plethodontidae disappear with the coming of cold weather, burrowing beneath surface materials or occupying crevices or other openings in shale, rock, or soil. The more aquatic plethodontids may be found throughout the year in or along the margins of springs or spring-fed streams. *Desmognathus o. ochrophæus,* which is essentially terrestrial during the warmer months, resorts to the saturated soil of springy banks for the winter, where they may remain more or less active even in extremely cold weather. Newts may hibernate in upland terrestrial situations or in the mud of a drying pond, or may remain active and feed all winter in ponds and springs. *Necturus* is sometimes found beneath the ice of bays and ponds, crawling about on the bottom, and is occasionally captured by fishermen. *Cryptobranchus,* although less active, can be found in winter beneath sheltering objects on the stream bottom and will swim away leisurely if disturbed.

In exceptionally dry seasons in the North, some salamanders burrow and remain hidden until the return of the rains. *Plethodon glutinosus* is particularly susceptible to drying and will burrow deeply during a drought. *Plethodon cinereus,* while less sensitive than *P. glutinosus,* will sometimes disappear for weeks in the absence of rainfall. In arid regions of the Southwest, some species are found only during the rainy season and will either burrow to reach moist situations or æstivate during the hot dry periods. *Siren* has been found æstivating in burrows in a sandy field in Texas.

Some salamanders perform seasonal migrations with considerable regularity. This is true of most ambystomids which enter the water in late winter or early spring for mating and egg-laying. The common spotted newt, which usually has a terrestrial stage in its life history, may return to the water in the fall or in the spring, often in concerted movements involving many individuals. Ambystomids which retire from the breeding ponds after the eggs are laid often migrate in the fall seeking suitable places in which to pass the winter. *A. maculatum* and *A. tigrinum* are sometimes discovered at considerable depths, but *A. jeffersonianum* may pass the winter beneath surface materials and withstand freezing temperatures.

In striking contrast to the frogs and toads, whose calls may serve for recognition marks as characteristic as those of birds, the salamanders are essentially voiceless. It is true that some are capable of producing various clicking noises, squeaks, yelps, or subdued bleats, but these can scarcely be regarded as true voice. Lacking vocal cords, the production of sound is, in some instances at least, mechanical, as when air is forced from the throat under stress of handling. *Siren* and its smaller relative *Pseudobranchus* sometimes voluntarily produce a faint yelping which may betray them in their retreats in the muck of a drying pond bottom. How this sound is produced has not been explained. It is *not* true that the common spotted newt *Triturus viridescens,* or the red salamander *Pseudotriton ruber,* calls like a *Hyla,* the several published records to the contrary notwithstanding.

COURTSHIP PATTERNS. Much remains to be learned about the courtship of salamanders, and only a few general statements can be made concerning certain groups. It is known that under favorable circumstances several species of *Ambystoma* participate in a kind of nuptial dance, the so-called *liebesspiel.* Usually males precede the females to the breeding waters and sometimes congregate in large numbers in limited areas. Here they engage in swimming and diving, frequently rising to the surface to gulp in air. They rub and nose one another and thresh around generally until the water fairly boils. The males, which often predominate in numbers at the beginning of the dance, are frequently joined by females until a hundred or more individuals may be participating. The purpose of the dance seems to be to stimulate the sexes for the mating activities which normally follow. The aroused male deposits a series of spermatophores, small stump-shaped structures of jelly, each surmounted by a cap of sperm, and any interested female may secure the sperm by removing the cap from the base of the spermatophore with the lips of her cloaca. The male of *Ambystoma jeffersonianum* may actually grasp the female, holding her in firm embrace for a time, usually employing his fore legs in front, or immediately behind, those of the female. After this action, the sexes separate, the male to deposit spermatophores and the female to recover the sperm-sac, if she has been sufficiently stimulated. Among the newts of the genus *Triturus,* mating activities may be even more elaborate. The male develops dark, horny ridges on the thighs and tips of the toes of the hind legs, pursues the female, grasps her firmly in the region of the fore limbs or neck, and wrestles her about vigorously. At times he presses his neck glands against her snout and vibrates the tip of his tail, which is directed forward, apparently to waft the secretions of his cloacal glands toward the head of his mate. As in the Ambystomids, the male newt produces spermatophores, and these are broad of base but with a slender spine-like apex on which the sperm-sac rests. In certain species of Plethodontidae, e. g. *Eurycea bislineata,* the male rubs and noses the female much as in *Triturus,* and also presses the side of his head against her snout; but in *Eurycea* and some other plethodontid genera, the glands which provide the stimulus are not limited to the side of the neck but are found on the snout, about the eyes, and on the tail. These are brought into play in various ways. In *Eurycea* the stimulated female straddles the tail of the male and presses her chin against the gland at the base of the tail of the male. Then follows a curious walk which Noble (1929, p. 3) has called the "tail-walk"; the pair waddle off together, the male wagging his tail from side to side and the female following along with her head turning in the opposite direction at each bend of the tail of the male. Essentially the same kind of walk is practiced by pairs of *Hemidactylium,* in this case on land instead of in water as in the case of *Eurycea.* Only a single observation has been made on *Gyrinophilus.* In this instance, the male grasped the female by the side of her head with his jaws and attempted to present his cloaca to her snout. The female in turn held the body of the male in front of the right hind leg and both turned and twisted, often as not upside down. The male *Necturus m. maculosus* engages in a kind of courtship which involves swimming and crawling about the female, frequently passing over her tail and between her legs. The female, in the only recorded observation, remained rather passive, held herself erect on the hind legs and tail, and made no attempt to escape.

FERTILIZATION AND CARE OF THE EGGS. As indicated in a paragraph above, Cryptobranchidae and Hynobiidae practice external fertilization, and anatomical evidence indicates that this is true of the Sirenidae. All other salamanders, so far as is known, have internal fertilization. In the external method, the eggs are inseminated as they are extruded; internal fertilization involves the production of spermatophores by the male and the storage of sperm in receptacles of the female cloaca. Eggs of salamanders are of two general types, pigmented and with a small vitellus, and non-pigmented with a large vitellus. The anatomically primitive members of several groups lay small, pigmented eggs which are abandoned in water, and this method, very likely, is the primitive one. A possible exception is found in Sirenidae, which, although specialized in some respects structurally, nevertheless retains the apparently primitive method of reproduction. The pigmentless eggs with large vitellus are usually deposited in "nests" or hidden in some way and attended by one parent or the other; the pigmented eggs are usually in more open situations. Here again the sirenids are an exception, for the eggs have been found in clumps in cavities in the mud of a pond bottom and attended by one of the parents. The more terrestrial plethodontids lay a comparatively small number of eggs which are usually accompanied by the female. The eggs may be laid in small clusters and attached by a pedicel to some support (*Plethodon cinereus* and *Desmognathus o. ochrophæus*), or deposited in a compact mass about which the parent coils (*Desmognathus f. fuscus* and *Hemidactylium scutatum*). The more aquatic species usually produce a larger number of eggs and attach them singly to the lower surface of a support in running water. This method is characteristic of *Eurycea b. bislineata, Gyrinophilus p. porphyriticus, Gyrinophilus d. danielsi* and *Pseudotriton r. ruber* of the family *Plethodontidae,* and of *Necturus m. maculosus* of the Proteidae. *Pseudotriton m. montanus,* on the other hand, which commonly lives in slow, muddy streams or leaf-filled trickles, lays her eggs singly, attaching them to leaves or debris on the bottom; and this method is also employed by *Manculus quadridigitatus.* A number of salamanders deposit the eggs in long rosary-like strings, one from each oviduct (*Cryptobranchus* in water, the male in attendance, and *Amphiuma* on land); others, *Stereochilus marginatus* and *Triturus,* attach the eggs singly among the leaves of water plants, often hiding them effectively. *Siren* and *Pseudobranchus* tuck the eggs away in pockets in the mud or muck of ponds of southern prairies. The eggs of the arboreal *Aneides lugubris* are frequently placed some distance above the ground in cavities in trees. The members of *Ambystoma* dispose their eggs in a variety of ways, the majority having an early-spring egg-laying season and depositing more or less globular masses, attached below the surface in quiet waters. The masses are distinguishable from those of frogs by the presence of a common envelope of jelly which encloses the whole. *Ambystoma opacum* deposits her eggs on land in the fall, in loose clusters not surrounded by a common envelope, instead of in the water during the early spring months. Some of the subspecies of *Ambystoma tigrinum* deposit the eggs in large clusters, while others attach them singly or in small groups.

LARVAE. Larvae of salamanders differ from those of frogs and toads in their possession of true teeth. They are of several types which are associated with particular environments. The larvae of the terrestrial species of *Plethodon,* and some others, may pass the entire period of development within the egg, emerging as fully transformed individuals; or they may retain the gills for a short time after hatching. These and certain other wholly terrestrial larvae are characterized by the possession of stag-horn gills, the three branches arising from a common base

(Fig. 1b), and the absence of fins on tail and back. *Plethodon cinereus, P. glutinosus* and *P. vandykei* provide familiar examples. Larvae living in quiet waters are usually provided with large, bushy gills and fins on

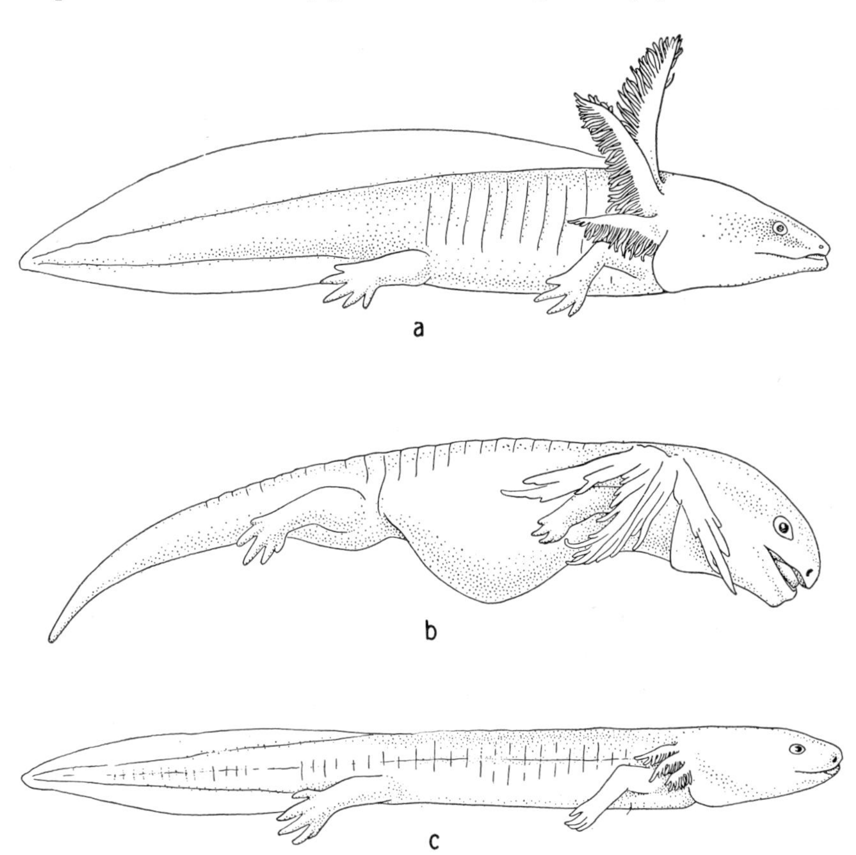


FIG. 1. (a) Pond-type larva of *Ambystoma tigrinum*. (b) Larva of terrestrial *Plethodon cinereus*. (c) Stream-type larva of *Gyrinophilus porphyriticus*. [M.L.S. *del.*]

the back and tail. This type is characteristic of *Ambystoma* (Fig. 1a), *Triturus,* and *Hemidactylium,* the last-named a terrestrial plethodontid with an aquatic larval period. Species which live in running water, and some more terrestrial forms which have evidently been derived from mountain stream ancestors, usually have short gills and the fin confined to the tail. *Dicamptodon ensatus* among the ambystomids, some species of *Desmognathus, Pseudotriton, Gyrinophilus* (Fig. 1c), and *Eurycea* of the family Plethodontidae, and *Cryptobranchus* fall into this category. *Necturus* is somewhat intermediate in character, possessing large bushy gills but having the fin confined to the tail.

NEOTENY. The retention of juvenile structures and habits by animals which become sexually mature is known as neoteny. The members of Sirenidae and Proteidae have been mentioned as exhibiting this condition and to these may be added the Cryptobranchidae and the Amphiumidae. Among the Ambystomidae the most notable example is the axolotl, *Siredon mexicana.* This larval type attains a large size and breeds regularly, but retains the gills, larval dentition, and fins on back and tail. While this is the usual condition in the axolotl, several species of *Ambystoma,* and *Dicamptodon ensatus,* which normally transform, will, under certain conditions, remain as larvae throughout life but carry on all adult functions. In certain of the high mountain ponds and lakes of Colorado, *Ambystoma* may transform into normal adults or continue as larvae, the percentage of transforming individuals varying from year to year as local conditions may dictate. In the family Salamandridae, some individuals of the common spotted newt, *Triturus v. viridescens,* are neotenic in the Coastal Plain from Massachusetts to New Jersey and probably elsewhere. Neotenic individuals of *T. perstriatus* and *T. v. louisianensis* are also known. In the family Plethodontidae, *Eurycea neotenes* and *E. nana* from Texas and *E. tynerensis* from Oklahoma are normally neotenic, while occasionally individuals of other species, which usually transform, will breed as larvae. The only known specimen of *Haideotriton* and all *Typhlomolge,* blind white salamanders of the underground waters of Georgia and Texas, are also neotenic.

### WHERE TO LOOK FOR SALAMANDERS

Salamanders should be looked for in all kinds of places. The terrestrial species will usually be found in greatest numbers where there is some moisture, but the apparently unsuitable situations should not be wholly neglected. Many species may be found by turning logs, pieces of bark, stones, and rock fragments. Well rotted logs should be pulled apart and the loose bark stripped from the trunks of standing and prostrate trees. Some species, such as *Aneides æneus* and *Plethodon glutinosus,* are often found in the crevices of rocky cliffs or shale banks, and *Plethodon wehrlei* beneath loose stones on the surface of the ground. *Aneides lugubris* occurs both on the ground and in cavities of trees. Many terrestrial species, which have an aquatic larval stage in their life history, congregate in large numbers in or near the breeding waters and advantage should be taken of such assemblages. In this category are many of the species of *Ambystoma. Hemidactylium scutatum,* although scattered widely during the greater part of the year, migrates in the spring to the vicinity of bogs, larch meadows, and woodland pools, where they may be found in moss, among the roots of grass, and occasionally beneath the bark of stumps or logs. Certain more or less aquatic species are adapted to life in or along the margins of springs and running streams, and in this kind of situation some species of *Eurycea, Pseudotriton, Gyrinophilus, Leurognathus,* and *Desmognathus* are often abundant. Strictly aquatic types, such as *Necturus* and *Cryptobranchus,* are most easily collected in the comparatively shallow water of running streams, where they may be found by day hiding beneath sheltering objects on the bottom; at night they often crawl openly over the bottom. In muddy streams, ditches, and bayous they may sometimes be taken by hook and line, and this is true of the voracious *Amphiuma.* The waters of slow streams, woodland pools, ponds, and bogs may be seined for larvae and adults of certain *Ambystoma, Triturus, Siren,* and *Pseudobranchus.* In the semi-arid regions of the Southwest, the "tanks" and reservoirs provide the only open water available to salamanders over wide areas. In the shallow, weed-choked waters of the Florida "prairies," *Pseudobranchus* is often extremely abundant and may be collected by rolling masses of aquatic plants upon the sloping shores. When other methods fail, Siren and some other aquatic types may be taken in wire traps such as are used for turtles and fishes. *Eurycea lucifuga, E. longicauda,* and *E. melanopleura* are often found living in or about the entrances of caves, and several other species normally found in the open enter caves occasionally. The larvae of *Typhlotriton spelæus* are present in greatest abundance in streams in the open, while the adults are partial to underground waters and caves. *Gyrinophilus p. duryi* has been found both in caves and in the open. The blind, white *Typhlomolge* and *Haideotriton,* so far as is known, are found only in underground waters. If one is interested in the breeding habits of salamanders under natural conditions, recourse must be had to night collecting; and this method is often the most effective for securing large numbers of certain species in the shortest time. During the breeding season, some species migrate by night and may be encountered on the way to or from the breeding waters.

## EQUIPMENT FOR COLLECTING

Effective equipment for collecting salamanders may be very simple. Cloth bags of various sizes, made of muslin or other stout cloth, will serve usefully in the field, if damp moss or leaves are provided to prevent drying of the specimens. Small cans, jars, or boxes will be needed if eggs or larvae are to be kept alive and in good condition. For the larger aquatic species, metal containers with tight-fitting perforated covers are desirable, although many will survive if kept moist in a grain bag. A useful tool for the collector is the common, four-tined potato hook. These are made in various sizes and weights and serve admirably in turning logs and stones, stripping bark, and raking aquatic plants and other debris from ponds and bogs. A long-handled aquatic dip net is almost a necessity, and with it one may secure *Necturus, Cryptobranchus, Siren, Amphiuma,* and other mainly aquatic forms. A small seine of fine mesh occupies little space and is extremely useful in collecting larvae of various types in both running and quiet waters. Small forceps will be found of service in handling eggs and larvae. Wide, deep pockets or a knapsack will carry the accessories conveniently.

The most important item for the field collector is the notebook. In it should be recorded all the pertinent facts relative to the specimens, the locality, the date, the collector's name, and notes on habitat and associated species. The notes relating to a particular species or collection should be identified by a number or symbol of some kind which should correspond to that assigned to the field label placed with the collection. Field labels should be written *at the time of collection* and placed in the container to avoid the inevitable confusion that will result if reliance is placed upon memory alone.

### HOW TO PRESERVE SALAMANDERS

Salamanders in many collections are poorly preserved and badly distorted. This may be avoided if care is used in killing and hardening. Chloretone for killing specimens may be purchased in crystal form and dissolved in water to make a saturated solution. An ounce of the solution added to a quart of water in a tightly covered container will kill specimens and leave them perfectly relaxed. In the absence of chloretone, a 20 per cent solution of alcohol may be used. The specimens should then be arranged in trays for hardening. One part of commercial 40 per cent formalin added to 16 parts of water will provide the proper strength for fixing the tissues of salamanders, and 48 hours is usually sufficient for all but the largest specimens. Small individuals may be placed in formalin solution without injecting them, but large specimens of any species should be injected or slit in such a way that the preservative may enter the abdominal cavity. A hypodermic syringe of 2- or 3-ounce capacity and needles of various sizes should be available, or, in their absence, a sharp-pointed knife or a pair of scissors. For the best permanent storage of specimens they should be transferred to a 60 per cent solution of ethyl (grain) alcohol. Alcohol denatured with formaldehyde will serve if pure alcohol is not available, but wood alcohol should be avoided. In the absence of alcohol, the specimens may, of course, be left in the dilute formalin solution. Formalin solution is hard on the skin of most persons and acts as a mild poison on some. It is best to avoid long contact with the solution, and specimens may be handled safely with long forceps or rubber gloves. Some protection to the hands is afforded by lanolin if it is spread generously over the surface of the skin.

## ADDITIONS TO THE CHECKLIST

One hundred and twenty-six species and subspecies are considered in this Handbook. This number includes the 102 forms recognized in the fourth edition of the Stejneger and Barbour "Checklist," 1939, with the exception of *Manculus quadridigitatus remifer* and *Necturus alabamensis,* which I do not regard as valid. Of this total I have seen living examples of 115, or 90 per cent. Additions to the Checklist are listed below, and, for convenience, references follow the names:

| | |
|---|---|
| PROTEIDAE: | *Necturus maculosus stictus* Bishop<br>Occ. Papers Mus. Zool. Univ. Mich. No. 451, pp. 9–12, Pl. 2, Figs. 3–4. 1941. |
| CRYPTOBRANCHIDAE: | *Cryptobranchus bishopi* Grobman<br>Occ. Papers Mus. Zool. Univ. Mich. No. 470, pp. 5–9. 1943. |
| SALAMANDRIDAE: | *Triturus granulosus mazamae* Myers<br>Copeia, No 2, pp. 80–81, 1 fig. 1942. |
| | *Triturus granulosus twittyi* Bishop<br>Occ. Papers Mus. Zool. Univ. Mich. No. 451, pp. 16–18, Pl. 1, Figs. 4–5. 1941 |
| | *Triturus perstriatus* Bishop<br>Occ. Papers Mus. Zool. Univ. Mich. No. 451, pp. 3–6, Pl. 1, Fig. 2. 1941. |
| | *Triturus sierrae* Twitty<br>Copeia, No. 2, pp. 65–67, Pls. 1, 4, c, c'. 1942. |
| AMBYSTOMIDAE: | *Ambystoma tigrinum diaboli* Dunn<br>Copeia, No. 3, pp. 160–161. 1940. |
| | *Ambystoma tigrinum mavortium* Baird<br>Jour. Acad. Nat. Sci. Phila. (2) 1: 284, 292. 1850; Dunn, Copeia, No. 3, p. 158. 1940. |
| | *Ambystoma tigrinum nebulosum* Hallowell<br>Proc. Acad. Nat. Sci. Phila. 6: 209. 1853; Dunn, Copeia, No. 3, pp. 158–159. 1940. |

*Ambystoma tigrinum melanostictum* (Baird) [*Siredon lichenoides melanosticta*]
Rept. Pacif. R.R. Survey, vol. 12, book 2, p. 306. 1860; Bishop, Copeia, No. 4, p. 256. 1942.

PLETHODONTIDAE: *Desmognathus quadramaculatus amphileucus* Bishop
Occ. Papers Mus. Zool. Univ. Mich. No. 451, pp. 12–14, Pl. 1, Fig. 3. 1941.

*Ensatina eschscholtzii picta* Wood
Univ. Cal. Pub. Zoöl. 42 (10): 425–428. 1940.

*Eurycea griseogaster* Moore and Hughes
Copeia, No. 3, pp. 139–142, Figs. 1–2. 1941.

*Eurycea nana* Bishop
Occ. Papers Mus. Zool. Univ. Mich. No. 451, pp. 6–9, Pl. 1, Fig. 1. 1941.

*Eurycea tynerensis* Moore and Hughes
Amer. Midland Nat. 22 (3): 696–699, Figs. 1–2. 1939.

*Gyrinophilus dunni* Mittleman and Jopson
Smiths. Miscel. Coll. 101 (2): 1–5, Pl. 1. 1941. (Regarded as a subspecies of *G. danielsi* in the Handbook.)

*Gyrinophilus porphyriticus inagnoscus* Mittleman
Proc. New Eng. Zoöl. Club. 20: 27–30, Pl. 6, Fig. F. 1942.

*Haideotriton wallacei* Carr
Occ. Papers Bost. Soc. Nat. Hist. 8: 333–336, Pls. 11–12. 1939.

*Plethodon clemsonae* Brimley
Copeia, No. 164, pp. 73–75. 1927; Bishop, Occ. Papers Mus. Zool. Univ. Mich. No. 451, p. 20. 1941.

*Plethodon hardii* Taylor
Proc. Biol. Soc. Wash. 54: 77–79. 1941.

*Plethodon idahoensis* Slater and Slipp
Occ. Papers Dept. Biol. College Puget Sound. No. 8, pp. 38–43, Figs. 1–2. 1940.

*Pseudotriton ruber vioscai* Bishop
Occ. Papers Boston Soc. Nat. Hist. 5: 247–249, Pl. 15. 1928.

*Pseudotriton montanus diastictus* Bishop
Occ. Papers Mus. Zool. Univ. Mich. No. 451, pp. 14–16, Pl. 2, Figs. 1–2. 1941.

| | |
|---|---|
| | *Pseudotriton montanus floridanus* Netting and Goin<br>Ann. Carnegie Mus. 29: 175–183, Pl. 1, Fig. 5. 1942. |
| SIRENIDAE: | *Pseudobranchus striatus axanthus* Netting and Goin<br>Ann. Carnegie Mus. 29: 183–193, Pl. 1, Figs. 1–2. 1942. |
| | *Siren intermedia nettingi* Goin<br>Ann. Carnegie Mus. 29: 211–217. 1942. |

## ON THE USE OF THE KEYS

The keys to aid in the identification of salamanders have been made as simple and usable as is consistent with accuracy and completeness. Of necessity, certain technical terms have been employed, but these may be understood by reference to the labeled drawings. It is not to be expected that all specimens of all the many species can be identified without trouble, and the detailed descriptions and ranges of the various types should be taken into account if the key should prove inadequate.

The keys are arranged so that alternative sets of characters appear in pairs. With the specimen in hand, examine the structures or characters indicated in the key, selecting those which fit the specimen. The number at the right-hand side of the page, following the alternative selected, will indicate the next set of characters to be examined. In several of the keys it will be found that the same generic or specific name appears more than once. This is to take into account different color phases, neotenic individuals, or other departures from normal.

In the paragraphs dealing with family characteristics, the sizes indicated, the costal-groove counts, the number of toes, and other items refer to American species, and no account is taken of representatives found elsewhere. The size limitations given apply to normal adults, although under the accounts of particular species the measurements of neotenic individuals will usually be found where they differ materially. It is, of course, to be expected that individuals may be found that exceed, in some degree, the limits of size indicated.

## MAPS

The maps showing the distribution of the various species and subspecies are based on records from the literature and museum collections. Only those believed to be reliable have been included, but it is scarcely possible that errors have been avoided in all cases. The peripheries of the ranges of a number of species will doubtless be modified as workers in particular regions study the local fauna more intensively.

In a number of instances the maps indicate the range of a form which, it is realized, is a complex of species or subspecies requiring additional study before its components can be properly delimited. This is certainly true of *Ambystoma jeffersonianum, Plethodon cinereus, P. glutinosus,* and *Siren intermedia nettingi,* and probably true of *Ambystoma texanum, Cryptobranchus alleganiensis,* and *Pseudobranchus* from extreme southern Florida.

In the case of subspecies occupying contiguous ranges, areas of known intergradation are indicated on the maps by overlapping of the symbols representing the subspecies. In a few instances, e. g. *Gyrinophilus danielsi danielsi* and *G. d. dunni,* where the subspecies are ecologically isolated within the same general range, the overlapping of symbols is unavoidable and does not necessarily indicate intergradation. The spotty distribution of some species, e. g. *Eurycea lucifuga,* is due to its cave-loving habits; in other instances, *Ambystoma talpoideum,* for example, it is probably due to our lack of knowledge. In the case of *Necturus maculosus maculosus,* the artificial introduction into several isolated river systems will account for the discontinuous distribution.

The distribution as shown on the maps should indicate, in many instances, where profitable studies may be prosecuted.

PART II

# ACCOUNTS OF SPECIES

PART II

# ACCOUNTS OF SPECIES

## KEY TO FAMILIES

### ADULTS

1. Body eel-like .......................................... 2
   Body not eel-like .......................................... 3
2. With 2 pairs of diminutive limbs; a single pair of gill slits; aquatic .......................................... AMPHIUMIDAE p. 49
   With anterior limbs, only, present; external gills present; aquatic .......................................... SIRENIDAE p. 457
3. Adult animals fully transformed; no gills, gill rudiments, or open gill slits .......................................... 4
   Adult animals not fully transformed; permanent larval types, the animals reaching sexual maturity while retaining some larval characteristics such as gills or gill rudiments, open gill slits, and larval dentition .......................................... 6
4. With a groove (nasolabial) extending from nostril to lip; vomerine and parasphenoid teeth present; lungs absent; aquatic and terrestrial .......................................... PLETHODONTIDAE p. 183
   Without a groove from nostril to lip; no parasphenoid teeth; lungs present .......................................... 5
5. Costal grooves not developed; vomero-palatine teeth in 2 lines converging anteriorly; aquatic and terrestrial stages ...... SALAMANDRIDAE p. 67
   Costal grooves present, indistinct in some; vomerine teeth in transverse series; mainly terrestrial except in breeding season .......................................... AMBYSTOMIDAE p. 111
6. With external gills in sexually mature adults .......................................... 7
   Without external gills in sexually mature adults; a single pair of gill slits opening ventrally; body much depressed; a wrinkled fleshy fold on either side of trunk; size large, to 27″; aquatic .......................................... CRYPTOBRANCHIDAE p. 59

7. Toes 4–4; with 3 pairs of gills, bushy, red in life; lungs present; aquatic .......... PROTEIDAE p. 25
   Toes 5–4; gills short, or long and slender, but not bushy and red; lungs present or absent .......... 8
8. Body without pigment; lungs absent; legs long, slender; eyes vestigial; aquatic, underground waters. Snout much depressed, gills small (*Typhlomolge*); or, snout not greatly depressed, gills long (*Haideotriton*) .......... PLETHODONTIDAE p. 183
   Body with pigment; eyes normal; lungs present or absent .......... 9
9. Costal grooves distinct; gills well developed .......... 10
   Costal groves not developed; gills often reduced to stubs or lacking, but with one or more open gill slits; size small, to 3½″; vomeropalatine teeth in 2 lines converging anteriorly; aquatic (*Triturus*) .......... SALAMANDRIDAE p. 67
10. Size large, 6″–14″; body stout; lungs present; aquatic; dorsal fin broad, confined to tail (*Dicamptodon*); or, dorsal fin extending nearly to head (*Ambystoma*) .......... AMBYSTOMIDAE p. 111
   Size small, to 3″; body slender; tail narrowly keeled above and below at tip; lungs absent; aquatic (*Eurycea*) .......... PLETHODONTIDAE p. 183

# Family PROTEIDAE

Aquatic; permanent larvae with three pairs of bushy red gills; lungs present; no nasolabial grooves; toes 4–4 in American species; no ypsiloid cartilage; no maxillae; eyelids lacking; pterygoid and palatine teeth in continuous series; size moderate to large, $6\frac{5}{8}''$ (168 mm.) to 17″ (432 mm.). The single genus, *Necturus,* in the area considered.

## Genus NECTURUS

### KEY TO THE SPECIES AND SUBSPECIES OF NECTURUS

1. Ground color above deep brown or russet with few or no large black spots or black dots . . . . . . . . 2
   Ground color above rust-brown to dark-brown or dark-gray, with many large rounded black spots; or, with few or no large black spots but with numerous small black dots . . . . . . . . 3
2. Dark brown to nearly black above, sometimes with a few small light spots on back and sides; throat dull white, belly bluish-white (yellowish in preservative); young larvae colored like adults; length to $5\frac{19}{32}''$ (142 mm.). Atlantic Coastal Plain streams in North and South Carolina and Georgia . . . . . . . . *punctatus* p. 46
   Ground color above russet to dark brown, sometimes with a few small dark spots on sides of trunk and tail; belly bluish-white (yellowish in preservative); chin and sides of throat grayish-brown; young larvae unknown; size small, to $6\frac{19}{32}''$ (167 mm.). Known only from the vicinity of Mobile, Alabama . . . . . . . . *lödingi* p. 35
3. Belly light . . . . . . . . 4
   Belly pigmented and more or less mottled and blotched . . . . . . . . 7
4. Belly whitish or yellowish without dark spots; dark bar on side of head conspicuous . . . . . . . . 5
   Belly lightly pigmented, at least along sides; round dark spots of lower

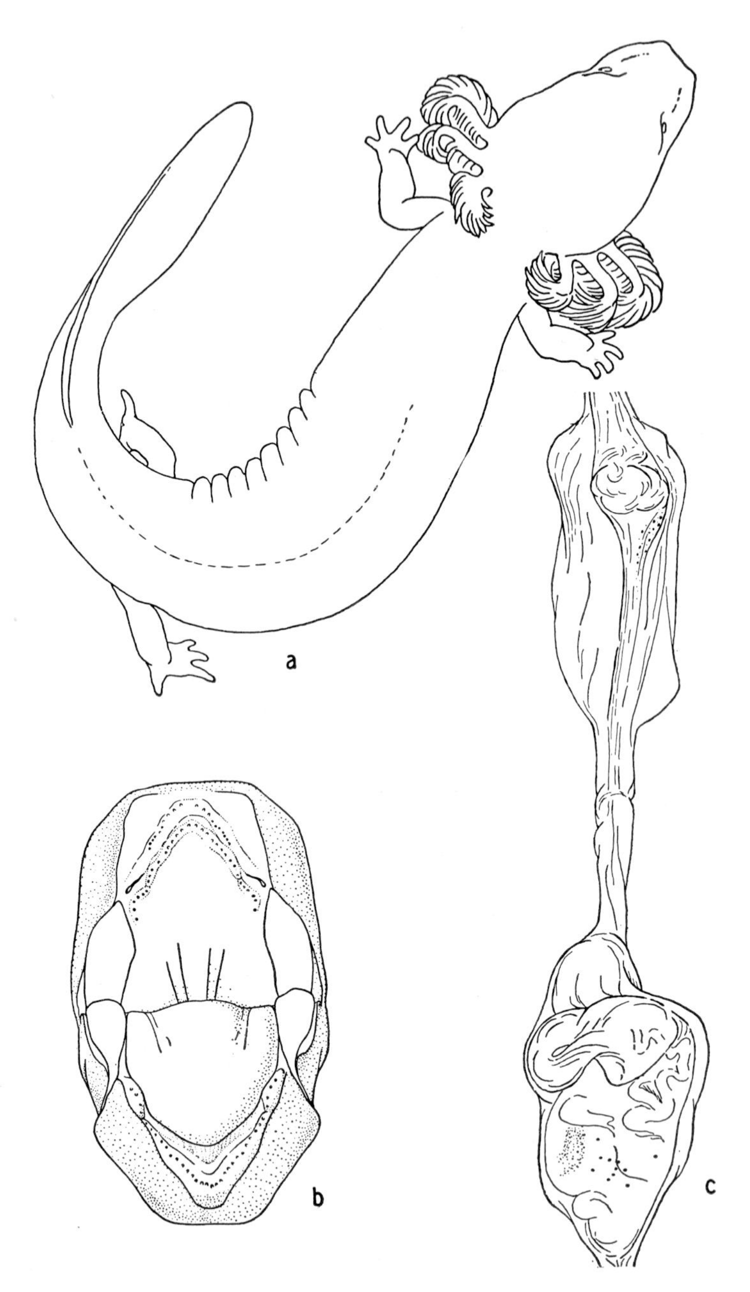


FIG. 2. (a) Outline of *Necturus m. maculosus* to show the body form, persistent gills, and four toes on both fore and hind feet. [H.P.C. *del.*] (b) Open mouth of *Necturus m. maculosus* to show the character of the tongue and teeth. [M.L.S. *del.*] (c) Spermatophores of *Necturus m. maculosus* deposited after stimulation with sheep anterior-lobe substance. [H.M.Z. *del.*]

sides usually encroaching on sides of belly; dark bar behind eye, well or poorly developed . . . . . . 6

5. Color above brown, with black spots forming more or less linear series on back and sides; on either side of mid-line of back a broad longitudinal band with irregular edges, lighter than adjacent areas; sides of tail with large, diffuse black spots; larvae with dorsolateral light stripes, the bordering dark areas with short, light dashes; size moderate, to 9 9/16″ (243 mm.). Known only from Big Creek, near Pollock, Louisiana . . . . . . *louisianensis* p. 37

6. Ground color above with many small tan spots forming a finely reticulated pattern; back, sides, and tail with many small to large, rounded black spots; light brown ground color of lower sides encroaching at least on sides of belly; dark bar on side of head through eye poorly developed or lacking; chin and at least sides of throat and belly with separate black spots; small larvae without longitudinal light stripes; larva 80 mm. long, spotted; length to 8 25/32″ (223 mm.). Southern Louisiana, Mississippi, and Alabama . . . . . . *beyeri* p. 28

   Ground color above rust-brown, with rounded blue-black spots scattered over back and sides; central area of belly light, sides of belly mottled and with a few dark spots; dark bar on side of head well developed; size large, to 17″ (432 mm.); larvae with light dorsolateral stripes; adjacent dark areas with few or no light dashes. Mississippi River system from Arkansas River northward; Great Lakes and St. Lawrence River and tributaries; rivers and lakes of New York; Delaware and Susquehanna Rivers and tributaries . . . . . . *maculosus maculosus* p. 40

7. Above with few to many small black dots on a tan-specked, dark-gray ground; no large black spots . . . . . . 9

   Above with many large black spots on a rust-brown ground color . . . 8

8. Belly with a finely reticulated pattern of brown and light tan and scattered separate black spots; throat and area between fore legs lightly mottled with few or no dark spots; dark bar on side of head narrow; larvae colored much like adults; size small to moderate, average 8″ (203 mm.), extreme length to 10 7/8″ (257 mm.). Known only from the Neuse and Tar River systems in North Carolina . . . . . . *lewisi* p. 32

   Belly light brown, with many small, tan or light yellow flecks and scattered dark spots most abundant along sides; dark bar on side of head broad; size large, to 17″ (432 mm.). Range as above under 6 . . . . . . *maculosus maculosus* p. 40

9. Ground color above dark gray, almost black, with many small tan flecks scattered generally on the dorsal and lateral surfaces; many small black dots, at least on head; occasionally a few large, widely separated black blotches; size large, to 12 5/8″ (320 mm.). Known from Lake Winnebago, Wisconsin, and Mackinac County, Michigan . . . . . . *maculosus stictus* p. 43

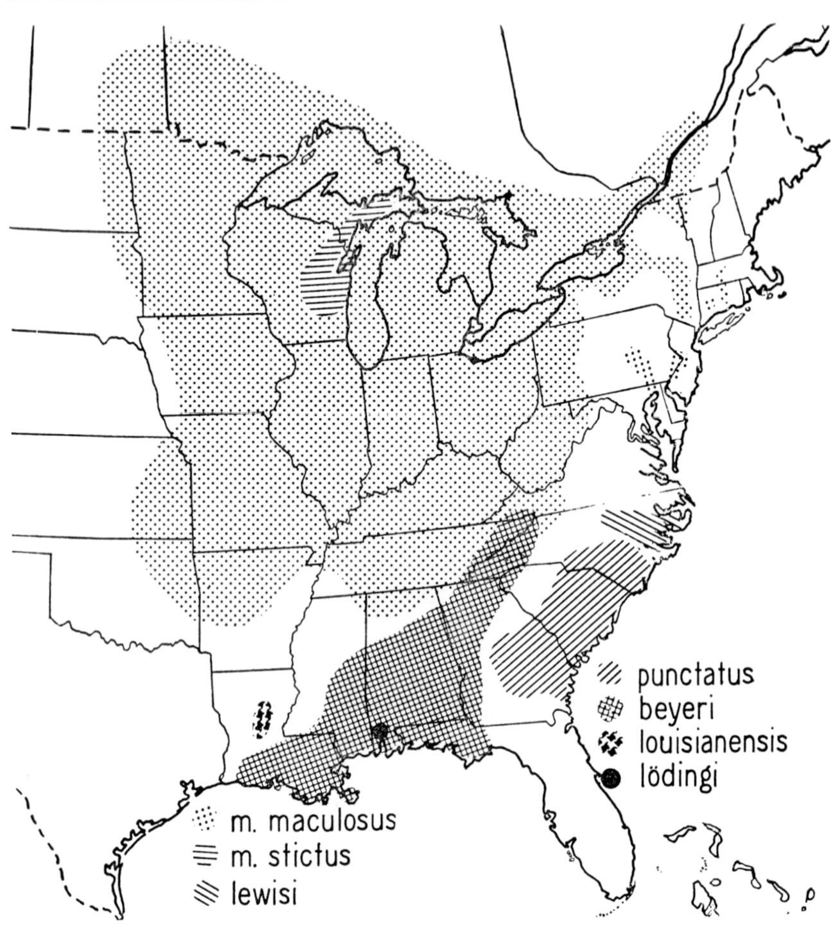


MAP 1.—Distribution of the subspecies of *Necturus maculosus*, *N. beyeri*, *N. lewisi*, *N. punctatus*, *N. louisianensis*, and *N. lödingi*.

**BEYER'S MUDPUPPY.** *Necturus beyeri* Viosca. Fig. 3. Map 1.

TYPE LOCALITY. Upper Calcasieu River near Oakdale, Louisiana.

RANGE. Gulf Coast streams from Mississippi to western Florida, northward through western Georgia to Virginia.

HABITAT. Viosca (1937, p. 136) writes that this species has been taken only in sandy, spring-fed streams which have an average temperature of approximately 70° F. While it has been found on both sides of the

Mississippi Valley, it has not been recorded from the sluggish and muddy waters of the bayous, lakes, and rivers of the Mississippi system in Louisiana.

SIZE. A series of adults measured by Viosca (1937, p. 124) varied in

FIG. 3. *Necturus beyeri* Viosca. (1) Adult female, actual length 6$^{15}/_{16}$″ (173 mm.). Bogue Falaya River at Covington, Louisiana. (2) Adult male, actual length 5$^{9}/_{32}$″ (135 mm.). (3) Same, ventral view. Bayou Lacombe Creek near Lacombe, Louisiana.

length from 6$^{9}/_{32}$″ (160 mm.) to 8$^{25}/_{32}$″ (223 mm.) and averaged 8″ (203 mm.) The proportions of an adult female 6$^{29}/_{32}$″ (175 mm.) long are as follows: head, length 1$^{3}/_{16}$″ (30 mm.), width $^{7}/_{8}$″ (22 mm.); head and trunk to posterior end of vent 4$^{3}/_{4}$″ (120 mm.); tail 2$^{5}/_{32}$″ (54 mm.). In this specimen the tail comprises 30.8 per cent of the total length.

DESCRIPTION. This is a species of moderate size and slender form. The head is widest immediately in front of the gills, tapers gently to the angle of the jaws, then more abruptly to the truncated snout. In side view, the snout is obliquely truncated, the swollen upper lip protruding beyond the lower. The eye has the pupil black and the iris silvery except where crossed by an indistinct dark bar from the nostril. The head and trunk are less noticeably depressed than in *Necturus maculosus* and *N. punctatus*. There is an impressed median dorsal line extending from the base of the tail to the back of the head, and the body in cross section is roughly subquadrate. The tail keel extends above to a point opposite the posterior end of the vent and is continued forward as a low fleshy ridge to a point above the insertion of the hind legs. The hind legs are a little larger than the fore; toes 4–4, normally 1–4–2–3 in order of length from the shortest, but in all adults examined some of the toes are curiously thickened and distorted. The costal grooves are not very prominent, but usually 16 may be counted, including 1 each in the axilla and groin.

COLOR. The general ground color above is a dark cocoa-brown (Natal Brown, Ridgway), with many small, irregular, tan-colored spots which form a finely reticulated pattern. The net-like pattern is continued over the sides of the head, trunk, and tail, the light spots on some individuals being fewer and farther apart on the lower sides and venter. The brown ground color fades somewhat on the lower sides of some, and on the venter becomes even lighter. Superimposed on the ground color are many deep brown, almost black, spots which are generally circular or roughly oval in outline. The dark spots vary in size from 2 to 5 mm. and are disposed in very irregular rows or in some individuals apparently at random. They tend to be smaller on the head, larger and more widely separated on the trunk and tail, and decrease somewhat in size on the lower sides. In specimens from west of the Mississippi Valley, the belly is strongly spotted on a dull flesh-colored ground (Viosca,

1937, p. 123, Fig. 2); in specimens I have examined from east of the Mississippi, the venter in some lacks the large dark spots and has the central area flesh color, with tinges of gray encroaching irregularly from the sides; in one individual the large dark spots are continued on the sides of the venter, leaving the central area brownish-gray, with separated, irregular, flesh-colored blotches along the mid-line. Some specimens in preservative take on a purplish cast above and are dull yellowish or grayish below.

BREEDING. Very little has been published on the breeding habits of this species. Viosca (1937, p. 136) reports that all adults of *N. beyeri* have been taken during the winter and early spring months and were in gravid condition. This suggests a spring nesting period.

LARVAE. I have not seen recently-hatched larvae of this species, but in a series of specimens sent me by Mr. Viosca are some apparently in their second season. These, taken in May and June 1937, vary in length from 52 to 64 mm. These larvae resemble the adults in general form. They differ in having the dorsal tail fin continued as a low ridge to a point halfway between the hind and fore legs, and in details of color and pattern. In addition to the fine reticulations, there are many small spots of dull yellow distributed over the head and trunk and aggregated in linear series along the margins of the tail fin, where they are in strong contrast to the more highly pigmented central portion. The venter is very light, slightly pigmented along the sides in some, in others immaculate. The larger, conspicuous, dark spots of the adults are barely indicated in the larger larvae and are absent in the smaller.

In a juvenile specimen 91 mm. long, the colors are essentially those of the adults except that the net-like pattern of tan spots is less well developed. There are many small dash-like, dull yellow marks on the sides and top of the head and trunk, arranged in fairly regular linear series. Two of these lines are continued on the tail, where they converge near the tip. The margins of the tail fins are mottled with dull gold and brown. The ventral surface of the head and trunk is lightly pigmented with brown.

Löding, 1922, p. 9, specimens from Montgomery County and Warrior River, Alabama; Viosca, 1937, p. 123.

**LEWIS' MUDPUPPY.** *Necturus lewisi* Brimley. Fig 4. Map 1.

TYPE LOCALITY. Neuse River near Raleigh, North Carolina.

RANGE. Known only from the Neuse and Tar River systems in North Carolina.

HABITAT. The majority of specimens of this species have been reported by Dr. C. S. Brimley of Raleigh, North Carolina. Little has been written concerning the habitat preferences. Many adults have been taken on hook and line in the early spring, and in addition Brimley and Mabee captured some specimens by dipnetting in the Neuse River near Raleigh. These individuals were among trash, in small backwaters of the main stream, in which there was a strong current. Specimens have also been taken in the Little River where the banks and bottom were sandy and muddy.

SIZE. A series of adults measured by Viosca (1937, p. 125) varied from $6\frac{5}{8}''$ (168 mm.) to $10\frac{7}{8}''$ (276 mm.) and averaged $7\frac{31}{32}''$ (202 mm.). The proportions of an adult male from Raleigh are as follows: total length $7\frac{5}{8}''$ (193 mm.); head length $1\frac{3}{8}''$ (34 mm.), width $\frac{31}{32}''$ (24 mm.); head and trunk to posterior end of vent $5\frac{1}{2}''$ (139 mm.); tail $2\frac{5}{32}''$ (54 mm.).

DESCRIPTION. *N. lewisi* differs from *N. maculosus* in its smaller size, in having relatively larger and fewer dark spots, and in details of color, pattern, and structure. The sides of the head are nearly parallel to the eyes, then converge rather abruptly to the truncated snout. Head and snout depressed, the trunk somewhat flattened dorsally, and with an impressed median line that extends from the base of the tail to the snout. The tail is less distinctly keeled above than in *N. maculosus* and extends forward only to the posterior end of the vent. In an adult male, the tail comprises 27.9 per cent of the total length. The legs are about equally developed. Toes 4–4, in order of length from the shortest 1–4–2–3. The 14 costal grooves are not well developed. The upper teeth are in

2 V-shaped series, the blunt apices directed forward. The premaxillary teeth, 10–12 in number on each side, form 2 short divergent series. The vomerine series are parallel with them and vary in number from 9 to 12. Continuing in line with the vomerine series but separated by a short interval are the 5–7 pterygoid teeth, which form short curved series, convex outwardly. Opposite and outside the break between the vo-

FIG. 4. *Necturus lewisi* Brimley. (1) Adult female, actual length 7⅝″ (194 mm.). (2) Same, ventral view. Little River, Wake County, North Carolina. [Mrs. R. C. Simpson, collector.]

merine and pterygoid series the inner nares open in an oblique groove.

COLOR. In the living animal, the general impression is that of a rust-brown salamander with large black spots scattered over the back and sides. The ground color actually is an intense yellow-brown and the detailed pattern consists of an irregular network of small, lichen-like, chocolate-brown spots which give the generally rusty appearance. Scattered over the dorsal surface and on the sides of the trunk and tail are deep brown or bluish-black blotches. In some individuals the large dark blotches form a fairly regular dorsolateral series, in others they are irregularly scattered. The lower sides and belly are heavily pigmented and have the yellow-brown in more scattered flecks. The ground color is dull brown or pale slate. The black blotches are generally fewer on the ventral surfaces and may be considerably smaller than those of the back and sides. The region between the fore legs and the under surface of the head is usually quite uniformly mottled, but lacks the black blotches in all specimens I have examined. The dorsal surface of the head generally has the dark markings of irregular size and shape but not rounded. The legs are marked above like the back, the hind usually with a few rounded black spots, the fore usually without them. A dark bar through the eye extends on the side of snout but does not reach the nostril. The tail keels are sometimes strongly tinged with orange.

BREEDING. Nothing has been published on the breeding habits of this species. A large male taken March 24, 1920, by F. Lewis, has the cloacal glands greatly swollen, the vent everted and exposing a large number of slender, thread-like papillae and, at the posterior margin, 2 short, widely separated nipples directed backward. This condition may indicate an early-spring mating season for this species, although some males of *N. maculosus,* which has a fall mating season, are known to retain the swollen glands until spring.

A female, 168 mm. long, taken April 8, 1919, by F. Lewis, had the belly distended by eggs apparently ready for deposition.

LARVAE. Brimley's smallest larvae were about 2½ inches long and were colored like the adults (Brimley, 1925, p. 14). Viosca (1937, p. 125) describes a 55-mm. formalin specimen as follows: "The entire dorsal region is a light grayish brown with a few ill-defined dark spots and vermiculations; this area broadens on the head, and in the rear it terminates at the origin of the upper dermal border of the tail. The sides are dark brown with more or less elongated light spots tending to be arranged in longitudinal rows. A faint darker band extends backward from the gills in a groove. A more or less irregular dark line extends along the canthus and through the eye to the anterior border of the gills. The tail is dark with scattered light spots, a light or spotted upper dermal border, and a light lower dermal border. The lower parts are light with occasional pigment spots, especially along the costal grooves, near the hind legs, and bordering the lower jaw."

**LOEDING'S MUDPUPPY.** *Necturus lödingi* Viosca. Fig. 5. Map 1.

TYPE LOCALITY. Eslava Creek near Mobile, Alabama.

RANGE. Known only from Eslava Creek (type locality) and Hall's Mill Creek near Mobile, Alabama.

FIG. 5. *Necturus lödingi* Viosca. (1) Adult female, actual length $5\frac{3}{4}$″ (147 mm.). (2) Same, ventral view. Hall's Mill Creek near Mobile, Alabama.

HABITAT. Nothing has been published on the habitat relationships of this species.

SIZE. The type, a female, has the following measurements as given by Viosca (1937, p. 126): length $6\frac{3}{16}$″ (157 mm.), tail $2\frac{1}{16}$″ (51.5 mm.); head length $1\frac{3}{16}$″ (30 mm.), width $\frac{25}{32}$″ (20 mm.). An adult female from Hall's Mill Creek has a total length of $6\frac{19}{32}$″ (167 mm.). Another sexually mature female from the same locality measures only $5\frac{17}{32}$″ (140 mm.). A pair of adult specimens give the following comparative measurements: male, length $5\frac{19}{32}$″ (142 mm.), tail $1\frac{11}{16}$″ (43 mm.); head length ⅞″ (22 mm.), width $\frac{13}{16}$″ (17 mm.). Female, length $5\frac{26}{32}$″ (147 mm.), tail $1\frac{15}{32}$″ (46 mm.); head length $\frac{31}{32}$″ (23 mm.), width ¾″ (18 mm.).

DESCRIPTION. This is a small species and in life is the lightest colored. The head is widest just in front of the gills, tapers gradually to the eyes, then more abruptly to the truncated snout. The gills are of moderate size and noticeably less bushy than in the other species. The trunk is roughly cylindrical, flattened slightly above, where there is an impressed median line, and more strongly below. The tail comprises 26–30 per cent of the total length in the male, and 31–32 per cent in the female. The tail fin of the male is noticeably broader than that of the female and extends to a higher level above the back. The ventral fin in both sexes arises as a fleshy ridge just back of the vent and becomes compressed into a thin keel about at mid-length. Legs are of moderate size, the toes rather slender and widened at the tips, where they are lighter in color. Toes in order of length 1–4–2–3; costal grooves 15, counting 1 each in the axilla and groin. Teeth, premaxillary, 9–9; vomero-palatine 9–11; pterygoid 7–7, in specimen from Hall's Mill Creek.

COLOR. In living specimens, the ground color above varies in different individuals from russet to dark brown (Buckthorn Brown to Mummy Brown, Ridgway). The ground color is developed as a fine network enclosing many small, irregular light spots. Scattered irregularly over the back and sides are black blotches which vary greatly in number in different individuals and in size from 1 to 3 mm. The black blotches in some tend to be larger and more diffuse on the sides of the tail, which may then be cloudy or marbled, the margins lighter, mottled with tan and reddish-orange blotches. The dark spots may be continued over the dorsal surface of the head, where they are fewer, smaller, and farther apart. In one specimen the dark spots are scarcely evident on the trunk, where they are few in number, and are entirely lacking on the head. A dark bar extends from the nostril through the eye, becoming somewhat broader and diffuse on the side of the head. The iris is tinged with silvery above and below the dark bar. The legs are colored above like the back and may have 1 or 2 dark spots. The margins of the pigmented areas on the belly may be nearly straight or highly irregular. In life the light ventral surfaces are bluish-white, with tinges of flesh on the center of the throat, between the legs, and in front of the vent. The chin and sides of throat are broadly pigmented with grayish-brown.

BREEDING. Nothing has been published about the breeding habits. Males taken Feb. 13, 1938, in Hall's Mill Creek near Mobile, Alabama, by Percy Viosca, Jr., have the cloacal glands greatly enlarged, the vent protuberant and everted to present many small slender fleshy filaments, and posteriorly a pair of broad nipples directed backward. In a female taken at the same time and place the ovarian eggs are very small, indicating that the season's complement had already been deposited or that the egg-laying season comes at a much later period in the year.

LARVAE. In a larva 84 mm. long, the ground color is light brown in a reticulated pattern, the network enclosing tan spots of irregular size and shape but tending to be elongate on either side of the back and forming an indistinct line from the head onto the sides of the tail. There is no evidence of larger black spots in this individual. The tail, and particularly the fins, are strongly mottled with light yellowish blotches which are in strong contrast to the darker central areas of the tail sides. The gills are well developed and bushier than in the adults I have examined. The tips of the toes are light, as in the adults. The dark bar through the eye is quite prominent but does not quite attain the nostrils. The legs are mottled tan and brown. The venter is dull flesh color, the lower lip, only, margined with dusky.

Viosca, 1937, p. 126.

**LOUISIANA MUDPUPPY.** *Necturus louisianensis* Viosca. Fig. 6. Map 1.

TYPE LOCALITY. Big Creek, near Pollock, Louisiana.

RANGE. Known from the type locality and from Indian Creek near Forest Hill, Louisiana.

HABITAT. Aside from the fact that the specimens have been found in streams, nothing has been published as yet on the habitat of this form.

SIZE. Viosca (1938, p. 144) writes that the adults average 8¾" (222 mm.), with the extremes 7²³⁄₃₂" (196 mm.) and 9⁹⁄₁₆" (243 mm.). I have measured sexually mature adult males from the type locality 7½" (190 mm.) and 9¹¹⁄₁₆" (246 mm.) in total length. The proportions of an adult male are as follows: total length 9¹¹⁄₁₆" (246 mm.), tail (tip lost) 2¾" (70 mm.); head length 1⅝" (41 mm.), width 1⁹⁄₃₂" (32 mm.). An adult female has the following measurements: total length 9¹³⁄₁₆" (239 mm.), tail 2²³⁄₃₂" (69 mm.); head length 1⅝" (41 mm.), width 1⁷⁄₃₂" (31 mm.).

DESCRIPTION. The head is depressed, with the sides back of the eyes slightly diverging, in front converging strongly to the bluntly rounded snout. The eyes are small, the long diameter about 3 times in the snout. Gills short, 1–2–3 in order of length, highly pigmented; filaments numerous, slender, and longest at the middle of the rachis. Trunk rounded on the sides, flattened above and below. Costal grooves usually 16, and about 8 intercostal folds between the toes of the appressed limbs. Tail subquadrate at base, strongly compressed distally, the dorsal keel arising above the vent as a low ridge and becoming thin and knife-edged at about the basal fourth. Ventral tail fin a rounded ridge behind the vent, becoming thin-edged at about the middle of the length. Legs about equally developed; toes 4-4, 1–4–2–3 in order of length from the shortest. The upper teeth form 2 V-shaped series, the blunt apices directed forward. The premaxillary teeth 9–12 on each side, thick and bluntly pointed on the anterior part of the bone, decreasing in size posteriorly; vomerine series parallel with the premaxillary, 9–12 on each side; pterygoid series short, each with 5 or 6 teeth. The vent of the female is a simple slit with oblique internal folds. The vent of the male is limited behind by two bluntly triangular papillae.

COLOR. The general color above and on the sides is light yellowish-brown or tan, disposed in large irregular blotches which form fairly distinct dorsolateral bands. These light bands are made up of aggregates of small tan spots in a reticulated pattern, the interspaces being darker brown. Scattered over the lighter ground color are large dark brown or blue-black spots, which on some individuals form more or less regular longitudinal lines. In specimens that are sparsely spotted above, there are usually only 2 rows of dark spots on the sides; individuals strongly spotted above may have several rows on the sides. There is a dark bar on the side of the head extending from the nostril through the

FIG. 6. *Necturus louisianensis* Viosca. (1) Adult female, dorsal view; actual length 8⅜″ (213 mm.). (2) Adult male, ventral view; actual length 7½″ (190 mm.). Big Creek near Pollack, Louisiana (type locality).

eye to the base of the gills. The head is mottled or blotched above, but the dark spots are smaller than on the trunk. The pigment of the lower sides encroaches on the sides of the venter, leaving a narrow, relatively clear, flesh-colored area extending from the posterior part of the throat nearly to the hind legs. The large dark spots of the sides tend to coalesce along the center of the sides of the tail to form dark areas of considerable extent, the margins of the tail being blotched with black and tan or with tinges of orange. The legs are marked above like the back and below like the sides of the belly.

BREEDING. Viosca (1938, p. 145) states that specimens taken in February are in mature sexual condition, indicating an early spawning period. Some specimens sent me by Mr. Viosca in January had fully developed eggs apparently ready for deposition in early April, and in nature might have been deposited before that date.

LARVAE. A larva 62 mm. long, taken Jan. 25, 1941, in Indian Creek near Forest Hill, Louisiana, has a broad, dorsolateral, light band extending on either side from a point behind the eye well onto the tail, the area between the bands being light brown. A dark band on the side extends from the nostril through the eye to the base of the gills and is continued along the sides of the trunk and tail to the tip. On the sides a fairly regular row of light dashes extends from the fore legs nearly to the tip of the tail, passing above the insertion of the hind legs. Dorsally there are a few small rounded black spots irregularly disposed. The ventral surfaces, except for the distal third of the tail and lower lip, are immaculate.

**MUDPUPPY. WATERDOG.** *Necturus maculosus maculosus* (Rafinesque). Figs. 2, 7. Map 1.

TYPE LOCALITY. The Ohio River.

RANGE. The Mississippi River system from the Arkansas River and northern Alabama northward; the Great Lakes and St. Lawrence and their tributaries in Manitoba, Ontario, and Quebec; Lake George and Lake Champlain, Mohawk River and Barge Canal, middle Hudson (probably *via* the Erie Canal), Susquehanna River, Delaware River (perhaps *via* the old Delaware and Hudson Canal), Connecticut and Farmington Rivers in New England (introduced).

HABITAT. Thoroughly aquatic but found under the most diverse conditions. Abundant in the clear waters of lakes and streams, but occurs in muddy and weed-choked bays and coves, and in canals and drainage ditches.

SIZE. Attains a maximum length of 17″ (432 mm.). Twenty-five adults from New York, Pennsylvania, and Kentucky average 11 5/16″ (287.4

mm.). The proportions of an adult male are: total length $11\frac{1}{2}''$ (292 mm.), tail $3\frac{9}{32}''$ (84 mm.); head length $1\frac{31}{32}''$ (50 mm.), width $1\frac{13}{32}''$ (36 mm.). The measurements of an adult female are: total length $11\frac{5}{8}''$

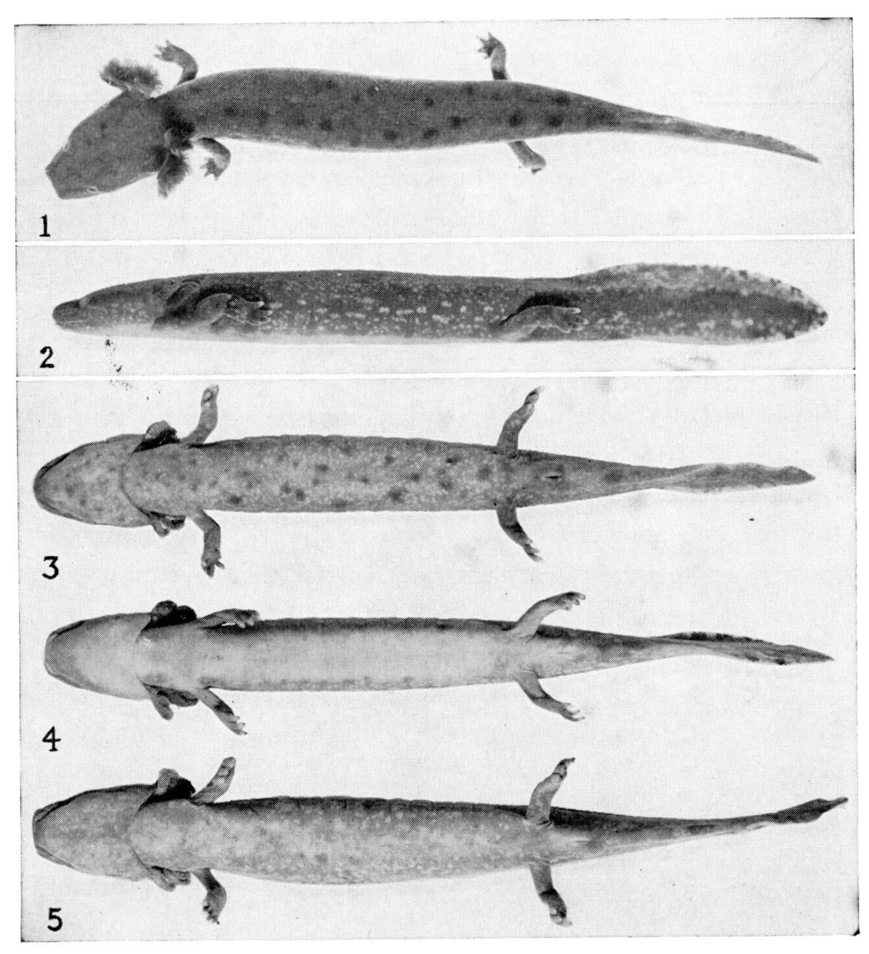


Fig. 7. *Necturus m. maculosus* (Rafinesque). (1) Young adult, dorsal view. (2) Same, lateral view. (3–5) Ventral views of three individuals to show variation in pattern. French Creek, Crawford County, Pennsylvania.

(295 mm.), tail $4\frac{1}{8}''$ (105 mm.); head length $1\frac{11}{16}''$ (43 mm.), width $1\frac{15}{32}''$ (38 mm.).

Description. The common Mudpuppy attains the largest size of any of the several forms of *Necturus* and has by far the widest distribution. The head behind the eyes is broad and swollen but tapers in front to the truncated snout. The trunk is somewhat depressed and has a distinct median, dorsal groove. The tail is strongly compressed and keeled above and below. The eyes are small, the horizontal diameter about 4 in the snout. There are usually 15 costal grooves. Legs short and stout; toes 4-4, in order of length from the shortest, 1-4-2-3. The upper teeth are in 2 V-shaped series with the sides nearly parallel and the apices directed forward. The premaxillary teeth average 11.5 and vary from 10 to 12; vomerines average 12.3 and vary from 10 to 14; pterygoids average 6.7 and vary from 5 to 9.

COLOR. The color is somewhat variable, but in general the impression is that of deep, rust-brown ground color, with rounded spots of blue-black scattered over the back and sides. Rarely the spots may be distributed in 2 fairly regular rows along the back, and there is sometimes a lateral series. The upper surface of the legs is usually spotted and the free margins of the tail are often tinged with orange or reddish. On each side of the head a dark bar extends from the snout through the eye. The belly may be light and relatively free from darker pigment, uniformly pigmented so that the general tone is grayish, or with rounded dark spots on a lighter ground color. Occasionally individuals may be found that are uniformly dark above and light in a narrow midventral area.

SEX DIFFERENCES. The sex of adults may be determined by the size and form of the vent. In the male this is a longitudinal slit, wrinkled at the margins, crossed behind by a crescentic groove and provided with 2 nipple-like papillae directed backward. During the breeding season it is surrounded by a swollen ridge which marks the limits of the cloacal gland beneath. The vent of the female is a simple slit.

BREEDING. The mating season is in the fall. In western New York and Pennsylvania the sexes may be found together late in September and through October and November; more rarely throughout the winter months and in the early spring. Often both sexes share the same shelter during the mating season. Fertilization is internal, males depositing spermatophores which consist of a gelatinous basal part, on the summit of which the sperm are aggregated in a whitish mass. After mating the sexes usually separate. In late May or early June, in the latitude of western New York, the female deposits her eggs and remains with them through the period of incubation, and often longer. She may be found occupying a shallow excavation beneath some stone or log with an entrance on the down-stream side. The eggs, attached singly to the lower surface of the sheltering object, are light yellow spheres ¼″ (5–6 mm.) in diameter and provided with 3 envelopes. The inner envelope is a thin, tough, and elastic membrane enclosing a jelly-like substance; the second layer is slightly thicker, clear and transparent; the third layer is thick, of jelly-like consistency, and forms the outer cover and attachment stalk. The eggs hatch in 38–63 days, depending on the temperature of the water.

LARVAE. The newly hatched larvae average a little less than 1″ in length (22–23 mm.). The fore legs and toes are fairly well developed, the hind legs short and directed backward, the toes scarcely indicated. The tail is keeled above and below. The color pattern of the larva differs markedly from that of the adult. There is a strongly pigmented median dorsal stripe which originates on the snout and extends the length of the trunk and on the tail, fading toward the tip. This is bordered on each side by a narrower yellow band extending from the base of the gills along the upper sides and basal third of the tail. Below the lateral light bands the sides are dark from the gills to the tip of the tail. The lower sides are somewhat lighter and fade to the yellow, yolk-distended belly. Sexual maturity is attained in about 5 years at a length of about 8″ (200 mm.).

**WISCONSIN MUDPUPPY.** *Necturus maculosus stictus* Bishop. Fig. 8. Map 1.

TYPE LOCALITY. Lake Winnebago, Winnebago County, Wisconsin.

RANGE. Known from the type locality and probably from Mackinac County, Michigan.

HABITAT. Found in the comparatively shallow parts of the lake.

SIZE. The average length of ten adults of both sexes from Lake Winnebago is $11\frac{1}{2}''$ (292 mm.), the extremes $9\frac{3}{8}''$ (238 mm.) and $14\frac{1}{4}''$ (362 mm.). The proportions of an adult male are as follows: total length $12\frac{5}{8}''$ (320 mm.), tail $3\frac{17}{32}''$ (90 mm.); head length $1\frac{31}{32}''$ (50 mm.), width $1\frac{11}{16}''$ (42 mm.). An adult female has the following

FIG. 8. *Necturus maculosus stictus* Bishop. (1) Adult female, actual length $14\frac{1}{4}''$ (363 mm.). (2) Same, ventral view. Lake Winnebago, Wisconsin (type locality).

measurements: total length $12\frac{1}{8}''$ (308 mm.), tail $3\frac{5}{8}''$ (92 mm.); head length $1\frac{13}{16}''$ (46 mm.), width $1\frac{11}{16}''$ (42 mm.).

DESCRIPTION. The general impression is that of a very dark Necturus, heavily pigmented above and below. The head is broad and flat, widest immediately in front of the gills, the sides converging slightly to the angle of the jaws and more abruptly from this point to the broadly truncated snout. The head width averages 86.5 per cent of the head length as compared with 81 per cent in specimens from the Allegheny River, Cattaraugus County, New York, and 82 per cent in a series from the tributaries of Lake Ontario. The eye is relatively small, $4\text{–}4\frac{1}{2}$ in the length of the snout, the iris blotched or flecked with yellow above and below the pupil. Gills short and bushy, the 3rd longest. The trunk is stout, flattened above and below, rounded on the sides. There are usually 15 costal grooves, occasionally 16, and approximately 6 intercostal folds between the toes of the appressed limbs. The tail is short, comprising 27–30 per cent of the total length in the males and 29–32 per cent in the females. It is broadly oval in section at base and with a dorsal keel that arises opposite the hind margin of the vent as a thick cord-like ridge and becomes compressed and thin-edged at about ½ the length; ventral fin thin distally, becoming thickened at about the distal third. The legs short and stout; toes 4–4, short, broad and blunt-tipped, usually 1–4–2–3 in order of length from the shortest, sometimes 1–2–4–3. The teeth differ somewhat in number from those of the typical subspecies. The pterygoid teeth average 6.2 and range from 5 to 8; the premaxillary average 10.7 and range from 9 to 15; the vomero-palatine vary from 11 to 15 and average 12.

COLOR. The general color above is dark gray, almost black. In life there is a superficial layer of tan chromatophores which obscures the ground color in varying degrees in different individuals. Scattered generally over the dorsal surfaces, but usually most abundant on the head, are many small deep brown or black dots, which are characteristic of this subspecies and absent in typical *maculosus*. In some, these dots are also present on the sides and ventral surfaces. In addition to the dark dots, a few individuals have a few large dark blotches on the back and sides, where the superficial light chromatophores are lacking. These large blotches, when present, are usually larger and fewer than in typical *maculosus*. The ventral surfaces are very dark, with only an occasional individual showing a lighter area along the midventral line. In some there are a few large dark blotches on the belly, and in many a scattering of small tan flecks, most abundant along the sides. The upper surface of the legs is colored like the back, the lower surface like the belly. The margins of the tail are sometimes tinged with ochre. Small individuals have a dark bar through the eye, but this is usually lost in the general coloration of the side of the head in old, dark individuals.

SEXUAL DIFFERENCES. The sexes may be distinguished by the form of the vent, which in the male is lined with slender finger-like papillae and limited behind by a transverse groove and a pair of fleshy triangular lobes directed mesially.

BREEDING. Nothing is known of the breeding habits which would distinguish this subspecies from the typical form, and it is likely that they are essentially similar in this respect. This subspecies has been figured by Viosca (1937, p. 129) and apparently by Pope and Dickinson (1928, Pl. I).

**SOUTHERN WATER-DOG. CAROLINA WATER DOG.** *Necturus punctatus* (Gibbes). Fig. 9. Map 1.

TYPE LOCALITY. Schoolbred's plantation on the South Santee River, a few miles from its mouth, South Carolina.

RANGE. Atlantic Coastal-Plain streams in North Carolina, South Carolina, and Georgia.

HABITAT. Permanently aquatic. Generally found in the lower and slower parts of streams having mud or sandy banks and bottoms; or in ditches where water is diverted from such streams. On Feb. 10, 1936, we collected in the Little River near Raleigh, North Carolina, and took specimens from among leaves and plant rubbish raked from the borders of the stream in a quiet backwater.

SIZE. This is one of the smallest of the several species of Necturus, adults averaging about 5⅙″ (125 mm.) and reaching an extreme length of about 6³⁄₁₆″ (157 mm.).

DESCRIPTION. *N. punctatus* is a slender species with the head relatively long and gently tapering from the gills to the bluntly rounded snout. The head and snout are depressed, the trunk slightly depressed and with an impressed median dorsal line. The tail is keeled above from a point opposite the vent and below nearly to the vent where it runs into a fleshy ridge. In the fully adult male the tail comprises about 35.6 per cent of the total length. The legs are about equally developed, with the hind a little stouter and shorter than the fore. Toes 4–4, in order of length from the shortest 1–4–2–3. The costal grooves are well developed and conspicuous in juvenile specimens and vary in number in different individuals from 14 to 16. The gills are neither as long nor as bushy as in *N. m. maculosus,* but the filaments are slender and the rachises mottled. The eye has the pupil black and the iris brassy. A dark horizontal bar passes across the eye but is not conspicuous. The upper teeth

FIG. 9. *Necturus punctatus* (Gibbes). (1) Larva, dorsal view. Upper Little River, North Carolina. [C. S. Brimley and Roxie Collie Simpson, collectors.] (2) Larva, lateral view; actual length 3³⁄₁₆″ (81 mm.). Middleton Mill Creek, North Carolina. [E. C. Raney, photograph.]

are in 2 V-shaped series with the sides nearly parallel and the apices directed forward.

Viewed from the side, the angle of the mouth reaches a point opposite the hind margin of the eye. The fleshy margins of the sides of the upper jaw are turned upward and inward to form a thin flap. The margins of the lower jaw are turned first inward and upward, then outward and downward. When the mouth is closed, the two flaps interlock to form a tongue and groove.

COLOR. The general color above is uniform dark brown (between Olive and Clove Brown, Ridgway). This color fades out somewhat on the lower sides at about the lower level of the legs. The head and the snout, below the level of the eyes, are dirty white. The tail is lightly marked with small, rounded, tan or buff spots scattered along the low dorsal keel. The ventral tail keel has the margin largely free from dark pigment and is yellowish or pinkish. The legs have several pale spots above. The throat is dull dirty white, the belly bluish-white without pigmented spots. Specimens long in preservative are usually uniformly dull brown above and dull whitish or yellowish below. A few pale spots may be evident along the margins of the tail fins.

BREEDING. Very little is known about the breeding habits of this species.

LARVAE. The smallest larva examined was 28 mm. (1⅛″) long. It is uniformly colored like the adults and is without dark spots or stripes. The tail is slightly mottled, owing to the presence of irregular pigment-free spots, and there is a row or two of pale spots of small size along the sides, marking the position of the lateral-line sense organs. The upper tail fin of young larvae extends forward on the trunk about halfway between the fore and hind legs. The color of the upper parts and sides is sharply cut off from the light lower sides and venter at a line which extends through the lower margin of the eye, the legs, and the ventral third of the tail. The 28-mm. larva mentioned above was taken by Mr. C. S. Brimley in Buffalo Creek, Wendell County, North Carolina, Nov. 30. At the same time and place other larvae 40 mm. long were taken. These obviously are young of the year, and the difference in length between these and the 28-mm. larva indicates a rather extended breeding season. An additional year group is apparently represented by 2 larvae taken in the Little River near Raleigh, North Carolina, in December 1934. These measure 65 and 88 mm. respectively.

Bishop, 1927, p. 187; Brimley, 1907, p. 155; 1915, p. 195; 1924, p. 166; 1926, p. 76; Brimley and Mabee, 1925, p. 14; Brimley and Sherman, 1908, p. 19; Cope, 1889, p. 27; Corrington, 1929, p. 61; Fowler and Dunn, 1917, p. 8; Pickens, 1927, p. 107; Viosca, 1937, p. 128; Wright, 1926, p. 82.

# Family AMPHIUMIDAE

Aquatic; body large, eel-like to 40″ (1015 mm.); lungs present; limbs diminutive, toes reduced; gills lacking; a single pair of open gill slits; no nasolabial groove; no ypsiloid cartilage; eyelids lacking, dentition larval; vertebrae amphicœlous; a posterior projection of the prevomer. The single genus, Amphiuma, in the family.

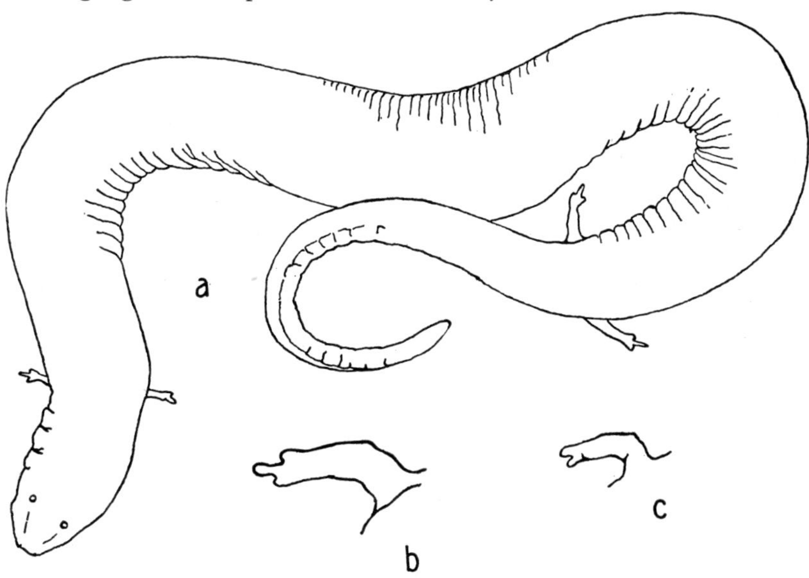


FIG. 10. (a) Outline of *Amphiuma means tridactylum* to show the eel-like body, diminutive limbs, and reduced toes. (b) Toes of *A. m. tridactylum* enlarged. (c) Toes of *A. m. means* enlarged. [H.P.C. *del.*]

## GENUS AMPHIUMA

### KEY TO THE SUBSPECIES OF AMPHIUMA MEANS

1. Toes normally 2–2; costal grooves 57–60, average 58; no sharp line of demarcation between color of back and lower sides; length to 35″ (889 mm.). Virginia south to Florida, intergrades with *A. m. tridactylum* in extreme western Florida and in parts of Alabama, Mississippi, and Louisiana .................... *means means* p. 50
2. Toes normally 3–3; costal grooves 60–64, average 62; a sharp line of demarcation between color of back and lower sides; length to 40″ (1015 mm.). Gulf States, Alabama to Texas, northward in the Mississippi valley to Missouri ................ *means tridactylum* p. 54

**TWO-TOED CONGO EEL. TWO-TOED AMPHIUMA.** *Amphiuma means means* (Garden). Figs. 10c, 11. Map 2.

TYPE LOCALITY. Not given, but apparently from Charleston or East Florida.

RANGE. Virginia south to southern Florida. Intergrades with *A. m. tridactylum* in extreme western Florida and in parts of Alabama, Mississippi, and Louisiana.

HABITAT. This is a nocturnal, lowlands form, commonly found in drainage ditches, muddy pools, and swamps, in wet swampy meadows and along the margins of muddy sloughs. Often found hiding in crayfish burrows and beneath debris, logs, and bark at the water's edge. That the species may occupy a definite retreat is indicated in the account of Harper (1935, p. 275), who writes, on the basis of a statement by Farley Lee, that a "large 'Snake Eel' . . . used to have a lair in the water beneath a tussock . . . on the Big Water" in the Okefinokee Swamp, Georgia.

SIZE. This species attains a maximum length of nearly 3 feet. Goin (in litt., 1939) has given me measurements of 3 large individuals: Gainesville, Florida, 35″ (889 mm.); Wallacetown, Virginia, 33⅜″ (848 mm.); and Charleston, South Carolina, 32″ (812 mm.). Harper (1935, p. 276) mentions an example from a cypress bay adjoining Chesser's Island in the Okefinokee Swamp, Georgia, that measured 33¼″

(844 mm.). The proportions of a moderate-sized individual from Floyd's Island Prairie, Okefinokee Swamp, sent me by Dr. Francis Harper, are as follows: length, 19″ (484 mm.); tail 5¼″ (133 mm.); head length, 1⅜″ (35 mm.), width ⅞″ (22.5 mm.).

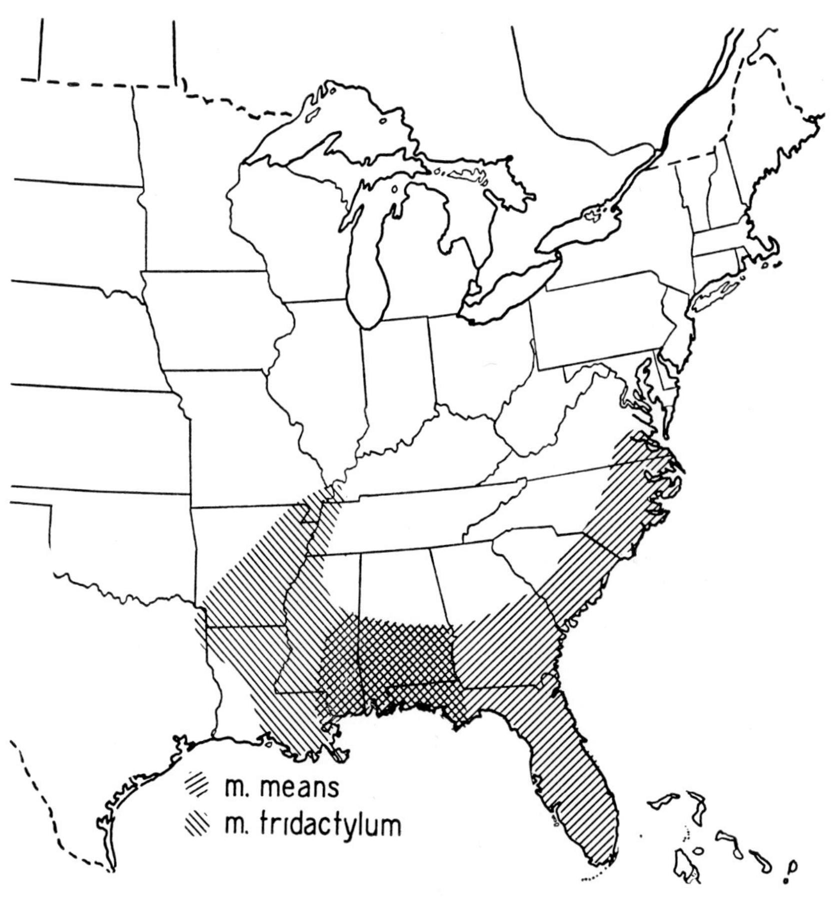


MAP 2.—Distribution of the subspecies of *Amphiuma means*. (*A. means tridactylum* also recorded from Clark County, Indiana.)

DESCRIPTION. This is a uniformly colored, eel-like form, with diminutive limbs and normally with but 2 toes and a single pair of open gill slits. The head is widest just in front of the gill slits, then tapers evenly to the pointed and depressed snout. The eyes are small and without lids,

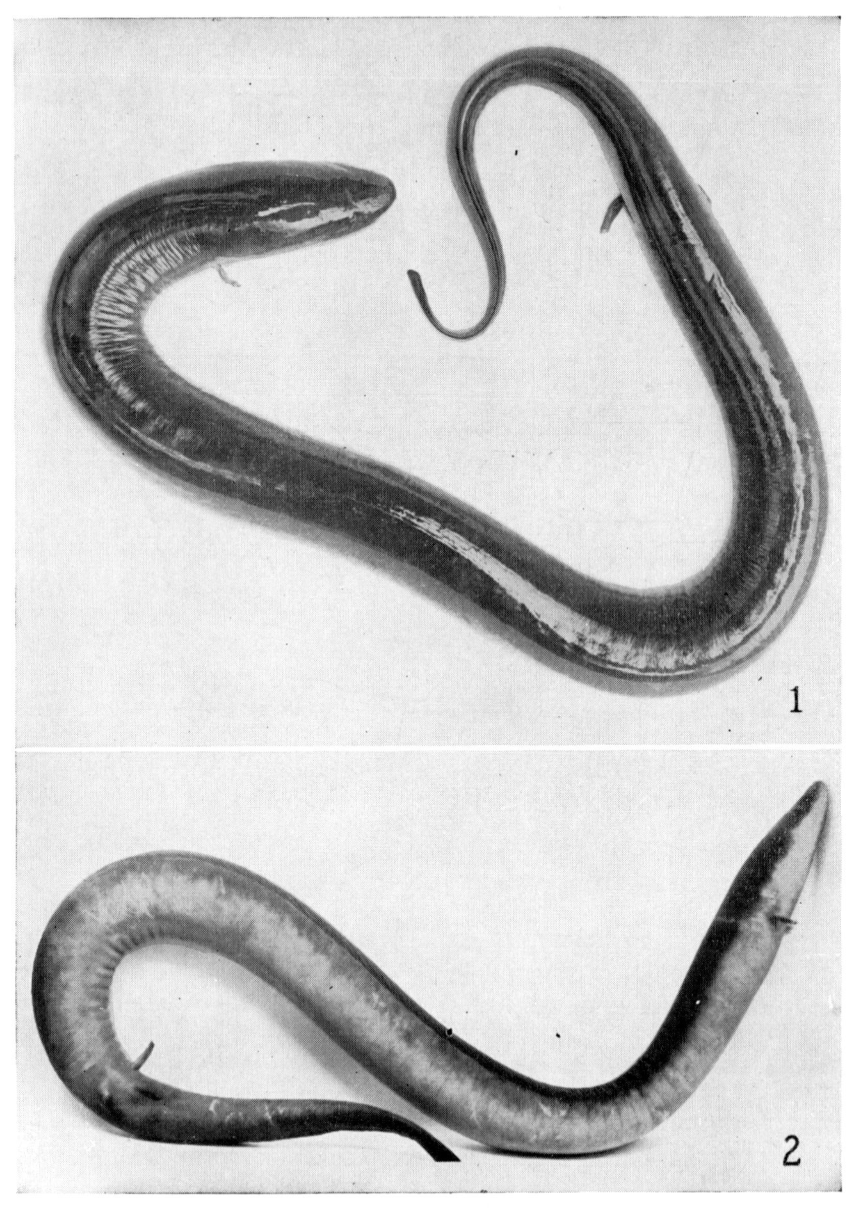


FIG. 11. *Amphiuma means means* Garden. (1) Adult male, actual length 23⅛" (588 mm.). (2) Same, ventrolateral view. North Biloxi, Mississippi.

the horizontal diameter about 4 times in the snout, the latter about 3½ in the head, measured from the tip of the snout to the gill opening. Viewed from the side, the snout projects beyond the lower jaw, and the fleshy margins of the upper jaw are thin and flange-like and overlap those of the lower jaw. The angle of the mouth is about midway between the tip of the snout and the insertion of the fore leg. There is no definite gular fold, but often a fairly distinct one on the side of the head just in front of the gill slit. The trunk is nearly cylindrical in life. Costal grooves 57–60 and average 58, according to Goin (1938, p. 128). In the specimen from the Okefinokee Swamp mentioned above the count is 59. The tail is nearly circular in section at the base, becoming keeled above and rounded beneath, the extreme tip strongly compressed and pointed. Legs very small and slender; in the 19″ specimen the fore legs are 9⁄32″ (7 mm.) long, the hind legs 3⁄8″ (9.5 mm.). Toes usually 2–2, the inner one shorter. The vomerine teeth are in 2 long series which arise at the inner margin of the posterior nares and converge at a point just behind the premaxillaries; approximately 20 teeth in each series.

COLOR. The general ground color in life is deep brown (Mummy Brown, Ridgway) on the dorsal surface of the head, changing to very dark grayish-brown on the trunk and tail, and only slightly paler (slate) on the sides and venter. The color of the dorsal surface fades gradually into the color of the lower sides, so that there is no sharp line of demarcation as in *A. m. tridactylum*. After being in preservative, the general color becomes dull bluish-brown, with the margins of the jaws, the throat, and the venter dirty grayish-white.

BREEDING. Little has been published. Cope's account (1889, p. 218), quoted from Hay, refers to *A. m. tridactylum*. Brimley (1910, p. 10) reported 4 lots of eggs from Hastings, Florida, found in July beneath logs in the partially dried mud of drying pools, and there are eggs in the National Museum collection (No. 61239) taken by S. F. Hildebrand, July 24, 1918, at Hamburg, South Carolina. The following notes are from the unpublished account of Mr. J. A. Weber of Miami, Florida, who has generously permitted their use. On Feb. 3, 1933, Mr. Weber found a nest of this species beneath a board at the edge of a dried pool in the Everglades south of Royal Palm Park, Florida. The nest was an elliptical depression from which a burrow led to subterranean crayfish holes and contained 49 eggs disposed in a single long string at intervals of 5–10 mm., the whole disposed in a flattened ball-like mass and attended by a female 39 cm. long. When discovered the embryos were visible through the egg envelopes and by Feb. 22 had attained a length of 36 mm. Another lot of eggs successfully hatched June 21, 1935, was believed to have been deposited during January, thus giving an incubation period of about 5 months.

LARVAE. Larvae at hatching had an average length of 55 mm. and were black above with a tan belly, dusky chin, grayish throat, and white gills. The legs were functional, the feet each with 2 toes. The young apparently transform at a small size, examples less than 3″ long but without gills having been found (Harlan, 1825, p. 269).

**THREE-TOED CONGO EEL. THREE-TOED AMPHIUMA.** *Amphiuma means tridactylum* Cuvier. Figs. 10a-b, 12. Map 2.

TYPE LOCALITY. New Orleans, Louisiana.

RANGE. Alabama, Mississippi, Louisiana, and Texas on the Gulf, up the Mississippi to Arkansas, Tennessee, and Missouri. Intergrades with *A. means means* in extreme western Florida and parts of Alabama, Mississippi, and Louisiana.

HABITAT. This subspecies is found in lakes, open-spring streams of running water, and streams flowing over calcareous rocks. They are also reported in drainage ditches, bayous, and wooded alluvial swamp lands.

SIZE. Attains a somewhat larger size than *A. means means,* adults reaching an extreme length of about 40″ (1015 mm.). The proportions of a male from Plaquemine, Louisiana, are as follows: length 33¼″ (844 mm.), tail 6¼″ (158 mm.); head length, 2½″ (63 mm.), width 1½″ (37.5 mm.). A female from St. Bernard Parish, Louisiana, measures: length, 20¼″ (515 mm.), tail 4¼″ (108 mm.); head length 1⅞″ (47.5 mm.), width 1⅛″ (28 mm.).

DESCRIPTION. This is a bicolored subspecies which, in typical examples, has 3 toes. The head is widest a short distance anterior to the gill slits, then tapers evenly to the pointed and depressed snout. The eyes are small and circular in outline, flush with the surface of the head and without lids, the horizontal diameter 6–7 times in the snout. The snout about 3

FIG. 12. *Amphiuma means tridactylum* Cuvier. Adult male, actual length 33⅛″ (842 mm.). Plaquemine, Louisiana.

times in the head, measured from the tip to the gill opening. Viewed from the side, the head is wedge-shaped, the tip of the snout obliquely truncated and projecting a short distance beyond the lower jaw. The fleshy margins of the upper jaw are thin, and overlap those of the lower jaw. The angle of the mouth is about midway between the tip of the snout and the posterior margin of the gill slit. The costal grooves vary from 60 to 64, according to Goin (1938, p. 128), and average 62. In the large male from Plaquemine, Louisiana, the count is 63, and in a female from near New Orleans the count is 60. The trunk is cylindrical, the tail oval in section at base, somewhat compressed distally, and more strongly keeled than in many specimens of *A. m. means*. The legs are small and slender; in the 20″ specimen mentioned above the fore legs are only 7/16″ (11 mm.), the hind legs 21/32″ (16 mm.). Toes normally 3–3, the 1st and 3rd about equal, the 2nd 2 or 3 times as long. The vomerine teeth in 2 long series which arise at the inner margin of the inner nares and converge at a point a short distance behind the premaxillaries; approximately 16–22 teeth in each series. The skin is smooth and well supplied with mucous glands. On the head the openings of the sense organs form double series, one on either side above and below the eye, converging on the snout. They also form irregular series on the sides of the head above the angle of the jaws and are continued in indistinct lines along the sides of the trunk.

COLOR. The general ground color of this form is lighter than that of *Amphiuma m. means,* browner above and much lighter below. The dorsal color, which extends on the sides to about the level of the legs, is Benzo Brown or Clove Brown (Ridgway) and is rather sharply cut off from that of the lower sides, which are light gray (Light Mouse Gray, Ridgway). The snout is lighter than the rest of the head and the light color of the sides extends to a line that passes above the gill slit but below the eye.

SEXUAL DIFFERENCES. The sexes may be distinguished by the form of the vent. In the female the vent is a simple slit, the margins bordered by a flat flange-like ridge. In the male the vent is bordered by a narrow raised ridge, and the inner surface anteriorly is lined by oblique serrated ridges.

BREEDING. Hay (1888, p. 316) was apparently the first to give an account of the breeding habits of this subspecies. On Sept. 1, 1887, while collecting in the drying bed of a cypress swamp near Little Rock, Arkansas, he discovered a large specimen coiled about a mass of eggs in a shallow depression beneath a large log. The eggs, about 9 mm. in diameter, contained highly developed embryos and were connected one to another by a slender cord. The entire mass was estimated to contain about 150. The young were coiled within the capsules and upon release were found to be about 45 mm. long, dusky above, with indications of a darker dorsal stripe, and with a dark band on each side. Gills, legs, and toes were well developed. A large male specimen, with sperm held in a viscid mass in the vent, was found during May in northern Tennessee (Davison, 1895, p. 395). Parker (1937, p. 61) found nests in Tennessee under logs in fairly dry soil in August and September and in midwinter. Some were near water and others a considerable distance from it.

Baker (1937, pp. 211–212) reported that the females guard the eggs, nests having been found beneath logs in August and in late September or early October in Tennessee. Adults kept under observation in the laboratory left the water in early July to burrow in the muck and there carry on the mating activities.

On Nov. 16, 1941, a female and her eggs were found below the surface of the ground in a garden at Baton Rouge, Louisiana, by Mr. O. J. Wenzel, and sent to the Museum of Zoology at the University of Michigan. Many of the eggs hatched *en route* to Ann Arbor, Nov. 17, and these specimens a few days later averaged about 64 mm. in total length. A single individual removed from the egg envelopes had a total length of 60 mm. and a tail length of 11 mm.

COLOR. The color in life shortly after hatching was dark brown above and on the sides, with many small, round, light dots scattered generally over the surface; venter pigmented but lighter than the side, gray over flesh; legs well developed, grayish; toes 3–3, fully differentiated; gills very short, white; eye black; dorsal tail fin rather broad and originating above the posterior end of vent; tip of tail rounded, ventral fin confined to the distal half.

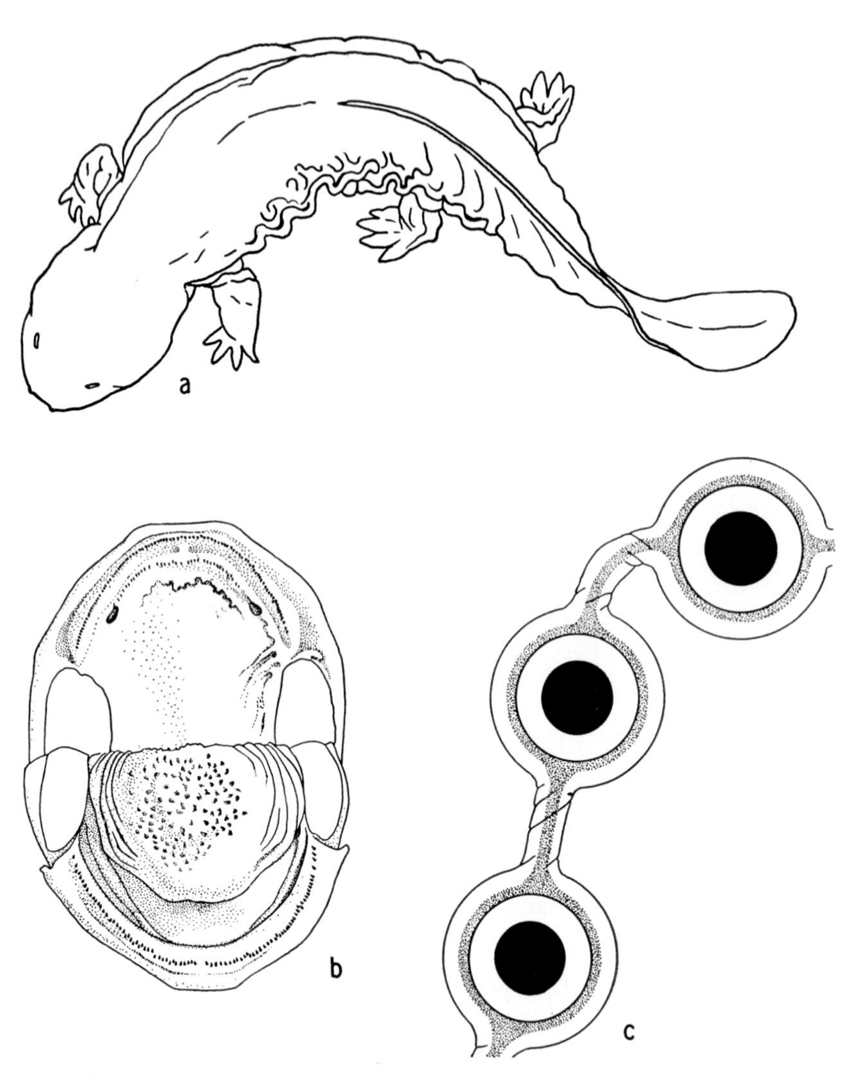


FIG. 13. (a) Outline of *Cryptobranchus alleganiensis* to show the form of the body, lateral folds on trunk and base of tail, and keels on the hind margin of the legs. [H.P.C. *del.*] (b) Mouth parts to show the character of the tongue and teeth. [M.L.S. *del.*] (c) Part of egg-string to show the rosary-like arrangement of eggs. [M.L.S. *del.*]

# Family CRYPTOBRANCHIDAE

Aquatic; body large, depressed, length to 27″ (686 mm.); gills absent, a single pair of open gill slits, sometimes reduced or lost on one side or the other; eyelids lacking; lacrimal and septomaxillary bones lacking; teeth larval; vertebrae amphicœlous; fertilization external. The single genus *Cryptobranchus* in the area considered.

## GENUS CRYPTOBRANCHUS

### KEY TO THE SPECIES OF CRYPTOBRANCHUS

Spiracular opening large, the long diameter contained twice in the distance between the external nares; gular region not noticeably spotted with black; dark pigment on body and tail usually in spots. Susquehanna River, Mississippi River and tributaries from New York and Iowa southward to Missouri and Georgia ........ *alleganiensis* p. 59

Spiracular opening small, often reduced to a pore-like opening, the long diameter contained in the distance between the external nares not less than 2½ times; gular region usually spotted with black; dark pigment of trunk and tail usually in large blotches. Current River and Eleven Point River in Missouri and probably Spring River in Arkansas ........ *bishopi* p. 63

**HELLBENDER. GIANT SALAMANDER. "ALLIGATOR."** *Cryptobranchus alleganiensis* (Daudin). Figs. 13–14. Map 3.

TYPE LOCALITY. Probably in the vicinity of Davenport's plantation, North Toe River, North Carolina.[1]

RANGE. The Susquehanna and tributaries in New York and Pennsylvania; the Ohio and tributaries including the Allegheny; the Mississippi

[1] Harper, F. Amer. Midland Nat. 23 (3): 720. 1940.

River southward to Missouri, Arkansas, and Georgia. Also recorded from Iowa.

HABITAT. Perennially aquatic and usually confined to running waters of fairly large streams and rivers, where it hides by day beneath sheltering objects on the bottom.

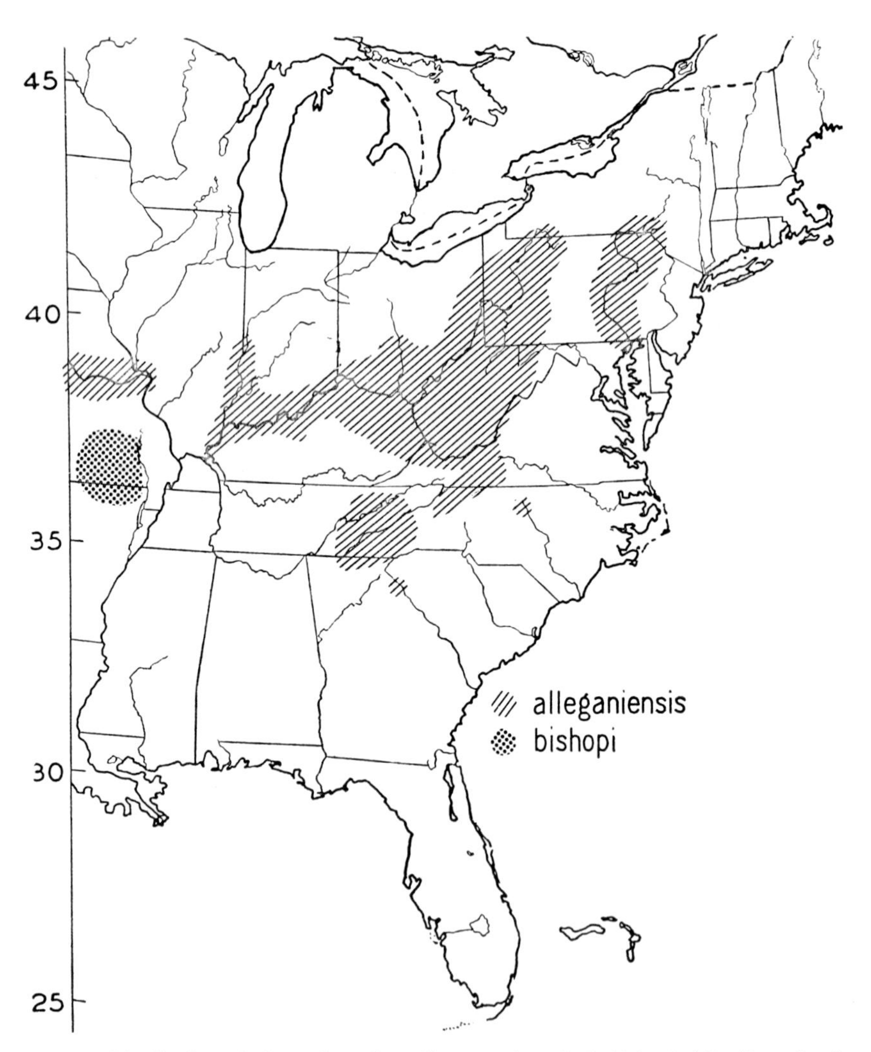


MAP 3.—Distribution of *Cryptobranchus alleganiensis* and *C. bishopi*. (*C. alleganiensis* also recorded from Iowa.)

SIZE. Adults on the breeding grounds average about 17⅞″ (454 mm.) but mature males may vary from 11⅝″ (295 mm.) to 22″ (560 mm.). Females attain a larger size, the maximum length about 27″ (686 mm.).

DESCRIPTION. The Hell-bender is a great soft-bodied creature with a wrinkled fleshy fold along each side of the trunk and a fleshy keel on

FIG. 14. *Cryptobranchus alleganiensis* (Daudin). (1) Adult female, actual length 23″ (585 mm.). (2) Same, ventral view. French Creek, Saegertown, Pennsylvania.

the hind margin of each leg. The head is broad and much depressed, widest back of the eyes and rounding in front to the truncated snout. The trunk is somewhat depressed and there is a median impressed line most strongly developed anteriorily. The tail is very strongly compressed and provided dorsally with a deep, thin keel. The throat has many longitudinal wrinkles that converge on the chin. The gill slits are reduced to a single pair, the opening guarded by narrow lips and concealed by folds of skin on the throat. The eyes are small and the nostrils are placed at the angles of the snout. The costal grooves are not well developed on the sides and are generally lost in the wrinkled skin. The legs are relatively short and broad. There are 5 toes on the hind feet, rounded and flattened at the tip, and with free fleshy margins. The fore feet have 4 toes. The upper teeth are in 2 arched series, those of the maxillary extending to the angle of the jaws, the prevomerine in series of 32–48.

COLOR. The ground color varies from yellowish-brown through dull brick-red to deep brown, almost black. Scattered over the dorsal surfaces there are usually lighter or darker spots irregular in size, shape, and disposition. The ventral surfaces are usually lighter than the back, sometimes spotted, often uniformly colored except for light areas on the throat, in front of the vent, and a few scattered spots on the belly. The sexes are essentially alike in color, but the males are broader and heavier than females of the same length. The vent of the male in the breeding season is surrounded by a broad, swollen ridge.

BREEDING. The mating season in western Pennsylvania and New York begins about the third week in August and extends well into September. At this season the male may be found guarding the nest and accepting females that have not deposited their eggs. Fertilization is external, the exception among American salamanders, and is accomplished by the emission of a white, cloudy, or ropy mass consisting of the seminal fluid and secretions of the cloacal glands. This is discharged into the water as the eggs are deposited by the female. The nest is a saucer-shaped excavation in the stream bottom beneath some large sheltering object, usually a flat rock. The entrance is on the side away from the direct current and the male often lies among the eggs with his head guarding the opening.

EGGS. The eggs are laid in long rosary-like strings, one from each oviduct, and lie in a tangled mass at the bottom of the nest; 300–450 eggs may be deposited by a single female, and several females may lay in a single nest. The individual egg is pale to bright yellow and unpigmented, about ¼″ (6 mm.) in diameter. It is surrounded by a thin, transparent vitellus and floats in a thin fluid enclosed by a thick, laminated outer layer. The fresh egg and its envelopes measure about ¾″

(18–20 mm.) in diameter. The incubation period varies from about 68 to 84 days. In the Allegheny River near Wolf Run, fresh eggs may be found from the third week in August through the first week in September. Some may hatch as early as Oct. 19, others as late as Nov. 23.

LARVAE. At hatching the larvae average about 1 3/16″ (30 mm.) in length. The fore legs are short and stubby, the hind margins keeled and with 2 toes indicated. The hind legs are short, thick, paddle-shaped lobes directed backward. The tail is strongly compressed and with a prominent fin above and below. The larvae are strongly and rather evenly pigmented above, the color extending over the head, trunk, and tail fins but thinning out on the sides to the clear, yellow, yolk-colored belly. The gills are provided with short, flattened filaments. The larvae lose the gills when 4–5½″ (100–130 mm.) in length and at an age of approximately 18 months.

**OZARK HELLBENDER.** *Cryptobranchus bishopi* Grobman. Fig. 15. Map. 3.

TYPE LOCALITY. Current River at Big Spring Park, Carter County, Missouri.

RANGE. Known from the Current River and from Eleven Point River in Missouri, and probably from Spring River, Arkansas.

HABITAT. So far as known confined to larger streams and rivers.

SIZE. The average length of the 12 specimens comprising the type series is 12¾″ (373 mm.), the extremes 10 15/16″ (278 mm.) and 17⅝″ (447 mm.). The proportions of an adult female from Eleven Point River, near Greer Spring, Missouri, are: total length 17⅝″ (447 mm.), tail 4⅜″ (112 mm.); head length 2¾″ (70 mm.), width 2 17/32″ (65 mm.).

DESCRIPTION. The head is depressed, broadly oval in outline, widest about midway between the eye and the spiracle, the sides in front of the eyes converging to the truncated snout. The fleshy folds of the sides of the trunk are continued on the sides of the head nearly to the angle of the mouth. The trunk is depressed, the lateral fleshy folds prominent, in 1–3 somewhat overlapping series on the posterior part of the trunk, but usually with only a single fold anteriorly. Tail strongly compressed, bluntly rounded at tip, the dorsal margin with a thin keel which is continued on the trunk as a low rounded ridge. Legs short, strongly depressed, margined posteriorly by thin, fleshy keels. Toes 5–4, those of the hind feet 1–5–2–3–4 in order of length from the shortest, occasionally 1–5–4–2–3; toes of the fore feet 1–4–3–2. Throat with many wrinkles which converge at the chin. Gill slits small, one or the other often reduced to a pore-like opening, much smaller than in *alleganiensis.* Eyes

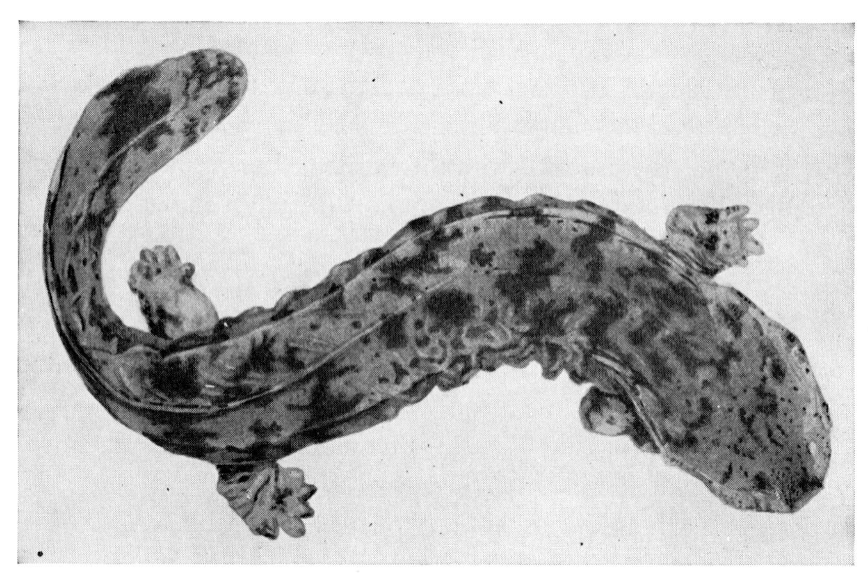

FIG. 15. *Cryptobranchus bishopi* Grobman. Adult female, actual length 15⅞″ (403 mm.). Eleven Point River near Greer Spring, Missouri. [Drawn by Hugh P. Chrisp.]

small, without lids. Nostrils at the anteriolateral angles of the snout, the opening often larger than in *alleganiensis.* The teeth of the upper jaw in 2 broadly arched series, the premaxillary extending to the angle of the jaws, the prevomerine series shorter, with 28–51 teeth in the series counted.

COLOR. Judging only by the condition in preserved specimens, the general ground color of *bishopi* is lighter than that of *alleganiensis,* but the dark blotches are usually noticeably larger. The dark blotches commonly reach their greatest development on the lateral surfaces of the tail, but in some individuals are continued on the dorsal surface of the trunk and head. The gular region in most *bishopi* is rather heavily spotted with black; immaculate or lightly flecked in *alleganiensis*. The venter may be nearly immaculate or with scattered small black spots and larger blotches.

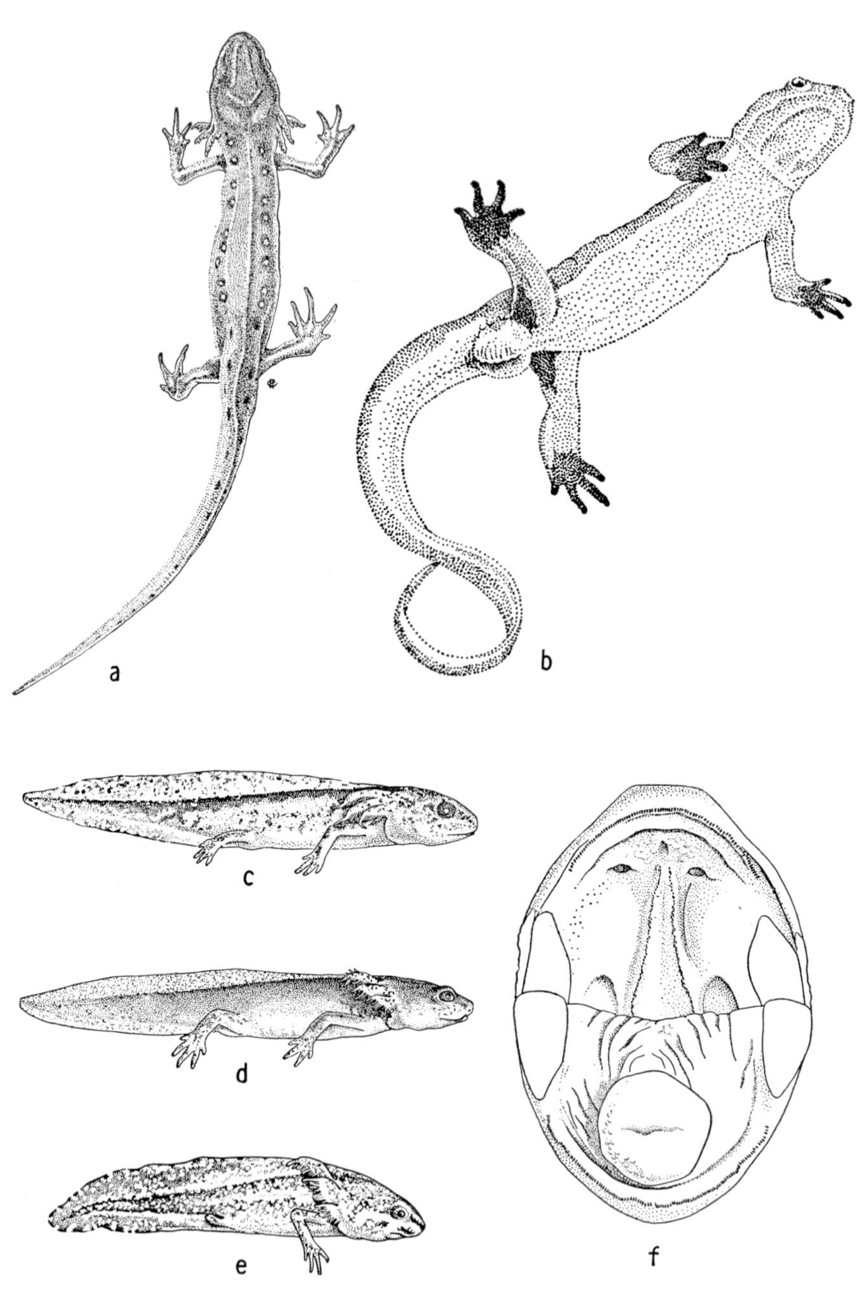


FIG. 16. (a) Neotenic *Triturus v. viridescens,* Coram, New York. (b) *Triturus granulosus twittyi,* Saratoga, California; adult male in breeding condition showing swollen cloacal glands and sexual excrescences on toes, feet, and thighs. (c) Larva of *Triturus torosus.* (d) Larva of *Triturus rivularis.* (e) Larva of *Triturus g. granulosus.* (f) Mouth parts of *Triturus g. granulosus* to show the character of the tongue and teeth. [a, b, H.P.C. *del.;* c–f, M.L.S. *del.*]

# Family SALAMANDRIDAE

Aquatic and terrestrial stages; size small to large, 2½" (64 mm.) to 8½" (216 mm.); lungs present; no nasolabial groove; adults normally without gills; eyes well developed, lids present; toes 5-4; vomero-palatine teeth in long series divergent posteriorly; tongue small, only slightly free at margins; ypsiloid cartilage present; a small, bony pterygoid; vertebrae opisthocœlous. The single genus, *Triturus,* in the area considered.

## Genus TRITURUS

### KEY TO THE SPECIES AND SUBSPECIES OF TRITURUS

1. Uniformly colored above and on sides, no spots, stripes or other markings; venter sometimes with brown-tipped tubercles and dark pigment in large irregular blotches, but never with separate round black dots or spots of pigment; size moderate to large, length to 8½" (216 mm.). West of Rocky Mountains . . . . . . . . . . . . . . . . . . . . . . . . 2

   Color above and on sides not uniform; variously spotted, dotted or striped; belly usually speckled or spotted with black pigment, but no brown-tipped tubercles; size small to medium, length to 4$^{11}/_{16}$" (120 mm.). East of Rocky Mountains . . . . . . . . . . . . . . . . . . . . . . . . 8

2. Color above deep brown to almost black, belly tomato-red; adults usually with a broad dark band crossing vent in male, occasionally in female; skin of breeding individuals smooth; head narrow, snout bluntly pointed; eyes dark brown; length to 7$^{9}/_{32}$" (185 mm.). Northwestern California . . . . . . . . . . . . . . . . . . . . . . . . *rivularis* p. 89

   Color above brown, reddish-brown, or umber; belly yellow to orange but never tomato-red; sometimes a pointed or rounded lobe of pigment on sides of vent, or dark band crossing posterior end of vent; eyes not dark brown; skin of aquatic adults moderately smooth to strongly roughened; head broad, snout broadly rounded . . . . . . . . . . 3

3. Adults with dark pigment of lower sides encroaching irregularly on venter, most pronounced on throat and in area between fore legs, occasionally on belly; tail of breeding male with narrow keels; lower surface of legs and tail mainly black or blotched with black; posterior end of vent crossed by a black band in both sexes; length to 8½" (216 mm.). Crater Lake, Oregon . . . . . . . . . . . . *granulosus mazamae* p. 74

   Dark pigment of lower sides not encroaching irregularly on venter; lower surface of legs never wholly black; posterior half of vent not crossed by a black band . . . . . . . . . . . . . . . . . . . . . . . . . . . . . . . . . 4

4. Above dark brown to nearly black, reddish-brown, dark yellowish-brown to burnt umber; skin of aquatic adults sparsely to densely studded with small, brown-tipped tubercles; sides of vent of male, less frequently of female, with rounded lobes of pigment; vent of breeding female on a conical elevation with opening directed obliquely downward; vomero-palatine teeth in 2 lines diverging evenly posteriorly . . 5

   Above dilute chocolate-brown, dark seal-brown, or reddish-brown to burnt umber; skin of aquatic adults relatively smooth, with brown-tipped tubercles, when present, mainly limited to margin of lower jaw and anterior part of belly; or, skin roughened by numerous warts; vent of male sometimes with a pointed lobe of pigment; vent of female not on a conical elevation; vomero-palatine teeth in 2 lines nearly parallel anteriorly, diverging abruptly and curving outward posteriorly . . . . . . . . . . . . . . . . . . . . . . . . . . . . . . . . . . . . . . . . . . . 6

5. Color above reddish-brown to deep brown, venter yellow to bright reddish-orange; skin of aquatic adults sparsely, if at all, studded with brown-tipped tubercles, within small reticulated areas on back and venter; pigment of lower sides sharply cut off from lighter venter along a nearly straight line; pigment on legs extending to ventral surface; fingers and toes short and wide; length to 7$^{15}/_{16}$" (201 mm.). Mendocino County, California, northward through western Oregon and Washington to British Columbia and Alaska . . . . . . . . . . . . . . . . . . . . . . . . . . . . . . . . . . . . . . . . . . . . *granulosus granulosus* p. 70

   Color above dark yellowish-brown to burnt umber, venter yellow to orange but less bright than in *T. g. granulosus;* skin of aquatic adults usually densely studded with brown-tipped tubercles; pigment of lower sides somewhat diffuse and meeting lighter venter along an irregular line; pigment on legs involving only upper half; fingers and toes longer; length to 7$^{15}/_{32}$" (196 mm.); average of males 6$^{11}/_{16}$" (170 mm.), average of females 5$^{13}/_{16}$" (140 mm.). Saratoga County, California, northward to Navarro River, Mendocino County, California . . . . . . . . . . . . . . . . . . . . . . . . . . . . . . . . . . *granulosus twittyi* p. 77

6. Skin with many warts above and below, particularly noticeable around eyes, lips, fingers, toes, and vent; snout short and blunt; length to 6$^{5}/_{32}$" (157 mm.); average of males 6$^{1}/_{16}$" (154 mm.), average of fe-

male 5 7/16" (139.5 mm.). San Diego County, California .... *klauberi* p. 80

Skin without warts and relatively smooth in aquatic adults; sometimes with low, rounded tubercles tipped with brown around margins of lower jaw and on belly .......................................... 7

7. Above chocolate-brown, venter burnt orange; skin over eyes pale; with many small brown-tipped tubercles on margin of lower jaw, between fore legs and at least on sides of belly and frequently along mid-line; tail fin of breeding male of moderate width; tail 56 per cent of total length in male, 52 per cent in female; length to 7 5/8" (194 mm.), average of males 7" (178 mm.), of females 6 1/8" (156 mm.). Vicinity of Chico, California .......................................... *sierrae* p. 93

Above dilute chocolate to dark seal-brown, breeding individuals paler; with few or no brown-tipped tubercles on margin of lower jaw, when present on belly, restricted usually to sides; tail fins of breeding male wide; tail 53 per cent of total length in male, 50 per cent in female; length to 7 3/4" (197 mm.). Northern Monterey County to middle Mendocino County, California .................... *torosus* p. 95

8. With a dorsolateral series of small, round, black-bordered red spots; or, with dorsolateral red stripes which may be entire or broken at intervals .......................................................... 9

Without black-bordered crimson spots or red stripes, occasionally with a few small dorsolateral pinkish or yellowish spots which are without black borders .............................................. 11

9. Above brown to olive-green; a dorsolateral series of small, round, black-bordered red spots; belly yellow with black specks or dots. Skin smooth, tail strongly compressed, keeled (aquatic adults); skin rough, tail oval in section, ground color reddish-brown to bright orange-red, with dorsolateral series of black-bordered red spots (terrestrial stage); length to 4 3/4" (120 mm.). Eastern North America, Hudson Bay, and northern Ontario to Georgia and Alabama, northward to Wisconsin and Michigan ............................ *viridescens viridescens* p. 99

With dorsolateral red stripes .................................... 10

10. Above olive-green; a dorsolateral, black-bordered red stripe interrupted at one or more points and usually limited to the trunk and posterior part of head; lower sides and venter yellow, speckled with black. Skin smooth, tail compressed (aquatic adults); skin rough, tail oval in section, ground color reddish (terrestrial stage). Adult males with 3–3 pits on side of head; adult females with 3–3 or fewer, smaller, pits; length to 3 19/32" (92 mm.). North and South Carolina, on the Coastal Plain and Piedmont .................. *viridescens dorsalis* p. 103

Ground color above uniformly brown to olive-green; a narrow, usually unbroken dorsolateral stripe on head and trunk, bright red to red strongly suffused with dusky; stripe often broken on tail; often a narrow middorsal light stripe; sometimes a row of red spots on lower sides; belly light yellow with small black flecks, occasionally immaculate; adult male with 3–3 pits on sides of head, female without pits; skin smooth, tail keeled (aquatic stage); skin rough, tail oval in section, color reddish (terrestrial stage); length to 2 15/16" (75 mm.). Southeastern Georgia and northern Florida ............ *perstriatus* p. 86

11. Ground color above light yellowish-brown to olive-brown or olive-green with scattered small black dots; sometimes with a few small dorsolateral light specks, yellowish or pale red but not black-bordered; belly straw-yellow, tinged with orange in the region of the vent and with a few small scattered black dots; adult male with 3–3 pits on sides of head, female without pits; length to 3½" (89 mm.). South Carolina to Florida; west to Texas, Oklahoma, and Kansas; northward to Wisconsin, Minnesota, and Michigan ......... *viridescens louisianensis* p. 106

Olive-green above, marked with numerous large round black spots and smaller, dull yellow spots or dashes and a narrow yellow line on either side of mid-line of back; sides tinged with bluish-green; venter bright orange with large rounded black spots; length to 4⅛" (105 mm.). Southeastern Texas ......................... *meridionalis* p. 82

**OREGON NEWT.** *Triturus granulosus granulosus* (Skilton). Figs. 16e-f, 17. Map. 4.

TYPE LOCALITY. Oregon.

RANGE. From Mendocino County, California, northward through western Oregon, Washington and adjacent islands, British Columbia, and possibly Alaska.

HABITAT. The aquatic adults are common in roadside ditches and borrow pits and in the streams, lakes, ponds, and woodland pools. The terrestrial newts are found wandering about on the forest floor or hiding beneath logs and bark.

SIZE. The average length of 18 adult males from Oregon is 7⅛" (181 mm.), with extremes of 6 5/16" (160 mm.) and 7 15/16" (201 mm.). A series of 14 adult females average 5½" (140 mm.), with extremes of 5⅛" (130 mm.) and 6" (152 mm.). The proportions of an adult male from Newport, Oregon, are as follows: total length 6 15/16" (176 mm.), tail 3⅞" (98 mm.); head length ¾" (19 mm.), width 19/32" (15 mm.). An adult female from the same locality has the following measurements: total length 5¾" (146 mm.), tail 3 5/32" (80 mm.); head length 21/32" (16 mm.), width 17/32" (13.5 mm.).

DESCRIPTION. The head is broad and slightly convex above, widest at the angle of the jaws, the sides converging gently to the lateral extensions of the gular fold and the sides of the snout tapering abruptly to the truncated tip. Eyes moderate, crossed by a dark horizontal bar, the

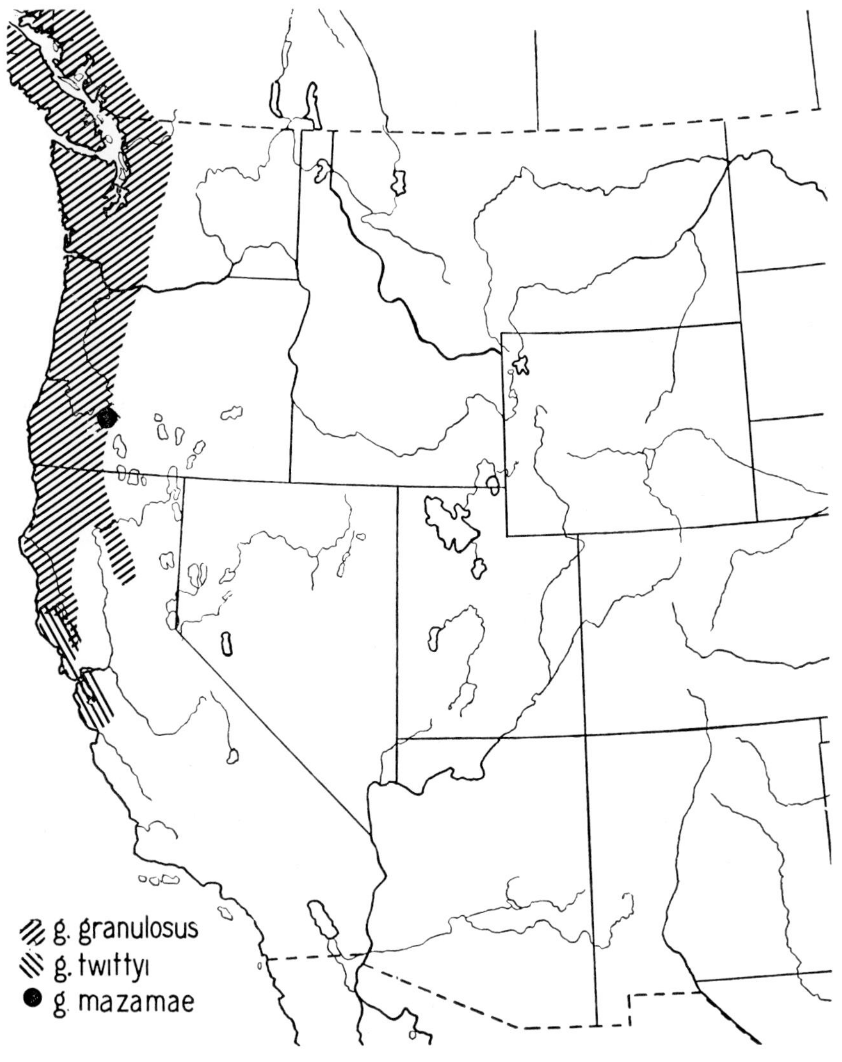


MAP 4.—Distribution of the subspecies of *Triturus granulosus*. (*T. g. granulosus*, or a form closely related to it, extends as far north as Admiralty Island, Alaska.)

iris above whitish, below dark with a whitish spot or bar immediately below the pupil. Interorbital distance about 2⅓ in the head, measured along the side below the eye from the gular fold to nostril. Trunk well rounded above and on the sides, belly flattened. Costal grooves not de-

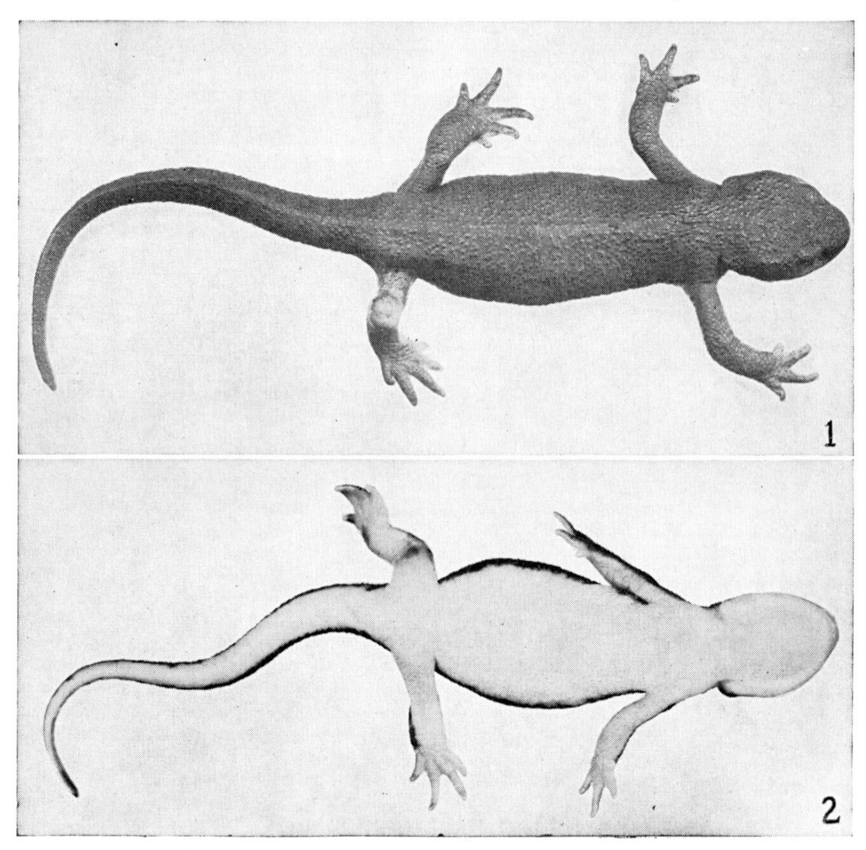


FIG. 17. *Triturus granulosus granulosus* (Skilton). (1) Adult male, actual length 5⁹⁄₁₆″ (142 mm.). (2) Same, ventral view. Robinson Creek, Ukiah, California. [V. C. Twitty, collector.]

veloped. Tail broadly oval in section at base, becoming sharp-edged above a short distance beyond the vent and strongly compressed distally; in breeding males the dorsal tail keel is widened so that its highest point is slightly above the level of the back; the ventral keel is narrow and does not reach the vent. Legs and feet strongly developed; toes short, flattened, and wide at base, abruptly tapering, those of the hind feet 1–5–2–4–3 in order of length from the shortest; toes of the fore feet 1–4–2–3. Palmar and plantar tubercles, when present, poorly developed. Skin tuberculate within small reticulated areas, sometimes a few tubercles tipped with brown. Males in breeding season with skin relatively smooth. Tongue small, broadly oval in outline, and slightly free at the sides. The palatine teeth in 2 rows nearly in contact anteriorly, evenly diverging posteriorly and becoming separated by a distance about equal to twice the diameter of an inner naris.

COLOR. The general ground color above varies from deep brown to reddish-brown. The dark pigment involves all dorsal surfaces and the sides of the head to the level of the nostril, leaving the upper lip light; and also the sides of the trunk to a level which involves the entire dorsal and lateral surfaces of the legs and all but a narrow stripe along the ventrolateral surface of the tail. In the male there is frequently a pointed lobe of pigment extending toward or reaching the vent, and in both sexes the dark pigment encroaches on the ventral surface of the forearms. The color of the ventral surfaces varies from yellow to bright reddish-orange.

SEXUAL DIFFERENCES. The sexual differences are expressed in the greater size of the males, and by the form of the vent, which, in the female, is small and cone-like, with the opening directed obliquely downward, while in the male it is larger and longer and, in the breeding season, may be greatly swollen and strongly protuberant, the sides ridged and grooved, the opening with short free filaments. Soles of the feet and a spot at base of thighs below often pigmented in breeding males.

BREEDING. The breeding season is apparently an extended one. Slater and Brockman (1936, p. 113) reported that eggs are laid soon after the ice melts from the ponds in northern Washington, and Gordon (1939, p. 51) remarked that eggs are laid in late winter in the vicinity of Corvallis, Oregon. In our own experience, mated pairs collected June 1 and 2, 1936, at Miranda, California, deposited many eggs a day or so later in captivity, and freshly deposited eggs were collected June 21, 1936, in a pond 2 miles west of Lake Crescent, Washington. The eggs are deposited singly or in small clusters and have a diameter of 1.85–2 mm.; with their envelopes they average about 3.3 mm., but may reach 4 mm. There are 3 conspicuous envelopes in addition to the vitelline membrane, the 2 inner ones about equal in thickness, and together about ½ the diameter of the outermost.

LARVAE. The young and half-grown larvae are strongly blotched and mottled on the fins, the pigment somewhat concentrated along the margins. The dorsal fin arises at the back of the head and reaches its greatest width about opposite the vent; the ventral tail fin extends to the vent. There is a conspicuous dark bar from the eye to the nostril, and usually 2 rows of larval light spots on the trunk, the upper next the dorsal fin, the lower along the sides above the legs, where, in older individuals, they often form a light stripe. In some large larvae the mottling of the fins is largely lost, but remnants of larval blotches remain in the form of rounded dark spots on a more uniform ground color of reddish-brown. Larvae may attain a length of 55 mm. and transform the fall of the year in which they were hatched, or spend an additional winter and spring in the water and reach a length of 70 mm. before transforming in June and July.

**CRATER LAKE NEWT.** *Triturus granulosus mazamae* Myers. Fig. 18. Map 4.

TYPE LOCALITY. Mt. Mazama, Crater Lake National Park, Klamath County, Oregon.

RANGE. Vicinity of Crater Lake, Oregon.

HABITAT. At present known only from the higher slopes of Mt. Mazama. Nothing has been published on the character of the habitat.

SIZE. The average length of thirteen adult males is 6 9/16″ (166 mm.), the extremes, 5½″ (139 mm.) and 8 7/16″ (214 mm.); nine adult females average 5⅞″ (149 mm.), the extremes 5¼″ (133 mm.) and 6¾″ (171 mm.). The proportions of a male are as follows: total length 8″

(202 mm.), tail 4 7/16" (113 mm.); head length 25/32" (20 mm.), width 5/8" (16 mm.). An adult female has the following measurements: total length 5 13/16" (147 mm.), tail 3 1/16" (78 mm.); head length 5/8" (16 mm.), width 17/32" (13 mm.).

DESCRIPTION. The head is broadly oval in outline, widest opposite the angle of the jaws, the sides converging to the lateral extensions of the gular fold, in front more abruptly to the truncated snout. The eyes are

FIG. 18. *Triturus granulosus mazamae* Myers. (1) Adult male, actual length, 7 7/8" (200 mm.). (2) Same, ventral view. Crater Lake, Oregon. [Drawn by Hugh P. Chrisp.]

of moderate size, the horizontal diameter equal to the distance from the anterior angle of the eye to the nostril of the same side. Trunk fairly stout, rounded above and on the sides, flattened below. Legs stout, toes 5-4. Toes rather short and bluntly tipped, 1-5-2-4-3 or 3-4 in order of length from the shortest, the first much reduced; toes of fore feet 1-4-2-3. Tail long, slender, compressed, becoming sharp-edged above a short distance behind the vent; ventral tail keel narrow and not reaching vent. Tongue small, broadly oval in outline. Vomero-palatine teeth in 2 series, in contact anteriorly, and only slightly diverging posteriorly for the anterior ⅔ of their length, then more abruptly. In some specimens nearly V-shaped (apex forward), evenly diverging posteriorly.

COLOR. This is a highly pigmented form, the preserved specimens being uniformly dark brown above and on the sides, the light areas on the venter yellow (possibly orange in life). The dark pigment extends on the sides of the head to involve the upper lip (rarely a light line along the extreme lower margin of upper lip), on the sides of the trunk to involve the under surface of the limbs, and on the tail so that only the extreme ventral margin of the basal part is light in color. In many individuals there are large dark blotches on the throat, on the area between the fore limbs, and in varying degrees on the sides of the venter. A broad dark bar crosses the posterior half of the vent in most individuals of both sexes.

SEXUAL DIFFERENCES. The sexes may be distinguished not only by average-size differences, as indicated above, but by tail-length, that of the males averaging 55 per cent of the total and of the female 53 per cent. The form of vent is also characteristic of the sex, that of the mature male being large and strongly protuberant, the sides marked by oblique ridges separated by narrow grooves which converge at the slit, the ridges studded by a few dark-tipped tubercles. Internally there is, at the anterior end of the vent, a triangular lobe with the point directed caudally and bearing, ventrally, a round, flat, disk-like structure which in some individuals is highly pigmented. On either side of this triangular lobe is a fleshy fold, with oblique grooves, which is expanded into a rounded lobe behind and meets its fellow of the opposite side at the middle line. In the female the vent is smaller, the sides with oblique grooves and folds which are continued within the margins.

BREEDING. Nothing has been published on the breeding habits, eggs, or larvae of this race. An adult female taken July 7, 1938, at Crater Lake, possessed ovarian eggs of large size, but in the majority of specimens taken at the same time and place the eggs were small.

Four of 26 specimens examined retained rudiments of the gills. These varied in length from 4⅞″ (124 mm.) to 6″ (152 mm.) and involved both sexes. For the loan of a fine series for study, from the type locality, I am indebted to Dr. G. S. Myers.

**TWITTY'S NEWT.** *Triturus granulosus twittyi* Bishop. Figs. 16b, 19. Map 4.

TYPE LOCALITY. Saratoga, California.

RANGE. Santa Clara, Santa Cruz, and Marin Counties, California. Intergrades are known from Marin, Napa, and southern Mendocino Counties.

HABITAT. This subspecies is common in streams, ponds, and pools, and is occasionally found on land.

SIZE. Twenty adult males from Saratoga and Boonville, California, average $6\frac{11}{16}''$ (170 mm.), the extremes $5\frac{21}{32}''$ (145 mm.) and $7\frac{15}{32}''$ (196 mm.). Twelve adult females from the same localities measure as follows: total length $5\frac{13}{16}''$ (149 mm.), extremes $5\frac{1}{4}''$ (134 mm.) and $6\frac{1}{4}''$ (159 mm.). The proportions of an adult male from Saratoga are as follows: total length $7\frac{3}{32}''$ (180 mm.), tail 4″ (102 mm.); head length $\frac{26}{32}''$ (21 mm.), width $\frac{23}{32}''$ (19 mm.). An adult female from the same locality measures: total length $6\frac{1}{4}''$ (159 mm.), tail $3\frac{3}{8}''$ (86 mm.); head length $\frac{23}{32}''$ (19 mm.), width $\frac{21}{32}''$ (17 mm.).

DESCRIPTION. The sides of the head back of the eyes are broadly rounded to the lateral extensions of the gular fold, where there is a definite constriction. In front of the eyes the sides converge rather abruptly to the bluntly pointed snout. The eyes are of moderate size, the long diameter about twice in the snout. The eye is crossed by a broad, dark, horizontal bar, the iris above silvery, below mottled silvery and black. The interorbital distance about $2\frac{1}{2}$ in the length of the head measured along the side from the tip of the snout to the lateral extension of the gular fold. The trunk is well rounded above and on the sides, flattened below. Costal grooves not developed. The tail is broad at base, subquadrate in section, becoming sharp-edged a short distance behind the vent and strongly compressed distally. Legs and feet stout. Toes 5–4, moderately long, depressed and widened basally, pointed at tip, those of the hind feet 1–5–2–4–3 in order of length from the shortest; toes of the fore feet 1–4–2–3. Outer palmar tubercle developed, inner poorly developed or lacking. Skin roughened and finely tuberculate, the brown-tipped tubercles usually very numerous on the belly, often single, sometimes in little clusters of 5 or 6 within small reticulated areas,

FIG. 19. *Triturus granulosus twittyi* Bishop. (1) Adult female, actual length 6″ (153 mm.). (2) Same, ventral view. Saratoga, California. [V. C. Twitty, collector.]

fewer and larger on the back and sides. Tongue small, nearly circular in outline, and slightly free at the sides. The palatine teeth in a V-shaped series, the apex directed forward, slightly diverging for ½ the length, then more strongly diverging so that the rearmost are separated by about the distance between the inner nares.

COLOR. The general ground color above varies from dark yellowish-brown to burnt umber, the dark pigment covering the dorsal surfaces and the sides to involve the upper ½ or ⅔ of the legs, the upper ¾ of the tail and the sides of the head to below the nostril and eye. The ventral surfaces vary from yellow to orange but are less bright than in *T. g. granulosus*. In a few males the dark pigment encroaches on the sides of the vent in the shape of blunt lobes.

SEXUAL DIFFERENCES. During the breeding season, sexual differences are well marked. The vent of the male is large, strongly protuberant, the sides with oblique broken folds, the opening lined with fleshy lobes tipped with short filaments. The soles of the feet and a patch on the under side of the thigh near the base often pigmented. The vent of the female is small and cone-like, the opening directed obliquely downward.

BREEDING. The eggs have been found, usually in quiet water, as early as Feb. 25 near Saratoga, California (Twitty, 1935, p. 76), with the height of the breeding season probably in March. There is some evidence, as indicated by Twitty, that breeding may be extended over a considerable period of the year. On May 20, 1936, we collected adults of both sexes at Saratoga, and on May 23 eggs were found in the containers. The eggs were deposited singly and in small clusters of 2 or 3. Individual eggs I have measured have a diameter of 2 mm., or, with the envelopes, 2.5 to 3 mm. In addition to the vitellus there are only 2 well defined envelopes, the inner one thin and having the appearance of being finely laminated, the outer about twice as thick, softer and more homogeneous.

LARVAE. Larvae at hatching are provided with balancers as in *T. torosus* and have a length of approximately 12 mm. at the time of independent feeding (Twitty, 1935, p. 79). At first there is a poorly defined dark dorsal stripe on each side, but this soon becomes diffuse and the pigment becomes distributed over the back and sides to form a very light reticulated pattern. The dorsal fin arises back of the head and extends to the tip of the pointed tail; the ventral fin extends to the vent. The larvae may attain a length of 75 mm. before transforming. Large larvae may have the fins lightly mottled and blotched, and there are a few small, scattered, dark blotches of irregular size and shape on the sides of the trunk and tail. It is evident that not all larvae transform in the fall of the year in which they were hatched, for we have collected individuals up to 70 mm. in length during May, when some adults were just depositing eggs.

**WARTY NEWT.** *Triturus klauberi* (Wolterstorff). Fig. 20. Map 6.

TYPE LOCALITY. Boulder Creek, San Diego County, California.

RANGE. San Diego County, California.

HABITAT. The breeding populations are known chiefly from Boulder and Cedar Creeks, in San Diego County, California. Klauber (1934, p. 4) writes: "This newt is found in localities where permanent water is available. While usually present in the pools of streams during the spawning season, it spends most of its life on land. As in the case of most amphibians, night is the time of its greatest activity; in the daytime, unless in water, it will usually be found hidden in leaves and under logs."

SIZE. The average length of 27 sexually mature females from the type locality, sent me by Mr. L. M. Klauber, is $5\frac{1}{2}''$ (139.5 mm.), with extremes of $5\frac{3}{32}''$ (126 mm.) and $6\frac{1}{32}''$ (153 mm.). Three males from the same locality average $6\frac{1}{16}''$ (154 mm.), with extremes of $5\frac{29}{32}''$ (150 mm.) and $6\frac{3}{16}''$ (157 mm.). The proportions of a male and a female having the same total length are as follows: Male, total length $5\frac{29}{32}''$ (150 mm.), tail $3\frac{9}{32}''$ (83 mm.); head length $\frac{11}{16}''$ (17 mm.), width $\frac{9}{16}''$ (14 mm.). Female, total length $5\frac{29}{32}''$ (150 mm.), tail $3\frac{3}{16}''$ (81 mm.); head length $\frac{21}{32}''$ (16 mm.), width $\frac{5}{8}''$ (15.5 mm.).

DESCRIPTION. This species is characterized by its short, broad head and the extreme wartiness of the skin. The head is widest opposite the eyes, the sides behind tapering quite strongly to the lateral extensions of the gular fold and in front more abruptly to the short, blunt snout. Eyes moderate, the interorbital width a little less than the distance from the nostril to the angle of the mouth, and only twice in the length of the head measured from the nostril to the lateral extension of the gular fold. Costal grooves not developed. Gular fold strongly developed. Trunk well rounded, flattened below. Legs stout, toes short, 5–4, those of the hind feet 1–5–2–4–3 in order of length from the shortest; toes of the fore feet 1–4–2–3, the 2nd and 3rd about equal. Tail subquadrate in section at base, becoming somewhat compressed and sharp-edged

Fig. 20. *Triturus klauberi* (Wolterstorff). (1) Adult female, actual length 5³⁄₁₆″ (132 mm.). (2) Same, ventral view. San Diego, California. [L. M. Klauber, collector.]

above and below distally. Tongue small, roughly circular in outline, free at the sides only, and often with a few rounded tubercles. Palatine teeth in 2 series, parallel and narrowly separated anteriorly for ½–⅔ the length, abruptly diverging posteriorily, or, the posterior ⅓ or ½ curved outward and backward. Skin everywhere beset with many wart-like tuberosities, often most conspicuous around the eyes, along the margins of the mouth, toes, and vent.

COLOR. In life, according to Klauber (1934, p. 4), the general color is burnt orange above and yellow below. The dark pigment extends on the side of the head to the lower margin of the eyes, on the trunk to the lower level of the legs and the exposed surfaces of the tail. The eyelids, the tip of the snout, and the toes are frequently lighter in color than adjacent areas.

SEXUAL DIFFERENCES. The sexes may be distinguished by the form of the vent, which in the male is large and quite strongly protuberant and in the female smaller and directed obliquely downward.

BREEDING. Little is known of the breeding habits. A female taken April 8 in Boulder Creek, San Diego, California, by L. M. Klauber, has ovarian eggs 2 mm. in diameter. The eggs are brown above and dull yellow below after being in preservative. The sides of the vent in this individual are swollen and somewhat protuberant, broken into ridges and reticulated areas, and studded with a few brown tubercles.

**TEXAS NEWT.** *Triturus meridionalis* (Cope). Frontispiece. Fig. 21. Map 5.

TYPE LOCALITY. Matamoros, Tamaulipas, Mexico . . . "tributaries of the Medina River and southward."

RANGE. East central Texas from Houston south to Brownsville, west to San Diego, San Antonio, and Helotes, north to Waco; Mexico, vicinity of Matamoros.

HABITAT. The aquatic adults are found in the lagoons, bayous, and other quiet waters. Strecker (1922, p. 6) found them in numbers in Laguna Lake, Falls County, Texas. The land form has also been reported by Strecker.

SIZE. In a series of specimens from Kingsville and the Rio Grande Valley in Texas, collected by Mr. Clyde T. Reed, and in specimens from Brownsville, Texas, the females vary in length from 95 to 105 mm. and the males from 86 to 95 mm. In the females the tail comprises 46–50 per cent of the total length and in the males 46–53 per cent.

DESCRIPTION. This is a stout little newt with a relatively broad and flat head, rounded trunk, and somewhat compressed tail. The skin is slightly roughened even in the aquatic adults, and in this respect differs from most other newts. The head above has 2 narrowly separated and nearly parallel ridges which originate on the snout and extend nearly to the back of the head. In some individuals these ridges fade out be-

MAP 5.—Distribution of *Triturus perstriatus* and *T. meridionalis.*

hind, in others they converge and continue as a low median ridge along the back. In living specimens these ridges may not be conspicuous. The eyes are crossed by a horizontal dark bar with the iris, above and below, brassy yellow. The tail is oval in cross section near the base, but becomes more flattened with narrow keels above and below, beginning a short distance behind the vent. The costal grooves are very poorly defined. The fore legs are rather slender, the 4 toes 1–4–2–3 in order of length from the shortest; hind legs stout, the 1st and 5th toes short, about equal in length, others 2–4–3. The tongue is small and fleshy, and slightly free only at the lateral margins. The teeth on the roof of the mouth are in

FIG. 21. *Triturus meridionalis* (Cope). (1) Adult female, actual length 4 1/16" (104 mm.). (2) Same, lateral view. Harlingen, Texas. [From Mrs. L. I. Davis, Harlingen, Texas.]

2 long series, widely separated behind, converging abruptly for half their length, then more gradually to a point between the inner nares.

COLOR. The ground color varies from light blue-green on the sides to light olive-green on the back. Scattered at random over the dorsal surface are rounded black spots of moderate size and more irregular, somewhat elongate, diffuse yellow spots. There are no red spots. Originating on the snout and continuing on the head and trunk are 2 narrow yellow lines, one on each side of the middorsal line. These lines are irregular and broken on the head, but become well defined on the trunk, fading out on the basal half of the tail. The legs are marked above with small round black spots and a few diffuse, yellow blotches; the toes are pale green at the base with tinges of yellow toward the tips. The lower sides have the black spots arranged in irregular lines and between them some yellow blotches. The black spots are slightly larger along the mid-line of the side of the tail, where they form a fairly regular series; a less well defined series of smaller black spots follows the edges of the tail keels. The sides of the head below the level of the eyes are yellowish. The entire ventral surface is bright orange, over which are scattered many rounded black spots. The black spots are small on the throat and legs, larger on the belly and tail. The orange of the ventral surface of the tail extends upward onto the lower third of the sides of the tail and halfway on the sides of the legs.

SEXUAL DIFFERENCES. Sexual differences are quite strongly marked. The head of the male is widest behind, tapers gently to the eyes and then abruptly to the pointed snout; in the female, the sides of the head behind the eyes are parallel. On the side of the head of the male, in a line behind the eye, there are 2 or 3 elongate pits, partially covered above by a well defined ridge. The hind legs of the male are stouter than those of the female, and during the breeding season are armed with transverse horny ridges on the under surface of the thighs and with horny tips to the toes. The tail fins of the male are broader than those of the female, especially during the breeding season. The vent of the male is larger and more protuberant, but in both sexes the margins of the vent are thrown into folds.

BREEDING. Strecker (1922, p. 6) reports that this species breeds in March and April. Single eggs were found enclosed by immersed leaves in the smaller sloughs, and larvae were seined from small arms of these sloughs. Strecker found the young larvae indistinguishable from larvae of Triturus from Raleigh, North Carolina.

Boulenger, 1888, p. 24 (*Molge meridionalis*); Cope, 1880, p. 30 (*Diemyctylus miniatus,* subsp. *meridionalis*); 1888, p. 395; Fowler and Dunn, 1917, p. 28; Strecker, 1908, p. 56; 1908a, p. 80: 1908b, p. 48 (*Die-*

*myctylus meridionalis*); 1915, p. 54 (*D. v. meridionalis*); 1922, p. 6 (*Notophthalmus meridionalis*); Wolterstorff, 1925, p. 293 (*Diemictylus meridionalis*).

**STRIPED NEWT.** *Triturus perstriatus* Bishop. Fig. 22. Map 5.

TYPE LOCALITY. Dedge Pond, 2 miles east of Chesser's Island, Charlton County, Georgia.

RANGE. Lakeland, Lanier County, Georgia, and the Okefinokee Swamp in southeastern Georgia southward to central Florida, westward to Leon County, Florida.

HABITAT. Aquatic adults have been found in the shallow ponds of the piney woods in the Okefinokee, and efts sometimes in fairly dry situations on the islands. In Florida, Carr (1940, p. 45) reports larvae and adults in ponds and drainage ditches.

SIZE. The average length of 86 adults of both sexes is 2½″ (63.4 mm.), the extremes 2¹⁄₁₆″ (52 mm.) and 3³⁄₃₂″ (79 mm.), each sex attaining the same average length. The proportions of an adult male from near Chesser's Island, Okefinokee Swamp, Georgia, are as follows: total length 2¾″ (70 mm.), tail 1³⁄₃₂″ (36 mm.); head length ⅜″ (10 mm.), width ³⁄₁₆″ (5 mm.). A female from the same locality measures: total length 2¹¹⁄₃₂″ (60 mm.), tail 1³⁄₃₂″ (28 mm.).

DESCRIPTION. The head is widest at the eyes, the sides behind nearly straight and slightly converging posteriorly, in front tapering abruptly to the bluntly pointed snout. Cranial ridges not well developed; when present they arise on the top of the snout about midway between the anterior angle of the eye and the nostril, and extend, nearly parallel, to a point opposite the posterior angle of the eye, diverging somewhat at this point and broadly rounded behind. Eye with a dark horizontal bar, the iris above silvery, the long diameter about equal the distance from the anterior angle to the nostril. On each side of head of male a series of 3 elongate pits partially overhung by a low ridge; on the female these pits are lacking. The trunk is slender and somewhat compressed, costal grooves and gular fold not developed. The legs are generally slender except the hind legs of the male in the breeding season, which then become enlarged and develop a thin keel on the hind margin. Toes 5–4, those of the hind feet 5–1–2–4–3, the 5th short and webbed

FIG. 22. *Triturus perstriatus* Bishop. (1) Adult female, actual length $2^{13}/_{32}$″ (61 mm.). (2) Same, lateral view. (3) Same, ventral view. Dedge Pond, near Chesser's Island, Okefinokee Swamp, Georgia. [Photographs of a preserved specimen.]

internally at base, the 1st thickened and rounded at tip; toes of the fore feet 1–4–2–3, long and slender. Tail strongly compressed, in the aquatic adults with a dorsal keel that arises above the insertion of the hind legs, reaches its greatest width at a point just behind the vent, and tapers evenly to the tip; vent of the male large, strongly protuberant, the opening directed obliquely down and backward, and lined with many slender filaments. Tongue small, oval. Vomero-palatine teeth in 2 long series which arise opposite the hind margin of the inner nares and extend, narrowly separated, in parallel lines for ⅔ their length, then diverge abruptly.

COLOR. Ground color above uniform dark brown to olive-green. On either side an unbroken narrow dorsolateral light stripe, bright red to red, strongly suffused with dusky. The stripes arise on the head between the eyes and continue the length of the trunk and often on the basal half of the tail; on the distal half, they may be broken into separate spots or lacking. Often there is a middorsal stripe lighter than adjacent areas but not red. The dorsolateral red stripes are not heavily black-bordered, as in *dorsalis,* but may have a slight concentration of dark pigment on either side. In some individuals there are a few small red spots on the lower sides between the legs. In most specimens there are a few small black flecks scattered over the dorsal pigmented areas; in some, the black flecks are mainly restricted to series which extend along either side of the middorsal line and the bases of the tail fins. The dark ground color extends on the sides of the head to the level of the eye, on the trunk to a line which passes just above the legs, and on the tail to different levels in different individuals, but mainly restricted to the upper ⅔. The ventral surfaces are light yellow, usually with a few small, scattered, black flecks; an occasional individual may be more heavily flecked, and, in about 25 per cent of all, the belly is immaculate. The upper surface of the limbs is colored like the sides.

EFTS. The terrestrial efts are orange-red, the dorsolateral red stripes present as in the adults, the skin roughened and the tail less compressed than in aquatic adults. Individuals I have measured varied in length from 43 to 51 mm.

BREEDING. Little is known of the breeding habits. There is apparently an extended period of mating and egg-laying, for we have taken adults in breeding condition near Chesser's Island, Okefinokee Swamp, Georgia, in February, and adults from the same general locality in the Cornell University collection were in breeding condition when taken during June.

LARVAE. The larvae I have examined were marked dorsolaterally with a series of pale spots, which may have been red in life, and the sides of the tail with scattered dusky spots; they varied in length from 25 to 37 mm. Often neotenic individuals are found. These are usually marked dorsolaterally with a series of pale spots or stripes and on the lower sides with a series of small light spots. These attain the size of aquatic adults.

**WESTERN RED-BELLIED NEWT.** *Triturus rivularis* Twitty. Figs. 16d, 23. Map 6.

TYPE LOCALITY. Gibson Creek, near Ukiah, California.

RANGE. Known from Sonoma, Mendocino, and Humboldt Counties in northwestern California.

HABITAT. Usually found in or near mountain streams (Twitty, 1935, p. 73). On May 31, 1936, we collected at the type locality of *T. rivularis,* Gibson Creek near Ukiah, California. Gibson Creek has its course in part through a wooded glen and over a rough, bouldery bottom. Here in the more quiet pools *Triturus granulosus* and larvae of *Dicamptodon* were abundant, but no larvae and only a single adult of *T. rivularis.* The scarcity of adults at this time of the year was perhaps to be expected, since Twitty has found that they disappear shortly after spawning in April.

SIZE. A series of 10 adult males from the type locality, measured by Twitty (1935, p. 74), varied in length from $6\frac{1}{2}''$ (165 mm.) to $7\frac{5}{16}''$ (185 mm.) and averaged $7\frac{1}{16}''$ (179.6 mm.); 5 adult females varied from $5\frac{23}{32}''$ (145 mm.) to $6\frac{15}{16}''$ (173 mm.) and averaged $6\frac{1}{4}''$ (158.6 mm.).

DESCRIPTION. This species may generally be distinguished from *T. torosus* and *T. granulosus* by its smoother skin, darker dorsal coloration, tomato-red venter, and dark iris, and by the arrangement of the teeth. The head is narrower than in *T. t. torosus,* widest opposite the angle of the jaws, the sides converging slightly behind to the lateral extensions of the gular fold, in front abruptly to the depressed and pointed snout. The eyes dark brown. Costal grooves not developed; gular fold a transverse crease in some, in others fairly well defined. Legs stout, the

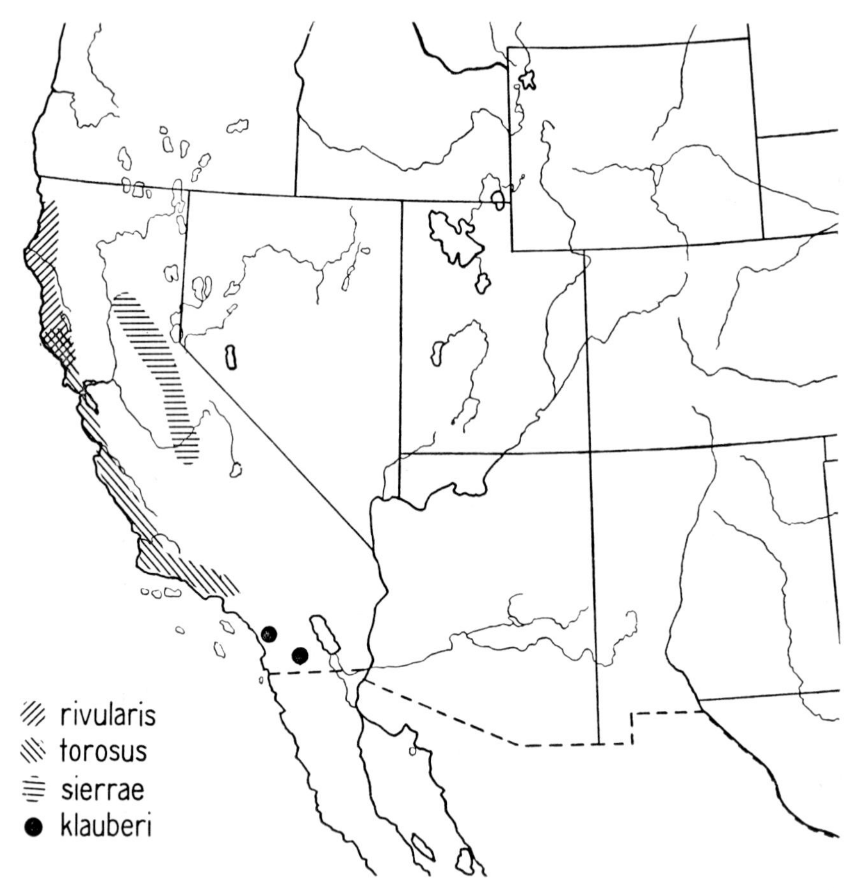


MAP 6.—Distribution of *Triturus rivularis, T. torosus, T. sierrae,* and *T. klauberi.*

toes 5–4, those of the hind feet usually 1–5–2–4–3 in order of length from the shortest; toes of the fore feet 1–4–2–3 or 3–2. The trunk is well rounded; the tail long and slender, oval in section at the base but without wide keels even in the breeding season. The tongue is small, roughly oval in outline, free only narrowly at the sides and behind. The palatine teeth are in 2 rows, parallel and narrowly separated anteriorly for ⅔ the length, abruptly diverging posteriorly.

COLOR. The general color above is a deep, dark brown, almost black. The feet are slightly lighter and the toes, especially the 1st and 2nd, reddish-orange suffused with brownish. The dark color of the dorsal

FIG. 23. *Triturus rivularis* Twitty. (1) Adult male, actual length 6¾″ (172 mm.). (2) Same, ventral view. Ukiah, California. [V. C. Twitty, collector.]

surfaces extends on the sides to a well defined line which runs from the nostril along the sides of the snout to the eye, where it involves only the upper half of the lower lid, and from the posterior corner of the eye slants downward to the level of the jaws at the lateral extension of the gular fold. The dark dorsal color involves the exposed surfaces of the limbs and the entire tail except for a narrow ventral area which becomes line-like posteriorly. The ventral surface of the throat, belly, and tail is a bright tomato-red; the soles of the feet dusky orange-red with the tips of the toes darker. In the male a broad, dark brown band extends directly across the vent; in the female the band may or may not be present. In some individuals the bright color of the underparts extends on the tip of the snout above the level of the nostril and faintly upon the upper eyelid. The surface of the skin when seen under low magnification appears to be broken up into small angular areas, within which are low, rounded tubercles; but the skin is smooth as compared with that of *T. torosus* and *T. granulosus*.

SEXUAL DIFFERENCES. Sexual differences are not well marked. The male apparently attains a larger average size than the female and during the breeding season develops a few rugosities on the lower surface of the thighs, while the skin of the rest of the body is generally quite smooth.

BREEDING. Twitty (1935, p. 73) found that breeding occurred only in streams of typical mountain-brook character. The eggs, in distinctive flat packets, were generally attached to the lower surface of stones in swift water. Several hundred egg clusters were collected or observed in Gibson Creek and Robinson Creek near Ukiah, California, in April 1935. Active spawning had apparently been confined to approximately the first 3 weeks of April, with the peak near the beginning of the month. The eggs in the small flattened clusters varied from 5 to 15 in number and averaged 9 in a series of 20 counted. Individual eggs in the blastula stage averaged 2.75 mm. in diameter, as compared with 2.25 mm. for those of *torosus*. The eggs are variously pigmented, dark gray or grayish-brown, and are deposited in great abundance, in one instance a count revealing approximately 70 masses attached to the lower surface of a single stone and crowded within an area 12″ x 8½″.

LARVAE. The recently hatched larva has the balancer either absent or rudimentary, never attaining half the normal size of that of *T. torosus* (Twitty, 1935, p. 76). The dorsal fin is less prominent and does not extend as far forward as in *torosus*. The head is flatter and longer than in the other described species and the body is uniformly pigmented over the back and sides. The larvae transform at a length of 45–55 mm., essentially the same as in *T. torosus*.

**SIERRA NEWT.** *Triturus sierrae* Twitty. Fig. 24. Map 6.

TYPE LOCALITY. Cherokee Creek near Chico, Butte County, California.

RANGE. Known from the vicinity of the type locality and from Eldorado, Mariposa, Tuolumne, Placer, and Fresno Counties, California.

HABITAT. The western Sierran slopes, where, during the breeding season at least, it tends to occupy streams with considerable current.

SIZE. The average length of 15 adult males from the type locality is 7″ (178 mm.), the extremes 6⅞2″ (158 mm.) and 7⅝″ (194 mm.); 6 adult females from the same locality average 6⅞″ (156 mm.), the extremes 5²⁷⁄₃₂″ (150 mm.) and 6¹³⁄₃₂″ (163 mm.). The proportions of an adult male are: total length 7½″ (190 mm.), tail 4⁵⁄₁₆″ (110 mm.); head length ²⁶⁄₃₂″ (21 mm.), width ⅝″ (16 mm.). In the males the tail averages 56 per cent of the total length; in the females 52 per cent.

DESCRIPTION. The head is widest immediately back of the eyes, the sides behind rounding gently to the lateral extensions of the gular fold, in front tapering abruptly to the truncated snout. The eye is crossed by a dusky horizontal bar, the iris silvery with a few dark flecks. The trunk is slightly depressed, rounded on the sides, flattened below. Legs strongly developed; toes 5–4, those of the hind feet 1–5–2–(4–3) in order of length from the shortest; toes of the fore feet 1–4–2–3; palmar and plantar tubercles not developed. The tail is subquadrate in section at base, becoming sharp-edged above a short distance behind the vent, and slender and strongly compressed distally; in breeding males about intermediate in width between that of *granulosus* and *torosus*. Tongue small, broadly oval in outline, the surface spongy. Palatine teeth in 2 long series nearly parallel or slightly diverging for ⅔ the length, abruptly diverging posteriorly.

COLOR. The color above varies from chocolate-brown to burnt umber, with the skin over the eyes and the feet lighter, dusky orange. The dark pigment extends on the sides to involve the upper ½ or ⅔ of the legs, on the cheeks to the lower margin of the eye, and on the tail to the lower third. The ventral surfaces are burnt orange, brighter on the throat, tail, and sides of the belly than along the mid-line of the belly. The skin is relatively smooth and somewhat shining, and with a few small, soft, brown tubercles which are most apparent against the orange of the belly but present on all surfaces.

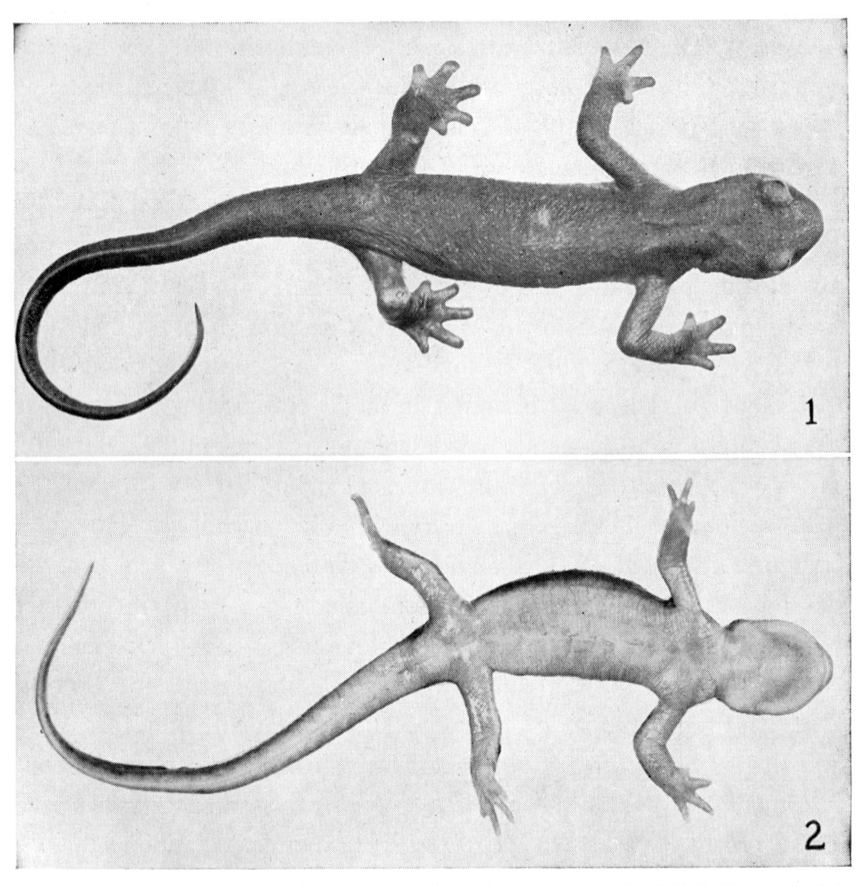


FIG. 24. *Triturus sierrae* Twitty. (1) Adult male, actual length 6⅝″ (168 mm.). (2) Same, ventral view. Chico, California. [V. C. Twitty, collector.]

SEXUAL DIFFERENCES. The sexes, aside from average differences in size, may be distinguished by the form of the vent. In the male it is large, strongly protuberant, and lined with ridges topped by many papillae. In the female, the vent is smaller and less protuberant, bordered either side by a swollen ridge which is transversely grooved and armed on the surface by blunt tubercles. The sides of the vent proper are thrown into folds which converge to the slit and the inner surface is somewhat papillate.

BREEDING. Twitty (1942, p. 66) found large numbers of adults mating and the females depositing eggs on April 9, 1937, in Cherokee Creek near Chico, California. The eggs were deposited in nearly spherical clusters containing 11–22 and were attached to the lower surface of stones in fairly swift water. Individual eggs averaged about 2.8 mm. in diameter.

LARVAE. Several larvae from Antelope Creek, Placer County, California, taken August 26, 1899, varied from 55 to 62 mm. in total length. In the smaller specimens there is a strong suggestion of the pattern of *Triturus t. torosus,* in the development of a dark stripe on either side of the middorsal line. These narrow dark stripes originate on the back of the head and extend well on the base of the tail. The back between the stripes is sparsely marked with small dark spots, and the sides of the head, trunk, and tail have larger spots disposed somewhat in lines. The venter is immaculate. The dorsal fin arises just behind the insertion of the fore legs and reaches its greatest width about at the mid-length of the tail. The ventral fin extends to the vent. Recently hatched larvae possess balancers, as in *torosus,* and young larvae have numerous melanophores extending farther ventrally on the sides than in that species.

## CALIFORNIA NEWT. *Triturus torosus* (Rathke). Figs. 16c, 25. Map 6.

TYPE LOCALITY. Surroundings of the Bay of San Francisco, California.

RANGE. From Monterey Bay north to middle Mendocino County, California. South of Monterey Bay a related form extends south at least to Los Angeles County and possibly to extreme northern Lower California.

HABITAT. The adults enter the water for the breeding season and may be found in streams, ponds, and reservoirs. The males frequently remain in the water for a considerable time, but transforming individuals leave the water if possible, and many females become more or less terrestrial after the egg-laying season.

SIZE. The measurements of the series that follow are from Twitty (1935, p. 78). The average length of 7 adult males from Ukiah and Stanford University, California, is $6\frac{3}{8}''$ (161.4 mm.), the extremes $7\frac{11}{16}''$ (195 mm.) and $6\frac{13}{32}''$ (163 mm.); 10 females from the same localities average $6\frac{1}{4}''$ (158.2 mm.), the extremes $6\frac{31}{32}''$ (177 mm.)

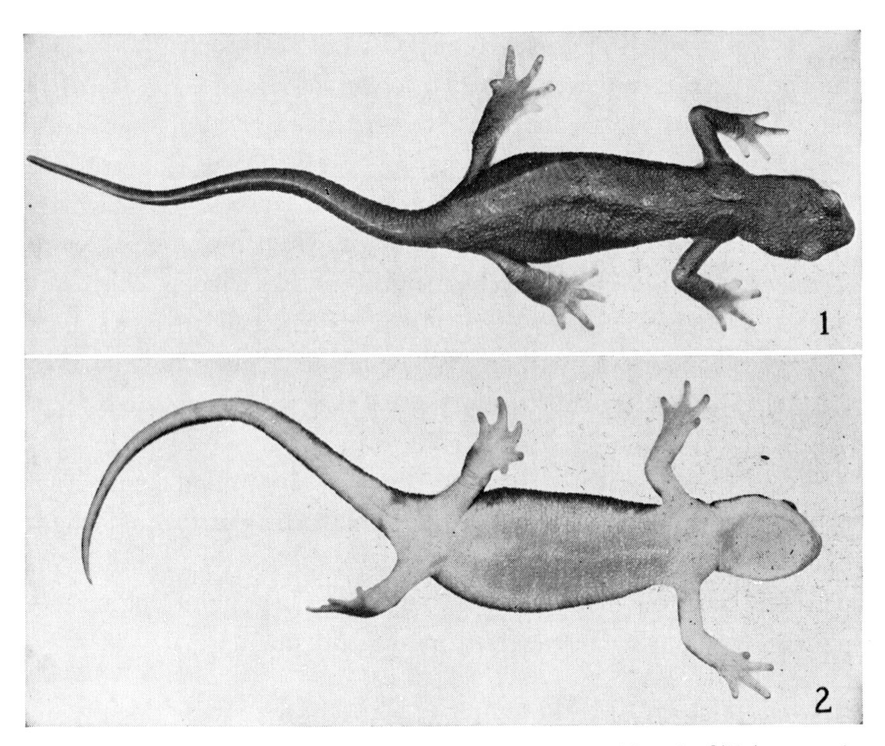


FIG. 25. *Triturus torosus* (Rathke). (1) Adult female, actual length $6\frac{3}{4}''$ (171 mm.). Mount Hamilton, California. [V. C. Twitty, collector.]

and $5\frac{21}{32}''$ (145 mm.). The proportions of an adult male from Ukiah, California, are as follows: total length $7\frac{1}{8}''$ (181 mm.), tail $3\frac{27}{32}''$ (98 mm.); head length $\frac{27}{32}''$ (22 mm.), width $\frac{23}{32}''$ (19 mm.). An adult female from the same locality measures: total length $5\frac{23}{32}''$ (146 mm.), tail $2\frac{27}{32}''$ (73 mm.); head length $\frac{23}{32}''$ (19 mm.), width $\frac{19}{32}''$ (15 mm.).

DESCRIPTION. The head is widest immediately behind the eyes, the sides converging gently to the lateral extension of the gular fold, in front more abruptly to the bluntly pointed snout. The eyes are relatively large, the horizontal diameter about 1¼ in the snout. The interorbital distance is contained in the length of the head about 2½ times. The eye is crossed by a dark horizontal band, the iris above silvery-yellow to greenish-yellow, more or less flecked or diffused with dusky. The legs are large and stout. Toes 5–4, those of the hind feet 1–5–2–3–4 in order of length from the shortest, rather stout and depressed, bluntly pointed; toes of the fore feet 1–4–2–3. Tail subquadrate in section at base, becoming sharp-edged above a short distance behind the vent and compressed distally. The tongue is small, broadly oval in outline, slightly free at the sides and behind. The palatine teeth are in 2 series nearly parallel anteriorly, diverging abruptly at about the posterior third. Aquatic adults have the skin smooth and practically devoid of tubercles above and with relatively few below. Terrestrial individuals have the skin roughened by blunt tubercles which are larger and more numerous above than below.

COLOR. The general color varies somewhat in any local population, but the most pronounced differences are found when the aquatic breeding males are compared with animals that have assumed a terrestrial life. The adult males—and, to a less extent, the females—become much lighter in color, pale brown, as compared with the general dorsal color of terrestrial animals, which vary from dilute chocolate-brown to dark seal-brown. The pigmented areas include the sides of the head to the level of the nostril, the sides of the trunk to involve the upper ½ of the limbs, and the upper ¾ of the tail. The color of the ventral surfaces varies from yellow to bright orange, brightest on the throat, tail, under surface of the limbs, and along the sides of the belly. Often the sides of the body, especially in the males, become greatly darkened and in strong contrast with the lighter dorsal surfaces.

SEXUAL DIFFERENCES. Sexual differences become well marked in the breeding season. The tail fins of the male are widened, the sides of the vent become greatly swollen and crossed by prominent ridges, the opening lined with short filaments; the tips of the toes and a spot at the base of the thighs below become dark and horny. The sides of the vent of the female, while somewhat protuberant, never attain the size of those of the male and the inner margins are relatively smooth.

BREEDING. The breeding period, more or less influenced by the character of the season, may begin as early as December and continue to March. The male grasps the female with his fore limbs and may be seen carrying his mate as he swims about in the deeper pools or rests quietly on the bottom in shallow water. The eggs are deposited in small globular clusters containing on the average about 16 but varying from 7 to 29 (Storer, 1925, p. 54). The masses are attached to sticks, vegetation, or other objects in the water. Individual eggs average about 2 mm. in diameter but may reach 2.5 mm. They are enclosed in an oval or round, fluid-filled envelope which has an average size of about 4.5 x 5.3 mm. The thickness of the jelly envelope varies from .41 to 1.18 mm. (Storer, *ibid.*, p. 55).

LARVAE. The larvae at hatching have a length of about 11–12 mm. and are provided with balancers and short buds of the fore limbs. The general ground color is a light greenish-tan, the head slightly more yellowish. Beginning at a point just back of the eyes and extending the entire length of the trunk and tail there is a broad median light band which is limited each side by a narrow black line. The pigment within the light band is rather evenly distributed except on the dorsal fin of the tail, where it is concentrated to form dark blotches of considerable size. The sides immediately below the dorsolateral dark lines are evenly pigmented, but the lower sides and lower tail fin are blotched. The legs are pale greenish-yellow mottled with black. Larvae may attain a length of nearly 60 mm. before transformation, but the majority transform at a length of about 50 mm., beginning in early September and continuing through the several succeeding months.

**COMMON NEWT. SPOTTED NEWT** (aquatic adult). **EFT** (land stage). *Triturus viridescens viridescens* (Rafinesque). Figs. 16a, 26. Map 7.

TYPE LOCALITY. Lake George and Lake Champlain, New York.

RANGE. Height of Land in Ontario and Quebec, eastward to the Gaspé Peninsula and New Brunswick; southward through the eastern states to Georgia and Alabama; Tennessee northward to Indiana and Michigan.

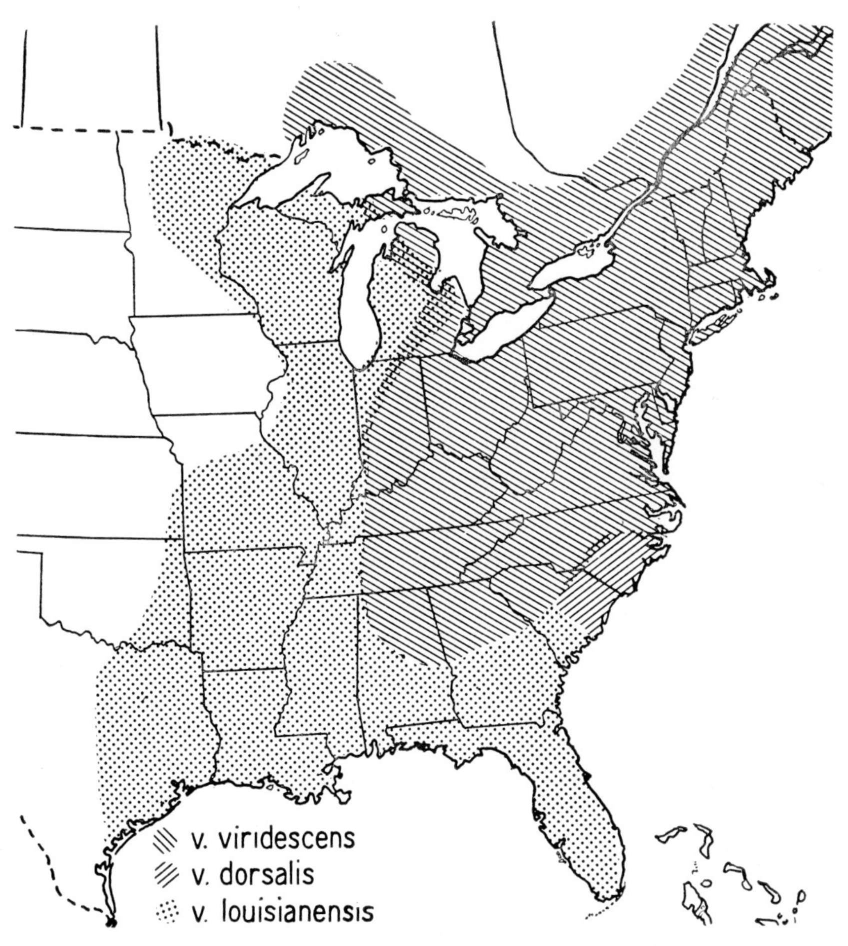


MAP 7.—Distribution of the subspecies of *Triturus viridescens*.

HABITAT. The aquatic adults are found in pools, ponds, swamps, and quiet reaches of streams both in woods and in the open. The usually immature, terrestrial "efts" are most abundant in wooded areas in fairly moist situations.

SIZE. The average length of 25 adults of both sexes from Clear Lake, Adirondack Lodge, New York, is $3\frac{1}{16}''$ (78 mm.), the extremes $2\frac{27}{32}''$ (73 mm.) and $3\frac{27}{32}''$ (98 mm.). In the mountains of western North and South Carolina this subspecies attains a larger size, 25 adults averaging $3\frac{25}{32}''$ (97 mm.), the extremes $3\frac{1}{32}''$ (77 mm.) and $4\frac{5}{16}''$ (110 mm.). The terrestrial efts vary from $1\frac{3}{8}''$ (35 mm.) to $3\frac{3}{8}''$ (86 mm.).

DESCRIPTION. Typical aquatic adults have a soft, smooth skin, strongly compressed tail, and a low dorsal ridge continuous with the fin of the tail. The head is moderately narrow and with prominent cranial ridges. The sides of the head converge gradually to the eyes, then more abruptly to the pointed snout. On the side of the head, at the level of the eye, a series of 3 pits in both sexes. Costal grooves and gular fold not well developed. Toes 5–4, those of the hind legs 1–5–2–4–3 in order of length from the shortest; toes of the fore feet 1–4–2–3. As indicated above, the tail is strongly compressed, that of the male noticeably widened during the breeding season. The tail of the female is always narrower, the dorsal keel never extending above the level of the back. The tongue is small, fleshy, broadly oval in outline. Vomero-palatine teeth in 2 long series which arise at about the level of the inner nares and diverge slightly posteriorly.

COLOR. The usual ground color is olive-green above, but varies from yellowish-brown to dark greenish-brown. The dorsal color extends on the sides to include the eyes, the upper $\frac{1}{2}$ of the legs, and $\frac{1}{2}$–$\frac{2}{3}$ of the tail. There is a dark line extending from the nostril through the eye and widening somewhat on the side of the neck. The throat, belly, lower surfaces of the limbs, and lower half of the tail are light to bright yellow. Scattered over both dorsal and ventral surfaces are many small, irregular, black spots, and on either side of the middorsal line is a series of bright red, black-bordered spots, variable in number and size. In

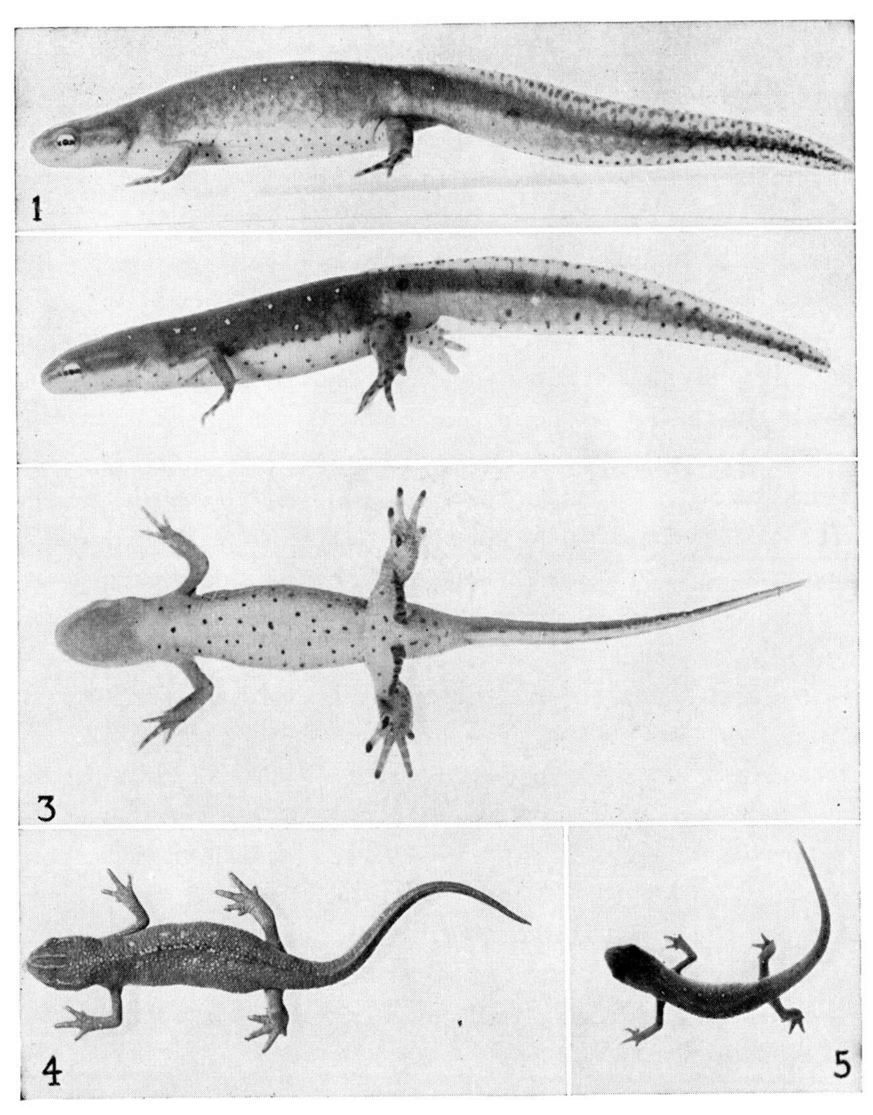


FIG. 26. *Triturus viridescens viridescens* Rafinesque. (1) Adult female, actual length 4 15/32″ (114 mm.). Thacher Park, New York. (2) Adult male, actual length about 4 1/8″ (105 mm.). (3) Same, ventral view. Thacher Park, New York. (4) Full-grown eft, actual length about 2 1/2″ (64 mm.). (5) Eft, year following transformation; actual length about 1 15/32″ (38 mm.).

typical efts the body has a more rounded contour, the tail is scarcely keeled, and the skin is rough. They are orange-red above, yellow to orange below, and provided with the black-bordered crimson spots as in the adults. In some localities the efts are dull red and never attain the brilliant color of specimens from moist upland situations. The land stage, eft, may be skipped in certain Coastal-Plain regions; in other localities the land stage may be limited to a single year or extended to three.

SEXUAL DIFFERENCES. Males in the breeding season, in addition to the broadly keeled tails, have a protuberant vent and the inner surface of the thighs and tips of the hind toes provided with black, horny excrescences.

BREEDING. The true mating and egg-laying season is in the spring. The males, however, may develop the sexual structures during the fall and winter and go through mating antics. Fertilization is internal, the males depositing small spermatophores, the sperm caps of which are picked up by the females.

EGGS. Eggs to the number of 200–375 are deposited singly and usually fastened to leaves or stems of aquatic plants in quiet waters, rarely to the surface of stones. The individual egg has a diameter of about 1.5 mm. and with its elliptical envelopes measures about 2.42 x 3.57 mm. The upper pole of the egg varies from light- to dark-brown, the lower pole is light yellowish-green. A thin vitelline membrane closely surrounds the yolk. The layer next without is elliptical and encloses a thin fluid in which the egg floats. Outside of this there are 2 thin layers, the outer covered with fine hair-like filaments. The period of incubation varies from 20 to 35 days.

LARVAE. Larvae at hatching average about 5⁄16″ (7.5 mm.) in length. The gills are developed but only slightly branched, and balancers are present, one on each side of the head. The front legs are short blunt buds, the hind legs not developed. The tail keel is continuous with the dorsal keel, which extends nearly to the head. The general color is light greenish-yellow. On each side of the middorsal line a gray band extends the length of the body. There is a poorly developed line through the eye. Larvae attain a length of $1\frac{3}{8}$–$1\frac{9}{16}$″ (35–40 mm.), average about 36 mm., and normally transform from August to October at a length of $1\frac{7}{16}$″ (36 mm.), a developmental period of 2–3 months. In the Woods Hole region, Massachusetts, and on Long Island, New York, some individuals remain in the water throughout life and retain vestiges of the gills.

**BROKEN-STRIPE NEWT.** *Triturus viridescens dorsalis* (Harlan). Fig. 27. Map 7.

TYPE LOCALITY. Vicinity of Camden, South Carolina.

RANGE. From Harnett and Onslow Counties, North Carolina, southward in the Coastal Plain to Kershaw and Georgetown Counties, South Carolina. Intergrades between this form and *T. v. viridescens* have been examined from the collection of Dr. C. S. Brimley from Moore, Harnett, Wake, and Craven Counties, North Carolina.

HABITAT. Aquatic adults are found in ponds, pools, and slow streams. Throughout its range, efts are occasionally found in terrestrial situations.

SIZE. The average length of 29 adults of both sexes is 3″ (76 mm.), the extremes $2\frac{23}{32}$″ (69 mm.) and $3\frac{3}{8}$″ (86 mm.). The proportions of an adult male from Wilmington, North Carolina, are as follows: total length $2\frac{15}{16}$″ (75 mm.), tail $1\frac{15}{32}$″ (38 mm.); head length $\frac{3}{8}$″ (10 mm.), width $\frac{9}{32}$″ (7 mm.); a female from the same locality measures: total length $2\frac{25}{32}$″ (71 mm.), tail $1\frac{3}{8}$″ (35 mm.); head length $\frac{3}{8}$″ (10 mm.), width $\frac{9}{32}$″ (7 mm.).

DESCRIPTION. This subspecies averages considerably smaller than typical *viridescens*. The sides of the head back of the eyes are nearly parallel, in front converging abruptly to the pointed snout. The eye is of moderate size, the horizontal diameter about equal the distance between the anterior angle and the nostril. On the sides of the head back of the eye, a series of small pits, usually 3–3 in the male; in the female the pits are a little smaller and may vary in number from 1 to 3; very rarely, they are absent. The trunk is slightly compressed, less stout than in

Fig. 27. *Triturus viridescens dorsalis* (Harlan). (1) Adult female, actual length about 3″ (77 mm.). (2) Same, lateral view. (3) Same, ventral view. Near Wilmington, North Carolina.

*T. v. viridescens.* Legs comparatively slender except hind legs of male in the breeding season; toes 5-4, those of the hind feet 5-1-2-4-3 or 1-5-2-4-3 in order of length from the shortest, the inner and outermost much reduced; toes of the fore feet 1-4-2-3. Tail compressed and keeled, but not so greatly widened as in typical *viridescens.* Tongue small, oval in outline. Vomero-palatine series long, the lines of teeth nearly parallel anteriorly, slightly diverging at the posterior third.

COLOR. The ground color above, of the aquatic adults, is usually some shade of olive-green. This darker color is sharply cut off from the yellow of the lower sides and venter, along a line which passes through the eye and along the trunk just above the legs. On the tail, the line of separation may continue that of the sides, jog upward to involve only the upper half, or turn downward to include the ventral fin. On each side of the mid-line of the back there is a narrow, broken, black-bordered red stripe which originates on the back of the head and extends to the base of the tail. In addition to the dorsolateral red stripes there is sometimes a row of small red spots on the lower sides between the fore and hind legs, and a light line along the ridge of the back. Scattered generally over the back and sides of the head, trunk, and tail are many small black dots or specks which are sometimes concentrated to form a fairly regular series along the line which separates the dark dorsal surfaces from the lighter ventral. On the sides of the tail the dark dots are sometimes larger and more diffuse. The lower sides and ventral surfaces vary from pale to bright yellow, with separate black dots or flecks larger and much more numerous than in *T. v. viridescens*.

SEXUAL DIFFERENCES. The sexes are easily distinguished, especially during the breeding season. The hind legs of the male are enlarged, the under surface of the thighs with horny ridges, the toes tipped with black. The vent of the male is large and strongly protuberant, the opening provided anteriorly with many large, soft, finger-like processes bearing short projections, and posteriorly with a group of longer unbranched filaments radiating from a common center. The vent of the female is small, the opening at the summit of a cone-like projection, the sides with ridges or series of low tubercles.

EFTS. The terrestrial efts are reddish-brown and have the red stripes not so strongly bordered by black as in the aquatic adults.

BREEDING. No detailed account of the breeding habits has appeared, but, so far as known, they do not differ materially from *T. v. viridescens*.

A male taken Dec. 12 at Wilmington, North Carolina, has the sexual excrescences well developed, which may indicate an earlier season than that of typical *viridescens.*

LARVAE. Larvae attain a length of at least $1\frac{19}{32}''$ (41 mm.), with the tail comprising about 50 per cent of the total length. Since terrestrial efts of equal size have been taken, the measurement indicated above may approximate the dimensions at transformation. Larvae I have examined had the dorsolateral red stripes faintly developed and not bordered by black, but with the belly faintly specked with black.

**LOUISIANA NEWT.** *Triturus viridescens louisianensis* (Wolterstorff). Fig. 28. Map 7.

TYPE LOCALITY. New Orleans, Louisiana.

RANGE. Southeastern Atlantic and Gulf States from South Carolina to Florida, west to Texas, eastern Oklahoma and Kansas, north to Wisconsin, Minnesota, and Michigan.

HABITAT. On March 19, 1936, we collected adults of both sexes in a small pond in pine woods at Natalbany, Louisiana. Larvae were taken in a roadside ditch choked with vegetation near Lake Okeechobee, Florida. In Louisiana the species is fairly common in pools, ponds, and slow waters of river bottoms of the Coastal Plain.

SIZE. Smaller and more slender than typical *viridescens,* the aquatic adults vary from $2\frac{9}{16}''$ (65 mm.) to $3\frac{1}{2}''$ (89 mm.) and average about 3″ (76 mm.). The terrestrial "red efts" I have examined vary from $1\frac{11}{16}''$ (43 mm.) to $3\frac{1}{16}''$ (77.5 mm.). An aquatic adult with gill rudiments, from Bayou Boeuf, La Fourche Parish, Louisiana, is $3\frac{1}{16}''$ (77.5 mm.); two larvae from near Okeechobee, Florida, measure 2″ and $2\frac{3}{16}''$ (50.5 and 55 mm.) respectively.

DESCRIPTION. The aquatic adults have a smooth, soft skin and compressed tail. The head is narrow, the sides parallel back of the eyes, the snout bluntly pointed. Cranial ridges prominent, nearly straight, slightly converging anteriorly. Eyes with iris brassy, tinged with gray and black. Males normally with 3 pits on each side of head, on a level with and behind the eye; females without pits. Gular fold poorly developed in some, though normally well defined. Costal grooves not developed. Legs

FIG. 28. *Triturus viridescens louisianensis* (Wolterstorff). (1) Adult male, actual length about 3″ (76 mm.). (2) Same, dorsal view. (3) Same, ventral view. Louisiana.

generally smaller and more slender than in typical *viridescens,* hind legs enlarged in male. Toes 5–4, those of the hind feet, 5–1–2–4–3 in order of length from the shortest; fore feet, 1–4–2–3, the inner rudimentary. Aquatic adults of both sexes, not in breeding condition, with tail fins narrow and not extending anteriorly beyond the vent. In breeding males the dorsal fin arises abruptly above the insertion of the hind legs, reaches its greatest width a short distance behind the vent, and tapers gradually to the tip; the ventral fin arises immediately behind the vent and continues to the tip without much variation in width. Sexual excrescences on the ventral surfaces of hind legs of male, forming short, broad, transverse ridges on thighs and, in some, smaller and flatter patches on the lower leg and sole of foot; toes with horny tips. Vent of male large, protuberant, with many slender papillae; vent of female smaller, somewhat constricted above the base and with the sides thrown into ridges.

COLOR. The general ground color above varies from light yellowish-brown to olive-brown or olive-green, and is sharply cut off from the straw-yellow venter at a line which extends from the nostril along the side of the head through the eye and above the legs to the base of the tail; in some individuals along the tail at the base of the ventral fin. In males taken in the breeding ponds at Natalbany, Louisiana, the belly was tinged with orange in the region of the vent. Scattered over the dorsal surfaces are many small black specks. In some individuals, especially males, the spots are larger and more diffuse on the sides of the tail. The black specks of the ventral surfaces are generally fewer and smaller than in typical *viridescens,* although there is wide variation. In general, females are colored like the males, but are sometimes darker, with the lower half of the tail lighter, and dark pigment of the tail aggregated to form one or two longitudinal bands. Throughout the greater part of the range this form *lacks* the dorsolateral series of black-bordered crimson spots so characteristic of *T. v. viridescens.* Where the range reaches that of *viridescens,* individuals are found with faint dorsal red spots, but they are incompletely ringed with black. I have seen intergrades from Illinois and South Carolina. "Specimens from the coastal plain of East Louisiana, north of the lakes, show small red spots, which are not seen on specimens south of Lake Pontchartrain or west of the Mississippi river" (Viosca, in litt.). Contrary to statements of several writers, the red land stage appears regularly in various parts of the range. It resembles that of *T. v. viridescens* except that the red spots are lacking or greatly reduced. I have specimens from Arkansas, Mississippi, and South Carolina. Some individuals apparently omit the land stage from the life history and may retain vestiges of the gills. I have specimens from Louisiana, and Schmidt and Necker (1935, p. 62) report that in ponds at Miller, Indiana, transformation takes place on floating vegetation and a considerable proportion of adults retain traces of the external gills.

BREEDING. Little has been reported concerning the breeding habits. Males and females in breeding condition were collected March 19, 1936, at Greensboro and Natalbany, Louisiana, and Mr. Percy Viosca, Jr., collected breeding pairs near Pollock, Louisiana, Feb. 22, 1938. The few larvae I have collected resemble closely the aquatic adults. I have not seen the eggs or recently hatched young, but the eggs are reported to be like those of typical *viridescens* (Turtox News, 20 (1): 9. 1942). In the collections of the University of Michigan are four neotenic individuals from Gainesville, Florida, which vary in length from 2 3/32" (53 mm.) to 2 5/8" (67 mm.).

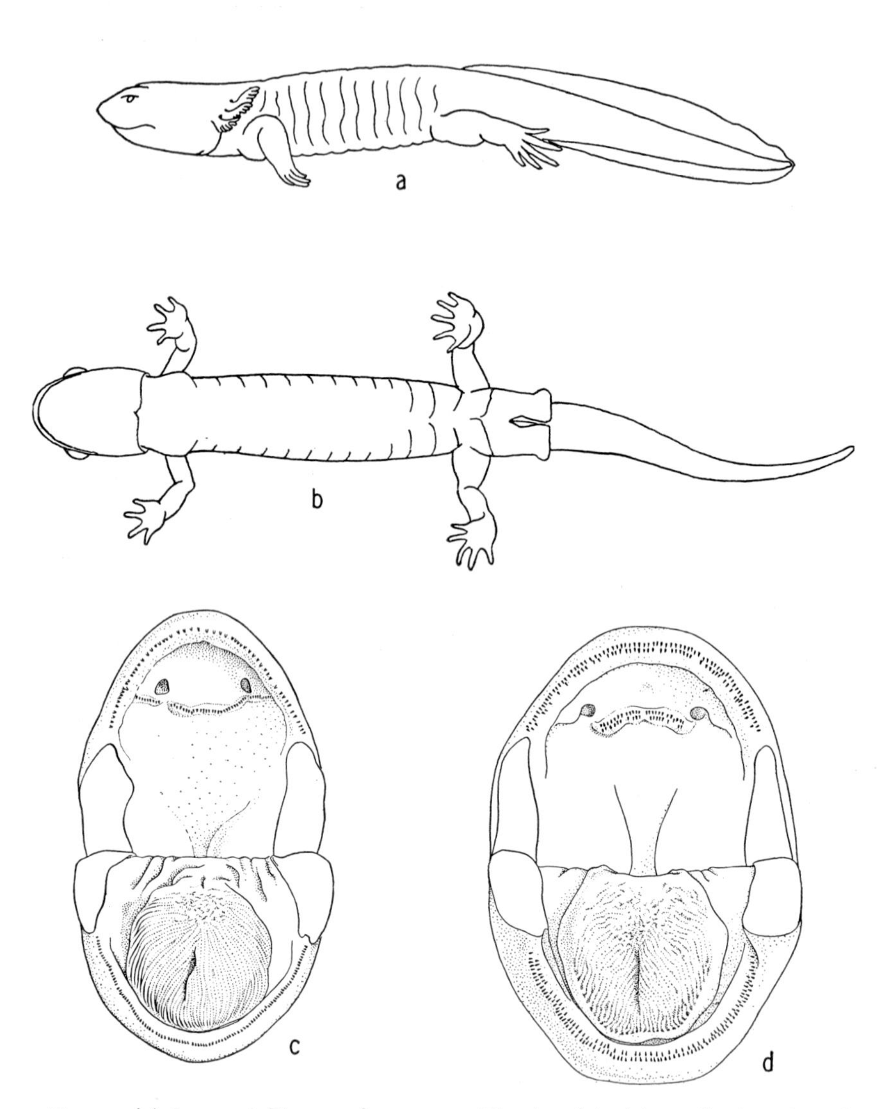


Fig. 29. (a) Larva of *Dicamptodon ensatus*. The dorsal keel is confined to the tail. (b) Adult male of *Rhyacotriton olympicus*. The squarish lobes at the vent are characteristic. (c) Mouth parts of *Ambystoma maculatum* to show the character of the tongue and teeth. (d) Mouth parts of *Ambystoma texanum* to show the character of the tongue and teeth. [a, b, H.P.C. *del.;* c, d, M.L.S. *del.*]

# Family AMBYSTOMIDAE

Mainly terrestrial except in the breeding season; size small to large, usually stout; lungs present; adults normally without gills;[1] no naso-labial grooves; eyes well developed, eyelids present; legs usually stout; toes 5–4; vomerine teeth in transverse series, usually interrupted one or more times; no parasphenoid teeth; tongue large, fleshy, free at sides; gular fold present; vertebrae amphicœlous; tarsus and carpus ossified; ypsiloid cartilage present. Three genera in the area considered.

The structural characters distinguishing the genera are internal and not readily usable in the field. In the following key they are included for the sake of completeness and may be used in connection with the external features indicated.

KEY TO THE GENERA OF AMBYSTOMIDAE

1. Eye large and strongly protuberant, the horizontal diameter equal to or greater than, length of snout; size small, to 4 1/16″ (103 mm.); ground color above seal-brown variegated with small white flecks; belly yellow or orange; tail short; vomerine teeth in 2 short, curved series widely separated; no nasal bones; lungs and ypsiloid cartilage reduced. Only a single species, western Oregon and Washington ................................................ *Rhyacotriton olympicus* p. 178

 Eye small to moderate, the horizontal diameter not equal to length of snout; size small to large, 4″ (102 mm.) to 12″ (305 mm.); ground color above not seal-brown with white flecks and belly yellow; vomerine series variable but usually in well developed transverse series,

[1] The larvae of Dicamptodon and of several species of Ambystoma sometimes attain a large size, retain the gills, remain permanently aquatic, and breed without the assumption of all adult characteristics. Such individuals are termed neotenic.

entire or interrupted one or more times; nasal bones present; lungs and ypsiloid cartilage normal . . . . . . . . . . 2

2. Tail strongly compressed and thin-edged above distally; costal grooves not well defined; size large, to 12" (305 mm.); ground color above brown, mottled and marbled with darker brown or black, venter lighter without dark markings. Only a single species, Santa Cruz County, California, north to southern coastal region of British Columbia; Idaho and Montana (Rocky Mountains) . . . . . . . . . . *Dicamptodon ensatus* p. 175

   Tail often glandular at base above, oval in section or rounded, sometimes compressed but less sharp-edged above; costal grooves well defined; lacrimals lacking; size small to large; 17 species and subspecies from Alaska to Mexico . . . . . . . . . . *Ambystoma* p. 112

## Genus AMBYSTOMA

### KEY TO THE SPECIES AND SUBSPECIES OF AMBYSTOMA

1. Tongue with a median groove from which the plicae extend obliquely outward . . . . . . . . . . 2

   Tongue without a median groove, the plicae radiating from the posterior part to the margins; or, plicae narrow, nearly parallel and extending to anterior margin of tongue . . . . . . . . . . 5

2. Vomerine teeth in 2 short series wholly between inner nares, each series consisting of 2 or 3 rows of small teeth; teeth on margins of jaws in 3 to 5 rows . . . . . . . . . . 3

   Vomerine teeth not in multiplie series and wholly between inner nares; teeth on margins of the jaw in a single row; color above deep brown, the lower sides with many grayish flecks; 13 costal grooves, counting 2 in groin; length to 4" (102 mm.). North and South Carolina and Georgia . . . . . . . . . . *mabeei* p. 136

3. Ground color above black, with narrow white or yellow cross bands which extend on sides of trunk to level of the legs and on tail to ventral edge; belly dark with small, scattered, light spots; 15 costal grooves; length to 8" (203 mm.). Arkansas and Missouri . . . . . . . . . . *annulatum* p. 115

   Ground color deep brown to black; light markings not in the form of narrow white or yellow cross bands; costal grooves 13 or 14 . . . . 4

4. Ground color deep brown to black, blotched above and on the sides with grayish or yellowish, lichen-like patches; costal grooves usually 14; length to 6" (153 mm.). Iowa, Ohio, Indiana, and Illinois, west to Nebraska and Kansas, south to Texas and Louisiana; North Carolina. Also recorded from South Carolina, Tennessee, Kentucky, and West Virginia . . . . . . . . . . *texanum* p. 155

   Ground color black, frosted above and on sides with many light gray fleckings which may form a reticulated pattern on the back, and narrow vertical lines on the sides between the costal grooves; 13–14 costal grooves; length to 4″ (102 mm.). South Carolina to northern Florida and Alabama ............................... *cingulatum* p. 123

5. Head back of eyes with large, raised, kidney-shaped parotoid glands; dorsal ridge of tail strongly glandular, thicker than ventral edge; color dark brown above, head and dorsal surface of tail lighter; venter lighter than back; dorsum rarely (in Olympic Mountains) with small yellow spots or blotches above; 11 costal grooves; length to 7¾″ (195 mm.). Northern California to British Columbia ....... *gracile* p. 128

   Head back of eyes without conspicuous parotoid glands; dorsal edge of tail not noticeably thickened ............................. 6

6. A broad, middorsal light band, tan, mustard-yellow or greenish-yellow, with irregular edges; sides brown to black; usually 12 costal grooves; length to 5″ (128 mm.). California to British Columbia and Alberta; Montana and Idaho; Iowa ...................... *macrodactylum* p. 139

   No broad middorsal light band ................................. 7

7. Above marked with 4 to 7 silvery-gray (female) or bright white (male) cross bands narrow dorsally, widened on the upper sides, where they sometimes unite to enclose a series of large, fairly regular black spots which extend along the mid-line of the trunk; costal grooves usually 11; length to 4 11/16″ (120 mm.). Massachusetts to Florida, west to Louisiana and Texas, Mississippi Basin north to Arkansas, Missouri, Indiana, and Illinois .................................. *opacum* p. 147

   Not with white or silvery-gray cross bands ....................... 8

8. Ground color above dark, with lighter spots or flecks; or with vertical yellow bars ............................................. 9

   Ground color above dark, with small round black spots ............ 16

9. Ground color above brown to black, with vertical yellow bars or blotches which often extend from the middorsal to the midventral line; light markings sometimes confluent on lower sides; venter often irregularly blotched with dark; costals usually 13, occasionally 12 or 14; length to 8⅜″ (213 mm.). Kansas, Oklahoma, central and western Texas, eastern Colorado, central and eastern New Mexico .........
   ...................................... *tigrinum mavortium* p. 165

   Ground color above brown to black, without vertical yellow bars but with light spots, flecks or blotches of yellow, orange, whitish or bluish .................................................. 10

10. With yellow, orange, olive-yellow or brownish-yellow spots or blotches on a dark ground ......................................... 11

    With pale bluish or whitish spots or flecks on a dark ground ...... 14

11. Light spots above dull yellow in life and tending to fuse to form irregular blotches of considerable size, the light areas sometimes predominating; usually a light blotch on dorsum of head extending on to snout; usually 13 costal grooves; tongue with narrow, parallel plicae;

length to 8⅝" (219 mm.). British Columbia, Alberta, Washington, Oregon, Idaho, Montana, Wyoming, North and South Dakota, and Nebraska ........ *tigrinum melanostictum* p. 172

Light spots on back yellow, orange, olive-yellow or brownish-yellow, showing little tendency to fuse, mostly well separated ........ 12

12. A dorso-lateral series of rounded yellow or orange spots only; sides dark gray or slate; belly pale slate without spots or blotches; usually 12 costal grooves; tongue with plicae radiating from the posterior field; length to 8⅛" (206 mm.). Nova Scotia to Wisconsin, southward to Florida, Louisiana and Texas ........ *maculatum* p. 143

Light spots above not limited to a dorsolateral series ........ 13

13. With a few large, dorsolateral yellow or orange spots (rarely invading middorsal region) and large yellow spots or blotches on the sides; belly generally grayish with a few small dull yellow spots along midline; costals usually 12; tongue with plicae nearly parallel except on sides; size to 8$\frac{3}{16}$" (207 mm.). California ........ *californiense* p. 119

With olive-yellow or brownish-yellow spots or blotches on back, sides, and belly; 12 costal grooves; tongue with plicae diverging from posterior field; length to 10" (254 mm.). New York to northern Florida, to Minnesota, to Missouri ........ *tigrinum tigrinum* p. 159

14. With 10 costal grooves; body short and stout; deep brown above in a broad dorsal band, lighter on tail; above with many small dull bluish-white flecks aggregated in lichen-like patches; lower sides gray instead of brown, with light markings more numerous and forming larger blotches; venter like lower sides but with fewer light markings; length to 3$\frac{25}{32}$" (97 mm.). North Carolina to Louisiana, north to Illinois ........ *talpoideum* p. 151

With more than 10 costal grooves; body not short and stout ........ 15

15. Body comparatively long and moderately slender; above dark-brown to almost black, with pale blue flecks on lower sides, old adults sometimes with blue spots lacking; usually 12 costal grooves; toes long and slender; length to 7¼" (185 mm.). Hudson Bay southward through New England and New York to Virginia, westward to Wisconsin, Illinois, and Arkansas ........ *jeffersonianum* p. 133

Body large and stout; dark chocolate-brown above, with many dull whitish spots irregular in shape and size; belly light brown; 11 costal grooves; length to 8½" (216 mm.). Alaska and British Columbia to the Olympic Peninsula in Washington ........ *decorticatum* p. 126

16. Olive-green or dark gray above, often with small scattered black spots on back, sides, and tail in recently transformed individuals, becoming uniformly dark above with age; ventral surfaces lighter, mottled and blotched with dark spots of irregular size and shape; usually 13 costal grooves; tongue with plicae narrow and parallel, or plicae broken into rows of small papillae; length to 9$\frac{1}{16}$" (230 mm.). Interior Basin and Colorado Plateau in Utah, western Colorado, northwestern New

Mexico, and northern Arizona .............. *tigrinum nebulosum* p. 168
Gray, brown, or light-olive above, with many small rounded black spots, sometimes elongate on sides and below; sides lighter than back; dull yellow ventrally; usually 12 costal grooves, or 13 if 2 in groin are counted; tongue with narrow plicae extending to anterior margin; length to 12¼″ (312 mm.). North Dakota east of Altamount Moraine into Alberta and Saskatchewan ............ *tigrinum diaboli* p. 162

**RINGED SALAMANDER.** *Ambystoma annulatum* Cope. Fig. 30. Map 8.

TYPE LOCALITY. Not known.

RANGE. Vicinity of Hot Springs, Arkansas, and Stone County, Missouri.

HABITAT. Strecker (1908, p. 86), reporting the observations of B. L. Combs, writes that specimens were found in October, beneath a log by the side of a creek, near Hot Springs, Arkansas. The log was partially imbedded in the mud, very much decayed, and almost covered by pine needles. In March of the year following, additional specimens were found beneath the same log. Noble and Marshall (1929, p. 1) report that on Sept. 30, 1927, a number of adults were secured from a cistern in a sweet-potato patch in Stone County, Missouri, and several from beneath piles of vines in the field and brush and leaves at the margin of a small pond. This species evidently burrows and spends most of the year below the surface of the ground.

SIZE. One of the specimens measured by Combs had a length of 8″ (203 mm.). The proportions of a pair of specimens from Missouri are are follows: Male, total length 177 mm., tail 88 mm.; head length 17 mm., width 14 mm. Female, total length 165 mm., tail 79 mm.; head length 16 mm., width 12 mm.

DESCRIPTION. A strikingly marked salamander with a small head and long, slender tail. The body is well rounded but slender. The head is narrower than the widest part of the trunk, widest just in front of the lateral extensions of the gular fold, then tapering slightly to the eyes. The snout is depressed, evenly and bluntly rounded. There are 15 costal grooves, counting 1 each in the axilla and groin. The legs and toes are slender and when appressed to the sides are separated by 3½–4 intercostal spaces. Toes 5-4, those of the hind feet 1-5-2-3-4 in order of length from the shortest; fore feet 1-4-2-3. The vomerine teeth are in 2

Map 8.—Distribution of *Ambystoma annulatum.*

short series entirely between the inner nares, each series consisting of 3 rows of about 7-11 small blunt teeth. The lines of teeth are directed obliquely forward and form a wide, inverted V. Teeth on the margins of the jaws in 4–5 rows. The tongue is of moderate size, attached at the front, free along the sides, and provided with a median groove to which the plicae extend obliquely.

COLOR. The general ground color varies from Clove Brown to Blackish

FIG. 30. *Ambystoma annulatum.* (1) Male, lateral view; actual length 6⅜″ (162 mm.). (2) Same, dorsal view. Reed's Spring, Stone County, Missouri.

Brown No. 3 (Ridgway), and the light cross bars and spots from deep Olive Buff to Baryta Yellow. In other words, the ground color may be deep seal-brown to almost black, and the lighter markings tan to deep sulphur-yellow. There is some variation in the intensity of markings on the same individual. The intercostal spaces on the lower sides are clouded with pale grayish-white, which forms an indefinite, broad band between the fore and hind legs. The belly is slate, tinged with flesh, and marked with many small irregular whitish spots. The legs are grayish above, uniformly colored in some and spotted in others. There is usually a short transverse light bar between the eyes, and in some specimens this is continued below the eyes as short bars directed diagonally backward; usually there is a bar crossing the head back of the eyes, but this may be interrupted at the middle line or reduced to a single median spot; 2–4 complete bars may cross the trunk, the anterior one above the insertion of the front legs, or the bars may be narrowly or widely interrupted along the mid-line and appear as short vertical side bars or elongate spots; from above, the tail may appear to be completely ringed with 4–7 fairly wide bars; in a few specimens the sides of the tail are blotched and mottled irregularly. The light cross bars terminate on the lower sides, fading into the generally lighter areas. In no instances do they cross the venter.

BREEDING. The observations of Combs (reported by Strecker, 1908, p. 86) indicate that on occasion this salamander may deposit its eggs on land in the spring. In March 1895, Combs turned over a log and discovered a female in the act of depositing eggs. The female was accompanied by a male, which evinced interest in the process and crawled after her as she moved leisurely along; 35 eggs had been dropped singly on the ground, and the female, when placed in a container, added to the complement until 150 had been deposited. These eggs began to hatch on the 10th day after laying. On Sept. 30, 1927, Noble and Marshall (1929, p. 1) found eggs abundant in a pond between Reed's Spring and Marvel Cave, Missouri. These eggs were deposited in clusters averaging about 10 but varying from 2 to 45. Some were attached to vegetation, but the majority were scattered over the bottom.

EGGS. The eggs are clustered together by a thin, common envelope which follows the outline of the eggs. In addition, each egg is surrounded by a thin, adherent, vitelline membrane and 3 definite capsules. The inner capsule encloses a milky jelly; the layer next without is thinner, denser, and somewhat laminated; the outer capsule has about twice the diameter of the middle and is thin and clear.

LARVAE. Larvae taken April 18, 1936, by B. C. Marshall, average about 48 mm. in length and vary from 46 to 50 mm. Larvae of this size have the legs and toes well developed, a dorsal fin that extends to the head, and rather uniform pigmentation over the dorsal surfaces and lower sides. There is a definite and fairly broad, pigment-free band on the sides of the trunk from the gills to the basal third of the tail.

Brimley and Brimley, 1895, p. 168 (*Linguelapsus annulatus*); Cope, 1886, p. 525; 1889, p. 115; Hurter and Strecker, 1909, p. 11 (*Ambystoma annulatum*); Noble and Marshall, 1929, p. 1; Stejneger, 1894, p. 599; Strecker, 1908, p. 85.

**CALIFORNIA TIGER SALAMANDER.** *Ambystoma californiense* (Gray). Fig. 31. Map 9.

TYPE LOCALITY. Monterey, California.

RANGE. Found in the Great Central Valley and adjacent valley and foothill districts, from Sonoma and Sacramento to Monterey and Kern Counties, California.

HABITAT. The region where this species is found is generally dry except during the winter rainy season, and the salamander is obliged to lead a subterranean existence during much of the year. Specimens have occasionally been found during ditching operations, in cellars, on irrigated lands, or in other damp situations. During the breeding season they resort to ponds usually of a temporary character, for egg-laying (Storer, 1925).

SIZE. Several females measured by Storer (1925, p. 61) varied from 6 7/16 to 8 3/16″ (163–207 mm.) in total length. In these the tail comprised 40.7–49.2 per cent of the total. The proportions of a female from Fresno, sent me by Dr. A. H. Wright, are as follows: total length 6 9/16″ (166 mm.), tail 2 7/8″ (73 mm.); head length 1 1/32″ (26 mm.), width

$1\frac{3}{16}''$ (20 mm.). A male from Palo Alto has a total length of $8\frac{1}{2}''$ (206 mm.), of which the tail comprises $3\frac{1}{2}''$ (89 mm.) or 43.2 per cent of the total.

DESCRIPTION. This is a well marked species, distinct in a number of

MAP 9.—Distribution of *Ambystoma californiense.*

respects from the typical eastern *Ambystoma tigrinum*. The head is depressed, the eyes small but prominent, the snout broadly rounded. The skin is generally smooth, with little indication of large gland openings. The mouth is large, the angle of the jaws well behind the posterior angle of the eyes. Eye with the pupil small and black, the iris with a tinge of brassy. The gular fold is prominent with slight vertical extensions. A depressed line extends backward from the eye and bends downward back of the angle of the jaws. Costal grooves variable from 11 to 13, usually 12 if 1 each is counted in the axilla and groin. Toes of the hind feet 1-5-2-3-4 in order of length from the shortest; fore feet 1-4-2-3. Tips of toes pinkish. Vomerine teeth in a transverse line which

FIG. 31. *Ambystoma californiense* Gray. (1) Female, lateral view; actual length 6½″ (165 mm.). (2) Same, dorsal view. [J. Westman, photograph.]

may be interrupted at the mid-point; the line arching forward between the inner nares, the apex slightly in advance of the nares, the ends curving outward behind them. The tongue is large, filling the floor of the mouth, the plicae forming narrow parallel ridges except on the sides, where they diverge slightly.

COLOR. The ground color in living examples which I have examined is lustrous black. Specimens from Fresno, California, were marked dorsally with a double row of small, rounded, lemon-yellow spots, disposed on the snout, on the head back of the eyes, and rather irregularly on the trunk. A secondary row of larger, more irregular spots is developed on the lower side of the head and trunk and is continued on the sides of the tail, where a few toward the tip cross the middorsal line. The spots of this secondary series fuse in some instances and encroach on the sides of the belly. The belly in general is grayish and in some individuals there are a few small, dirty yellowish spots along the mid-line. The legs are spotted on the sides, more rarely across the upper surfaces; toes mottled.

BREEDING. Storer (1925, p. 65) writes that breeding normally occurs during the height of the winter rainy season in January or February. In some years eggs may be deposited as early as December 10 (Twitty, 1941, p. 1). The ponds regularly visited are usually shallow and of temporary character. The eggs are attached to stems of grass or weeds below the surface, usually deposited singly, although small groups of 2–4 are not infrequent. Individual eggs are surrounded by 3 coats of jelly, the outermost thick-walled and soft, the 2 inner thin and of apparently greater density. The eggs hatch after a short developmental period of about 20 days.

LARVAE. Larvae at hatching measure about 10.5 mm. in total length, but individuals may be 13 mm. In color they are pale yellowish-brown, with dark pigmented spots along either side of the tail (Storer, 1925, p. 67). Larvae transform after a growth period of approximately 4 months and at a size considerably smaller than that attained by *A. tigrinum* in the East, Storer having metamorphosed individuals only 60 mm. in total length. Some larvae from Palo Alto, presented to me by Dr. V. C. Twitty, had attained a length of 76 mm. when captured late in March 1935. These have a noticeably depressed head and bluntly rounded snout. Storer (1935, p. 62) describes the larvae in life as follows: "Body coloration dull dark green and light yellow, mixed without obvious pattern; dorsal and caudal fins with many large diffuse melanophores; belly semi-iridescent, of a yellowish tinge; legs and feet almost colorless; iris golden yellow, surrounded by black."

**RETICULATED SALAMANDER.** *Ambystoma cingulatum* Cope. Fig. 32. Map 10.

TYPE LOCALITY. Grahamville, South Carolina.

RANGE. South Carolina through eastern Georgia to northern Florida and Alabama. Also recorded from New Orleans, Louisiana.

HABITAT. Little is known of the habitat preferences of this species. Two specimens were taken in the Okefinokee Swamp, Georgia: the

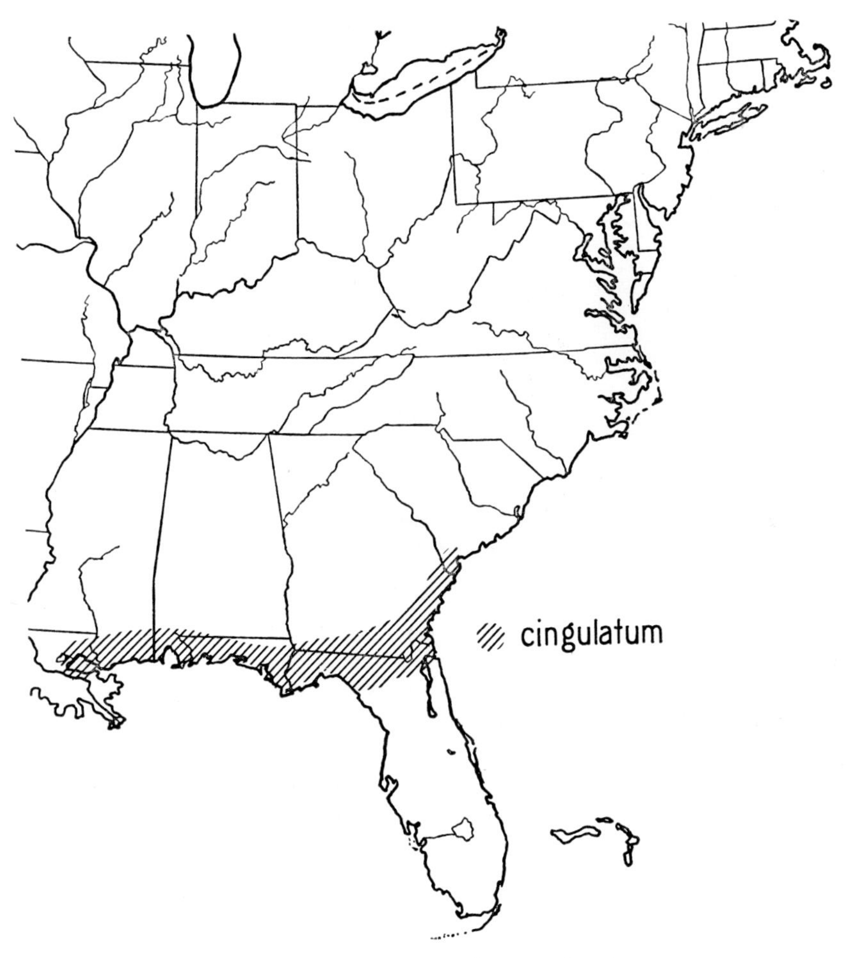


MAP 10.—Distribution of *Ambystoma cingulatum*.

first, June 15, 1922, was found by A. H. Wright, Francis Harper, and Miles Pernie beneath a log at the edge of a cypress swamp in a moist situation; the second was found by Thomas Chesser July 23, 1922, be-

FIG. 32. *Ambystoma cingulatum* Cope. (1) Adult, lateral view; actual length about $3\frac{1}{2}$″ (89 mm.). Chesser's Island, Okefinokee Swamp, Georgia. (2) Adult, dorsal view. Two miles east of Chesser's Island, Okefinokee Swamp, Georgia. [Photographs by A. H. Wright and Francis Harper.]

neath a log in the pine barrens of Chesser Island, a relatively dry habitat.

SIZE. Moderate among the species of Ambystoma. A series of adults of both sexes from Alabama and Florida average $3\frac{23}{32}$″ (95 mm.) in total length and vary from $3\frac{1}{2}$″ (89 mm.) to $3\frac{31}{32}$″ (101 mm.). The proportions of an adult female from Mobile, Alabama, in the collection of the U. S. National Museum (No. 57390), are as follows: total length 3 11/16″ (94 mm.), tail 1 17/32″ (39 mm.); head length 1/2″ (13 mm.), width 5/16″ (8 mm.). A male from the same locality (No. 42861) measures: total length 3 31/32″ (101 mm.), tail 1 27/32″ (47 mm.); head length 15/32″ (12 mm.), width 5/16″ (8 mm.).

DESCRIPTION. The head is oval in outline and convex above. It is widest at the angle of the jaws, curving gently behind to the lateral extension of the gular fold and somewhat more abruptly to the bluntly pointed snout. The eyes are small, the long diameter twice in the snout. An impressed line from the posterior angle of the eye to the gular fold and a short vertical groove from this line to the angle of the jaw. The trunk is rounded with an impressed median line above. Costal grooves 13 or 14, and 1 or 2 intercostal spaces between the toes of the appressed limbs. Tail broadly oval in section at base, becoming elongate oval at about mid-length and somewhat compressed and sharp-edged above toward the tip. Legs moderate. Toes 5–4, those of the hind feet 1–5–2–3–4 in order of length from the shortest, the 1st rudimentary. Toes of fore feet 1–4–2–3. Tongue elongate oval, free narrowly at the sides, and provided with a median groove from which the plicae diverge obliquely. Maxillary and premaxillary teeth small, in 2–5 rows. Mandibular teeth small and in 3–6 closely set rows. Vomerine series short, containing about 7 teeth in each of 2 or 3 closely set rows on either side of the mid-line. The series lie wholly between the inner nares and are separated by about the width of a naris.

COLOR. I am indebted to Dr. A. H. Wright for the following note on the color in life: "In general the color is black with many gray spots. On each side is a series of thin, gray, vertical stripes, bifurcate ventrally." The light spots are small and are aggregated on the dorsal surfaces of the head, trunk, and tail to form a distinctly reticulated pattern. This dorsal pattern extends on the sides to the level of and includes the dorsal surface of the legs and upper half of the tail. The ventral surfaces of the trunk, throat, and legs are light, with many small yellowish or tan specks. Ventral surface of the tail, slate with pale flecks.

BREEDING. Nothing has been reported on the breeding habits of this species, and the eggs and larvae are unknown.

**BRITISH COLUMBIA SALAMANDER.** *Ambystoma decorticatum* Cope. Fig. 33. Map 11.

TYPE LOCALITY. Port Simpson, British Columbia.

RANGE. Coastal region of southeastern Alaska and British Columbia to the Olympic Peninsula in Washington.

HABITAT. Apparently restricted to the coastal humid belt. Slater's

MAP 11.—Distribution of *Ambystoma decorticatum* and *Batrachoseps caudatus.*

specimen from the Olympic Peninsula was found beneath a plank on the bank of a stream.

SIZE. Slater's specimen, a large male, has the following dimensions: total length 8½″ (216 mm.), tail 4⁵⁄₁₆″ (110 mm.); head length 1″ (25

FIG. 33. *Ambystoma decorticatum* Cope. Adult male, actual length 8¹⁵⁄₃₂″ (215 mm.). Forks, Washington. [Preserved specimen; J. R. Slater, collector.]

mm.), width ¹³⁄₁₆″ (21 mm.). Measurements of the type as given by Cope (1889, p. 108) are: length 174 mm., to base of tail 60 mm., width of head 16 mm.

DESCRIPTION. This is a large, robust species with a short, blunt snout and sturdy limbs. It is closely related to *A. gracile*. The head is widest just behind the angle of the jaws, the sides behind the eyes nearly straight or slightly converging, in front of the eyes strongly converging to the bluntly pointed snout. The corners of the mouth only a short distance behind the eyes. Glandular areas on the head back of the eyes are large but not raised above the surface as in *A. gracile*. The eyes prominent. In the male from the Olympic Peninsula there is a short, deep groove extending from the eye to the corner of the mouth, a long, sinuous groove from the posterior angle of the eye to the vertical extensions of the gular fold, and a less prominent groove from the angle of the jaw to the long horizontal groove. The trunk is fairly short, the appressed limbs overlapping by about 5 costal folds; a median groove from the back of the head to the base of the tail. Eleven costal grooves, counting 1 each in the axilla and groin. The legs are large and strong, the feet broad; no plantar or palmar tubercles. Toes 5–4, those of the hind feet 5–1–2–4–3, 2 and 4 about equal; toes of fore feet 4–1–3–2. Tail long and slender, rounded above and somewhat glandular but not to the extent found in *A. gracile.* Vomerine teeth in 4 short transverse series, the inner ones entirely between the inner nares and narrowly separated at the mid-line, the outer series widely separated from the middle, placed closely behind the nares, which they border only on the outer half. Tongue broad, without a definite median groove but with the plicae radiating from a poorly defined center opposite the point where the tongue narrows conspicuously.

I have never seen this species in life. Slater (1930, p. 87) describes the color as "dark chocolate brown, closely studded with quite small whitish spots of irregular size and shape. The ventral coloration is lighter than the dorsal, but is distinct brown."

Apparently nothing is known about the breeding habits of this species, and the larvae have not been described

**NORTHWESTERN SALAMANDER.** *Ambystoma gracile* (Baird). Fig. 34. Map 12.

TYPE LOCALITY. Cascade Mountains, Oregon, latitude 44° N.

RANGE. From Humboldt and Del Norte Counties, northwestern California, to British Columbia, Vancouver Island.

HABITAT. Found under bark and logs in damp situations. Van Denburgh's specimen from Requa, California, was found beneath a stump in wet earth. On June 15, 1936, I found a fine large adult beneath a log on a damp hillside near Sandy, Oregon. During the breeding season the adults resort to ponds, swamps, and slow streams for the egg-laying.

SIZE. A large female has a total length of 7⅝″ (193 mm.) of which the tail comprises 4″ (102 mm.). The smallest fully adult female measures only 5¾″ (146 mm.) and has a tail length of 2⅜″ (61 mm.). In a series of males the total length varies from 5⅝″ (143 mm.), tail 2 13/16″ (71 mm.), to 7¾″ (195 mm.), tail 4¼″ (107 mm.).

DESCRIPTION. This species is of moderate size, smaller and more slender

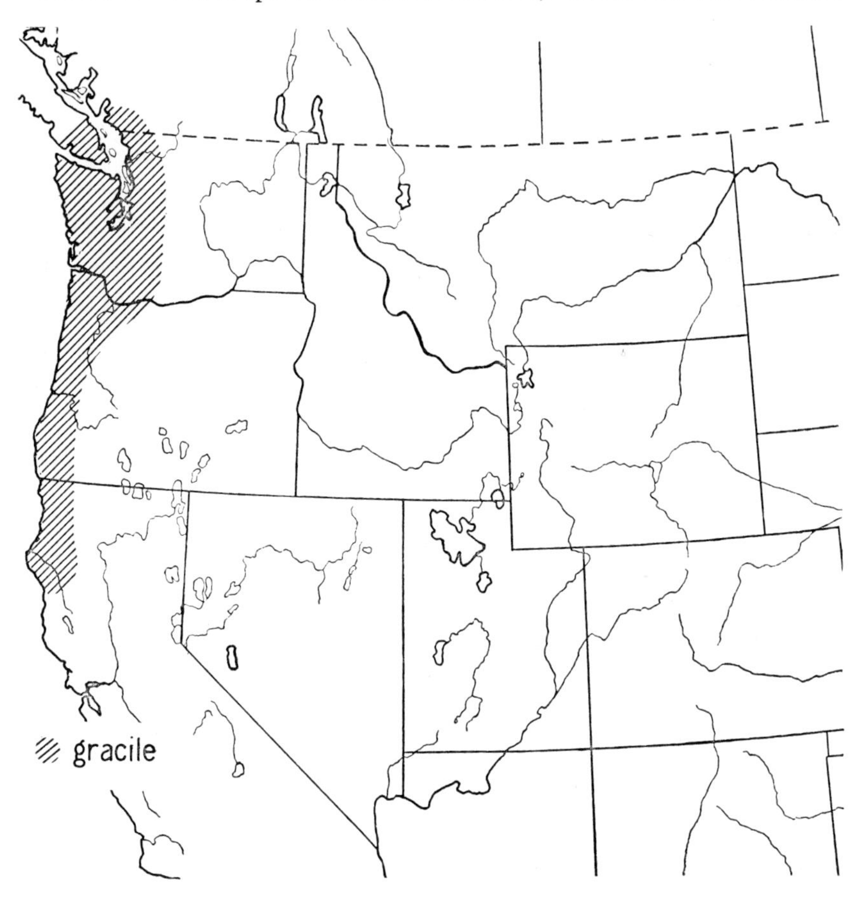


MAP 12.—Distribution of *Ambystoma gracile.*

than its near relative, *A. decorticatum.* The head is broad, widest just back of the eyes, and with a broadly rounded snout. On the head back of the eyes are large, raised, kidney-shaped parotoid glands. Other glandular areas lie along the jaws behind the eyes and form a conspicuous ridge along the dorsal surface of the tail. Eyes prominent, iris dark brown. A deep groove extending obliquely downward from the eye is met by a shallower one extending vertically from the angle of the jaw and then continues to the vertical extensions of the gular fold. Trunk well rounded, in the females frequently with a shallow groove along the middorsal line and in the males often with a low ridge. Costal grooves 11, counting 1 each in the axilla and groin. Legs large and strong, in the male with appressed toes overlapping 6 costal folds. Soles and palms without tubercles. Toes 5–4, those of the hind feet 5–1–2–4–3 in order of length from the shortest; fore feet, 1–4 about equal, 2–3. Tail long and slender, with the dorsal ridge rounded over, the ventral edge more knife-like, particularly on the distal half, in this respect differing from most Ambystoma. Vomerine teeth in a transverse sinuous line which may be broken into 3 or 4 series, those between the inner nares sometimes forming an uninterrupted, slightly arched series, or 2 series more strongly arched forward and narrowly separated at the mid-line. Lateral series short, arising behind the middle of the inner nares and extending slightly beyond their lateral borders. Tongue moderate, fleshy, the plicae diverging slightly on the sides from the posterior field.

COLOR. The color in life is rich seal-brown above, with the glandular areas in the parotoid region, sides of head below the eyes, and the dorsal surface of the tail lighter brown. The ventral surfaces are brown, lighter than the back, and varied with tinges of bluish and flesh. Legs and feet colored dorsally like the back and ventrally like the belly. Individuals from a small lake in the Olympic Mountains, about 6 miles south of Sol Duc Hot Springs, Washington, are marked above with yellow flecks and spots which vary in size from mere points to irregular blotches several millimeters in diameter.

BREEDING. On June 25, 1936, we collected the adults, larvae, and eggs of this species in Frog Haven Ponds, which lie at an elevation of about 4500′ on the slopes of Mt. Rainier, Washington. Another series was taken in ponds fed by melting snow water near Reflection Lakes, Mt. Rainier. The eggs are disposed in groups attached to sticks, stems of weeds, or grass, below the surface of the water, and vary in size, from

FIG. 34. *Ambystoma gracile* (Baird). (1) Adult, lateral view; actual length about 5¼″ (134 mm.). Estacada, Oregon. (2) Adult, dorsal view; actual length 7⅝″ (194 mm.). Sandy, Oregon. (3) Adult male, dorsal view; actual length 6⅝″ (168 mm.). Near Deer Lake, Olympic Peninsula.

small clusters containing 25–30 eggs to large elongate masses containing as many as 270, and measuring 2–3″ thick by 5–6″ long. The outermost jelly layer of the eggs lying near the surface of the mass fuses to form a common envelope which is thick and firm. Next within is a comparatively thin and transparent envelope having a diameter of about 6–7 mm., but which may widen before hatching to 10–15 mm. Separating this layer from the yolk is another envelope enclosing a thin jelly which is often milky. The yolk itself varies in diameter from 2.5 to 3 mm., and has the animal pole dark brown to black and the vegetal pole white to cream-gray. Slater (1936, p. 234) observed a migration of adults and sexually mature larvae and watched a 140-mm. larva deposit eggs in a small pond at South Tacoma Swamp, Washington, Feb. 23, 1931.

LARVAE. The larvae, 1–7 days after hatching, vary in length from 15 to 20 mm. At this age the fore legs and gills are well developed and about equal in length, the feet with 3 toes. The balancers are present in the younger larvae, lost in the older. At hatching, the pigment is quite uniformly distributed over the sides of the abdomen and head and is concentrated in a band along either side of the dorsal keel, where it is interrupted at short intervals by yellowish-white blotches. Recently hatched larvae have a very broad dorsal fin which arises at the back of the head, reaches its greatest width just beyond the vent, and tapers to the tip of the tail. The ventral fin is broad and continues on the belly a short distance anterior to the vent, which opens through the keel itself. Full-grown larvae attain a length in excess of 6″. In these the ground color varies, in different individuals, from deep brown to greenish- or olive-brown; the head lighter with reddish-brown tinges. Scattered over the sides and dorsum of the trunk and upper surfaces of the head and legs are many small, irregular, black flecks. The color fades slightly on the lower sides to the bluish-gray venter. In half-grown larvae the tail keels are moderately wide; in full-grown larvae they are quite narrow.

**JEFFERSON'S SALAMANDER. BLUE-SPOTTED SALAMANDER.** *Ambystoma jeffersonianum* (Green). Fig. 35. Map 13.

TYPE LOCALITY. Near Jefferson College, Canonsburg, Pennsylvania.

RANGE. From Manitoba and Ontario north to 53° N. latitude; Hull,

MAP 13.—Distribution of *Ambystoma jeffersonianum.* (Also recorded from Mammoth Cave, Kentucky, and Mitchell County, North Carolina. Scattered records south of the general range may be *A. texanum* or some other species.)

Quebec, eastward to Gaspé, New Brunswick, and southward through New England and New York to Virginia, westward to Wisconsin, Illinois, and Arkansas.

HABITAT. Mixed and deciduous woods with swamps, pools, and slow streams. Often extremely abundant on river flats, where they hide by day beneath old logs, bark, or other surface cover.

SIZE. In the eastern form large specimens attain a length of 7¼″ (185 mm.) but average about 6⅜″ (162 mm.), the males and females about equal. In western New York and in the north and middle West there is a smaller, darker form.

DESCRIPTION. This is a moderately slender species with a strongly glandular skin and faint pale bluish markings on the sides, in the adults. The head is only moderately broad and is widest just back of the prominent eyes. The trunk is well rounded and with a median impressed line. The tail is oval in cross section at the base, but strongly compressed distally, and marked with vertical grooves on the basal half. There are 12 costal grooves and the gular fold is prominent. The limbs are strong and the toes very long and slender, those of the front feet 1–4–2–3 in order of length. The hind feet have 5 toes, the 1st shortest, and the others, in order of length, 5–2–3–4. The vomerine teeth form a transverse series interrupted back of the inner nares and sometimes at the mid-line. The tongue is large and fleshy and nearly fills the floor of the mouth. In breeding specimens the color is quite uniform dark brown over the dorsal surfaces and upper part of the sides. On the lower sides are weakly developed, pale, irregular, bluish fleckings. The venter is lighter, with some bluish markings, most abundant on the throat. Sexual differences are not well marked, but in the breeding season the vent of the male is swollen and protuberant. The legs of the male are slightly larger and longer than those of the female and the trunk length is shorter. In western New York and in the Middle West there is a smaller, darker form, with light markings often better developed. The average length of 10 mature individuals from western New York is 4$\frac{9}{16}$″ (115.7 mm.).

BREEDING. In March or early April the adults migrate to breeding ponds for the egg-laying. Fertilization is internal by means of spermatophores—small, pyramidal structures with a gelatinous base and apical sperm mass. The male sometimes embraces the female, his fore legs behind hers. The females usually outnumber the males and often bid for attention during the mating season. The eggs are in small, cylindrical masses, containing on the average about 16, and are attached to slender twigs or other support below the surface of quiet pools or ponds. The number per mass varies from 7 to 40, and several masses complete the

Fig. 35. *Ambystoma jeffersonianum* (Green). (1) Adult female, actual length 6⅜″ (162 mm.). Albany, New York. (2) Adult male, actual length 6⅜″ (162 mm.). Albany, New York. (3) Egg mass, about ¾ natural size. Albany, New York, April 7. (4) Larva, actual length $2\frac{15}{16}''$ (75 mm.). Albany, New York.

complement, which may total over 200 eggs. Individual eggs have a diameter of 2–2.5 mm., and are enclosed by distinct envelopes in addition to the soft, thick jelly layer which surrounds the mass and gives the appearance of a common envelope. The upper pole of the egg varies from dark brown to black; the lower pole is lighter. Under field conditions, the incubation period varies from 30 to 45 days.

LARVAE. Larvae at hatching have an average length of 9/16″ (app. 13 mm.) but vary from 10 to 14 mm. The head is broad, twice the width of the trunk, and has the sides parallel. The snout is blunt and the balancers are attached midway between the eyes and the first pair of gills. The gills are provided with 2 rows of filaments. The fore legs are represented by elongate buds directed backward, the hind legs are not developed at hatching. The tail fin is continuous with that of the back, which extends almost to the back of the head. Larvae are strongly pigmented above, the color varying from olive-green to brown, with tinges of yellow on the sides of the head and neck and rarely along the sides of the dorsal fin. Black pigment is concentrated to form large paired spots on either side of the middorsal line. Mature larvae vary from 1⅞″ to 3″ (48–75 mm.). The length of the period is extremely variable, 56–125 days, and transforming young may be found from July through September. During the year following transformation a length of approximately 4¼″ (107 mm.) is attained, and the animals breed the following spring.

**MABEE'S SALAMANDER.** *Ambystoma mabeei* Bishop. Fig. 36. Map 14.

TYPE LOCALITY. Low grounds of the Black River near Dunn, North Carolina.

RANGE. Known from the vicinity of Dunn and Wilmington, North Carolina, Charleston County, South Carolina, and Liberty County, Georgia.

HABITAT. The first specimen of this species was secured by Mr. W. B. Mabee in the low grounds of the Black River near Dunn, North Carolina, May 12, 1923. The species was described from this preserved material and no additional living specimens became available for study until 1934. On Feb. 23, 1934, A. H. Wright collected an adult near Wilmington, North Carolina, and two days later 10 or 12 adults and one juvenile specimen in Charleston County, South Carolina, near the Edisto River. These specimens were found in or beneath semi-rotted logs and bark, in the low bottom lands bordering the highway. On Feb.

12, 1936, we visited the locality, in Charleston County, where A. H. Wright had taken specimens, but failed to find this species.

SIZE. The type specimen has a total length of 100 mm., of which the

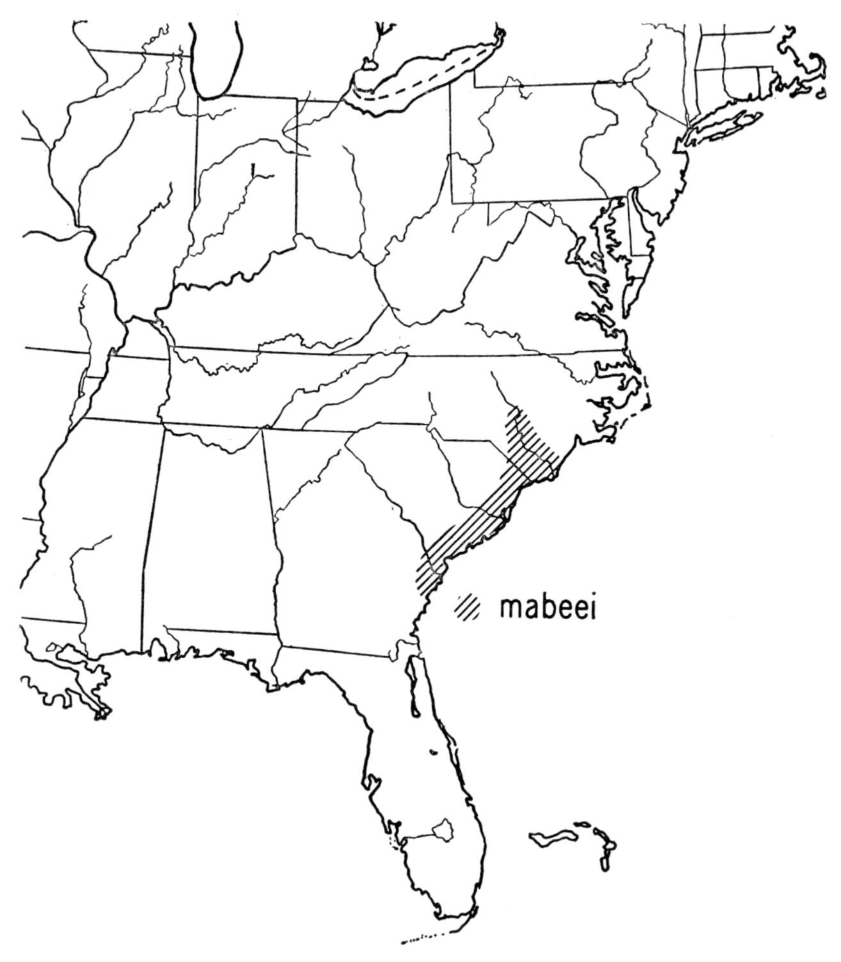


MAP 14.—Distribution of *Ambystoma mabeei.*

tail comprises 40 per cent. Other measurements are as follows: head length 13 mm., width 11 mm. A female from Charleston County, South Carolina, presented by A. H. Wright, has a total length of 95 mm., tail 31 mm.; head length 13 mm., width 8 mm.

DESCRIPTION. A small, stout species with the general conformation of *Ambystoma opacum.* The head is oval in outline, widest immediately behind the eyes, gently tapering posteriorly, and more abruptly to the bluntly rounded snout. The eyes are small but prominent, with a tinge

FIG. 36. *Ambystoma mabeei* Bishop. (1) Adult female, natural size, dorsolateral view. (2) Same, dorsal view. Near Wilmington, North Carolina. [A. H. Wright, collector.]

of brassy in the iris. There are 13 costal grooves, counting 2 that run together in the groin and 1 in the axilla. There are about 2 intercostal spaces between the appressed toes. The legs are stout and proportionally large for the size of the salamander. Toes 5–4, long and slender, those of the hind feet in order of length from the shortest, 4–3–2–5–1; fore feet, 3–2–4–1. The tail is noticeably short and comprises 33⅓–40 per cent of the total length. The tongue is oval in outline, fills the floor of the mouth, and is marked by a median impressed line, the plicae meeting it obliquely. The vomerine teeth form a transverse line interrupted back of the inner nares.

COLOR. In life the general ground color is deep brown in a broad band along the dorsal surfaces. On the lower sides are many small grayish-white flecks which become indistinct, dull tan as they approach the dorsal surface, where they seem to be overlaid by darker pigment. The light flecks are abundant on the head and legs, but few and scattered on the light brown venter. The throat is lighter, dull flesh color. In alcohol the color becomes almost uniformly deep brown or slate-gray, the lower sides and venter somewhat lighter. The pale spots mostly disappear in time, but a few may remain on the throat and along the sides of the belly.

BREEDING. Nothing has been reported on the breeding habits of this species and the eggs are unknown.

LARVAE. A larva approaching transformation at a length of 63 mm. was collected by Mr. C. S. Brimley at Andrews, North Carolina, in 1908. It may be this species. It has the costal-groove count and general coloration of *A. mabeei.*

This species bears a superficial resemblance to some specimens of *A. texanum* (*microstomum*), and the type specimen was first recorded under that name by Brimley (1927, p. 10). A specimen recorded as *A. microstomum* from Harnett, North Carolina, by Brimley (1926, p. 77) may also be this species. It is also likely that some of Cope's early records from South Carolina may refer to *A. mabeei.*

Bishop, 1928, p. 157 (*Ambystoma mabeei*); Brimley, 1926, p. 77 (*Ambystoma microstomum*); Brimley, 1927, p. 10.

**LONG-TOED SALAMANDER.** *Ambystoma macrodactylum* Baird. Fig. 37. Map 15.

TYPE LOCALITY. Astoria, Oregon.

RANGE. From Calaveras County, California, north to British Columbia and Alberta; Montana and Idaho; also recorded from Iowa.

HABITAT. At Crater Lake in Oregon this species was found in abundance beneath rocks just above water line (Evermann, 1897, p. 235). It is occasionally found beneath the loose bark of fallen trees. During the breeding season it resorts to ponds even though they be partly icebound.

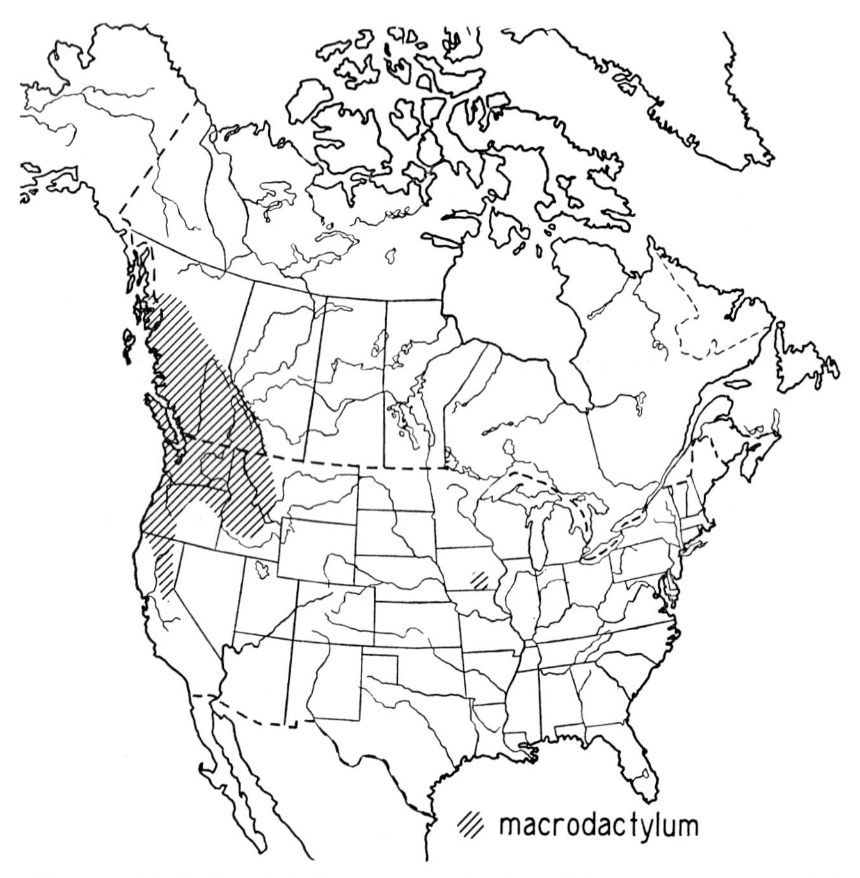


MAP 15.—Distribution of *Ambystoma macrodactylum.* (Also reported from Iowa.)

On June 25, 1936, we collected adults and eggs in ice-bordered pools on the slopes of Mt. Rainier. Here the adults crawled slowly over the bottom in shallow water and were easily captured.

SIZE. My largest specimen, a male, has a total length of 4⅝″ (122.5 mm.), tail 2⁹⁄₃₂″ (58 mm.); head length ⁹⁄₁₆″ (14 mm.), width ¹³⁄₃₂″ (10 mm.). An adult female has a total length of 4⅛″ (104 mm.) and a tail length of 2⅛″ (53.5 mm.). Slater (1936, p. 236) mentions a female 127 mm. long. These are perhaps exceptionally large individuals, for the majority of my specimens are considerably shorter.

FIG. 37. *Ambystoma macrodactylum* Baird. (1) Adult male, actual length 5″ (128 mm.). Mt. Rainier National Park, Washington. (2) Adult female, dorsal view; actual length 5¾″ (96 mm.). (3) Same, dorsolateral view. Pullman, Washington.

DESCRIPTION. This slender species of Ambystoma is most strikingly marked. Viewed from above, the head is broadly oval in outline, the snout bluntly rounded, and the sides behind the eyes only slightly converging to the strong constrictions marking the lateral extensions of the gular fold. The eyes are large and strongly protuberant, the iris tinged with dull brassy. A groove curves downward from the posterior angle of the eye to the angle of the jaw and from this an impressed line continues to the lateral extension of the gular fold. The trunk is slender, rounded above and on the sides, and flattened ventrally. Costal grooves 12 or 13, counting 1 each in the axilla and groin, the usual number 12. Toes of appressed limbs may overlap 2 costal interspaces. Tail oval in section at base, becoming compressed and with a more pronounced dorsal ridge toward the tip. Legs moderately stout, toes 5–4, those of the hind feet 1–5–2–3–4 in order of length; fore feet 1–4–2–3. Tongue with the plicae radiating from the posterior field. Vomerine teeth in 4 series which form a fairly regular transverse line. The middle series of 9 or 10 teeth each are longer than the outer ones, lie wholly between the inner nares, and curve outward and backward. They are separated at the mid-line and from the inner nares by a distance about equal to ½ the length of one series. The outer series is short, contains 3–5 teeth, and forms a line directed obliquely outward behind the inner nares.

COLOR. In life the ground color varies from deep chocolate-brown to black. The animal is strikingly marked above, with a median stripe which varies in color from dull tan to a bright mustard-yellow, with some individuals definitely greenish-yellow. The stripe usually originates at the back of the head, but sometimes involves the entire dorsal surface of the head and continues without interruption to the end of the tail. In some the stripe is broken one or more times on the trunk, and on the head may be represented by a number of small irregular blotches. The sides of the dorsal stripe are never entirely straight and usually are quite irregular. Immediately below the dorsal stripe the sides are black, the lower sides with many very small, light flecks in a broad band which originates on the side of the snout below the level of the eye and continues along the side to involve the exposed surfaces of the legs. The lower side of the tail may be flecked heavily with white or be nearly uniform black. Often, on the upper surface of the limbs, there are a few irregular tan or yellowish spots. The belly and lower surface of the legs are dark brown, everywhere minutely flecked with whitish points. The throat and lower surface of the tail have the ground color slightly lighter. The sexes may be distinguished by the form and size of the vent. In the male it is large and strongly protuberant, in the female, except in the breeding season, a simple slit.

BREEDING. On June 25, 1936, our party, collecting with Professor J. R. Slater in small ponds near Reflection Lakes on Mt. Rainier, was fortunate to find adults and eggs of this species in some numbers. The eggs were scattered singly and in small masses containing 8–10. Slater (1936, p. 235) gives additional information as follows: "The egg . . . is 2.5 mm. in diameter, black at the animal pole, with the lower two-fifths at the vegetal pole light gray. The two envelopes do not show very clearly; the inner is 6 to 7 mm., and the outer 12 to 17 mm. in diameter, depending on the age of the egg. They hatch in five to fifteen days and may transform at sea level in July, while in the high mountain ponds most of the larvae do not transform until the beginning of their second year. The size at transformation is 60 to 75 mm.". . . "From a female 127 mm. in length 184 mature eggs were taken."

LARVAE. Some fully developed larvae received from J. R. Slater on August 15 completed transformation August 21. The larvae were quite uniformly light greenish-gray to brownish-gray, flecked and mottled with dark brown and black. Ventral surfaces dirty white. In the recently transformed individual, the dorsal surfaces became greenish-yellow with metallic reflections, the head and legs less brightly colored than the trunk and tail. The ventral surfaces of the trunk and tail dark, finely speckled with whitish, yellowish, and silvery dots. Length 2⅞″ (70 mm.).

**SPOTTED SALAMANDER.** *Ambystoma maculatum* (Shaw). Figs. 29c, 38. Map 16.

TYPE LOCALITY. "Carolina."

RANGE. Nova Scotia, Prince Edward Island and mainland New Brunswick and Gaspé, Quebec; west to Lake Nipigon, Ontario, and Wisconsin; southward to northern Florida, Louisiana, and Texas.

HABITAT. Most abundant in areas of deciduous and mixed forests where ponds, slow streams, or temporary pools offer suitable breeding places.

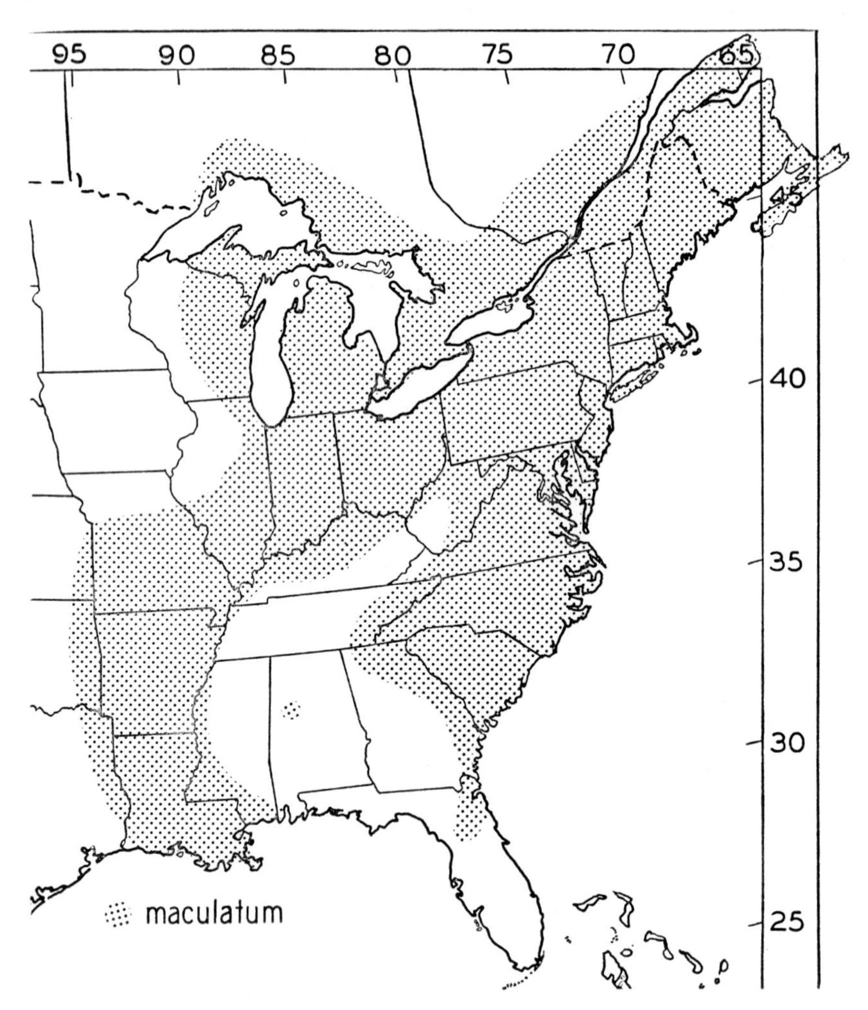


MAP 16.—Distribution of *Ambystoma maculatum.*

SIZE. Adult males vary from 5⁵⁄₁₆ to 7¾″ (150–196 mm.) and average about 6⅞″ (175.5 mm.). Females are a little smaller, averaging about 6¹¹⁄₁₆″ (170 mm.).

DESCRIPTION. The Spotted Salamander is stout-bodied and intermediate in size between *A. jeffersonianum* and *A. tigrinum*. The body is somewhat depressed, but the snout is broadly rounded and less flattened than in Jefferson's Salamander. The sides of the head back of the angle of the jaws are often swollen. The tail is oval in cross section at base and moderately flattened distally. The usual number of costal grooves is 12, sometimes 11 or 13. Gular fold prominent. The legs are large and strong, with toes 4–5, those of the front feet 1–4–2–3 in order of length; hind feet with toes 1–5–2–3–4. The vomerine teeth form a transverse sinuous line, continuous or interrupted at the middle and back of the inner nares. The tongue is broad, free only at the margin, and marked with plicae that radiate from the posterior part. The color above is deep bluish-black, the lower half of the sides, the venter, and under surface of the limbs lighter, pale slate. On either side of the middorsal line of the body there is a more or less regular row of large, round, yellow or orange spots varying in number from 24 to about 45. The males are slightly slimmer than the females and in the breeding season may be recognized by the protuberant vent. The females at this season have the ventral surfaces paler.

BREEDING. Adults of both sexes migrate to the breeding ponds in the spring, late March or early April in the North. Sometimes many individuals congregate in a limited area and indulge in a kind of nuptial dance, swimming about vigorously, rubbing and nosing one another until the water fairly boils. Fertilization is internal by means of spermatophores, stump- or vase-shaped gelatinous structures capped by whitish sperm masses. The eggs are deposited in large, compact masses 2½″–3½″ in diameter and attached to submerged objects. The external envelope of jelly may be clear or milky white. It is thicker than in *A. jeffersonianum* and much denser. The number of eggs per mass varies from a dozen to over 250, but the average is about 125. Total complement may be deposited in half a dozen small masses or limited to 1 or 2 large ones. The individual egg is 2½–3 mm. in diameter, with the upper pole dark brown or gray and the lower pole dirty white or dull yellow. In the field the incubation period varies from 31 to 54 days, depending on the temperature of the water.

FIG. 38 (legend at foot of opposite page).

LARVAE. Larvae at hatching have a length of ½″ (12–13 mm.). The ground color is dull greenish yellow, with darker areas of olive on the head, and small rounded black spots scattered over the dorsal surfaces and forming an indistinct band on either side of the middorsal line. There are no large paired dark spots as in *A. jeffersonianum*. Gills and balancers are present at hatching, and the fore legs are represented by elongate buds; the hind legs not evident. The broad tail fin is continuous with the dorsal fin, which extends to a point opposite the fore legs. Larvae attain a length of 2″–3½″ (40–75 mm.) and average about 2½″ (51 mm.) during a development period of 61–110 days. Transforming young may be found from August to September, rarely to October in colder waters. The yellow or orange spots are sometimes acquired within a week following transformation. During the year following transformation the Spotted Salamander attains a length of about 3½″–4″ (85–100 mm.) and may attain sexual maturity by the following spring.

**MARBLED SALAMANDER. BANDED SALAMANDER.** *Ambystoma opacum* (Gravenhorst). Fig. 39. Map 17.

TYPE LOCALITY. New York.

RANGE. From New Hampshire to Florida, west to Louisiana and Texas, Mississippi Basin north to Arkansas, Missouri, Indiana, Illinois, and Wisconsin.

HABITAT. Found in drier situations than are suitable for most species of *Ambystoma*. Abundant in sandy and gravelly areas and in the low grounds bordering ponds and slow streams.

SIZE. One of the smaller species of the genus, adults averaging about 4″ (100 mm.) in length but ranging to 120 mm. The females are a little longer and larger than the males.

FIG. 38 (see opposite page). *Ambystoma maculatum* (Shaw). (1) Adult male, ventral view; actual length 5⅝″ (144 mm.). Albany, New York. (2) Adult female, dorsal view; actual length 5¾″ (147 mm.). Albany, New York. (3) Adult male, dorsolateral view; actual length 5 13/16″ (149 mm.). Albany, New York. (4) Egg mass. (5) Spermatophores.

DESCRIPTION. In this strikingly marked salamander the body is thick, short, and more or less cylindrical or slightly depressed. In well-fed specimens there is no evidence of a dorsal groove. The head is moder-

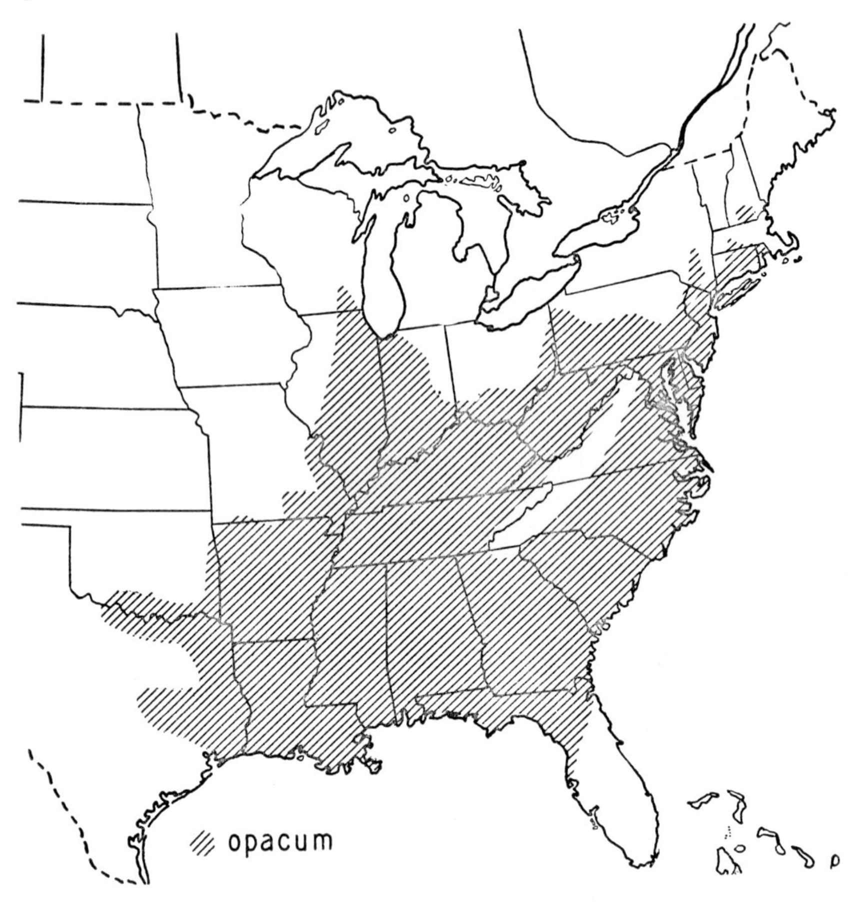


MAP 17.—Distribution of *Ambystoma opacum.*

ately broad, slightly rounded above, and widest at a point above the angle of the jaw. The tail comprises 34.8–42.3 per cent of the total length. It is not greatly compressed and is thicker above than below. Gular fold prominent. Costal grooves 11, or 12 if 1 immediately above the fore leg is counted. The legs are stout, the toes 5–4, those of the fore feet 1–4–2–3 in order of length; toes of the hind feet 1–5–2–3–4. The vomerine teeth are in a transverse sinuous line, widely interrupted back of the inner nares and more narrowly interrupted at the middle. The

FIG. 39. *Ambystoma opacum* (Gravenhorst). (1) Adult female, actual length 3¾″ (96 mm.). Raleigh, North Carolina. (2) Recently transformed individual, actual length 2⁵⁄₁₆″ (59 mm.). Coram, New York. (3) Larva, actual length 1¾″ (45 mm.). Coram, New York. (4) Adult female with eggs. Raleigh, North Carolina, Oct. 26. (5) Adult male, actual length 3½″ (89 mm.). Raleigh, North Carolina.

tongue is short, thick, and fleshy, with the plicae radiating from a V-shaped area at the back. The ground color is deep lustrous black, with brownish tinges on the under side of the head and on the legs and toes. The light markings, bright white in the males and dull grayish in the females, are extremely variable in size, shape, and distribution. In general they are restricted to the back and upper sides of the head, trunk, and tail, and are arranged in transverse bands, narrow dorsally and widened on the upper sides, where they sometimes unite with the ends of adjacent bands to enclose a series of fairly regular black spots along the mid-line of the back. Sometimes the light markings fuse to form longitudinal bands, one on each side of the back. On the tail the light markings are usually simple transverse bands. The sexes may be distinguished by the protuberant vent and brighter white of the males.

BREEDING. In striking contrast to the habits of the majority of the species of Ambystoma, which deposit their eggs in early spring in water, the Marbled salamander migrates to the breeding grounds in the fall and deposits her eggs on land. Fertilization is internal by means of spermatophores. There is a brief courtship, apparently initiated by the male. The female deposits her eggs in a little depression in the ground beneath surface materials, old bark, logs, moss, or weeds. The situation chosen is one likely to be flooded by late summer or fall rains. The eggs, 50–200 in number, are deposited singly, not aggregated by a common jelly covering as in other species of the genus, and are attended by the female. In the North the egg-laying season may start in September, farther south in October to December. The individual egg is about 2.7 mm. in diameter and is provided with two separate envelopes, in addition to the vitelline membrane, which give it an outside diameter of 4.2–5 mm. The length of the incubation period is variable, depending on the amount of moisture present and perhaps, to some extent, on temperature. The eggs usually hatch in the fall but may carry over to spring in the absence of rains.

LARVAE. Larvae at hatching have a length of about ¾″ (19 mm.), but individuals forced to remain in the egg may attain a length of 1″ (25 mm.). The larva is uniformly, lightly pigmented above and has a grayish tinge which soon changes to brown. The tail fins and sides are mottled with yellowish areas where pigment is lacking. The gills are long, slender, and tinged with gray and yellow. The fore legs are slender and delicate but with the toes well developed. The hind legs are elongate buds without indication of toes. The balancers may be present or lost before hatching. Larvae that hatch in the fall may attain a length of 23–45 mm. by the following March, and at the time of transformation in June vary from 63 to 74 mm. Sexual maturity is reached in the fall of the year following transformation or when 15–17 months old.

**MOLE SALAMANDER.** *Ambystoma talpoideum* (Holbrook). Fig. 40. Map 18.

TYPE LOCALITY. Sea islands off the coast of South Carolina.

RANGE. South Atlantic and Gulf States, North Carolina to Louisiana and northward to Illinois. Also recorded from Oklahoma, Arkansas, and Texas.

HABITAT. Except during the breeding season, the adults are much given to burrowing and are only occasionally found beneath rotten wood in damp situations. In Louisiana, confined to areas of loose sand not affected by salt water (Viosca, 1938, in litt.). Strecker and Frierson (1935, p. 33) found an adult occupying a mole-like burrow on the east bank of Wallace Bayou, De Soto Parish, Louisiana. Van Hyning (1933, p. 3) found larvae abundant in ditches and ponds near Gainesville, Florida.

SIZE. Adults I have measured vary in length from 85 to 97 mm. The proportions of a large female are as follows: length 97 mm., tail 36 mm.; head length 18 mm., width 14 mm.

DESCRIPTION. This is a small, stout species with a broad, depressed head. The head is widest opposite the angles of the jaws, the sides slightly converging behind to the lateral extensions of the gular folds, more abruptly from opposite the eyes to the bluntly rounded snout. The trunk is well rounded, with a scarcely evident, median impressed line. The tail is short, comprising 34–37 per cent of total length, elongate oval in cross section at base, compressed toward the tip and slightly keeled above, rounded below. Costal grooves 10, counting 1 each in axilla and groin.

Legs proportionally stout, toes 5–4, those of the hind feet 1–5–2–3–4 or 1–5–2–4–3 in order of length; fore feet, toes 1–4–2–3. Toes meeting or slightly overlapping when legs are appressed to the sides. Tongue rela-

MAP 18.—Distribution of *Ambystoma talpoideum*. (Also reported from French Broad River, Henderson County, North Carolina, and from Transylvania County, North Carolina.)

tively narrow, about as wide as distance between the eyes, smooth behind in a triangular area and with narrow plicae radiating from this posterior field. Vomerine teeth in 3 distinct series, the middle series some distance posterior to and wholly between the inner nares, the lateral series short, with 5–7 teeth, placed behind the inner nares and extending a short distance outside of the lateral margins of the nares.

COLOR. The ground color above is deep brown in a broad band which involves the entire dorsal surface of the head and trunk, and a lighter brown on the dorsal surface of the tail. Scattered over the dorsal surfaces are many small bluish-white flecks which are aggregated in small clusters to give a lichen-like effect. The light flecks are more abundant and smaller on the tail, fewer and more widely scattered on the head. The lower sides of the head, trunk, and tail have the general ground color bluish-gray instead of brown, and the light fleckings are much more numerous and form larger blotches. The legs above are colored like the back, brown with light fleckings. The ventral surfaces are colored like the lower sides except that the light markings are fewer and farther apart.

BREEDING. Allen (1932, p. 3) found this species breeding in Harrison County, Mississippi, during February 1930. Eleven specimens were found in a hole in which there were 2–3 inches of water. In St. Tammany Parish, Louisiana, Mr. Henry B. Chase found eggs and early larval stages during or following warm rains in January, when air temperatures varied from 50° to 70° F. The eggs were deposited in temporary pools in stands of sour gum, *Nyssa biflora,* and red gum, *Liquidambar styraciflua,* and in ponds in pine woods in the sandy valleys of spring-fed rivers. The eggs were aggregated in small, fragile masses containing 4–20 and measuring 30–57 mm. in length by 24–34 mm. in width.

LARVAE. A series of larvae collected April 21, 1935, at Mt. Pleasant, South Carolina, and presented to me by Mr. E. B. Chamberlain, vary in length from 1⅜″ (35 mm.) to 2¼″ (57 mm.). The head is broad and depressed, the snout bluntly pointed. The dorsal fin arises just back of the head, reaches its greatest width about opposite the vent, and continues to the tip of the tail. The ventral fin is narrower and extends anteriorly to the vent. The fins are speckled and lightly mottled with brown pigment. On either side of the dorsal fin is a broad, dark brown band which is broken into a series of quadrate spots by lighter interspaces. Below these dark bands, a narrow light line extends from the gills to a point opposite the vent. The sides below the light line are

FIG. 40. *Ambystoma talpoideum* (Holbrook). (1) Adult female, actual length 3⅜″ (86 mm.). (2) Same, ventral view. Alton, Louisiana.

brown, somewhat mottled, and lighter than the dorsal stripes. Throat and belly light, the former slightly mottled. Sides of head usually with a dark band through the eye from snout to base of gills, below this mottled.

According to Mr. Henry B. Chase and Mr. Percy Viosca, Jr., the larvae transform at the end of April or early in May, having attained nearly the size of adults. M. J. Allen (1932, p. 4) found this species breeding Feb. 12, 1930, in Mississippi, and larvae in the water until May 15. Individuals lost the gills at lengths varying from 50 to 65 mm., while transformed specimens, taken May 14, varied from 55 to 70 mm.

**TEXAS SALAMANDER. SMALL-MOUTHED SALAMANDER.** *Ambystoma texanum* (Matthes). Figs. 29d, 41. Map 19.

TYPE LOCALITY. "Das erste Exemplar fand ich unter einem faulen Baumstamme im Urwalde am Rio Colorado, das zweite ebenfalls . . . inn Cumming's Creeck [*sic*] Bottom, Fayette County" [Texas].

RANGE. Ohio west to southern Iowa, Kansas, and Nebraska; south to Texas and Louisiana; northward to Tennessee, Kentucky and western West Virginia. Also reported from Mississippi, North Carolina, and South Carolina.

HABITAT. In Ohio reported by Morse (1904, p. 110) to be common in hilly regions, where they are found beneath logs and partially buried in damp turf. In Texas, Strecker (1922, p. 7) reports the species from lowlands as well as uplands and on the prairie farms and flats. They are commonly found in hiding beneath old logs and drift material along river bottoms and sloughs. The species is much given to burrowing and frequently occupies deserted crayfish holes.

SIZE. Strecker (1922, p. 6) says that the species attains a length of 6″ or more. The females average a little larger than the males. The largest I have measured, a female, has a total length of 5 13/16″ (147 mm.), tail 2 3/8″ (60 mm.). The proportions of a moderate-sized female from Waco, Texas, are as follows: total length, 5 1/2″ (139 mm.), tail 2 1/4″ (57 mm.); head length 19/32″ (15 mm.), width 13/32″ (12 mm.). A male from the same locality measures: total length 5 1/8″ (130 mm.), tail 2 1/4″ (57 mm.); head length 19/32″ (15 mm.), width 14/32″ (11 mm.). Measurements of a considerable number of specimens indicate that the tail of the male is proportionally longer than that of the female.

DESCRIPTION. A species of moderate size, characterized by the slender head and small mouth. The head is roughly oval in outline when viewed from above, the sides behind the eyes gently converging to the lateral extensions of the gular fold and in front abruptly to the short blunt snout. Eyes small, the iris tinged with brassy. A sinuous groove from the

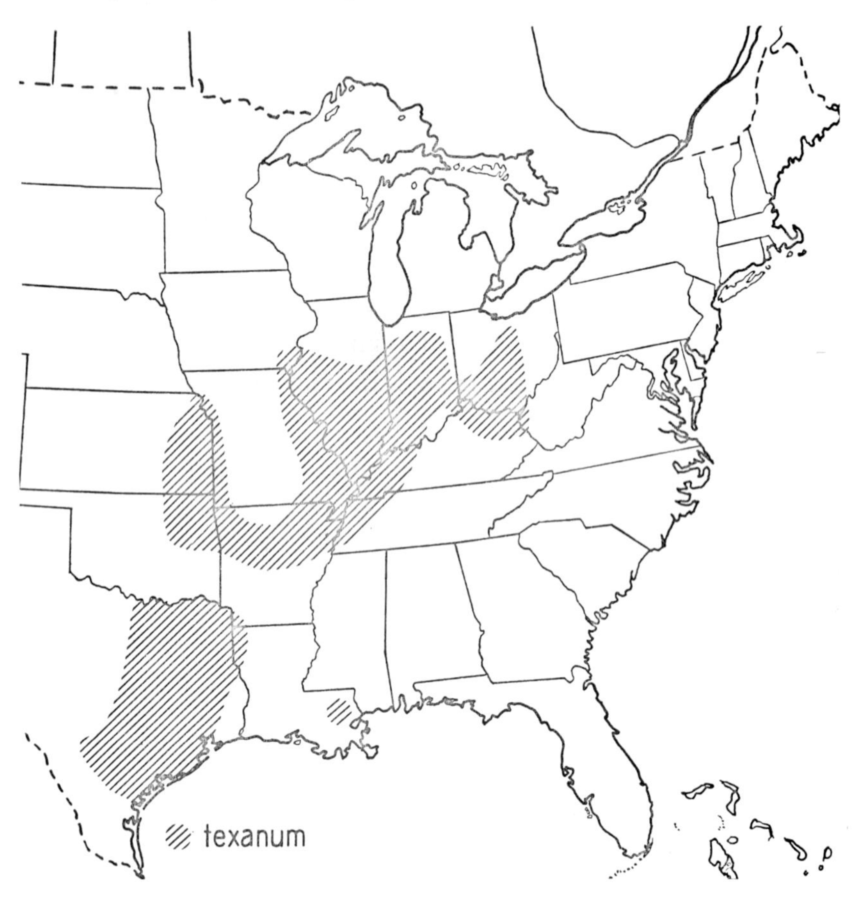


MAP 19.—Distribution of *Ambystoma texanum*. (Also recorded from Richmond County, Georgia, and Rutherford County, North Carolina.)

posterior angle of the eye to the lateral extensions of the gular fold, poorly developed in some; a short vertical groove from this line passing just behind the angle of the mouth. The head is shorter and wider in the males than in the females. The trunk is rounded on the sides and above, flattened below. A slight median ridge from the back of the head to the base of the tail and frequently an impressed line along the mid-line of the belly from the fore to the hind legs. The tail is oval in section at the base, slightly rounded below on the basal third, and compressed above,

FIG. 41. *Ambystoma texanum* (Matthes). (1) Adult female, actual length 5″ (128 mm.). Shreveport, Louisiana. (2) Adult male, actual length 4 11/16″ (120 mm.). Gayle, Louisiana. (3) Adult, dorsal view. Niles Township, Delaware County, Indiana. [Photograph by Leo T. Murray.]

becoming definitely flattened distally. Costal grooves 14 or 15, the usual number being 14 when 1 each is counted in the axilla and groin; 2–3½ intercostal spaces between the toes of the appressed limbs. Legs moderate, the hind pair in the males stouter. Toes 5–4, those of the hind feet

1-5-2-3-4 or 4-3 in order of length from the shortest; toes of the fore feet, 1-4-2-3. Tongue moderate in size, the extreme edges smooth, the center marked with a longitudinal groove from which the plicae diverge obliquely. Vomerine teeth very slender, 10-12 in double rows behind and entirely between the inner nares, where they usually form 2 slightly separated series. On the margins of the jaws the teeth are small, foot-shaped and in 2 or 3 rows.

COLOR. In living adults, the ground color above is a deep brown, almost black, blotched above and on the upper sides with irregular, lichen-like markings of grayish-brown. On the lower sides the superficial lichen-like markings are usually much lighter, grayish-white, and frequently larger. The whitish markings are continued on the sides of the head below the level of the eyes and on the lower half of the tail to the tip. The basal segments of the legs are mottled above like the back; in front and behind the spots are lighter, more like the lower sides. The ground color of the ventral surfaces is brown, lighter than that of the back, and here the light markings are fewer in number, less irregular in shape, and more widely scattered on the belly. Strecker (1909, p. 18) records some examples having yellowish underparts and with light-yellow—instead of grayish-white—spots between the costal grooves on the lower sides and belly.

BREEDING. In Ohio this species has been observed as early as February (Morse, 1904, p. 110), and, in Indiana, Hay (1889, p. 602) found eggs on March 3. Strecker (1928, p. 6) reported that in northeastern Texas the breeding season is during the first 10 days of March.

EGGS. The eggs are deposited singly or in small clusters and are attached to stems and sticks below the surface of the water. Individual eggs are small, about 2 mm. in diameter and, according to Hay (*ibid.*, p. 603), are surrounded by gelatinous capsules composed of 2 principal layers separated by a very thin layer. Some eggs observed in the field March 3 had already hatched by March 28 and a few continued to hatch until April 10.

LARVAE. Larvae at hatching have a length of approximately 10-13 mm., bright olive-green above, with indications of squarish blotches along the back. Burt (1938, p. 377), who studied young material from Texas, remarks: "The youngest specimens exhibit a series of prominent vertical light bars or spots in addition to a distinctive light stripe on each side of the body and tail." In Ohio and Indiana the larvae are ready to transform by June. In the vicinity of Waco, Texas, transformation may take place in May at a length of approximately 2¾″ (70 mm.).

**EASTERN TIGER SALAMANDER.** *Ambystoma tigrinum tigrinum* (Green). Figs. 1a, 42. Map 20.

TYPE LOCALITY. Near Moorestown, New Jersey.

RANGE. From Albany and Long Island, New York, southward to northern Florida, westward to Texas, northward to Minnesota and Point Pelee, Ontario.

HABITAT. The Tiger salamander, as an adult, is essentially a burrowing species and spends most of its life below the surface of the ground. The breeding waters, to which it resorts in the spring, are found in semi-desert regions, pine barrens, and forests of plains and mountains, its ability to survive under such diverse situations probably being due to the burrowing habit.

SIZE. The largest of the species of Ambystoma. The males exceed the females in size, averaging about 8″ (202 mm.), but individuals of the eastern form attain a length of 10″ (254 mm.). The females average about 7⅟₁₆″ (177 mm.).

DESCRIPTION. The body is stout and muscular and the skin fairly smooth. The head is broad, widest back of the eyes, and tapering gradually to the narrow neck. The snout is broadly convex above and bluntly rounded. The tail is oval in section at the base, thicker below than above, and compressed toward the tip. There are 12 costal grooves, counting 1 each in the axilla and groin. Gular fold prominent. The legs are stout and moderately long, the toes 5–4; those of the hind feet 4–3–2–5–1 in order of length; fore feet 3–2–4–1. The toes are much depressed, short, sharply tapering to blunt tips. The vomerine teeth are usually in a single continuous series, but sometimes interrupted at the middle point. The tongue is broad and fleshy, with only the margins free and thin.

COLOR. The ground color above and on the sides is deep brown or dull black. On the lower sides the dark color meets the olive-yellow of the

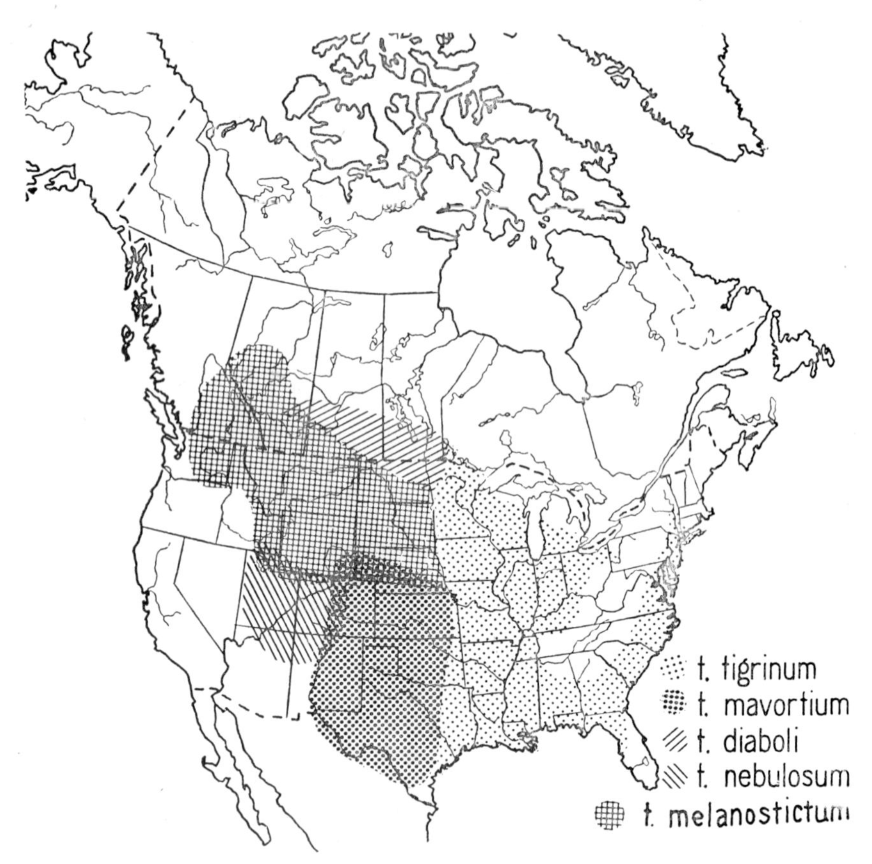


MAP 20.—Distribution of the subspecies of *Ambystoma tigrinum*.

venter along an irregular line, with the light markings often extending upward between the costal grooves. Scattered over the dark ground color are many pale, brownish-olive or brownish-yellow blotches, irregular in size and shape, and tending to fuse and form bands on the lower sides and tail in some individuals. The under surface of the head is often uniformly yellow or yellow tinged with olive. The venter is olive-yellow blotched and marbled with darker. The males are larger than the females and during the breeding season the vent is swollen and

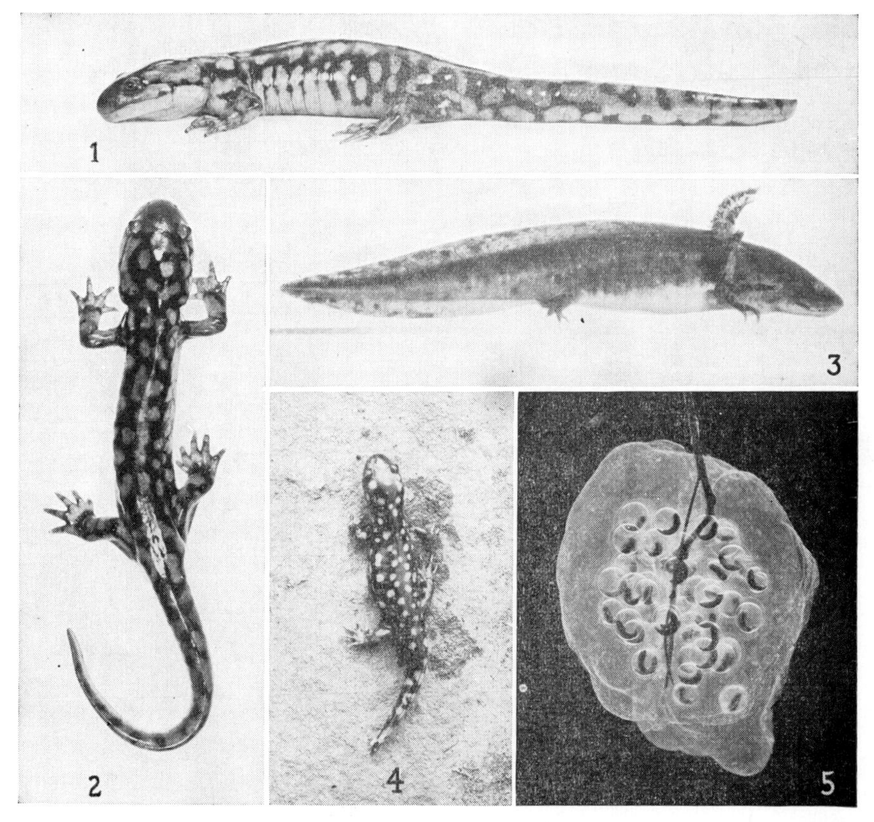


FIG. 42. *Ambystoma tigrinum tigrinum* (Green). (1) Adult female, actual length 7½″ (190 mm.). (2) Same, dorsal view. Setauket, Long Island, New York. [L. G. Barth, collector; J. Westman, photograph.] (3) Larva, fully grown, actual length 4¾″ (120 mm.). Syosset, Long Island, New York, June 25. (4) Recently transformed individual, actual length 3½″ (89 mm.). Syosset. (5) Egg mass. Syosset.

protuberant. The tail is usually longer and the hind legs longer and stouter in the male.

BREEDING. In the early spring, March in the North, January or February farther south, the Tiger salamanders resort to the breeding ponds. There is a courtship in which the sexes nose and push one another with the snout, and this is followed by the deposition of spermatophores. The eggs are deposited in masses and attached to twigs and weed stems below the surface of the water. The egg mass may be globular, oblong, or kidney-shaped, and measure from 2¼" x 2¾" (55 x 70 mm.) to 3 x 4" (75 x 100 mm.). The eggs average about 52 per mass but vary from 23 to 110. Compared with that of *A. maculatum,* the egg mass of the Tiger salamander has the common jelly envelope thicker but less dense, the eggs fewer in number, the individual envelopes with a greater diameter. The vitellus has a diameter of 3 mm., the upper pole light- to dark-brown and the lower pole pale cream to buff. There are 3 distinct envelopes in addition to the vitelline membrane. The extent of the incubation period is variable, 24–30 days or longer.

LARVAE. Larvae at hatching average about 9/16" (14 mm.) but vary from 13 to 17 mm. The head of the young larva is broad, widest in front of the gills, and rounding to the short blunt snout. The gills are as long as the head and with few filaments. The dorsal keel extends backward from the head and is continuous with that of the tail. The balancers and limbs are lacking at hatching, and the forelegs represented by low humps. The ground color above is yellowish-green with some darker markings. On either side of the middorsal line there is a dark band, usually broken to form about 6 pairs of dark spots. Normally a light band extends along each side from the eye well onto the tail. The larvae may attain a length of 3 9/16"–4⅞" (90–123 mm.) by June in the North, after a development period of 75–118 days. Sexual maturity may be reached the spring following transformation.

**DEVIL'S LAKE TIGER SALAMANDER.** *Ambystoma tigrinum diaboli* Dunn. Fig. 43. Map 20.

TYPE LOCALITY. Devil's Lake, North Dakota.

RANGE. North Dakota (north and east of Altamount Moraine) into Alberta, Saskatchewan, and Manitoba.

SIZE. Attains a maximum size of 12¼" (312 mm.). The average of 15 adults of both sexes from the vicinity of Turtle Mountain and Devil's

Lake, North Dakota, is 7½″ (190 mm.), with extremes of 11⅛″ (282 mm.) and 4¹¹⁄₁₆″ (119 mm.). The proportions of the Devil's Lake female are as follows: total length 11⅛″ (282 mm.), tail 5⁵⁄₃₂″ (131 mm.); head length 1²¹⁄₃₂″ (42 mm.), width 1⁷⁄₁₆″ (37 mm.). An adult male from near Lake Upsilon, North Dakota, measures: total length 8¾″ (222 mm.), tail 4⅜″ (111 mm.); head length 1⁷⁄₃₂″ (31 mm.), width 1¹⁄₁₆″ (27 mm.).

HABITAT. Both adults and larvae are known from the fresh-water lakes in North Dakota and from Devil's Lake, which has a high salt concentration; but the species has been reported as breeding only in fresh water.

DESCRIPTION. The head is broad, convex above behind the eyes, widest opposite the angle of the jaws; the sides behind the eyes converge slightly to the gular fold; the snout depressed and bluntly rounded. The eye is relatively small, its horizontal diameter about 3 times in the snout. An impressed line from the posterior angle of the eye to the lateral extension of the gular fold; a vertical groove from this line to the angle of the mouth. Gular fold strongly developed. Trunk widened ventrally, somewhat narrowed dorsally, and with an impressed median line. Usually 12 costal grooves, or 13 counting 1 in the axilla and 2 that run together in the groin, and 1–3 intercostal folds overlapped by the toes of the appressed limbs. Tail widest at about the basal third, basal half rounded below, compressed above; distal half strongly compressed. Legs large, stout; toes 5–4, those of the hind feet 1–5–2–3–4 in order of length from the shortest; toes of the fore feet 1–4–2–3. Toes flattened and pointed, with flange-like margins, the tips black and horny; hind toes one-half webbed, fore feet webbed at base. Plantar and palmar tubercles well developed. Tongue nearly circular in outline, the parallel plicae narrow and extending to the anterior margin from a posterior field which is crossed by several crescentic folds. Vomerine teeth small and in 2 or 3 irregular rows forming a wide V-shaped figure, the apex directed forward.

COLOR. I have not seen this species in life. In preserved specimens the ground color above varies from gray to brown and light olive, lighter on the sides and dull yellowish below. This species is marked with many small rounded black spots, which in some individuals may become somewhat elongate on the sides of the trunk; a few black spots on the throat,

FIG. 43. *Ambystoma tigrinum diaboli* Dunn. (1) Adult female, ventral view; actual length 8⅜″ (213 mm.). (2) Same, dorsal view. Near Lake Upsilon, North Dakota. [Photographs of a preserved specimen.]

sides of the belly and under surface of the limbs. In some the black spots on the belly may be elongate and arranged transversely. Some specimens from near St. Johns, North Dakota, have a strongly mottled appearance, the dark spots showing a tendency to run together on the sides and belly, and occasionally on the back.

BREEDING. A large female from Devil's Lake, North Dakota, unfortunately without the date of collection, has large ovarian eggs apparently ready to be deposited. The eggs are pigmented above (brown in preservative) and yellowish below. Apparently this subspecies is often neotenic.

LARVAE. The larvae attain a large size and when fully grown have the general color and pattern of adults. A series of 11 from Sweetwater Lake, North Dakota, varied in length from $5\frac{29}{32}''$ (151 mm.) to $7\frac{1}{8}''$ (181 mm.) and averaged $6\frac{5}{8}''$ (168 mm.). The broad dorsal tail fin arises at the back of the head and is widest at a point just behind the vent. The ventral fin extends to the vent. The gills are long and slender, 1–2–3 in order of length from the anterior. In some specimens there is a tendency for the dark spots to form bars on the tail.

Partly grown larvae from Grand Forks, North Dakota, vary in length from $3\frac{19}{32}''$ (92 mm.) to $4\frac{7}{32}''$ (108 mm.). These either exhibit a marbled pattern or have the dark pigment concentrated in fairly definite spots. The dorsal tail keel is strongly blotched, the ventral fin relatively free from darker markings. On some individuals there is a well defined row of light spots on the sides, extending from the head onto the basal half of the tail; in a few specimens a secondary row of light spots on either side of the dorsal fin.

**YELLOW-BARRED TIGER SALAMANDER.** *Ambystoma tigrinum mavortium* Baird. Fig. 44. Map 20.

TYPE LOCALITY. New Mexico.

RANGE. Kansas, Oklahoma, central and western Texas, eastern Colorado, central and eastern New Mexico (Dunn, 1940, p. 158).

SIZE. The average length of 16 adults of both sexes from Kansas and Oklahoma is $6\frac{23}{32}''$ (171 mm.). Five adult males average $7\frac{9}{16}''$ (192 mm.), the extremes $5\frac{19}{32}''$ (143 mm.) and $8\frac{3}{8}''$ (213 mm.). Eleven females average $6\frac{3}{8}''$ (162 mm.) and vary from $4\frac{13}{32}''$ (123 mm.) to $8\frac{9}{32}''$ (211 mm.). The proportions of an adult male from Riley County, Kansas, are as follows: total length $7\frac{11}{16}''$ (195 mm.), tail $3\frac{13}{32}''$ (87 mm.); head length 1 1/32″ (26 mm.), width 25/32″ (20 mm.). A female which is of the same length and which came from the same locality has a tail length of 3 1/8″ (80 mm.); head length 1 3/32″ (28 mm.), width 25/32″ (20 mm.).

HABITAT. During the dry season of the year the adults are seldom seen above the surface of the ground, but are occasionally dug out of burrows at a depth where the soil is at least slightly moist. During the first heavy rains of spring they often appear in large numbers and occupy the pools, ponds, "tanks," and slow streams for the breeding season. I have collected the larvae in temporary ponds, irrigation ditches, and backwaters of streams.

DESCRIPTION. The head is broad and somewhat convex above, widest at the angle of the jaws, the sides behind converging to the lateral extensions of the gular fold, in front abruptly to the broadly rounded and depressed snout. The eye is small, its horizontal diameter about 3 times in the snout; the iris brassy. An irregular slightly impressed line from the posterior angle of the eye to the lateral extension of the gular fold. Gular fold well developed. Trunk stout and with little indication of a median impressed line. Costal grooves usually 13, occasionally 12 or 14, counting 1 each in the axilla and groin, and the toes of the appressed limbs overlap 2 to 4 intercostal folds. The tail of the adult is rather long and slender, strongly compressed, rounded below and sharp-edged above. In the male it may comprise 42–50.7 per cent of the total length, and in the female 35–46.5 per cent, the higher percentages associated with the larger specimens. Legs large and stout. Toes 5–4, those of the hind feet 1–5–2–3–4 in order of length from the shortest; toes of the fore feet 1–4–2–3. Toes all webbed at base, rather long, flattened, tapering, and horny tipped. Plantar and palmar tubercles well developed. Tongue fleshy and broadly triangular, the narrow plicae parallel and extending to the margin from the broad posterior area. Vomerine teeth usually in a transverse, slightly arched series, which may be interrupted only at the mid-line or at this point and just inside the inner margin of each inner naris. Occasionally the teeth form a single series which arises just

Fig. 44. *Ambystoma tigrinum mavortium* Baird. (1) Adult male, actual length 8¼″ (210 mm.). (2) Same, lateral view [from life]. (3) Same, ventral view [preserved specimen]. Winfield, Kansas.

outside and behind the outer margin of the inner naris and arches strongly forward between the nares.

COLOR. The general ground color is deep brown or black. The light markings are developed in the form of irregular vertical bars or elongate blotches which vary in width and completeness but usually extend from the mid-line of the belly to the mid-line of the back. The light markings are sometimes confluent on the lower sides, forming a broad band between the legs and leaving the middle of the belly irregularly blotched with dark. Rarely the light markings unite at the mid-line of the back and more often cross the tail, giving it a ringed appearance. The light markings are yellow, varying in degree of intensity, but usually suffused with dusky on the back and clear yellow on the lower sides. The head and limbs are irregularly blotched.

BREEDING. In Lost Lake, Riley County, Kansas, Burt (1927, p. 2) found, on March 7, many eggs in cleavage stages attached singly or in small clusters to weed tops. Some eggs appeared to form a continuous series along the stem of a weed, and when removed to the laboratory began to hatch March 14.

LARVAE. The larvae attain a large size, to 8 29/32" (226 mm.), sometimes exceeding the adults, and some are certainly neotenic. They are quite uniformly colored, in some waters light greenish, and acquire the barred pattern of the adults at the time of transformation. The axolotls have a very broad head, deep body, and broad fins. The dorsal fin arises on the back behind the head and attains its greatest width a short distance behind the vent. The ventral tail fin extends to the vent. The gills on some individuals are very long, the uppermost extending nearly to the hind legs. The legs are stout, the toes strongly depressed, broad at base, and sharply pointed.

**CLOUDED TIGER SALAMANDER.** *Ambystoma tigrinum nebulosum* Hallowell. Fig. 45. Map 20.

TYPE LOCALITY. San Francisco Mountains, Arizona.

RANGE. Interior Basin and Colorado Plateau in Utah, western Colorado, northwestern New Mexico, northern Arizona (Dunn, 1940, p. 158).

HABITAT. Adults and larvae are often found in the mountain lakes and ponds and, occasionally, the adults on land, hiding beneath stones, logs, or bark. In the tanks and pools of ranches the axolotls are known as "water dogs." In the region of the type locality, San Francisco Mountains, Arizona, the salamander is said to be confined to the balsam-fir zone.

SIZE. The average length of 12 adults of both sexes from Utah and Boulder County, Colorado, is 7 19/32" (192.25 mm.), the extremes 6 3/32" (155 mm.) and 9 1/16" (230 mm.). The males are a little larger than the females in the series I have measured, averaging 8 1/8" (206 mm.), the females 6 17/32" (166 mm.). The proportions of an adult male from Boulder County, Colorado, are as follows: total length 8 3/16" (208 mm.), tail 3 23/32" (95 mm.); head length 1 1/4" (32 mm.), width 31/32" (25 mm.). An adult female from the same locality measures: total length 6 23/32" (171 mm.), tail 3 1/32" (77 mm.); head length 31/32" (25 mm.), width 26/32" (21 mm.).

DESCRIPTION. The head is broad and convex above, the sides nearly parallel back of the eyes, in front evenly and broadly rounded to the tip of the snout. The eye is of moderate size, about 2 1/2 in the snout. An impressed line from the posterior angle of the eye to the lateral extension of the gular fold, a short vertical groove crossing this line at the angle of the mouth. Gular fold strongly developed. Trunk flattened ventrally, somewhat narrowed dorsally, and with an impressed median line. Tail flattened above at base, becoming strongly compressed immediately behind the vent, thin distally. There are usually 13 costal grooves, counting 1 in the axilla and 2 that run together in the groin, occasionally 12 or 14, and 2–4 intercostal folds are overlapped by the toes of the appressed limbs. Legs large and stout, toes 5–4, those of the hind feet broad at base, flattened and pointed, 1–5–2–3–4 in order of length from the shortest; toes of the fore feet 1–4–2–3. Palmar and plantar tubercles well developed, 2 on each foot. Tongue large and fleshy, free at the sides and behind only, the plicae narrow, parallel and extending to anterior margin; or, plicae broken into rows of small papillae. Vomerine teeth variable, sometimes forming a continuous series which is strongly arched forward between the inner nares, commonly interrupted just inside the inner margin of the naris and again at the mid-line. When interrupted, 6–11 teeth in the outermost series and 12–20 in each of the inner series. During the breeding season the males have the cloacal glands greatly enlarged and the sides of the vent are swollen into 2 broad, rugose lobes, bluntly pointed behind. The vent of the female is much smaller, less protuberant, and has a narrow flange-like margin.

COLOR. The adults are olive-green or dark gray above, with a few small, scattered black spots on the back, sides, legs, and tail. The ventral surfaces are usually lighter than the dorsal and mottled and blotched with dark spots of irregular size and shape. There are from one to several dark spots on the throat, and often the blotches of the sides of the tail run together and form dark areas of considerable extent. Other races of the Tiger salamander have light spots above on a dark ground; *nebulosum* has black spots on a dark ground.

BREEDING. There is little known of the breeding habits of this subspecies and the only reference to the eggs I have found is a note by Dwight Franklin (Copeia, No. 21, 1915, p. 30). At an elevation of approximately 7000′ at Flagstaff, Arizona, eggs were said to have been found in June. The salamander breeds in many of the streams, lakes, and ponds in the mountains; the larvae are often extremely abundant and are usually associated with a few transformed individuals.

LARVAE. In the high lakes of Colorado and other western states this species is often neotenic and reaches a size comparable to that of the transformed adults, the average length of 5 being 7⁹⁄₁₆″ (192 mm.). In life the larvae are rather uniformly olive-brown above, with small dark irregular spots on the back, sides, and tail. The belly is a little lighter and sometimes has a few small dark spots. The larvae have broad, convex heads and stout legs, with much flattened and pointed toes. The dorsal keel arises at the back of the head and reaches its greatest width a short distance behind the vent.

FIG. 45. *Ambystoma tigrinum nebulosum* Hallowell. (1) Recently transformed adult, actual length about 5½″ (139 mm.). (2) Same, ventral view. Zion National Park, Utah. [A. H. Wright, collector.] (3) Old adult, female [preserved specimen]; actual length 6½″ (165 mm.). Wasatch County, Utah.

**NORTHWESTERN TIGER SALAMANDER.** *Ambystoma tigrinum melanostictum* (Baird). Fig. 46. Map 20.

TYPE LOCALITY. Between Fort Union and Fort Benton, Montana.

RANGE. British Columbia, Alberta, Washington, Oregon, Idaho, Montana, Wyoming, North and South Dakota, Nebraska. (Dunn, 1940, p. 159.)

HABITAT. Adults are occasionally found in damp situations beneath logs, bark, boards, and stones. The larvae are sometimes extremely abundant in fresh-water ponds and lakes, where by day they are mainly in the deeper areas but at night come into the shallower waters along shore.

SIZE. The transformed adults reach an extreme length of about $8\frac{5}{8}''$ (219 mm.). Thirteen transformed individuals of both sexes from Washington average $5\frac{7}{16}''$ (139 mm.), the extremes $3\frac{3}{8}''$ (86 mm.) and $8\frac{3}{32}''$ (205 mm.). Six sexually mature adults in this series average $6\frac{19}{32}''$ (167 mm.). The proportions of a male from Lincoln County, Washington, follow: total length $6\frac{1}{2}''$ (165 mm.), tail $3\frac{1}{8}''$ (80 mm.); head length $\frac{15}{16}''$ (24 mm.), width $\frac{11}{16}''$ (18 mm.). A female from Medical Lake, Washington, measures: total length $5\frac{27}{32}''$ (150 mm.), tail $2\frac{7}{16}''$ (62 mm.); head length $\frac{15}{16}''$ (24 mm.), width $\frac{11}{16}''$ (18 mm.).

DESCRIPTION. The head is rather broad and short, widest at the angle of the jaws, the sides behind the eyes rounding to the lateral extensions of the gular fold, in front converging more abruptly to the bluntly rounded snout. Eyes moderate, the horizontal diameter about twice in the snout. An impressed line extends obliquely from the posterior angle of the eye to the gular fold and a short vertical groove from this line to the angle of the jaw. The trunk is well rounded and with an impressed median dorsal line. The usual number of costal grooves is 13, counting 1 in the axilla and 2 that run together in the groin; sometimes only 12, rarely 14, and the toes of the appressed limbs overlap 2–4 intercostal folds. The tail is flattened above at base and compressed shortly behind the vent, where it is rounded below and narrow above, and tapers distally. A free dorsal keel is not well developed in the adults, but is evident in some recently transformed individuals, where it extends to a point a short distance behind the vent. The legs are large and stout, toes 5–4, those of the hind feet 1–5–2–4–3 in order of length from the shortest, the tips often pointed and horny, the bases slightly webbed; toes of the fore feet 1–4–2–3, slightly webbed at base. There are 2 plantar

FIG. 46. *Ambystoma tigrinum melanostictum* (Baird). (1) Young adult, dorsal view; actual length 5⅝″ (144 mm.). Medical Lake, Spokane County, Washington. [J. R. Slater, collector.] (2) Larva, lateral view; actual length 3⅛″ (80 mm.). Near Crystal Creek, Wyoming. [Photographs of preserved specimens.]

and 2 palmar tubercles. The tongue is large, broadly oval in outline, free slightly at the sides, and provided with narrow parallel plicae. The vomerine teeth may be disposed in 3 series, but sometimes the middle one is narrowly interrupted at the mid-line. The lateral series of 6–10 teeth usually arise outside the outer margin of the inner naris and extend obliquely inward and forward to a point just inside the inner margin of the naris. A short break separates this series from the central one, which is usually strongly arched forward and may have 16–20 teeth on each limb.

COLOR. This subspecies is characterized by the extreme development of the light areas, which often form connected blotches of considerable size on the dorsal surfaces, so that the darker ground color persists in irregular patches. The light areas in life are dull yellow, the dark ground color brown to black, often darkest bordering the light areas. The upper surface of the legs is mottled and blotched, the sides of the tail with dark-bordered light areas sometimes forming more or less complete vertical bands. Recently transformed individuals sometimes have the light areas small and scattered over the dorsal and lateral surfaces. The throat, belly, and ventral surface of the legs may be light, with only a few darker blotches, or the light and dark blotches may be about equally distributed. The dark blotches on the sides of the tail extend to the ventral edge.

BREEDING. This form is known to be neotenic at times, Slater (1937, p. 82) having found eggs deposited by larvae taken April 5, 1935, and kept in captivity. These eggs were found to be like others taken in Medical Lake, Washington, April 6, 1934. Under natural conditions the eggs are deposited singly or in small clusters of 3–5 and attached to vegetation in shallow water near the margin of the lake. Individual eggs have a diameter of 1.9 mm. and are normally provided with three envelopes, the inner 2.6 mm. in diameter, the middle 3.5 mm., and the outermost 4.5 mm. when freshly deposited (Slater, 1937, p. 82).

LARVAE. The larvae may transform at a length of only $3\frac{3}{8}''$ (86 mm.) or, as axolotls, attain a length of $10\frac{7}{8}''$ (276 mm.). Larvae only $2\frac{7}{16}''$ (62 mm.) long sometimes exhibit a color pattern suggestive of that of the adults. The light areas are extensive and the dark pigment is in the form of small irregular spots on the back and sides of the trunk, on the sides of the tail, and on the upper surface of the head and limbs. In Medical Lake, Washington, Slater found larvae transforming at the end of the summer.

## GENUS DICAMPTODON

**PACIFIC GIANT SALAMANDER.** *Dicamptodon ensatus* (Eschscholtz). Figs. 29a, 47. Map 21.

TYPE LOCALITY. Vicinity of the Bay of San Francisco, California.

RANGE. Santa Cruz and Santa Clara Counties, California, northward through Oregon and Washington to southwestern British Columbia; northwestern Idaho and Rocky Mountains of Montana.

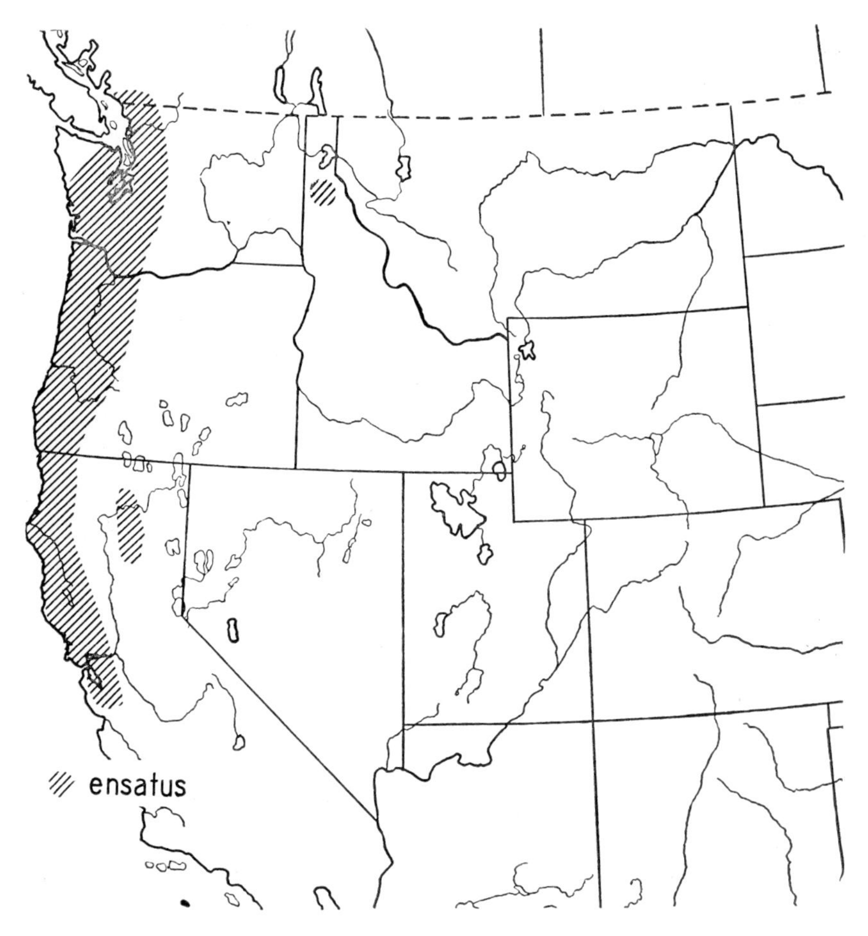


MAP 21.—Distribution of *Dicamptodon ensatus*. (Also recorded from the Rocky Mountains of Montana.)

HABITAT. This species is mainly limited to the humid coastal regions of the Northwest where there is abundant moisture throughout the year. On May 28, 1936, we collected adults of this species under logs and bark on the steep slopes of China Gulch at Gualala, California, and many larvae of various sizes in the stream below.

SIZE. This is the largest of the western species of salamanders, the transformed adults attaining an extreme length of nearly 12" (300 mm.). The measurements of an adult female from the south slope of Mt. Hood, Oregon, are as follows: total length 9 5/16" (236 mm.), tail 3 7/8" (98 mm.); head length 1 5/8" (41.5 mm.), width 1 1/8" (28 mm.). The smallest transformed individual I have measured is 5 5/16" (135 mm.) from Gualala, California, and the largest larva, from Oak Grove, Clackamas County, Oregon, sent me by Mr. Stanley G. Jewett, 11 1/4" (286 mm.). Storer (1925, p. 77) records a transformed individual only 122 mm. in total length.

DESCRIPTION. The body is stout, the limbs large and strong. The head is moderately broad, with the sides behind the eyes nearly parallel, in front tapering to the bluntly pointed and depressed snout. The mouth is large, the corner extending behind the eye a distance equal to the horizontal diameter of the eye. Eyes protuberant but of moderate size, iris deep brown, mottled with brassy. A horizontal impressed line from the posterior angle of the eye to the lateral extension of the gular fold. A short vertical line from the angle of the jaw to the horizontal groove. The trunk is well rounded above. Costal grooves 12, counting 1 each in axilla and groin, poorly defined. Tail thick and nearly subquadrate in section at base, tapering rapidly and becoming wedge-shaped at about 1/2 its length, the upper edge thinnest but without a definite keel. Toes 5-4, those of the hind feet 1-5-2-3-4 in order of length; front feet 1-4-2-3. Soles of the feet in recently transformed individuals sometimes with indistinct tubercles, in large adults smooth. Tongue large, fleshy, free along margins, plicae slightly diverging from the posterior field. Vomerine teeth behind inner nares and forming a transverse series which is narrowly interrupted where it bends forward at the mid-line.

Literally the teeth extend nearly to the outer margin of the inner naris.

COLOR. The ground color and markings vary considerably in intensity.

Fig. 47. *Dicamptodon ensatus* (Eschscholtz). (1) Adult male, actual length $9\frac{11}{16}''$ (245 mm.). (2) Same, lateral view. Sandy, Oregon. (3) Larva, actual length $11\frac{1}{8}''$ (286 mm.). Oak Grove, Clackamas County, Oregon.

In life a large female had the ground color above comparatively light (Dresden Brown, Ridgway), mottled and marbled with deep brown (Warm Sepia, Ridgway), fading to chocolate on the sides. The darker markings involve the entire dorsal and lateral surfaces of the head and encroach narrowly on the ventral surfaces of the lower jaws. On the sides of the trunk, the darker markings anteriorly extend to the level of the front legs, posteriorly the line of demarcation curves above the insertion of the hind legs and continues along the ventrolateral margins of the tail. The legs above are colored like the back. The ventral surfaces are lighter than the ground color of the lower sides; in the region of the throat, vent, and lower ridge of the tail, dull flesh color. In some individuals the mottlings may be very dark, almost black, and in strong contrast with the lighter ground color.

BREEDING. Eggs apparently of this species were reported by Henry and Twitty (1940, p. 247), attached singly to the surface of a timber in rapidly running water in San Mateo County, California, found June 19, 1937. The lot consisted of about 70 albino embryos in the "tail-bud" stage of development.

LARVAE. Larvae are characterized by the restriction of the dorsal fin to the tail, the absence of balancers, and the simultaneous development of the fore and hind limbs. In larvae nearing transformation the general ground color is Fuscous (Ridgway). Scattered over the dorsal surface of the body and limbs are aggregates of small, rounded, tan spots which form blotches of varying size. The lower sides are tinged with bluish-gray and the ventral surfaces light gray with tinges of flesh in the region of the vent. Gills short, bushy, and dull red. In larvae of all sizes the dorsal fin is never continued anteriorly beyond the insertion of the hind legs, and in large larvae often extends forward only to the posterior end of the vent.

## GENUS RHYACOTRITON

**OLYMPIC SALAMANDER.** *Rhyacotriton olympicus* (Gaige). Figs. 29b, 48. Map 22.

TYPE LOCALITY. Lake Cushman, Washington.

RANGE. From Rouge River Basin, Curry County and Clackamas County, Oregon, throughout western Washington.

HABITAT. On June 21, 1936, we collected along the Calawah River,

½-mile north of Forks, Washington, where springs flowed from beneath overhanging banks covered by tangles of thick vegetation. Here, among stones and gravel washed clean by the water, we found a fine series of specimens in company with *Plethodon vandykei, P. vehiculum*

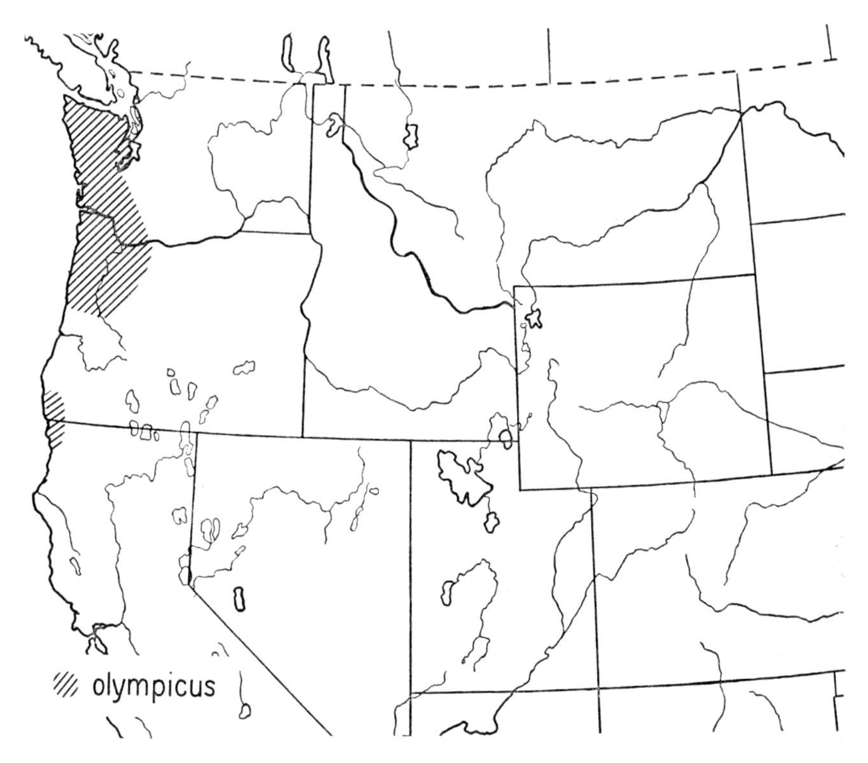


MAP 22.—Distribution of *Rhyacotriton olympicus.*

and *Ascaphus truei.* On June 20 we collected this species by the side of a small trickle on a steeply wooded slope near Shelton, Washington. Slater (1934, p. 140) reported this species in and near small streams which drain into Spirit Lake.

SIZE. The average length of 5 adult males is 3½″ (88.6 mm.); 5 adult females average 3$^{15}/_{32}$″ (87.8 mm.). The largest male, from ½ mile west of Forks, Washington, has a total length of 4$^{1}/_{16}$″ (103 mm.), tail 1$^{11}/_{16}$″ (42 mm.); head length ½″ (12 mm.), width ⅜″ (9 mm.). A female of exactly the same length from 7 mi. west of Shelton, Washington, has the tail 1 19/32″ (40 mm.); head length 17/32″ (13 mm.), width 3/8″ (9 mm.).

DESCRIPTION. This species is characterized by the short, wide head, very prominent eyes, and short tail. The head is widest at the eyes, tapering abruptly to the short, blunt snout, and with the sides back of the eyes straight and nearly parallel or slightly converging. The eyes are large and bulging, their horizontal diameter equal to or greater than the distance from the anterior angle of the eye to the end of the snout. The gular fold is strongly developed and has prominent lateral extensions. There is a deep decurved groove extending from the posterior angle of the eye to join the lateral extensions of the gular fold; a short groove from this to the angle of the mouth. The trunk is moderately stout, rounded on the sides, flattened beneath, and with an impressed median dorsal line. There are 14 costal grooves, counting 1 each in the axilla and groin, and 2–3 intercostal spaces between the toes of the appressed limbs. The tail is oval in section at the base, flattened or rounded below, compressed above, flattened distally, and produced into a blunt point. The legs are short and stout, toes 5–4, webbed at base, those of the hind feet 1–5–2–3–4 in order of length from the shortest; toes of the fore feet 1–4–2–3. Tongue moderate, longer than wide, the sides only free and narrowly depressed and smooth; the deep, narrow plicae diverge slightly from the posterior field. The vomerine teeth are in 2 strongly arched series narrowly separated at the mid-line. In the female, the series arise behind the middle of the inner naris and consist of about 11 teeth; in the male, the series arise nearer the outer margin of the inner naris, the teeth number 8–9 and are larger, longer, and more prominently hooked at tip.

COLOR. The ground color above is a rich seal-brown, variegated with fine white specks which become more numerous and somewhat larger on the sides. The upper surface of the tail is distinctly lighter brown, and there is a conspicuous depressed line on the back between the hind legs. The belly is bright yellow, nearly orange in some, except in the region below the liver, where there is a greenish tinge. The throat is marked with irregular grayish-white flecks and mottled and blotched with larger, light brown spots. The entire ventral surface is sparsely flecked with small whitish points and often with a few irregular brown spots. The dark dorsal coloration extends on the sides of the head to involve the upper jaw, and covers the upper surface of the limbs, the

FIG. 48. *Rhyacotriton olympicus* (Gaige). (1) Adult male, actual length 3 11/16″ (93 mm.). (2) Same, ventral view. Calawah River, Forks, Washington.

sides of the trunk to the sides of the belly, and the sides of the tail nearly to the ventral surface. The lower surface of the legs is yellow. Slater (1938, p. 136) mentions specimens from near Sandy, Oregon, which, in life, were light brown with small black spots on their lateral and dorsal surfaces and cream-colored below.

SEXUAL DIFFERENCES. The sexes may be distinguished by the form of the vent. In the male the sides of the vent are produced into squarish lobes behind, the outer angles light-tipped and visible from above.

BREEDING. Mr. Phillips G. Putnam, in the Olympic Mountains, Washington, discovered single salamander eggs during June under circumstances which led to the belief that they belonged to this species (Noble and Richards, 1932, p. 19). The eggs were attached to the lower surface of a stone in running water. Noble and Richards (*ibid.*, p. 20), employing the pituitary technique, subsequently induced a number of females to deposit eggs in the laboratory. The individual egg is large and without pigment, the yolk having an average diameter of 4.5 mm., without the 3 envelopes which surround it. Few eggs are deposited, the average being about 5.

LARVAE. The larvae may attain a length of 69 mm. before transformation (Slater, 1938, p. 136), but the usual size is considerably smaller. I have a single specimen with greatly reduced gills only 31 mm. long. The tail is narrowly keeled above, but the keel of the back is lacking as in *Dicamptodon ensatus*.

# Family PLETHODONTIDAE

Aquatic and terrestrial; size small to medium, 1⁹⁄₁₆″ (40 mm.) to 8⅝″ (219 mm.); lungs absent; a nasolabial groove; adults usually without gills, a few neotenic species; toes 5–4 or 4–4; vomerine and palatine teeth usually present; ypsiloid cartilage reduced or lacking; tongue free all around or attached anteriorly.

## KEY TO THE GENERA OF PLETHODONTIDAE

1. Toes 4–4 .......... 2
   Toes 5–4 .......... 5
2. Tail with a basal constriction; belly white with black flecks; size small, to 3½″ (89 mm.). Eastern states, southern Canada southward through Michigan to Gulf; only the single species .......... *Hemidactylium scutatum* p. 306
   Tail without a basal constriction; belly not white with black specks .......... 3
3. Tail compressed and slightly keeled above; a median dorsal band limited either side by a dark line; belly yellow; size small, to 3⁵⁄₁₆″ (84 mm.). Southeastern states; only the single species .......... *Manculus quadridigitatus* p. 446
   Tail not compressed and keeled above, nearly circular in section .......... 4
4. Body and tail elongate and worm-like; vomerine teeth in short series or patches widely separated from parasphenoids; costal grooves 17–21; belly not black with white spots; size to 6¾″ (172 mm.). Pacific Coast states and adjacent islands .......... BATRACHOSEPS p. 310
   Body and tail not exceptionally elongate and worm-like; vomerine series longer, 10–15, continuous with or narrowly separated from parasphenoids; above with a chestnut or reddish-brown band; belly black with white flecks; size to 3¹³⁄₁₆″ (97 mm.). Oregon; only the single species .......... *Plethopsis wrighti* p. 290
5. Tail with a basal constriction; palm with two tubercles; length to 5⅞″

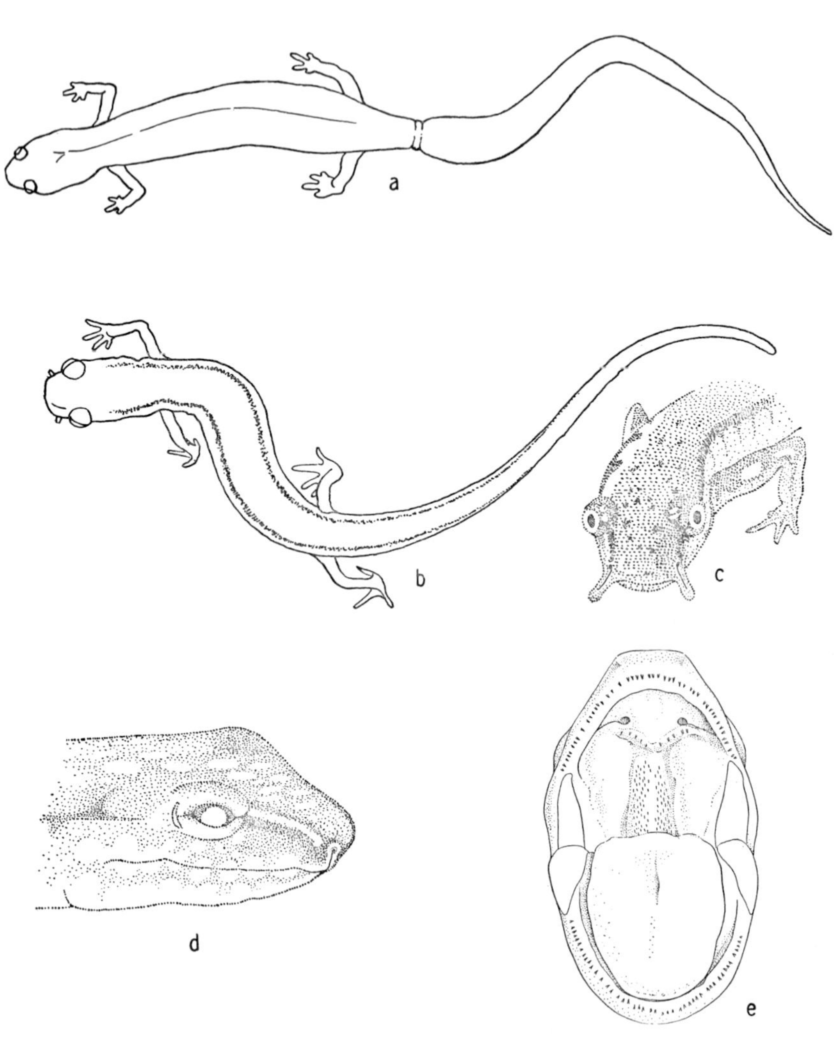


FIG. 49. (a) Outline of *Hemidactylum scutatum* to show the basal constriction of the tail and the four toes on both fore and hind feet. (b) Outline of *Eurycea bislineata cirrigera,* adult male, to show the cirri. (c) Same, head enlarged, front view. The naso-labial grooves extend along the cirri nearly to the tip. (d) Head of *Gyrinophilus* to show the canthus rostralis and the nasolabial grooves. (e) Mouth parts of *Plethodon jordani* to show the character of tongue and teeth. [a–d, H.P.C. *del.;* e, M.L.S. *del.*]

(150 mm.). Pacific Coast, southwestern California to British Columbia and Vancouver Island . . . . . . . . . . . . . . . . . . . . . . . . . . ENSATINA p. 293

Tail without a basal constriction; no palmar tubercles . . . . . . . . . . . 6

6. Tongue with a central pedicel, free all around . . . . . . . . . . . . . . . . . . . 7

Tongue free at sides and behind only . . . . . . . . . . . . . . . . . . . . . . . . . . . 10

7. Vomerine and parasphenoid teeth forming continuous series . . . . . . 8

Vomerine and parasphenoid teeth not in continuous series . . . . . . . . . . 9

8. Canthus rostralis marked by a light bar bordered above and/or below with black; dorsal pattern sometimes with dark flecks but without separate rounded black spots; size to 8⅝″ (219 mm.). Ontario, Canada southward through the eastern states to Georgia and Alabama . . . . . . . . . . . . . . . . . . . . . . . . . . . . . . . . . . . . . . . . . GYRINOPHILUS p. 360

No marked canthus rostralis bordered above and below by black; color above bright red to purplish or chocolate-brown, marked with black spots; size to 7″ (178 mm.). Eastern states, New York to Florida and the Gulf of Mexico . . . . . . . . . . . . . . . . . . . . . . . . . . . . . PSEUDOTRITON p. 376

9. Toes strongly webbed; above dark brown to black, marked with light flecks sometimes aggregated in lichen-like patches; costal grooves 13; tail short, nearly circular in section; size to 4³⁄₃₂″ (104 mm.). High Sierras in California; only the single species . . . . . . . . . . . . . . . . . . . . . . . . . . . . . . . . . . . . . . . . . . . . . . . . . . . . . . . . . . *Hydromantes platycephalus* p. 451

Toes not webbed, or at most only slightly at base; tail moderate to long, compressed; ground color usually some shade of yellow or orange; size to 7⁵⁄₃₂″ (182 mm.). New Brunswick and Quebec southward through the eastern states to Georgia, westward to New Mexico, northward through Texas, Oklahoma, and Kansas to southern Illinois . . . . . . . . . . . . . . . . . . . . . . . . . . . . . . . . . . . . . . . . . . . . . . . . . EURYCEA p. 402

10. Eyes normal; body well pigmented . . . . . . . . . . . . . . . . . . . . . . . . . . . 11

Eyes reduced, partly or completely hidden by fused lids, or eyes lacking; body with little or no dark pigment . . . . . . . . . . . . . . . . . . . . . . . . . 15

11. Posterior part of maxilla sharp-edged and without teeth; vomerine teeth present, usually 5–9; size to 6⅜″ (162 mm.). Cumberland Plateau in the east; British Columbia to Lower California . . . . . . . . . ANEIDES p. 327

Posterior part of maxilla with teeth; vomerine teeth often lacking in adult males . . . . . . . . . . . . . . . . . . . . . . . . . . . . . . . . . . . . . . . . . . . . . . . . . 12

12. Opening of inner nares conspicuous . . . . . . . . . . . . . . . . . . . . . . . . . . . 13

Opening of inner nares hidden in a fold at sides of mouth; vomerine teeth usually lacking in adult males, sometimes present but reduced in number in females; length to 5⅙″ (128 mm.). Western North Carolina . . . . . . . . . . . . . . . . . . . . . . . . . . . . . . . . . . . . . . . . LEUROGNATHUS p. 220

13 Sensory pits on head conspicuous; vomerine and parasphenoid series continuous; alternating light and dark longitudinal lines on sides; length to 3¾″ (95 mm.). From Dismal Swamp, Virginia, to Liberty County, Georgia, in the Coastal Plain; only the single species . . . . . . . . . . . . . . . . . . . . . . . . . . . . . . . . . . . . . . . . . . . . . . . . . . . *Stereochilus marginatus* p. 346

No conspicuous sensory pits on head; vomerine and parasphenoid series not continuous; no alternating dark and light longitudinal lines on sides . . . . . . . . . . 14

14. Tail trigonal or circular in section; a light bar from posterior angle of eye to angle of the jaw; body rather short and stout. . . DESMOGNATHUS p. 186

Tail always circular in section or nearly so; no light bar from eye to angle of the jaw; body usually long and slender . . . . . . . . PLETHODON p. 228

15. Adults fully transformed, without gills; white or very lightly pigmented; eyes with lids partly or completely fused; legs normal; length to 5 5/16″ (135 mm.). Ozark Plateau in Missouri, Oklahoma, Kansas, and Arkansas; only the single species . . . . . . . . . . *Typhlotriton spelæus* p. 351

Adults not fully transformed, with gills; pigment lacking; eyes lacking . . . . . . . . . . 16

16. Costal grooves 11; vomerine and palatine teeth in continuous series; snout wide but not greatly depressed; sides of head nearly parallel; length 3″ (75.5 mm.). Known only from a well at Albany, Georgia . . . . . . . . . . *Haideotriton wallacei* p. 355

Costal grooves 12; vomero-palatine teeth not forming a continuous series; snout narrower than head back of eyes and strongly depressed; length to 5 13/16″ (136 mm.). Eastern central Texas in wells and underground streams; only the single species . . . *Typhlomolge rathbuni* p. 357

# GENUS DESMOGNATHUS

## KEY TO THE SPECIES AND SUBSPECIES OF DESMOGNATHUS

1. Tail trigonal in section and often keeled above . . . . . . . . . . 2

Tail oval or nearly circular in section and not keeled above . . . . . . . . . 8

2. Snout and head to back of eyes, distal half of tail and legs white or very lightly pigmented; back deep brown, belly light brown; 14 costal grooves; one intercostal fold between toes of appressed limbs; length to 3 17/32″ (91 mm.). Known only from Demorest, Georgia . . . . . . . . . . *quadramaculatus amphileucus* p. 214

Snout and head, tail and legs not white or very light . . . . . . . . . . 3

3. Belly dark . . . . . . . . . . 4

Belly light . . . . . . . . . . 5

4. Belly uniformly pigmented, deep brown to blue-black; back and sides in old individuals black, sometimes with small rusty blotches; back of head and snout often rusty; juveniles with a row of small light spots dorsolaterally and a second row at juncture of lower sides and belly; vomerine teeth present in adults of both sexes, 5–12; light line from eye to angle of jaw often indistinct; usually 14 costal grooves and 2–3 intercostal folds between toes of appressed limbs; length to 6 7/8″ (175 mm.). West Virginia and Virginia southward in the mountains to northern Georgia . *quadramaculatus quadramaculatus* p. 210

Belly strongly mottled, deep brown and white; back dark brown (almost black); lower sides bluish-gray to grayish-brown; dorsal surface of tail basally often with a russet band with irregular edges; often with a dorsolateral row of small light dots and a second row on lower sides between the legs and on basal half of tail; light mark on side of head from eye to angle of jaw narrow to broad and bright orange-red to russet; length to $4^{19}/_{32}$" (118 mm.). Virginia south to central Florida, west to Louisiana .................... *fuscus auriculatus* p. 193

5. Belly uniformly but lightly pigmented .......................... 6
   Belly light but mottled .......................................... 7

6. Back without a longitudinal band or conspicuous markings, or, if band is present, then poorly defined and limited each side by faint light spots; usually a row of small light dots on sides above the legs and sometimes a second row at juncture of lower sides and belly; vomerine teeth lacking in adult males; usually 14 costal grooves and 3–5 intercostal folds between toes of appressed limbs; length to $5^{1}/_{4}$" (134 mm.). Northeastern Texas, Oklahoma, and Arkansas ........ ........................................ *fuscus brimleyorum* p. 196
   Back with conspicuous light spots or blotches heavily outlined with dark brown or black, or, back with heavy light and dark worm-like markings; lower sides usually mottled and sometimes with a row of light dots on sides between the legs; 13 costal grooves, occasionally 14, and 4–5 intercostal folds between toes of appressed limbs; vomerine teeth present in adults of both sexes; length to $5^{1}/_{8}$" (131 mm.). Southwest Pennsylvania southward through Virginia and West Virginia to Georgia and Tennessee in the mountains ............ *phoca* p. 206

7. Back yellowish-brown to nearly black, often without a dorsal band; if dorsal band present then with irregular edges or limited either side by dark-edged semicircular or worm-like markings; sides usually without a series of small light dots; usually 14 costal grooves and about 4 intercostal folds between toes of appressed limbs; length to $4^{23}/_{32}$" (121 mm.) in North, to $5^{7}/_{8}$" (149 mm.) in Harlan County, Kentucky. New Brunswick to northwestern Georgia, west to Mississippi, northward to Illinois ........................ *fuscus fuscus* p. 188
   Back deep brown to black, without a band but sometimes with a dorsolateral series of light, dark-bordered spots extending onto the tail; light bar from eye to angle of jaw conspicuous, reddish or yellowish; other characters as above under 4 .............. *fuscus auriculatus* p. 193

8. Median dorsal light stripe with straight edges, belly light; or, if uniformly black above, then belly dark; sides mottled; vomerine teeth lacking in adult male; length to $3^{3}/_{4}$" (96 mm.). New York southward to Virginia and West Virginia, west to Kentucky and northward to Ohio ......................... *ochrophæus ochrophæus* p. 199
   Median dorsal light stripe with irregular edges .................... 9

9. Size to $4^{13}/_{32}$" (113 mm.); dorsal light band variable in color but when

present always with irregular edges; belly lightly mottled; or, if uniformly black above, then belly also darker; upper sides usually dark; vomerine teeth lacking in adult males. Virginia and West Virginia south to Georgia in the mountains ........ *ochrophæus carolinensis* p. 203

Size small, to 2" (51 mm.); dorsal light band light-tan to reddish, often bronzy; many individuals with a definite herringbone pattern along the mid-line of back; light line from eye to angle of jaw often conspicuous; belly light; vomerine teeth 3–7, present in adults of both sexes; tail short, 34–43 per cent of the total length. High elevations in the mountains of North Carolina, Tennessee, and western Virginia ........................................... *wrighti* p. 216

**NORTHERN DUSKY SALAMANDER.** *Desmognathus fuscus fuscus* (Rafinesque). Figs. 50b, 51, 101b. Map 23.

TYPE LOCALITY. Northern parts of the State of New York.

RANGE. St. John's River, New Brunswick, to northern Georgia and Alabama, northward to Illinois, Indiana, Ohio, and Ontario opposite Buffalo, New York.

HABITAT. Frequents the margins of streams and springs, leaf-filled trickles, springy banks where the soil is constantly moist, and often the beds of partially dry streams in deep ravines. Occasionally enters the water but is essentially terrestrial.

SIZE. Ninety adult males from near Rochester, New York, average $3\frac{19}{32}$" (91.2 mm.), with extremes of $2\frac{9}{16}$" (65 mm.) and $4\frac{13}{32}$" (113 mm.); 65 adult females average $3\frac{7}{32}$" (82.5 mm.), the extremes $2\frac{19}{32}$" (66 mm.) and $3\frac{29}{32}$" (100 mm.). In this vicinity the males reach an extreme length of $4\frac{23}{32}$" (121 mm.). The proportions of an adult male are as follows: total length $4\frac{11}{32}$" (111 mm.), tail 2" (51 mm.); head length $\frac{19}{32}$" (15 mm.), width $\frac{3}{8}$" (10 mm.). An adult female measures: total length 4" (102 mm.), tail $1\frac{7}{8}$" (48 mm.); head length $\frac{1}{2}$" (13 mm.), width $\frac{5}{16}$" (8 mm.).

DESCRIPTION. This is an extremely variable species in color and pattern, and strongly aberrant individuals may be found in almost any locality. In the adult male the head is somewhat swollen between the angle of the jaw and the lateral extension of the gular fold and is widest posteriorly; from the angles of the jaw the sides curve evenly to the bluntly pointed snout. The eyes moderately large and protuberant and limited behind by vertical groove, the iris flecked with brassy, and with a narrow line of gold bordering the pupil. An impressed line from the posterior

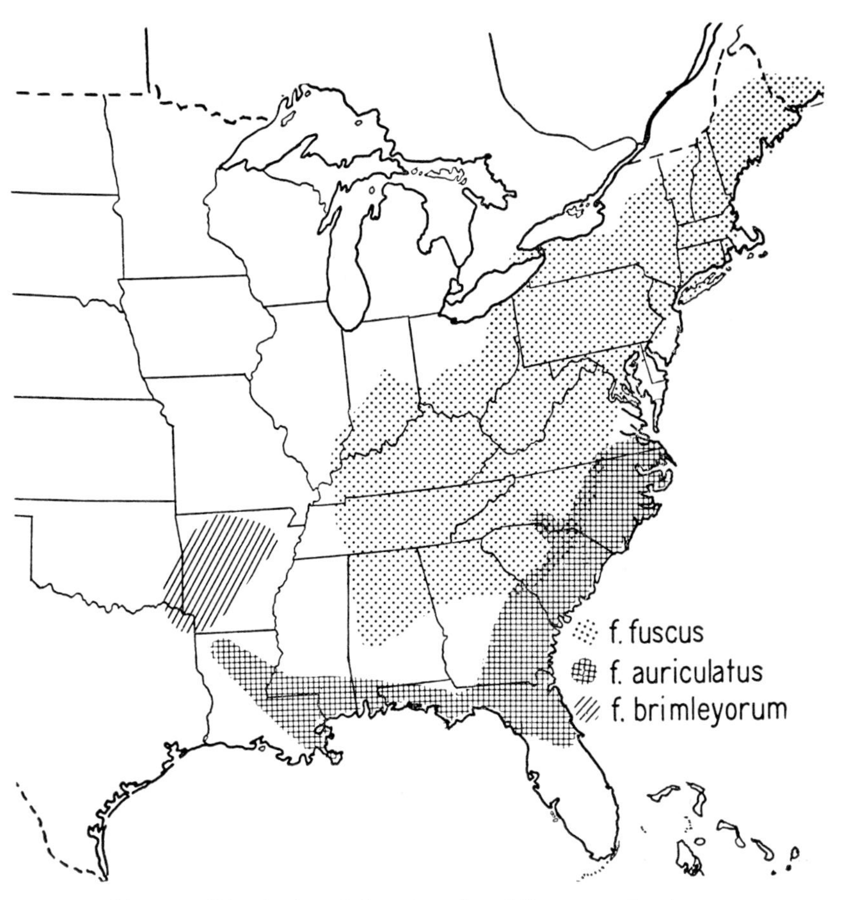


MAP 23.—Distribution of the subspecies of *Desmognathus fuscus.*

angle of the eye to the lateral extension of the gular fold; a short vertical groove from this line to the angle of the jaw. The trunk is well rounded above and on the sides and with an impressed median line above. There are usually 14 costal grooves, with 1 in the axilla sometimes forked above, and 4–4½ intercostal folds between the toes of the appressed limbs. The tail is subquadrate in section at base, trigonal in section beyond the base, and becoming knife-edged above and slender toward the tip. The legs are relatively short and stout, the hind noticeably larger than the fore. Toes 5–4, those of the hind feet 1–2–5–4–3 in order of length from

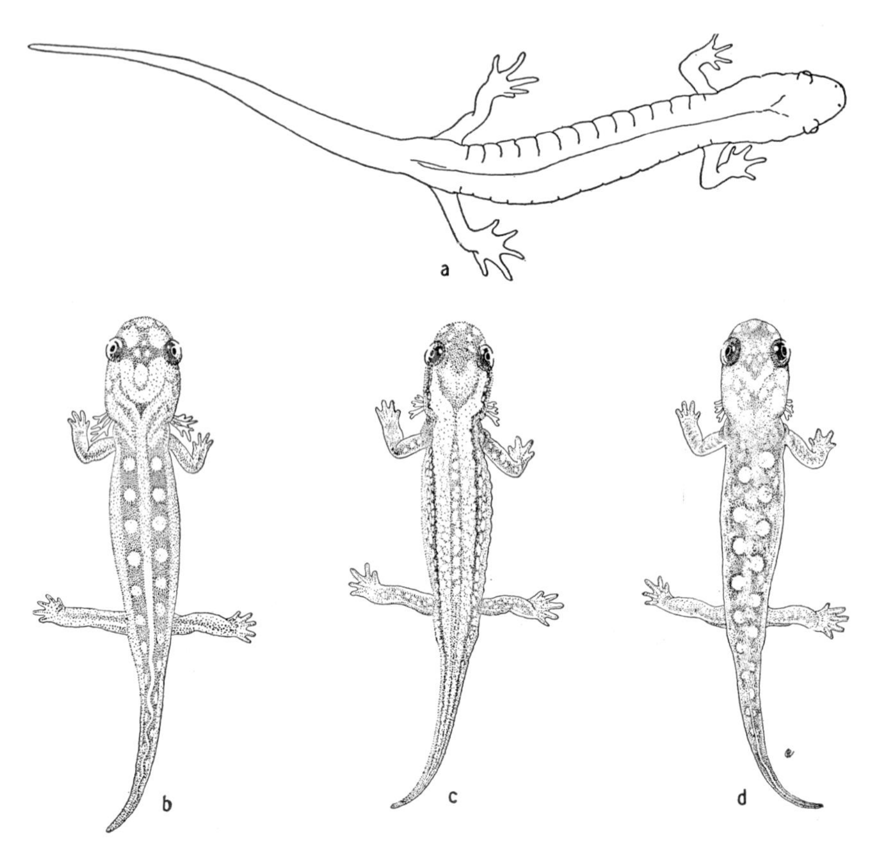


FIG. 50. (a) Outline of *Desmognathus ochrophæus carolinensis* to show the form of the body, legs, feet, and tail. (b) Recently hatched larva of *Desmognathus fuscus fuscus*. (c) Recently hatched larva of *Desmognathus ochrophæus ochrophæus*. (d) Recently hatched larva of *Desmognathus o. carolinensis*. [H. P. Chrisp, *del.*]

the shortest; toes of the fore feet 1–4–2–3. Tongue small, broadly oval in outline, free at the sides and back, the surface marked with narrow plicae which radiate from the posterior field. Vomerine teeth usually lacking in the adult male; in the female small, in short series of 5–7 which arise behind the inner margin of the inner naris and curve inward and backward toward the mid-line, where they are separated by about

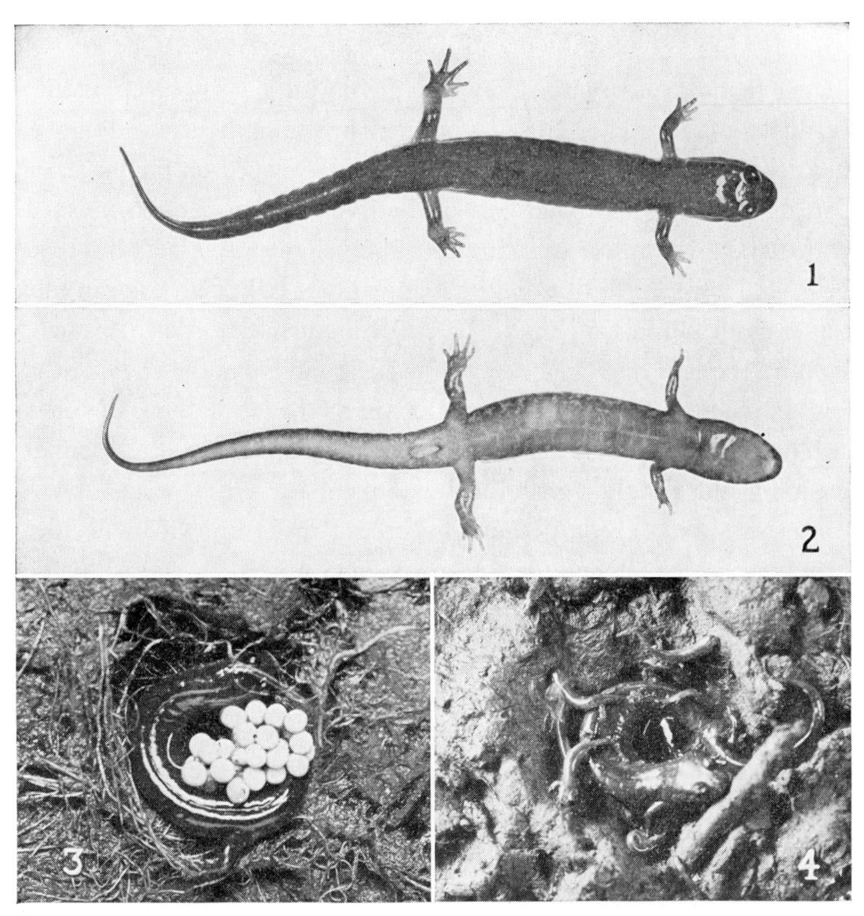


FIG. 51. *Desmognathus fuscus fuscus* (Rafinesque). (1) Adult male, actual length 3 5/16″ (84 mm.). (2) Same, ventral view. Allegany State Park, New York. (3) Female with eggs. Stamford, New York. (4) Female with young. Cornwall, Connecticut.

½ the diameter of a naris. Parasphenoid teeth in elongate patches which may be narrowly separated or united anteriorly and slightly divergent behind; widely separated from the vomerine teeth in the female.

COLOR. There is much variation in the intensity of pigmentation and in degree of development of the dorsal pattern. The ground color above varies from light yellowish-brown to deep brown and black. Typically there is a broad median band above with irregular dark edges, but the band is often obscured in the dark ground color. When present it is lighter than the adjacent sides, which are dark brown to black, fading gradually to the level of the legs and there becoming lighter and lightly mottled. Frequently the dorsal band is best developed on the posterior part of the trunk and basal part of the tail, where it is often invaded with darker worm-like markings enclosing lighter spots. The throat, belly, and lower surface of the limbs and tail are light flesh color in life, tinged with bluish, and varied by small pigment-free spots. A small, light bar extends from the posterior angle of the eye to the angle of the jaw. In the male the head is larger than in the female and has the sides swollen; the tip of the lower jaw is somewhat pointed and bears a small mental gland, and the vent is lined with papillae.

BREEDING. The breeding habits have been made known by Wilder (1913, 1923), Noble and Brady (1930), and others. In courtship the male applies the snout, cheeks, and mental gland to the snout of the female, who usually responds by picking up a spermatophore. The eggs are deposited in small, compact clusters of 12–26 (average 17), and cling to one another by extensions of the outer envelope. The eggs are found in June, July, and August, and are attended by the female in her nest beneath logs, stones, or bark, in the vicinity of water. Individual eggs are about 3 mm. in diameter and have 3 envelopes, the 1st of clear jelly and fairly thick, the 2nd thin, and the outer thin, tough, and elastic, and with a total diameter of about 4½ mm.

LARVAE. Larvae at hatching average about 16 mm. and are gray above in a broad band which extends from the head onto the basal half of the tail. Within the dorsal band, and separated by a median light stripe, are 7–9 pairs of round light spots. The sides of the trunk and legs above are lightly but distinctly mottled. Larvae may attain a length of 44 mm. and transform at an age of 7–9 months. Larval gills are slender and glistening white.

**SOUTHERN DUSKY SALAMANDER.** *Desmognathus fuscus auriculatus* (Holbrook). Fig. 52. Map 23.

TYPE LOCALITY. Riceborough, Georgia.

RANGE. Dismal Swamp, Virginia, south to central Florida, west to Louisiana in the Coastal Plain.

HABITAT. Found under logs, bark, and other surface debris, in or beside swamp streams and springs. Typical specimens were found in numbers in the bottom lands of the Tickfaw River near Greensburg, Louisiana, March 19, 1936, where they were hiding beneath logs and bark in damp situations.

SIZE. The average length of 13 adults of both sexes from Georgia and Florida is $3\frac{23}{32}''$ (95 mm.). Five adult males average $3\frac{29}{32}''$ (100 mm.) and vary from $3\frac{21}{32}''$ (93 mm.) to $4\frac{3}{16}''$ (107 mm.). Five adult females average $3\frac{1}{2}''$ (89 mm.), the extremes $3\frac{11}{32}''$ (85 mm.) and $3\frac{11}{16}''$ (94 mm.). The males reach an extreme length of about $4\frac{19}{32}''$ (118 mm.). The proportions of an adult male from the type locality, Riceborough, Georgia, are as follows: total length $4\frac{5}{16}''$ (110 mm.), tail $2\frac{3}{16}''$ (56 mm.); head length $\frac{1}{2}''$ (13 mm.), width $\frac{7}{32}''$ (6 mm.). A female from near Marianna, Florida, measures as follows: total length $3\frac{17}{32}''$ (90 mm.), tail $1\frac{5}{8}''$ (42 mm.); head length $\frac{13}{32}''$ (11 mm.), width $\frac{11}{32}''$ (8.5 mm.).

DESCRIPTION. The head is widest just in front of the lateral extensions of the gular fold, where it is somewhat swollen, the sides converging gently to the eyes and more abruptly to the bluntly pointed snout. In the female the head is narrower and the sides less swollen. The eye is of moderate size, its horizontal diameter about $1\frac{1}{2}$ in the snout, and limited behind by a vertical fold. An impressed line from the posterior angle of the eye to the lateral extension of the gular fold; a vertical groove from this line passing behind the angle of the mouth. Trunk well rounded and with a median impressed line. There are usually 14 costal grooves, counting 1 in the axilla and 2 that come together in the groin, occasionally 15, and 4–6 intercostal folds between the toes of the appressed limbs. Tail stout at base, sometimes flattened above; rounded below, sharp-edged above beginning behind the vent, and keeled above on the distal half. Legs rather short and stout. Toes 5–4, those of the hind feet 1–5–2–4–3 in order of length from the shortest; toes of the

Fig. 52. *Desmognathus fuscus auriculatus* (Holbrook). (1) Adult male, actual length 3⅞16″ (88 mm.). (2) Same, lateral view. (3) Same, ventral view. Ferry Pass, Pensacola, Florida.

fore feet 1–4–2–3. Tongue widest behind, bluntly pointed in front, free at the sides and behind. Vomerine teeth lacking in adult males; in the females 3–6, in short series which arise behind the inner margin of the inner naris and slant obliquely backward toward the mid-line, where they may be nearly in contact or separated by the width of a naris. Parasphenoid teeth in 2 slender patches widely separated from the vomerine.

COLOR. In life, above deep brown, almost black on the head, trunk and upper sides; bluish-gray on the lower sides, where there are many minute whitish specks. In old individuals there is scarcely an indication of a dorsal pattern on the trunk, but often, from the base of the tail well toward the tip, there is a conspicuous russet band with irregular edges. On the upper sides, a row of whitish or reddish spots extending from the back of the head onto the basal half of the tail, obscure in old, dark specimens. Often a second row of light spots on the lower sides between the legs. Belly dark, sometimes bluish-black and mottled with white, the ventral surface of the tail often brownish. On the side of the head from the eye to the angle of the jaw, a light mark which may be reduced to a line or may form an oval patch which varies in color from bright orange-red to russet. Half-grown individuals frequently exhibit a well developed dorsal light band with irregular dark edges.

BREEDING. Little is known of this subspecies.

LARVAE. The general ground color varies from light tan to dark brown. In most individuals there is an inconspicuous dorsolateral dark line or series of spots extending from the base of the gills onto the sides of the tail. The back between the dark lines is finely specked with brown pigment and a few larger, irregular brown spots disposed either in 2 much-broken lines or in a single median series. Along the upper edge of the dorsolateral dark line is a series of 7–9 small, pigment-free spots, 5–7 on the trunk, the rest on the tail. In many specimens there is a second series of light spots extending along the sides between the fore and hind legs and rarely onto the tail. Within most of these light spots there is a fleck of bright white pigment. The sides of the trunk below the dorsolateral dark lines are somewhat lighter than the back, and the head, legs, and sides of the tail are strongly pigmented and somewhat mottled. There is a light spot surrounding each nostril. The dark pigmentation of the sides encroaches in a narrow band around the margin of the lower jaw, on the sides of the throat and belly, and to the mid-line of the venter of the tail. Unlike the larvae of many species of *Desmognathus,* these have the gills fairly long and slender and highly pigmented. A series of larvae from the vicinity of Gainesville, Florida, collected by A. F. Carr during January 1941, varied from 21 to 29 mm. in total length, and one individual with reduced gills was approaching transformation with a trunk length of 22 mm., the tip of the tail being lost.

**BRIMLEY'S SALAMANDER.** *Desmognathus fuscus brimleyorum* Stejneger. Fig. 53. Map 23.

TYPE LOCALITY. Hot Springs, Arkansas.

RANGE. Northeastern Texas, southeastern Oklahoma, and Arkansas.

HABITAT. Under stones, rocks, logs, planks, and other debris in damp woods and in the vicinity of ponds and streams.

SIZE. The average length of 15 sexually mature individuals of both sexes from Rich Mountain, Oklahoma, is $4\frac{13}{32}''$ (113 mm.), the extremes $3\frac{1}{16}''$ (78 mm.) and $5\frac{13}{32}''$ (138 mm.). The proportions of an adult male from this locality are: total length $4\frac{3}{8}''$ (112 mm.), tail $2\frac{1}{32}''$ (52 mm.); head length $\frac{5}{8}''$ (16 mm.), width $\frac{3}{8}''$ (10 mm.). A female has the following measurements: total length $4\frac{3}{16}''$ (107 mm.), tail $2\frac{1}{32}''$ (52 mm.); head length $\frac{17}{32}''$ (14 mm.), width $\frac{5}{16}''$ (8 mm.).

DESCRIPTION. The male has the head somewhat swollen in the region between the angle of the jaw and the lateral extension of the gular fold, the sides of the head in front of the angle of the mouth tapering gently to the eyes, then more abruptly to the bluntly pointed snout. In the female the sides of the head back of the eyes are nearly parallel and the snout is more broadly rounded. The eyes are large and strongly protuberant, the horizontal diameter about $1\frac{1}{4}$ in the snout. Eye limited behind by a vertical fold. An impressed line from the posterior angle of the eye to the lateral extension of the gular fold; a short vertical groove from this line passing behind the angle of the mouth. The trunk is stout and subcylindrical, slightly flattened below. There are usually

14 costal grooves, counting 1 each in the axilla and groin, with 2 in the groin sometimes running together, and 3–5 intercostal folds between the toes of the appressed limbs. Tail subquadrate in section at base, becoming rounded below, sharp-edged and keeled above behind the vent,

FIG. 53. *Desmognathus fuscus brimleyorum* Stejnager. (1) Adult, actual length about 4″ (102 mm.). [Photograph by A. A. Heinze.] (2) Adult male, actual length 4⅜″ (112 mm.). (3) Same, ventral view. Rich Mountain, Le Flore County, Arkansas.

and strongly compressed distally. The tail is relatively short in this species, in old males comprising 43–47.5 per cent of the total length and in the females 43–46.3 per cent. Young individuals of both sexes have proportionally longer tails than adults, in the males attaining 50.2 per cent and in the females 48.6 per cent. Legs rather large and stout. Toes

5–4, those of the hind feet 1–5–2–(4–3) in order of length from the shortest and slightly webbed at base; toes of the fore feet 1–4–2–3. Tongue broadly oval in outline, free at the sides and behind only. Vomerine teeth lacking in old and large males, sometimes retained in individuals until a length of $4\frac{25}{32}''$ (122 mm.) is attained, or lost at a length of $3\frac{23}{32}''$ (95 mm.); when present in males 5–7 in short series which arise between and behind the inner nares and curve inward and backward toward the mid-line, where they are separated by a little more than the diameter of a naris. In the female the series are longer, usually 7–11; they arise behind the inner margin of the inner nares and curve inward and backward, forming rows which are narrowly separated at the mid-line. Parasphenoid teeth in long, slender, narrowly separated patches; or, in older individuals, the patches shorter and in contact anteriorly.

COLOR. This is a large and somewhat variable species with the general color above brown or grayish-brown, fading slightly on the lower sides. In some the back is mottled with fine, worm-like, gray markings on a black ground and the lower sides are lightly mottled. In all but the largest and oldest individuals there is a series of light dashes or rounded spots on the sides which extends from the head or shoulder along the trunk above the level of the legs and is often continued along the sides of the tail. Very large specimens tend to be uniformly brownish. In small, half-grown, and medium-sized individuals there is a tendency to retain, on the basal half of the tail above, a light tan or brownish-yellow band with irregular dark edges. In some light-colored individuals there is a dorsolateral series of small, pale, dark-bordered spots extending from the shoulders along the trunk and basal quarter of the tail. In these individuals there is frequently an additional series of small light dots on the lower sides between the legs. The light bar from the posterior angle of the eye to the angle of the jaw is well developed in the young, frequently diffuse in the older individuals. The throat, belly, and ventral surface of the tail are finely pigmented with brown but are light in general tone—flesh or tinged with yellow.

BREEDING. Strecker (1908, p. 88), quoting from the notes of Combs, has given an account of the egg-laying: "In the latter part of August or early in September the female triton deposits her eggs, which are from 30 to 36 in number, and attached in strings, in a crevice in the under side of a rotten log or in a mass of decaying wood near some small stream. The eggs are about an eighth of an inch in diameter. The female is much attached to her eggs and seldom goes far away from them. During a dry spell she will carry them down into her hole with her, and if it rains again before they are hatched, will again bring them to the surface."

LARVAE. A larva 31 mm. long has the general ground color, brown, disposed in broken streaks along the sides of the trunk and tail. There is a dorsolateral series of round light spots with irregular edges extending along the trunk and on the basal third of the tail. The pigment extends on the side of the head to involve the extreme margin of the lower jaw, on the trunk to the lower level of the legs, and the entire tail. The throat and a narrow area along the venter of the belly immaculate.

**ALLEGHENY MOUNTAINS SALAMANDER.** *Desmognathus ochrophæus ochrophæus* Cope. Figs. 50c, 54. Map 24.

TYPE LOCALITY. Susquehanna County, Pennsylvania.

RANGE. Adirondack Mountains, New York, south to Virginia and West Virginia, west to Kentucky and northward into Ohio.

HABITAT. Found under logs, bark, stones, and moss, usually in the vicinity of streams and springs, but not restricted to such situations. Often found on hillsides a considerable distance from the nearest water. More terrestrial than *D. f. fuscus,* with which it is sometimes associated along stream banks.

SIZE. The average length of 40 males from the vicinity of Rochester, New York, is 2¾″ (70.33 mm.), the extremes 2³⁄₃₂″ (53 mm.) and 3¾″ (96 mm.). Forty females averaged 2²⁷⁄₃₂″ (73 mm.) and varied from 1²⁵⁄₃₂″ (46 mm.) to 3¹⁵⁄₃₂″ (88 mm.). The proportions of an adult male are as follows: total length $3\frac{1}{16}''$ (78 mm.), tail $1\frac{15}{32}''$ (38 mm.); head length $\frac{13}{32}''$ (10.5 mm.), width $\frac{9}{32}''$ (7 mm.). An adult female has the

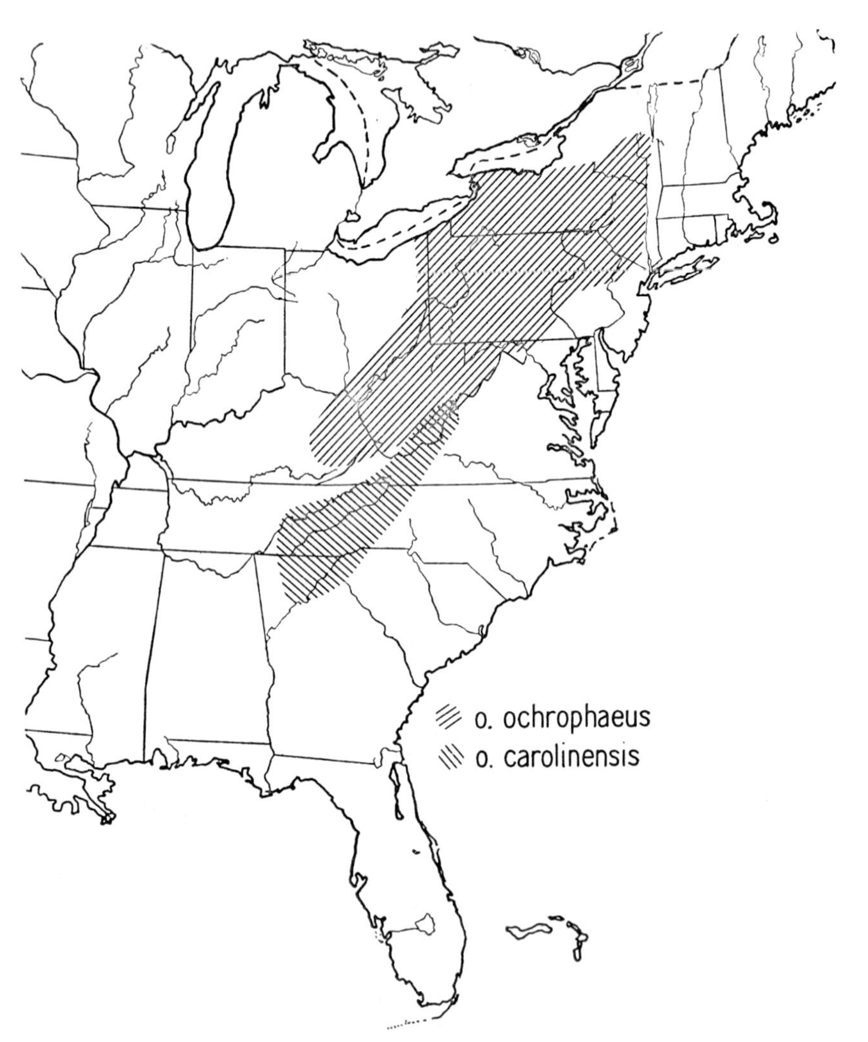


MAP 24.—Distribution of the subspecies of *Desmognathus ochrophæus.*

following measurements: total length $3\frac{1}{16}''$ (78 mm.), tail $1\frac{7}{16}''$ (37 mm.); head length $\frac{11}{32}''$ (9 mm.), width $\frac{7}{32}''$ (5.5 mm.).

DESCRIPTION. This is a slender species with a well rounded, slim, and tapering tail. The head of the male is widest just in front of the lateral extensions of the gular fold, slightly constricted at the angle of the jaws, the sides in front of this point converging to the bluntly pointed snout. In the female the sides of the head are nearly parallel back of the eyes. The eye is small, its long diameter about twice in the snout; limited behind by an oblique fold; the iris flecked with brassy. An impressed line from the posterior angle of the eye bends down posteriorly to the gular fold; a short vertical groove from this line to the angle of the jaw. A groove across anterior corner of upper lid. Trunk rounded above and on the sides, slightly flattened below. Costal grooves 14, counting 1 each in the axilla and groin, and 2½–4 intercostal spaces between the toes of the appressed limbs. Tail nearly circular in section, slender and tapering. No dorsal keel developed except when the tail has been lost and regenerated. Legs only moderately stout, the hind larger than the fore. Toes 5–4, those of the hind feet 1–5–2–4–3 in order of length from the shortest; toes of the fore feet 1–4–2–3. Tongue small, oval in outline, and without conspicuous plicae. Vomerine teeth lacking in the adult males; in the females the short series of 4–6 teeth arise behind the inner margin of the inner naris and curve gently inward and backward toward the mid-line, where they are separated by about the diameter of a naris. Parasphenoid teeth in 2 long slender patches divergent posteriorly and nearly or quite in contact anteriorly. Adult males with a small mental gland and pointed lower jaw; male vent with papillae.

COLOR. The color of the dorsal surfaces is extremely variable, but the pattern remains essentially constant. There is a broad, middorsal, light band with straight edges, below which the sides of the trunk and tail are dark, nearly black. The lower sides are generally mottled and fade into the lightly pigmented but scarcely mottled belly; throat and under surface of the limbs flesh color, lower surface of tail like belly. In old, dark individuals the dorsal light band may be almost obscured and the belly dark; in others the ground color may vary from light tan through various shades of gray, red, and brown. Within the band there is sometimes a median line of small black spots or scattered dark spots of various sizes and shapes. A small light bar extends from the posterior angle of the eye to the angle of the jaw.

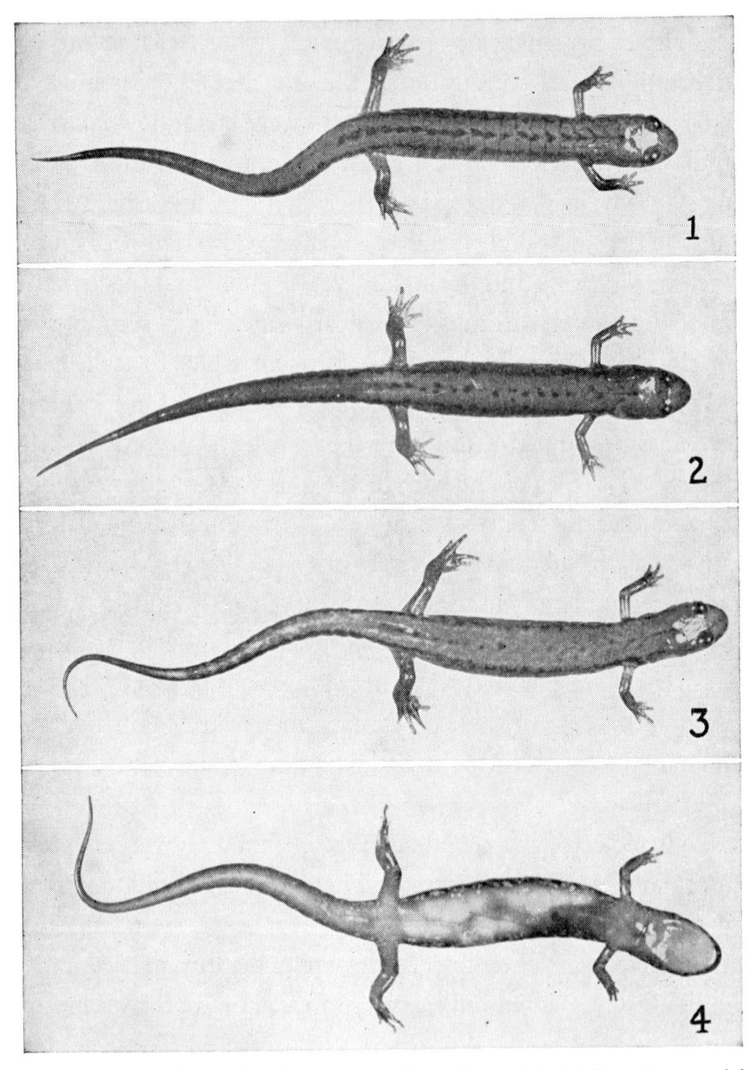


Fig. 54. *Desmognathus ochrophæus ochrophæus* Cope. (1) Adult male, actual length 3″ (76 mm.). (2) Adult male, another individual, actual length 3⁵⁄₁₆″ (84 mm.). (3) Adult female, actual length 2³¹⁄₃₂″ (75 mm.). (4) Same, ventral view; note the eggs. Allegany State Park, New York.

BREEDING. There is an extended late-summer-and-fall mating period, and considerable evidence to indicate an additional spring season of sexual activity. Eggs have been found beneath stones and logs on springy hillsides in August, September, and October, and recently hatched young throughout the late fall months and again in March. Spermatophores have been found both in the spring and fall (Bishop and Chrisp, 1933, p. 194). The eggs are deposited in small clusters of 11–14, occasionally in pairs or singly. The whitish, pigmentless eggs have a diameter, with the envelopes, of about 4 mm., and are attached to a common stalk by extensions of the outer envelope. The spermatophore is a small, stump-shaped mass of jelly surmounted by a whitish sperm cap.

LARVAE. At hatching the larvae average about 17 mm. They have a broad, median, light band extending from back of the eyes to the end of the tail; the band is limited on either side by a darker line of pigment; the sides of the trunk and tail lightly mottled. The dorsal light band lacks the conspicuous, dark-bordered light spots characteristic of *D. f. fuscus, D. o. carolinensis* and *D. phoca,* but is lightly mottled. Short white gills are present at hatching and are retained for a short time.

**CAROLINA MOUNTAIN SALAMANDER.** *Desmognathus ochrophæus carolinensis* Dunn. Figs. 50a, d; 55. Map 24.

TYPE LOCALITY. Mt. Mitchell, North Carolina.

RANGE. From Virginia and West Virginia south to Georgia in the mountains.

HABITAT. Generally terrestrial and found under logs, stones, and bark, often a considerable distance from water. During the egg-laying season most abundant in the vicinity of streams and springs.

SIZE. Forty adults of both sexes from North Carolina and Tennessee average $3\frac{1}{8}''$ (80 mm.) and range from $2\frac{5}{32}''$ (55 mm.) to $4\frac{3}{32}''$ (104.4 mm.). Dunn (1926, p. 109) records a male $4\frac{13}{32}''$ (113 mm.) in total length. The proportions of an adult male from Indian Gap, Tennessee, are as follows: total length $3\frac{25}{32}''$ (97 mm.), tail $1\frac{7}{8}''$ (48 mm.); head length $\frac{15}{32}''$ (12 mm.), width $\frac{5}{16}''$ (8 mm.). A female from the same locality: total length 3⅜" (86 mm.), tail 1¹¹⁄₁₆" (43 mm.), head length ¹³⁄₃₂" (11 mm.), width ⁹⁄₃₂" (7 mm.).

DESCRIPTION. In general conformation similar to *D. o. ochrophæus.* The head of the male is somewhat swollen between the lateral extensions of the gular fold and the angle of the jaw, where there is a constriction; in front of this point the sides converge to the bluntly pointed snout. In the female the sides of the head nearly parallel back of the eyes. Eyes moderate and limited behind by a vertical fold; a groove across the anterior corner of the upper lid. A sinuous impressed line from the posterior angle of the eye to the lateral extension of the gular fold, a short vertical groove from this line passing back of the angle of the jaw. Trunk rounded above, with an impressed median line. Costal grooves usually 14 but ranging from 13 to 15, counting 1 each in the axilla and groin, often the one in the groin forked above; 3–5 intercostal folds between the toes of the appressed limbs. Tail nearly round throughout its length, long, slender, and tapering; no keels. Legs stouter than in *D. o. ochrophæus;* toes 5–4, those of the hind feet 1–5–2–(4–3) in order of length from the shortest; toes of the fore feet 1–4–2–3. Tongue broadly oval in outline, free at the sides and behind, the surface spongy. Vomerine teeth lacking in the adult males; in the females the short series of 5–7 teeth lie between and behind the inner nares and slant inward and backward toward the mid-line, where they are separated by about the diameter of a naris. Parasphenoid teeth in 2 elongate patches which may be completely separated or narrowly in contact anteriorly, slightly diverging posteriorly. The vent of the male is lined anteriorly with short papillae and the tip of the lower jaw is pointed and bears a small mental gland. In the female the sides of the vent are thrown into folds and the tip of the jaw is broadly rounded.

COLOR. Extremely variable in color and pattern. Old individuals may be uniformly bluish-black above, with rusty mottlings on the head, and with the legs lighter than the back. In these the belly is finely reticulated with black pigment and a few scattered light flecks. Many individuals have a dorsal pattern consisting of a broad, median, light band with

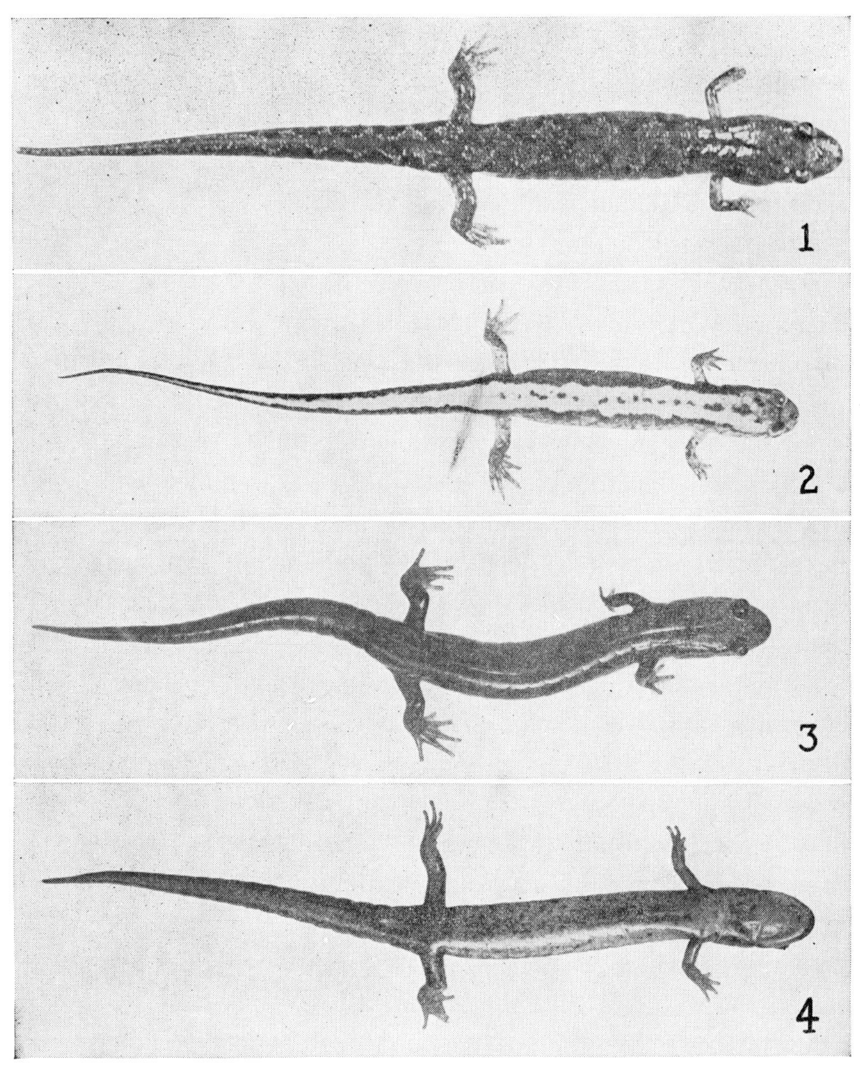


FIG. 55. *Desmognathus ochrophæus carolinensis* Dunn. (1) Adult male, actual length 2 31/32″ (75 mm.). Jocassee, South Carolina. (2) Adult female, actual length 3¼″ (82 mm.). Eight miles west of Blowing Rock, North Carolina. (3) Adult female, actual length 4 1/16″ (103 mm.). (4) Same, ventral view. Near Grandfather Mountain, North Carolina.

irregular or evenly scalloped edges. The upper sides bordering the light band are dark, fading gradually on the lower sides to the belly, which is always finely pigmented with black but lighter than the back. Other individuals may have a dorsal series of black-bordered light areas set in a zigzag pattern and representing the remnants of the larval light spots. The dorsal light areas may be of varying shades of yellow, brown, red, or gray.

BREEDING. During the period in which the eggs are found, July and August, the adults in attendance are to be found in the vicinity of shaded brooks and springy areas beneath dead leaves, in mud, and among moss-covered rocks. The eggs are deposited in small clusters, held together like a bunch of toy balloons by extensions of the outer envelope, and average about 10 in number but vary from 2 or 3 to 18. Individual eggs in early stages of development have a diameter of approximately 3 mm., increasing in size with development to 4.5 mm. (Pope, 1924, p. 4). Eggs preserved in formalin do not show the envelopes well and they have not been described.

LARVAE. Larvae at hatching have an average length of about 16 mm. and vary from 15 to 18 mm. There is considerable variation in the dorsal markings even in the young larvae. Often the dorsal dark-bordered light spots form a zigzag series; sometimes they are united in pairs which are connected by the middorsal light stripe. Occasionally there is a single broad median band with irregular edges. I have measured completely transformed individuals only 21 mm. long, and larvae with gills 23 mm. in length. The gills are short and stubby, pigmented at base, and with only a few short filaments.

**SEAL SALAMANDER.** *Desmognathus phoca* (Matthes). Fig. 56. Map 25.

TYPE LOCALITY. Taylor's Creek near Newport, Kentucky.

RANGE. From Pennsylvania south through Virginia, West Virginia, North Carolina, South Carolina, Georgia, Tennessee, and Kentucky, in the mountains.

HABITAT. Most abundant about the margins of mountain brooks and springs or in cool, well shaded ravines. They are to be found beneath logs, bark, and stones, and sometimes in shallow, muddy streams.

SIZE. The average length of 20 adults of both sexes from North Caro-

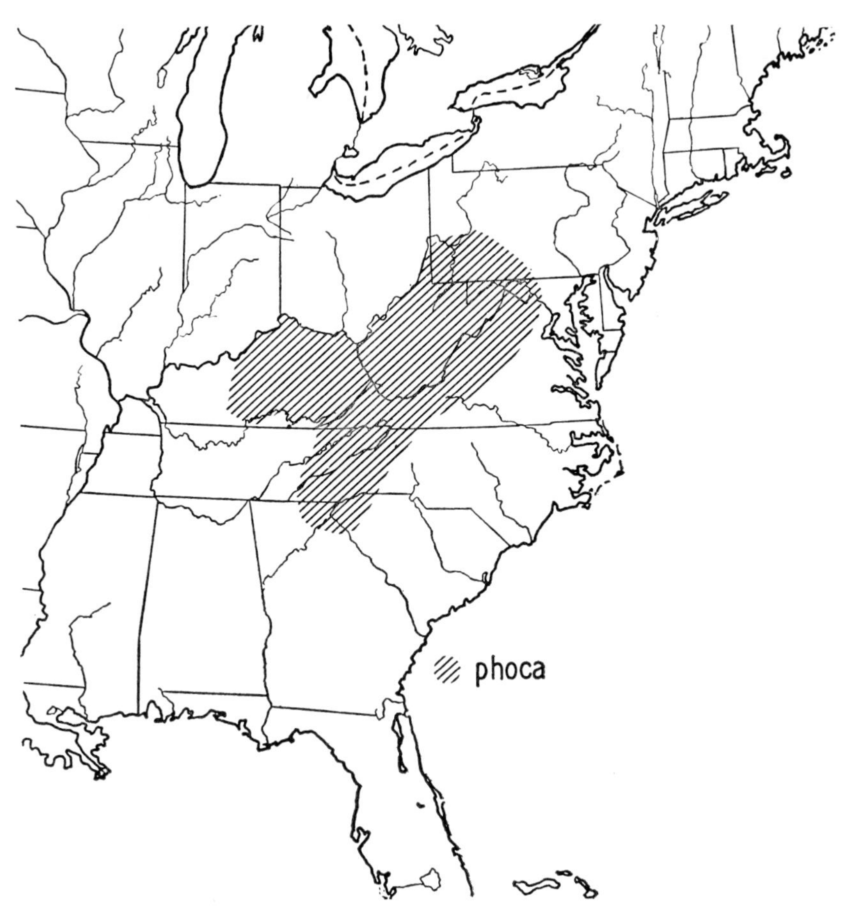


Map 25.—Distribution of *Desmognathus phoca.*

lina and Tennessee is 4 11/32″ (111 mm.), the extremes 3 1/32″ (77 mm.) and 5 1/8″ (131 mm.). The proportions of an adult female are as follows: total length 3 31/32″ (101 mm.), tail 2″ (51 mm.); head length 1/2″ (13 mm.), width 5/16″ (8 mm.). A male has the following measurements: total length 4 3/8″ (112 mm.), tail 2 1/8″ (54 mm.); head length 5/8″ (16 mm.), width 11/32″ (9 mm.). Dunn (1926, p. 77) recorded transformed individuals only 1 7/32″ (31 mm.) in total length.

DESCRIPTION. The head is widest about midway between the eyes and the lateral extensions of the gular fold, and from this point the sides behind converge slightly, and in front more abruptly, to the bluntly pointed snout. The eyes are large and strongly protuberant, the horizontal diameter about 1⅓ in the snout. A vertical fold immediately behind the eye; a light bar from the posterior angle of the eye to the angle of the jaw. An impressed line from the posterior angle of the eye to the lateral extension of the gular fold and a vertical groove crossing this line just behind the angle of the mouth. The trunk is rounded above and with a median groove, flattened beneath. There are usually 13 costal grooves, counting 1 in the axilla and 2 that run together in the groin, or 14 if the 2 in the groin are counted separately; but often the space between these and the next beyond is less than that between the other grooves. Intercostal folds 4–5 between the toes of the appressed limbs. Tail subquadrate in section at base, becoming sharp-edged above, rounded below beyond the vent, and compressed and slightly keeled above toward the slender, pointed tip. Legs stout; toes 5-4, those of the hind feet usually 1–5–2–4–3 in order of length from the shortest, the 3rd and 4th about equal; toes of the fore feet 1–4–2–3 or 1–2–4–3. The tongue is broad behind, narrowed and bluntly pointed in front, the surface spongy; free at the sides and behind. Vomerine teeth usually retained in adult males; when present in short, nearly transverse rows of 3–4 teeth, which may arise behind inner margin of inner nares or lie wholly between and slightly behind the nares, the series widely separated at the mid-line. Vomerine series in female longer, with 5–7 teeth. Parasphenoid teeth in 2 elongate, imperfectly separated patches, often in contact anteriorly and slightly divergent behind. In some males united for nearly the whole length.

COLOR. While variable in its dorsal markings, this salamander usually gives the impression of being very light below and strongly blotched above. The ground color of the light areas above varies from light buff to grayish-brown. The light areas may have dark borders and form fairly regular pairs along the back and become fused on the tail; or the light areas alternate with one another, with dark interspaces. Sometimes the light areas occupy the entire dorsal surface with only the outer margins limited by dark. The upper sides are usually more lightly mottled and fade on the lower sides to the light venter. The ventral surfaces may

FIG. 56. *Desmognathus phoca* (Matthes). (1) Adult male, actual length 4⅜″ (112 mm.). (2) Same, ventral view. Near Pine Ridge, Kentucky. [From a preserved specimen.]

be evenly but lightly pigmented with minute chromatophores, or the pigment may be aggregated to give a lightly mottled effect. The tip of the lower jaw of the male is slightly pointed and bears a mental gland, and the vent within is lined with papillae. The lower jaw of the female is broadly rounded, and the sides of the vent are thrown into folds.

BREEDING. The eggs of this species were first reported by Brady (1924, p. 29), who found a female coiled about a lot of 17 hatched eggs and 12 larvae near Harper's Ferry, Virginia, Sept. 5, 1923. Pope (1924, p. 7) reported additional lots found in the vicinity of Flat Rock and the base of Trenholm Mountain, North Carolina. The first group found by Pope, July 27, consisted of a cluster of 30 eggs attached to the lower side of a stone which was supported by other smaller stones above the bed of a clear brook. The second lot of 21 eggs, accompanied by a female, was taken August 1 in the cavity of a decayed, moss-covered log. Apparently in most instances the eggs are attached singly by extensions of the outer envelope. Brady recorded individual eggs as having a diameter of $2\frac{1}{4}$ mm., and Pope gave a measurement of 5 mm., apparently the diameter of the outer envelope.

LARVAE. I have not seen the recently hatched larvae, but Brady (*ibid.*) has described them as 18–20 mm. long and "having a row of well defined light spots each side of the dorsal groove from its beginning to the base of the tail." My smallest transformed specimen, 34 mm. in total length, has 6 pairs of narrowly separated light spots between the back of the head and the base of the tail and several offset confluent pairs on the basal half of the tail.

**BLACK-BELLIED SALAMANDER.** *Desmognathus quadramaculatus quadramaculatus* (Holbrook). Fig. 57. Map 26.

TYPE LOCALITY. Georgia and the Carolinas.

RANGE. From West Virginia and Virginia southward through the mountains to northern Georgia.

HABITAT. I have found this species in greatest numbers in and along the mountain streams above 2500′. They are particularly abundant where boulder-paved streams tumble in steep cascades down wooded ravines.

SIZE. Largest among the species of the genus, *D. quadramaculatus* attains an extreme length of $7\frac{5}{16}''$ (186 mm.) (Caesar's Head, South Carolina). The average length of 35 adults of both sexes from North Carolina and Tennessee is $4\frac{19}{32}''$ (118 mm.), the extremes $2\frac{13}{32}''$ (61 mm.) and $6\frac{7}{8}''$ (175 mm.). The proportions of an adult male from Mt. Le Conte, Tennessee, are as follows: total length $6\frac{5}{8}''$ (168 mm.), tail $2\frac{29}{32}''$ (74 mm.); head length $1\frac{1}{16}''$ (27 mm.), width $\frac{5}{8}''$

(16 mm.). An adult female from the same locality measures: total length 5 7/32″ (133 mm.), tail 2 7/16″ (62 mm.); head length 25/32″ (20 mm.), width 19/32″ (15 mm.).

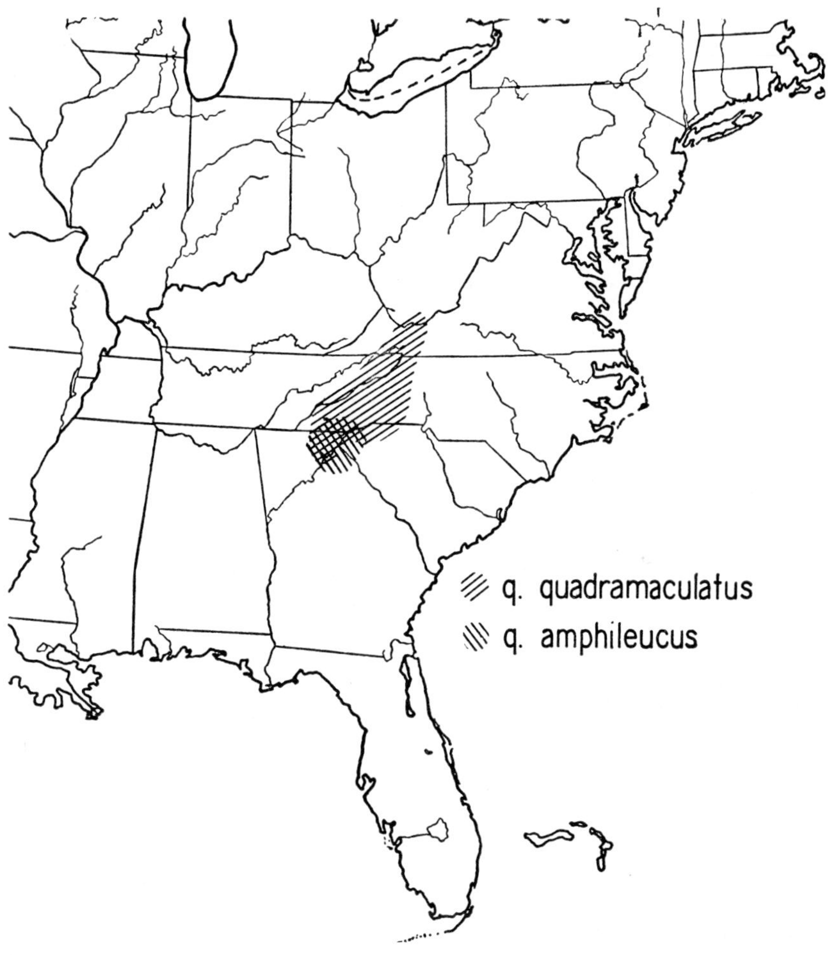


MAP 26.—Distribution of the subspecies of *Desmognathus quadramaculatus.*

DESCRIPTION. The head is large, somewhat swollen between the angle of the jaws and the lateral extensions of the gular fold, the sides in front of the angle of the mouth gradually converging to the bluntly rounded and depressed snout. The eyes are moderately large and protuberant, the long diameter about 1½ in the snout; limited behind by a vertical fold. A lightly impressed line from the posterior angle of the eye to the lateral extension of the gular fold; a deep groove from this line passing behind the angle of the mouth; a narrow light line from the posterior angle of the eye to the angle of the jaw. The trunk is large and strong and with an impressed median line above. There are usually 14 costal grooves, counting 1 each in the axilla and groin, or 15 if 2 that run together in the groin are counted separately; 2–3 intercostal folds between the toes of the appressed limbs. Tail subquadrate in section at base, rounded below, sharp-edged and keeled above beyond the vent. The dorsal tail keel arises as a low ridge immediately behind the vent and is thin and free beginning about at the basal third. Legs large and stout. Toes 5–4, those of the hind feet 1–5–2–3–4 in order of length from the shortest; toes of the fore feet 1–4–2–3. Tongue broadly oval in outline, free at the sides and behind, the surface broken into deep irregular folds. Vomerine teeth in series of 5–12, usually about 9, which arise back of the inner margin of the inner naris and curve inward and backward toward the mid-line, where they are separated by about the diameter of a naris; parasphenoid teeth in 2 long narrow patches narrowly separated or slightly in contact anteriorly, separated from the vomerine by 2 or 3 times the diameter of a naris.

COLOR. In old adults the ground color of the back and sides is black. Scattered over these areas and sometimes aggregated on the back to form irregular blotches are small greenish-yellow dots which give a rusty appearance. On the sides there is often a series of small light spots at the level of the eye and a second row between the legs at the juncture of the sides and belly. The belly and ventral surface of the tail and legs are black; the throat may be dark, strongly mottled black and dull yellow, or quite uniformly dusky yellowish. The back of the head and the snout are often rusty and sometimes there is a rusty spot at the base of the tail above. The dorsal surface of the legs is dark and quite uniformly pigmented, the soles of the feet light flesh color. In half-grown individuals the back and upper sides are dark, the lower sides light and mottled, the venter dark and lightly flecked with yellow. Re-

FIG. 57. *Desmognathus quadramaculatus quadramaculatus* (Holbrook). (1) Adult male, actual length 6⅝″ (168 mm.). Mount Le Conte, Tennessee. (2) Adult female, actual length 5⅝″ (143 mm.). Indian Gap, Tennessee.

cently transformed individuals are generally dark gray or brown above and light below, and many retain indications of the light larval markings as paired spots along the back and basal half of the tail.

BREEDING. In the disposition of the eggs, *D. quadramaculatus* resembles *D. phoca* and differs from the majority of the species of the genus in that they are attached singly to the lower surface of a support, usually a stone, and may be either submerged beneath the surface of the water or suspended above it. Pope (1924, p. 8) found the eggs and developing embryos in considerable numbers during July and early August in the mountain streams of North Carolina, and has indicated that normal complements may contain 25–40. The eggs are unpigmented and have a diameter of 5–6 mm. when well advanced in development. Each egg is attached by a slender pedicel, an extension of the outermost of the two jelly envelopes.

LARVAE. Larvae soon after hatching have a length of 20 mm. and are represented in Pope's figure (p. 10) as being uniformly pigmented on the back and sides except for a series of small pale spots along either side of the middorsal line of the back and basal half of the tail. The dorsal tail fin has its origin above the insertion of the hind legs and extends to the bluntly pointed tip; the ventral fin is narrower and does not reach the vent. Both fins are lightly mottled. Larger larvae are quite uniformly pigmented above and light below. Often the light larval marks are retained as paired spots narrowly separated along the mid-line of the back. Larvae may attain a length of 3½" (89 mm.) before losing the gills, or transform at less than 2" (51 mm.).

**WHITE-HEADED SALAMANDER.** *Desmognathus quadramaculatus amphileucus* Bishop. Fig. 58. Map 26.

TYPE LOCALITY. Demorest, Habersham County, Georgia.

RANGE. Known from the general vicinity of the type locality.

HABITAT. I know nothing of the circumstances under which the specimens were found.

SIZE. Only three specimens are known. The type has measurements as follows: total length 3 17/32" (90 mm.), tail 1⅝" (41 mm.); head length 17/32" (13 mm.), width 11/32" (9 mm.). Two paratypes measure respectively 3⅛″ (80 mm.) and 3 5/32″ (81 mm.), the smaller specimen with the tip of tail lost.

DESCRIPTION. A salamander of moderate size having the head, distal half of the tail, and the limbs white or lightly pigmented. The head has the sides back of the eyes gently converging to the lateral extensions of

FIG. 58. *Desmognathus quadramaculatus amphileucus* Bishop. (1) Adult female, actual length 3 17/32″ (90 mm.). (2) Same, ventral view. Type, from Demorest, Habersham County, Georgia.

the gular fold, in front more abruptly narrowing to the bluntly pointed snout. The eye is large and strongly protuberant, the horizontal diameter slightly less than the length of the snout. The trunk is slightly depressed and with a median impressed line, the sides rounded. There are 14 costal grooves, counting 1 in the axilla and 2 that run together in the groin and about 1 intercostal fold between the toes of the appressed limbs. Tail subquadrate in section at base, becoming compressed and keeled above immediately behind the vent; ventral tail keel narrow and limited to the distal third. Legs moderately stout; toes 5-4, those of the hind feet 1-5-2-4-3 in order of length from the shortest, webbed at base; toes of the fore feet 1-4-2-3. Tongue broadly heart-shaped, thin at the margins and free at the sides and behind, the plicae narrow and radiating from the center toward the sides and anterior margin. Vomerine teeth 7-10 in the females, in short series which arise slightly behind and inside the inner margin of the inner nares and curve inward and backward toward the mid-line, where they are separated by about the diameter of a naris. Parasphenoid teeth in 2 long, slender, club-shaped patches, slightly in contact anteriorly and separated from the vomerine by about twice the diameter of a naris.

COLOR. The general color above is dark brown, with the white of the head including the eyes and extending backward in a triangular point behind them, extending on the sides of the head to include the jaws and part of the neck, and on the ventral side halfway from the tip of the lower jaw to the gular fold. The distal half of the tail and the limbs white or lightly pigmented. The general brown color of the basal part of the tail and the upper sides is relieved by narrow light markings along the costal grooves and vertical grooves of the tail. The lower sides are mottled, yellowish-white and brown, the ventral surfaces lightly pigmented with brown except as noted above.

Intergrades between this subspecies and typical *quadramaculatus* have been noted in the general region about Demorest. They are generally larger than *amphileucus* and have the head, tail, and limbs lightly pigmented but suggesting the pattern as it is developed in this form.

BREEDING. Nothing is known of the breeding habits or larvae of this form.

**PYGMY SALAMANDER.** *Desmognathus wrighti* King. Fig. 59. Map 27.

TYPE LOCALITY. Mt. Le Conte, Sevier County, Tennessee.

RANGE. The spruce-fir forests of the Great Smoky Mountains in eastern

Tennessee; the mountains of western North Carolina northward to Whitetop Mountain, Virginia.

HABITAT. This is a terrestrial species found beneath stones, logs, bark,

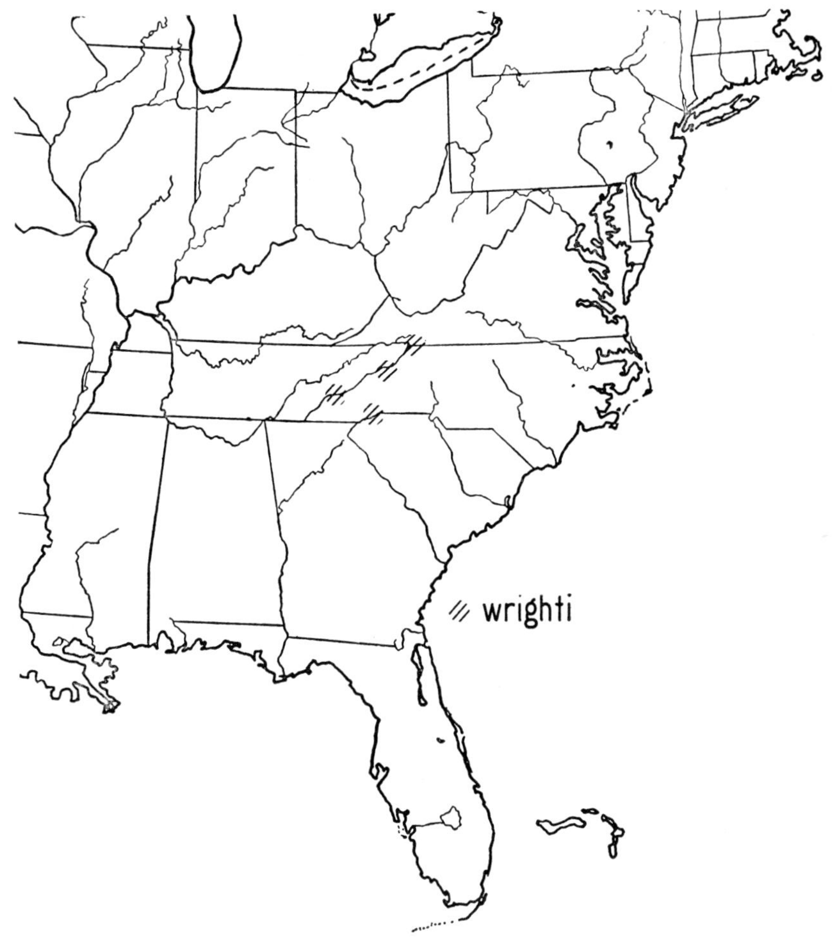


MAP 27.—Distribution of *Desmognathus wrighti,* a species limited to high elevations.

and moss, usually at elevations above 5000′ but occasionally as low as 3500′. On Clingman's Dome and Mt. Le Conte, Tennessee, reaches an elevation of 6500′.

SIZE. This is a diminutive species which attains an extreme length of about 2″. King (1936, p. 58) gives the measurements of 12 individuals from North Carolina and Tennessee which average 1 11/16″ (42.9 mm.) in total length and range from 1 7/16″ (37 mm.) to 1 29/32″ (49 mm.). The proportions of an adult male from Mt. Le Conte, Tennessee, are as follows: total length 1 5/8″ (42 mm.), tail 11/16″ (18 mm.); head length 7/32″ (6 mm.), width 5/32″ (4 mm.). A female from Grandfather Mountain, North Carolina, measures: total length 1 27/32″ (47 mm.), tail 23/32″ (19 mm.); head length 7/32″ (6 mm.), width 5/32″ (4 mm.).

DESCRIPTION. The sides of the head back of the eyes are nearly parallel, the snout short and bluntly rounded. Eye limited behind by an oblique fold, moderate, the long diameter about 1¼ in the snout; the iris tinged with brassy. A sinuous groove from the posterior angle of the eye to the lateral extension of the gular fold, a short vertical groove from this extending to the angle of the mouth. Trunk rounded, belly flattened. There are usually 14 costal grooves, counting 1 each in the axilla and groin, or 13 when 2 in the groin run together, and about 4 intercostal folds between the toes of the appressed limbs. The tail is flattened above at base, becoming circular in section beyond the vent, slender and tapering. Legs small but well developed. Toes 5–4, those of the hind feet 1–5–2–4–3 in order of length from the shortest; toes of the fore feet 1–4–2–3. Tongue wide behind, bluntly pointed in front, free at the sides and back. Vomerine teeth 3–7 in short series that arise between and behind the inner nares and curve inward and backward toward the mid-line, where they may be nearly in contact or separated by a distance equal to the diameter of a naris. Parasphenoid teeth in 2 slender, club-shaped patches in contact anteriorly.

COLOR. This small species has a light dorsal band and light belly. In most the light band arises on the snout and involves the full width of the head, narrowing down slightly at the shoulders and continuing to the tip of the tail. The light band varies in color from light tan to red, with the majority of specimens exhibiting a bronzy or reddish-brown tinge. In some individuals the sides of the band are fairly regular;

in others the remnants of the light larval spots persist as half-circles arranged opposite one another or alternating along the sides of the band. Many specimens have a definite herringbone pattern of black along the mid-line of the back, each segment separated from its fellow by a dash of yellow. Legs above reddish-orange mottled with brown and gray; feet tan and brown. On the side of the head a dark bar from the nostril through the eye and an oblique orange-yellow bar from the

FIG. 59. *Desmognathus wrighti* King. (1) Adult female, actual length 1⅝″ (41 mm.). (2) Same, ventral view. Grandfather Mountain, North Carolina.

posterior angle of the eye to the angle of the jaw. Sides of head behind jaw, and the trunk next to the dorsal band tan to brown lightly flecked with white, becoming lighter toward the belly. Belly lightly pigmented at sides, flesh color in life, yellowish-white in preservative; ventral surface of legs and tail tan, throat flesh with silvery reflections. Margins of lower jaw usually mottled with brown and tan.

BREEDING. Nothing has been reported on the breeding habits. King (*ibid.*, p. 59) dissected two females taken Sept. 23, 1934, each 45 mm. long, which contained 13 and 14 eggs respectively. The egg-laying season may be an extensive one, for a female taken July 22, 1937, on Clingman's Dome, and sent me by Dr. A. H. Wright, had large ovarian eggs which showed through the skin of the belly. On April 18 and 19, 1938, we collected this species in some numbers on Grandfather Mountain, North Carolina, at elevations above 5000′. Here they were in fairly dry situations, beneath stones and bits of logs and bark along the trail. At Indian Gap, Tennessee, they were in somewhat moister situations in the spruce-fir forests.

## Genus LEUROGNATHUS

### KEY TO THE SUBSPECIES OF LEUROGNATHUS MARMORATA

Above mottled with shades of buff and dark gray or black, the light and dark patches often forming a broad zigzag pattern; belly mottled and blotched along the sides, usually lighter centrally; vomerine teeth usually lacking in adults of both sexes, 2 or 3 sometimes present in the female; length to 5 1/16″ (128 mm.). Western North Carolina and Roan Mountain, Carter County, Tennessee ........ *marmorata marmorata* p. 220

Above dark brown with dorsolateral series of rounded, well separated light spots extending from the head onto the basal half of the tail; belly uniformly mottled and blotched and without a light central area; vomerine teeth usually lacking in adult male, occasionally 1 or 2 present; vomerine teeth in female variable, usually 1 or 2 on each side; length to 4 13/16″ (122 mm.). Known from the vicinity of Davis Gap, Waynesville, and Frying Pan Mountain, North Carolina ........ *marmorata intermedia* p. 224

**MOORE'S SALAMANDER.** *Leurognathus marmorata marmorata* Moore. Fig. 60. Map 28.

TYPE LOCALITY. Grandfather Mountain, North Carolina.

RANGE. Known from the following counties in western North Carolina: Watauga, Caldwell, Avery, Burke, McDowell, Yancey, Buncombe, Henderson, and Haywood, and from Roan Mountain, Carter County, Tennessee.

HABITAT. An aquatic species found in streams and spring runs, hiding beneath flat stones or boulders in the pools and riffles, or exposed and resting quietly on the bottom. Sometimes found in the open on the naked rock of stream beds where water flows in a thin sheet or in trickles.

SIZE. The average length of 10 adults of both sexes from the North Carolina mountains is 4 9/16″ (108.5 mm.), the extremes 3 9/16″ (90 mm.)

MAP 28.—Distribution of the subspecies of *Leurognathus marmorata.*

and 5 1/16″ (128 mm.). The proportions of an adult male from Grandfather Mountain, North Carolina, are: total length 4″ (101 mm.), tail 1 7/16″ (36 mm.); head length 21/32″ (16 mm.), width 14/32″ (11 mm.). A female from the same locality: total length 4 1/16″ (103 mm.), tail 1 17/32″ (38 mm.); head length, 21/32″ (16 mm.), width 13/32″ (10.5

mm.). A large female from near Linville Falls, North Carolina, has a total length of $5^{21}/_{32}$″ (145 mm.).

DESCRIPTION. This species has sometimes been confused with *Desmognathus quadramaculatus* and *D. phoca.* Viewed from above, the head is widest at the back, gently rounding or slightly tapering to the eyes, then more abruptly converging to the bluntly pointed snout. The snout is strongly depressed. The eyes are large, strongly protuberant, and limited behind by an oblique fold. An impressed sinuous line from the posterior angle of the eye to the lateral extension of the gular fold, a short vertical groove from this line passing behind the angle of the mouth. The trunk is somewhat flattened above and provided with an impressed median line; sides rounded. There are usually 13 costal grooves, rarely 14, counting 1 each in the axilla, which may be poorly developed, and 1 in the groin; 3 intercostal spaces between the toes of the appressed limbs. Tail subquadrate in section and flattened at base above, becoming strongly compressed and knife-edged above, about at the end of the proximal third. The tail is rounded below and without a free fin except at the distal fourth. Legs rather short and stout, toes 5-4, those of the hind feet 1–5–2–(4–3) in order of length from the shortest; toes of the fore feet 1–4–2–3; all toes tipped with black, and those of the hind feet slightly webbed at base. The tongue is large and fleshy and fills the floor of the mouth; it is broader than long, attached at the center in front, and with the margins free. Vomerine teeth usually lacking in adults of both sexes, 2 or 3 rarely developed in females. Parasphenoid teeth in 2 slender, elongate patches which taper and meet anteriorly and diverge posteriorly. Inner nares small, widely separated, and concealed in slit-like grooves at the sides of the mouth.

COLOR. In life this species is mottled above with shades of buff and gray that match the stream-bottom background of pebbles and small stones and render the animal inconspicuous until it moves. The light and dark patches alternate more or less regularly so that a broad zigzag pattern is formed, or 2 rows of light blotches with indefinite edges extend from the back of the head well onto the tail. The sides of the trunk and tail are grayish or brownish with light scattered flecks. The venter may be uniformly and lightly pigmented, or the darker colors of the lower sides may encroach irregularly on the sides of the venter, leaving a fairly light central area. In some individuals the venter may become very dark except for a narrow central spot. The legs are mottled above

FIG. 60. *Leurognathus marmorata marmorata* Moore. (1) Adult female, actual length 4 7/16″ (112 mm.). (2) Same, ventral view. Yonahlossee Road near Grandfather Mountain, North Carolina.

and on the sides, lightly blotched below. In the majority of specimens the anterior half of the throat, the sides of the vent, and the distal third of the tail below are lighter than surrounding areas. The vent of the male is larger than that of the female and has the margins ridged and grooved. Old individuals may be very dark above and below with only a faint indication of a dorsal pattern.

BREEDING. Dissection of adult females taken in October revealed 26–34 ovarian eggs, each approximately 1.6 mm. in diameter (Bishop, 1924,

p. 95). Nothing was learned of the egg-laying habits until Pope's discovery of a cluster of 28 eggs attached to the edge of a submerged stone in the running water of a small mountain stream. The eggs were grouped together in 2 layers and attached by means of short extensions of the outer envelope (Pope, 1924, p. 13).

LARVAE. Larvae 8–10 days after hatching have a total length of approximately 17 mm. (Pope, *ibid.*, Fig. 2). The larvae have a wide dorsal fin confined to the tail and rising in a broad curve above the line of the back. The gills are long, filamentous, and lacking in pigment, the eyes large. There is a row of small light spots on either side of the middorsal line, and the back and sides are pigmented to the lower level of the legs. Larvae may attain a length of 66 mm. before transformation, and the larger ones may be briefly characterized as follows: The general ground color is light brown (in preservative). On either side of the middorsal line a series of round tan spots extend along the trunk and basal quarter of the tail. The dark pigment extends on the sides of the head to involve the margin of the lower jaw and sides of the throat, on the trunk to the lower level of the legs, and on the tail encroaches narrowly on the ventral surface. The chin is narrowly margined with dusky. The inner nares are longitudinal slits which lie below the anterior end of the eye.

**POPE'S SALAMANDER.** *Leurognathus marmorata intermedia* Pope. Fig. 61. Map 28.

TYPE LOCALITY. Given as Davis Gap, Waynesville, North Carolina, but apparently from the Davis farm about 2 miles east and a little north of Waynesville, on Highway No. 276.

RANGE. In addition to the type locality, known from the Boone farm 3½ miles southeast of Waynesville and from Frying Pan Mountain, near Mt. Pisgah, at elevations of 4000′ and above.

HABITAT. In streams and springs, generally at high elevations. May be found hiding beneath stones or resting quietly at the pool bottoms. Essentially aquatic but occasionally on land.

SIZE. Five adults of both sexes average $4\frac{5}{32}''$ (105.7 mm.) with extremes of $3\frac{11}{16}''$ (94 mm.) and $4\frac{13}{16}''$ (122 mm.). The proportions of an adult female taken Oct. 13, 1926, on Frying Pan Mountain, Haywood County, North Carolina, are as follows: total length $4\frac{13}{16}''$ (122 mm.), tail $2\frac{1}{8}''$ (54 mm.); head length $\frac{11}{16}''$ (17 mm.), width $\frac{9}{16}''$ (14 mm.). An adult male from near Waynesville, North Carolina, has the following measurements: total length $4\frac{17}{32}''$ (115 mm.), tail $1\frac{10}{16}''$ (41 mm.); head length $\frac{21}{32}''$ (16 mm.), width $\frac{13}{32}''$ (10.5 mm.).

DESCRIPTION. Resembles some juvenile specimens of *Desmognathus quadramaculatus,* but may be distinguished by lack of well developed vomerine teeth and concealed inner nares. Viewed from above, the head is oval in outline, widest just in front of the lateral extensions of the gular fold, the sides curving broadly and evenly to the bluntly pointed and strongly depressed snout. The eyes are large and strongly protuberant, the iris tinged with brassy near the pupil. An oblique fold back of the eye and from this a sinuous line extending to the lateral extension of the gular fold. A short vertical groove crossing the sinuous line and extending to the angle of the mouth. In adults, the trunk somewhat flattened above, and with an impressed median line; the sides well rounded. There are 13 costal grooves, counting 1 in the axilla and 2 that run together in the groin, or 14 when the 2 in the groin remain separated, and 2½–3½ intercostal spaces between the toes of the appressed limbs. Tail flattened at base above and subquadrate in section, becoming compressed and sharp-edged and keeled above, beginning about at the end of the proximal fourth; ventral tail fin narrow and limited to the distal fourth. Legs short and stout; toes 5–4, those of the hind feet 1–5–2–(4–3) in order of length from the shortest, webbed at base; toes of the fore feet 1–4–2–3; toe tips usually black. Tongue large, attached medially in front, the side margins free. Vomerine teeth usually lacking in the male, although an occasional individual may develop 1 or 2; vomerine teeth in female variable, usually 1 or 2 on each side, or present on one side and lacking on the other; juveniles sometimes with fairly regular series of 4–4 teeth in slightly curved lines separated by about the length of a series. Parasphenoid teeth in 2 long slender patches tapering anteriorly where they unite, and narrowly diverging poste-

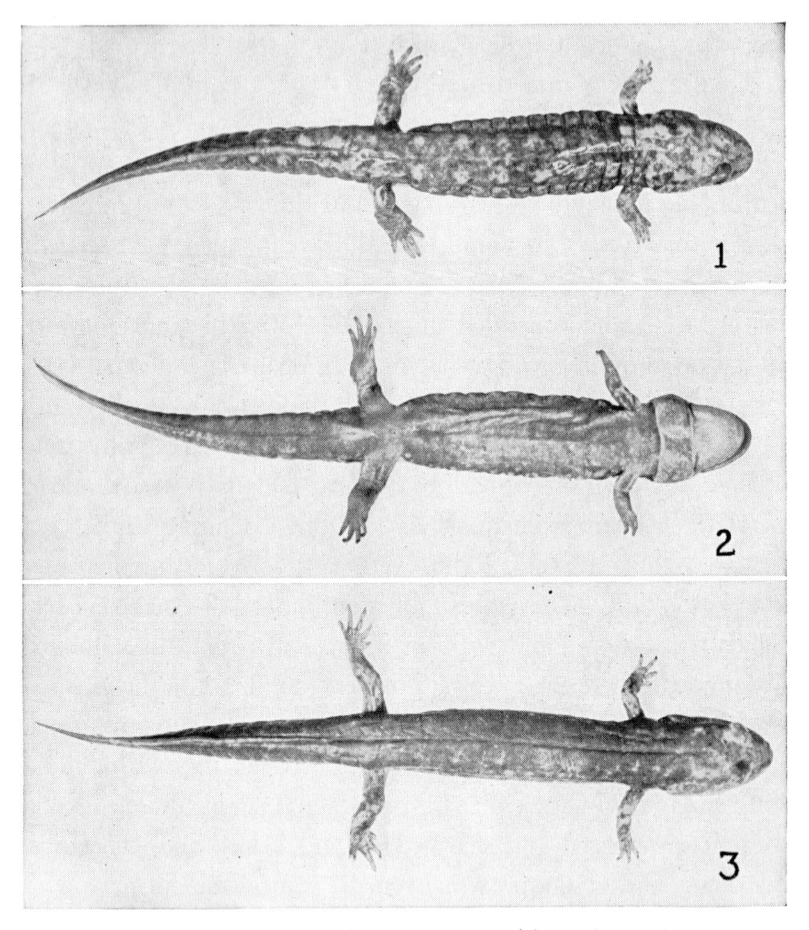


FIG. 61. *Leurognathus marmorata intermedia* Pope. (1) Adult female, actual length $4^{11}/_{16}$″ (120 mm.). (2) Same, ventral view. Frying Pan Mountain, near Big Pisgah, North Carolina. (3) Adult male, actual length $4^{3}/_{32}$″ (104 mm.). Near Waynesville, North Carolina. [Photographs from preserved specimens.]

riorly. Inner nares small and concealed in a groove at the side of the mouth.

COLOR. The color in life is variable, but the pattern is usually constant. In some the ground color above is dark walnut-brown, the sides lighter, grayish-brown. On each side above, a dorsolateral series of round light spots, well separated, extending from the head back of the eyes onto the basal half of the tail. The legs are lighter than the back and lightly mottled with yellow; upper margin of dorsal tail fin frequently lighter than sides. The venter may be much lighter than the back, speckled with small light yellow spots, or uniformly dark gray, nearly black, flecked with small light spots. The throat is lighter than the belly, although often lightly mottled, flesh-colored in the central area, and whitish on the chin. Tail beneath, colored like the belly at base, fading to yellowish toward the tip. On some individuals there are 2 secondary rows of small light spots on the sides, the upper extending from the eye well onto the tail, the lower forming a broken line between the limbs. In some dark specimens, a light line from the eye to the angle of the jaw. The vent of the male has a free flange-like margin and the sides within are covered by papillae. The vent of the female has the margins crossed by grooves and ridges but lacks the papillae. The tip of the lower jaw of the male pointed, that of the female evenly rounded.

BREEDING. Nothing has been published on the breeding habits of this species. The large female taken on Frying Pan Mountain, North Carolina, October 13, 1926, has large ovarian eggs apparently nearly ready to be deposited.

LARVAE. I have not found a published description of the larva. A single specimen from the type locality in my own collection is 1¾″ (45 mm.) in total length. It is uniformly Mummy Brown (Ridgway) above, very light, probably flesh color in life, below. Scattered generally over the surface of the head, back, and sides are many small, separate, black chromatophores, and dorsally a few scattered, rounded pigment-free spots. On the upper sides and ventrolaterally these light spots tend to form indistinct lines between the legs. The pigment of the lower sides encroaches on the sides of the throat, the sides of the belly and nearly to the mid-line of the venter of the tail. The gills are short and pigmented only at the base. The dorsal tail fin arises at a point opposite the posterior end of the vent and reaches its greatest width about midway the length of the tail. The ventral fin is narrower and confined to the distal half. The inner nares of the larva, as in the adult, are slit-like and lateral in position. The smallest fully transformed individual in my collection has a total length of 2 7/16" (62 mm.).

## Genus PLETHODON

### KEY TO THE SPECIES AND SUBSPECIES OF PLETHODON

Several species of Plethodon have distinctly different color phases, and most species undergo conspicuous changes of color in preservatives. To aid in the identification of both living and preserved material, a species may appear more than once in the key.

1. With a median, dorsal band having definite edges . . . . . 2
   Without a median dorsal band; or, if markings form a band, then not with definite edges . . . . . 13
2. Dorsal band gray, red, brown, or chestnut . . . . . 3
   Dorsal band dull white, tan, yellow, or greenish-yellow . . . . . 9
3. Dorsal band with zigzag margins; dorsal band red or red suffused with dusky (fades to dull red or yellow in preservatives); venter lighter than back, mottled; usually 17 costal grooves, occasionally 16 or 18; length to 3½" (89 mm.). Southern Ohio south through Kentucky and Tennessee to Alabama . . . . . *cinereus dorsalis* p. 236
   Dorsal band with margins straight or slightly irregular but never definitely zigzag . . . . . 4
4. Costal grooves 16 or less . . . . . 5
   Costal grooves 17 or more . . . . . 7
5. Intercostal folds between toes of appressed limbs 3 or less . . . . . 6
   Intercostal folds between toes of appressed limbs 5 or more; dorsal band red in various shades to red strongly suffused with dusky; if dorsal band indistinct, then its margins indicated by a narrow light line on each side; venter light, mottled; usually 15 costal grooves, occasionally 16; length to 4⅛" (105 mm.). Western Oregon, Washington, and British Columbia . . . . . *vehiculum* p. 278
6. Belly and throat lightly pigmented and sometimes blotched, lighter than back; median dorsal band chestnut-red; costal grooves usually 15, occasionally 16; intercostal folds between toes of appressed limbs, 0–1; length to 6½" (165 mm.). Blue Ridge Mountains in western North Carolina and southwestern Virginia . . . . . *yonahlossee* p. 287
   Belly as dark as back, throat light; above chestnut-brown overlying a darker ground color, with small brassy or golden flecks; upper sides chestnut varied with black; lower sides with a broad whitish band

with irregular upper edge; costal grooves usually 16, 15 to 17; intercostal folds between toes of appressed limbs, 1–3; length to 4¾″ (121 mm.). Oklahoma and Arkansas .................. *ouachitae* p. 269

7. Costal grooves 20–23; dorsal band narrow, poorly developed, scarcely lighter than adjacent sides; dorsal surfaces with minute silvery-white and bronze flecks; belly dark, throat somewhat lighter, mottled; costal grooves usually 21, occasionally 20 to 23; intercostal folds between toes of appressed limbs, 9–10; length to 5³⁄₃₂″ (130 mm.). Pennsylvania, West Virginia, Kentucky, and southeastern Ohio.... ........................................... *richmondi* p. 272

   Costal grooves 17–19 ........................................ 8

8. Parasphenoid teeth in a single patch notched behind; dorsal band brown or chestnut strongly suffused with dusky; upper sides dark, lower sides slate, often with minute whitish flecks; belly slate, throat lighter, mottled; usually 17 costal grooves, occasionally 18; intercostal folds between toes of appressed limbs 6–7; length to 5⅛″ (131 mm.). Extreme northwestern California and southwestern Oregon. .*elongatus* p. 246

   Parasphenoid teeth in 2 patches; dorsal band usually some shade of red, occasionally gray, often with flecks of black; sides gray or black; belly mottled, pepper-and-salt; 18–20 costal grooves; intercostal folds between toes of appressed limbs, 7–9; length to 4¹³⁄₁₆″ (122 mm.). Southeastern Canada south to Texas, Georgia, Missouri, and Arkansas ........................................ *cinereus cinereus* p. 232

9. Belly very light, flesh to yellow, sometimes faintly suffused with dusky; adults with parotoid glands; sides uniformly light brown; dorsal stripe yellow or tan; 13–14 costal grooves; 2–3 intercostal folds between toes of appressed limbs; length to 4¹⁷⁄₃₂″ (116 mm.). Western Washington .................................... *vandykei* p. 275

   Belly pigmented; adults without parotoid glands .................. 10

10. Costal grooves 18 or more; dorsal stripe tan or dull yellow; other characters as above under 8 ........................ *cinereus cinereus* p. 232

    Costal grooves 16 or less ........................................ 11

11. Costal grooves 13–14; belly black; 3 intercostal folds between toes of appressed limbs; dorsal stripe yellow, strongly suffused with brown on the head and tail tip; sides black with scattered gray flecks; length to 4″ (101.5 mm.). Kootenai County, Idaho ........... *idahoensis* p. 259

    Costal grooves 15 or 16; belly light, mottled .................... 12

12. Intercostal folds between toes of appressed limbs, about 4; dorsal band dull tan to yellowish-green; upper sides brown with many dull white and tan flecks and dashes; venter pale slate with many pale spots, whitish or yellowish; costal grooves usually 15, occasionally 16, rarely 14; vomerine teeth 8–13, rarely fewer; length to 5⅛″ (131 mm.). Southwestern Oregon to southwestern Washington ... *dunni* p. 242

    Intercostal folds between toes of appressed limbs 5 or more; dorsal band yellow or dull white (rarely living specimens and most preserved specimens on which the red has faded); other characters as above under 5; hind feet sparsely webbed ........ *vehiculum* p. 278

13. With red spots, flecks, or patches present ........ 14
    Without red spots, flecks, or patches ........ 15
14. Black above, with red spots or flecks on dorsal surface of legs (color fades to white or yellow in preservatives); lower sides light gray; belly light, bluish-white or pale gray; throat light; costal grooves 15–16; 1–2 intercostal folds between toes of appressed limbs; length to 6″ (153 mm.). Southwestern North Carolina ........ *glutinosus shermani* p. 253
    Black above, with a red patch on side of head (fades to white or yellowish in preservatives); lower sides and belly dull bluish-gray; throat anteriorly flesh color; costal grooves 15–16; 2–3 intercostal folds between toes of appressed limbs; length to 5$\frac{9}{32}$″ (135 mm.). Great Smoky Mountains in western North Carolina and eastern Tennessee ........ *jordani* p. 261
15. Uniformly dark gray or black above and on upper sides; no conspicuous spots, flecks, or other markings ........ 16
    Not uniformly dark gray or black above and on sides ........ 24
16. Costal grooves 16 or less ........ 17
    Costal grooves 17 or more ........ 20
17. Intercostal folds between toes of appressed limbs, 2 or less; bluish-brown to purplish-gray above; belly dull grayish; throat grayish-white; usually 16 costal grooves, occasionally 15; length to 6$\frac{1}{8}$″ (156 mm.). Southwestern Virginia through western North Carolina, northern South Carolina, northern Georgia, Roan Mountain, Tennessee, and possibly northeastern Alabama ........ *metcalfi* p. 264
    Intercostal folds between toes of appressed limbs 3 or more ........ 18
18. Belly strongly mottled; basal segments of legs above usually with some light spots; dorsal band lacking or indicated only by narrow light lines laterally; other characters as above under 5 (dark phase) ........ *vehiculum* p. 278
    Belly uniformly colored; basal segments of legs above without light spots ........ 19
19. Costal grooves 14–15; vomerine teeth 6–10, average 8; brown above, venter uniformly pigmented; throat lighter than belly; length to 3$\frac{11}{16}$″ (94 mm.). New Mexico ........ *hardii* p. 256
    Costal grooves 16; vomerine teeth 4–6, average 5; uniformly slate above; belly grayish-blue, lighter than back, uniformly pigmented or rarely slightly blotched and with a few light flecks anteriorly; under surface of legs frequently with a light mark at joint; length to 3$\frac{1}{16}$″ (78 mm.). Grandfather Mountain, North Carolina, and Whitetop Mountain, Virginia (preserved specimens) ........ *welleri* p. 285
20. Belly much lighter than back, mottled yellow and gray ........ 21

Belly dark, with few or no light flecks ........................ 22

21. Costal grooves 18–19, occasionally 20; uniformly dark brown or lead-colored above; lower sides, throat, and belly light, mottled; other characters as above under 8 (lead-backed phase) ........................................................ *cinereus cinereus* p. 232

Costal grooves usually 17, occasionally 16 or 18; uniformly brown or lead-colored above; vomerine teeth 4–6; other characters as above under 3 (dark phase) .......................... *cinereus dorsalis* p. 236

22. With 20–23 costal grooves; ground color uniformly dark above; belly dark, throat lighter, mottled; other characters as above under 7 (preserved specimens) ................................ *richmondi* p. 272

With 17–19 costal grooves ................................ 23

23. Usually with parasphenoid teeth in a single patch; 6–7 intercostal folds between toes of appressed limbs; other characters as above under 8 (dark phase) ................................ *elongatus* p. 246

With parasphenoid teeth in 2 patches; 5–6 intercostal folds between toes of appressed limbs; uniformly bluish-black above; venter slightly lighter and often with a few whitish flecks along sides of belly; throat lighter; costal grooves usually 18, 17 to 19; 5–6 intercostal folds between toes of appressed limbs; length to $3^{17}/_{32}''$ (90 mm.); West Virginia (preserved specimens) .................... *nettingi* p. 266

24. Dark gray or black above, with white or pale bluish-white spots or blotches on sides and sometimes on back; or, with light gray lichen-like patches on sides and often on back ...................... 25

Without silvery-white, bluish-white, or lichen-like gray patches on sides, but with small silvery, bronze, or golden flecks on back .... 29

25. Light markings forming lichen-like blotches on back and often on sides; belly and ventral surface of tail black; throat whitish; 15 costal grooves; 1–2 intercostal folds between toes of appressed limbs; iris deep brown without brassy flecks; length to $5^{13}/_{16}''$ (149 mm.). Vicinity of Jocassee, South Carolina .................. *clemsonae* p. 239

Light markings not forming lichen-like patches, often in separate spots but sometimes fused to form irregular bands on lower sides ...... 26

26. Entire ventral surface dark slate-color in life, blue-black in preserved specimens; light markings on sides, occasionally on back, silvery-white; usually 16 costal grooves; intercostal folds between toes of appressed limbs, 1–3; vomerine series long, 10–13; length to $7^{7}/_{16}''$ (188 mm.). New York westward to Wisconsin, southward to northern Florida, westward to Texas and Missouri ........ *g. glutinosus* p. 250

Entire ventral surface not dark ................................ 27

27. Belly dark, throat light ................................ 28

Belly and throat lightly pigmented and sometimes blotched, definitely lighter than back; lower sides with small white spots; costal grooves usually 15; other characters as above under 6 (preserved specimens) ........................................... *yonahlossee* p. 287

28. Parasphenoid teeth in a single broad patch; a few whitish flecks and faded blotches on sides; often with faded whitish spots or blotches on tail, above; vomerine teeth 7–13; other characters as above under 6 (preserved specimens) . . . . . . . . . . . . *ouachitae* p. 269

    Parasphenoid teeth in 2 patches; ground color above bluish-black; belly and ventral surface of tail grayish; throat light; light spots small, white to bluish-white, usually restricted to lower sides; costal grooves usually 16, occasionally 17; 2 or 3 intercostal spaces between toes of appressed limbs; vomerine teeth 6–9; length to 6″ (152 mm.). Southwestern New York, Pennsylvania, West Virginia, and southeastern Ohio . . . . . . . . . . . . *wehrlei* p. 281

29. Costal grooves 20–23; intercostal folds between toes of appressed limbs, 9–10; brown above on the trunk, with many small silver-white and bronze flecks; lower sides and venter with small light spots; other characters as above under 7 (dark phase) . . . . . . . . . . . . *richmondi* p. 272

    Costal grooves 19 or less . . . . . . . . . . . . 30

30. Costal grooves 16; 3–4 intercostal grooves between toes of appressed limbs; ground color above black, with many small dull golden blotches on the back and a few golden flecks on upper sides; sides mostly black, with a few whitish flecks near belly; belly slate, sometimes with a few scattered light spots; throat light, venter of tail darker; length to 3 1/16″ (78 mm.); Grandfather Mountain, North Carolina, and Whitetop Mountain, Virginia (living) . . . . . . *welleri* p. 285

    Costal grooves 17–19; above deep brown, almost black, with many fine, pale-brassy flecks over dorsal surfaces and upper sides of head and trunk; belly and ventral surface of tail dark slate; throat lighter; other characters as above under 23 (living) . . . . . . . . . . . . *nettingi* p. 266

**EASTERN RED-BACKED SALAMANDER. LEAD-BACKED SALAMANDER.** *Plethodon cinereus cinereus* (Green). Figs. 1b, 62. Map 29.

TYPE LOCALITY. New Jersey.

RANGE. Gaspesia and Cape Breton Island west to Fort William, Ontario, south to Dallas, Georgia, Missouri, and Arkansas (Stejneger and Barbour, 1939, p. 13).

HABITAT. The Red-backed salamander is completely adapted to terrestrial life and is abundant beneath old logs, bark, moss, leaf mold, and stones, in evergreen, mixed, and deciduous forests. May be found in fairly dry situations but is most abundant where there is indication of moisture.

SIZE. Reaches an extreme length of 122 mm. (Davis, 1942, p. 258).

The males are a little smaller than the females, the adults ranging from $2\frac{5}{16}''$ to $3\frac{5}{8}''$ (58–91 mm.) and averaging about $2\frac{3}{4}''$ (73 mm.). The females average $3\frac{1}{8}''$ (78 mm.) and vary from $2\frac{1}{2}$ to $3\frac{9}{16}''$ (64–90 mm.). In a series of specimens from Big Babson's Island, Maine, in the

MAP 29.—Distribution of the subspecies of *Plethodon cinereus.*

U. S. National Museum, there are individuals varying from 84 to 115 mm. in total length.

DESCRIPTION. The body is long and fairly slender, slightly flattened dorsally, and well rounded on the sides. The tail is nearly circular in section throughout its length. There are usually 18 or 19 costal grooves but the number may vary from 17 to 20. The gular fold is prominent.

The legs are small, the toes short and thick, those of the fore feet 1-4-2-3 in order of length; hind toes rather thick and fleshy, slightly webbed at base, 1-5-2-4-3 in order of length. The mouth is fairly large, with the angle of the jaws back of the eye. The small tongue does not fill the floor of the mouth. The vomerine series form 2 backward-curving lines of 5-7 teeth separated from each other and from the parasphenoid teeth, which are in 2 imperfectly separated patches.

COLOR. There are 2 distinct color phases, and individuals are encountered of an intermediate character. The red-backed phase is characterized by the possession of a broad, median, dorsal band that originates on the head and extends along the trunk and onto the tail. The dorsal stripe varies in color from light gray and dull yellow through shades of pink, brick-red to bright red, often with small flecks of black within the band. Below the dorsal band the sides are dark gray or black, becoming lighter and mottled toward the belly. The belly is strongly mottled with gray and white. In the lead-backed phase, the color above is uniformly dark gray to almost black, with the head and legs usually lighter. In some localities the color phases are about equally represented, in others one or the other phase may predominate.

SEXUAL DIFFERENCES. The male differs from the female in its more swollen snout, in having enlarged premaxillary teeth, and in having the legs proportionally longer. In body length the male is usually shorter.

BREEDING. There is a rather extended mating season in the fall, October to December, during which spermatophores are deposited by the male and recovered by the female. The egg-laying season extends from June through July, in the North. The eggs are deposited in little clusters, containing usually 8-10 but varying from 3 to 13, held together like a bunch of grapes and suspended by a common pedicel to the roof of the cavity chosen for the nest. Usually the eggs are found in well rotted logs, rarely in little cavities on the lower surface of stones, and attended by the female. The individual egg varies in size from 3.5 to 5 mm., depending somewhat on the amount of moisture present, and in addition to the vitelline membrane has 2 distinct envelopes. The outer envelope usually is covered by whitish mucous strands that converge to form the attachment stalk. The length of the period of incubation in the field

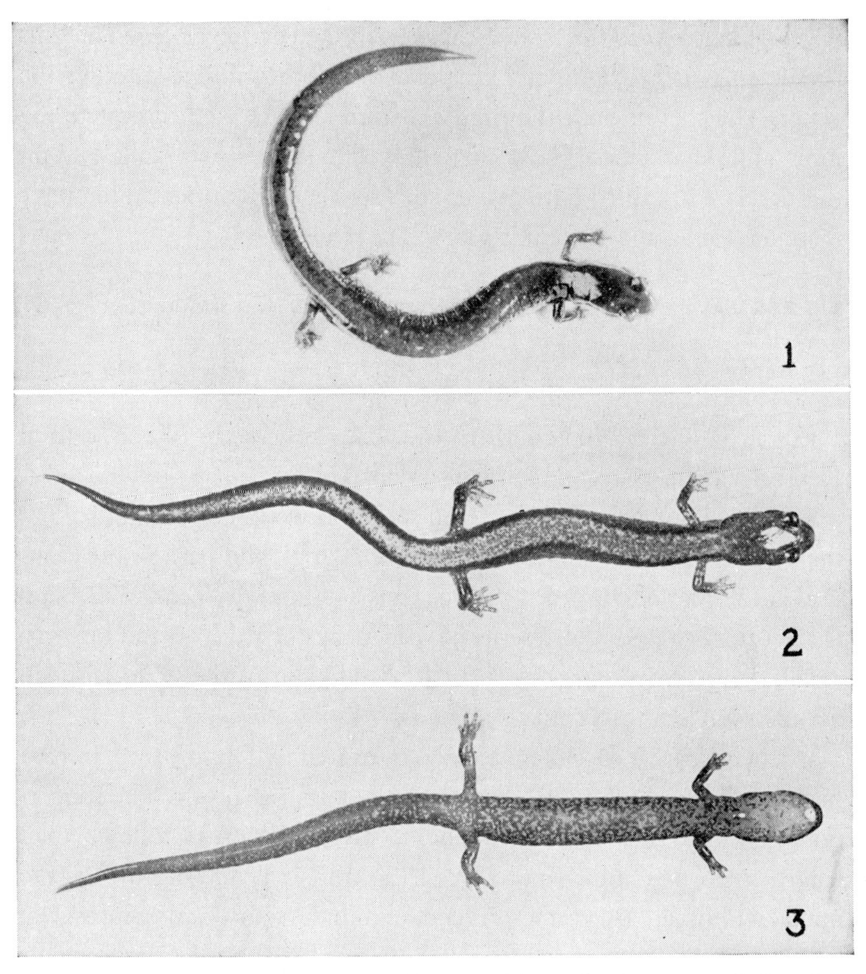


FIG. 62. *Plethodon cinereus cinereus* (Green). (1) Adult, about natural size. Kenwood, Albany County, New York. (2) Adult female, red-backed phase; actual length 2 29/43″ (74 mm.). (3) Same, ventral view. Rochester, New York.

has never been accurately determined, but recently hatched young may be found all through August.

LARVAE. The recently hatched larvae average about ¾″ (19 mm.) in length. In the red-backed phase, the dorsal band is well developed at the time of hatching and the upper sides are strongly pigmented. The broad, flat, leaf-like gills rise from a common base and often reach their full development before the larva escapes from the egg envelopes. The gills persist only a few days, and the larva develops without entering the water. The toes of both the hind and fore feet are well indicated, the inner and outer short. The young are well endowed with yolk, and on this they are sustained until able to shift for themselves. Sexual maturity is attained at an age of about 2 years.

**ZIG-ZAG SALAMANDER.** *Plethodon cinereus dorsalis* (Cope). Fig. 63. Map 29.

TYPE LOCALITY. Louisville, Kentucky.

RANGE. Southern Ohio, southern Indiana, southeastern Illinois, central Kentucky, Tennessee, and northern Alabama.

HABITAT. Found beneath logs and stones in woods, also in and about the mouths of caves. Common under stones, logs, and bark in the clearings at the foot of the Smoky Mountains at Elkmont, Tennessee. Blanchard (1925, p. 368) took a series of specimens under leaves and logs in a small, damp, wooded ravine near the flood-plain of the Ohio River in Henderson County, Kentucky, June 15.

SIZE. The largest specimen, a male, measured by Dunn (1926, p. 160) had a total length of $3\frac{7}{16}''$ (87 mm.), tail $1\frac{5}{8}''$ (41 mm.). The proportions of a male from Woodford County, Kentucky, are as follows: total length $3\frac{9}{32}''$ (83 mm.), tail $1\frac{17}{32}''$ (39 mm.); head length, $\frac{3}{8}''$ (10 mm.), width $\frac{7}{32}''$ (5.5 mm.). A female from the same locality measures: total length $3\frac{1}{4}''$ (82 mm.), tail $1\frac{15}{32}''$ (37 mm.); head measurements like those of male.

DESCRIPTION. This subspecies is smaller and more slender than its near relative, *Plethodon c. cinereus.* The head is slender, the sides behind the eyes slightly converging to the lateral extensions of the gular fold; snout short and bluntly pointed. Eyes strongly protuberant and relatively large. A sinuous groove from the posterior corner of the eye to the lateral extension of the gular fold; a short vertical groove from this to the angle of the jaw. Trunk slender, the sides well rounded, back and belly slightly flattened; back with a median depressed line from back of head onto basal third of tail. Tail nearly circular in cross section near base, gradually tapering to a slender tip. Costal grooves usually 17, counting

FIG. 63. *Plethodon cinereus dorsalis* Cope. (1) Adult female, actual length 3³⁄16″ (81 mm.). (2) Same, ventral view. (3) Same, from life. Woodford County, Kentucky.

1 each in axilla and groin, rarely 16, less rarely 18; 6–7 costal folds between toes of appressed limbs. Legs slender but well developed; toes 5–4, those of hind feet 1–5–2–3–4 or 4–3 in order of length from the shortest; fore feet 1–4–2–3, the innermost rudimentary; all toes slightly webbed at base. Tongue large, filling floor of mouth, the margins thin

and smooth. Vomerine teeth 5 or 6, in 2 short series beginning behind inner nares and curving gently inward and backward, separated by the width of a naris. Parasphenoid teeth in 2 closely approximated club-shaped patches, separated from the vomerine by about the length of a vomerine series.

COLOR. The pattern above consists of a zigzag band extending from the back of the head to the base of the tail, and usually of some shade of red. It is limited each side by a narrow dark line. The upper sides are brownish or reddish with small bluish-white points which give a shade only slightly darker than the dorsal band. In some, the sides may be quite uniformly grayish-white. The zigzag band may extend the entire length of the trunk and be sharply set off, or the zigzags may be limited to the anterior half of the trunk, the remaining part having straight sides. In some specimens the general color of the upper sides may be continued across the dorsum, to the almost complete obliteration of the dorsal band. In a few individuals, the dorsal band may have the sides straight, but in these the band is narrower than in specimens of *P. cinereus* of the same size. On the dorsal surface of the tail the band usually has straight sides; it may continue to the tip in light-colored individuals, or fade out on the basal third in dark specimens, the tip being silvery-gray. This species has a dark phase suggestive of that of *P. cinereus* but differing somewhat in color. Several specimens from Kentucky are dark reddish-brown above, fading on the sides to the mottled belly. The belly has a ground color bluish-white in life but changing to dull yellow in preservatives. It is strongly mottled with clusters of grayish chromatophores. The throat is usually lighter and there is a light area, sometimes red, on the sides of the trunk in front of each fore leg. The legs are colored above like the sides of the trunk, below like the venter.

SEXUAL DIFFERENCES. There are no well marked sexual differences. The tip of the lower jaw is slightly more pointed in the male, and the vent may have the sides thrown into inconspicuous ridges.

BREEDING. Little is known of the breeding habits of this species. Females taken April 24 in Woodford County, Kentucky, had a few ovarian eggs 1 mm. in diameter and many small ones. The larvae have not been described.

**FROSTED SALAMANDER.** *Plethodon clemsonae* Brimley. Fig. 64. Map 30.

TYPE LOCALITY. Jocassee, South Carolina.

RANGE. Known from the vicinity of Jocassee, South Carolina.

HABITAT. The type and several paratypes were taken at an elevation of

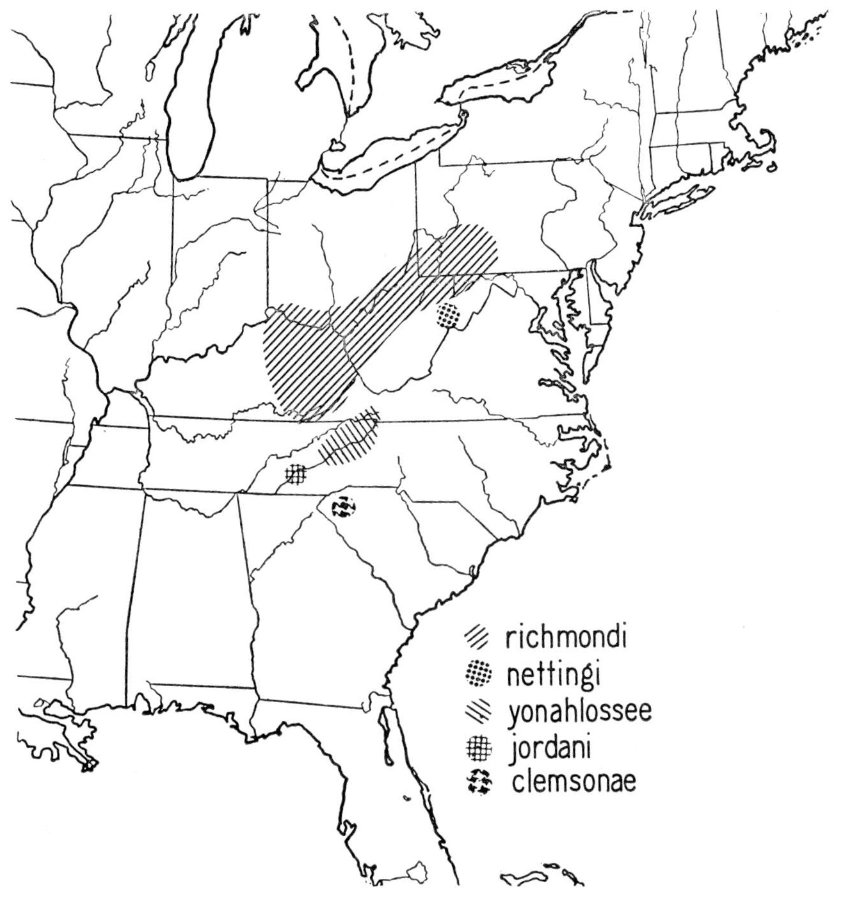


MAP 30.—Distribution of *Plethodon richmondi, P. nettingi, P. yonahlossee, P. jordani,* and *P. clemsonae.*

1200′ to 1500′ under logs in woods. On April 8, 1941, Arnold B. Grobman and M. B. Mittleman, collecting in the vicinity of the type locality, took 2 adults under logs on a heavily wooded slope.

SIZE. The largest specimen recorded, a male, has a total length of 5 13/16″ (149 mm.); tail 3 1/8″ (80 mm.). An adult female has the following proportions: total length 5 9/16″ (142 mm.), tail 2 23/32″ (69 mm.); head length 25/32″ (20 mm.), width 15/32″ (12 mm.).

DESCRIPTION. This species has the general proportions of *Plethodon metcalfi* but differs in its larger size and in color and pattern. The head is widest immediately behind the eyes, the sides gently converging to the lateral extensions of the gular fold and in front abruptly to the bluntly rounded snout. The eyes are large and strongly protuberant, about 1 1/4 in the snout; iris deep brown without brassy flecks. A deep sinuous groove from the posterior angle of the eye to the lateral extension of the gular fold; a short vertical groove from this line to the angle of the jaw. Trunk rounded above and on the sides and with a slightly impressed median dorsal line, best developed above the fore and hind legs. There are 15 costal grooves, counting 1 each in the axilla and groin, and 1–2 intercostal folds between the toes of the appressed limbs. Tail long, slender, nearly circular in section throughout its length, and comprising 47.46–53.69 per cent of the total length in the series measured. Legs stout, toes 5–4, those of the hind feet slightly webbed at base, 1–5–2–3–4 in order of length from the shortest; toes of the fore feet 1–4–2–3. Tongue large, nearly filling the floor of the mouth, free at the sides and behind, the narrow plicae parallel but not reaching the front margin. Vomerine teeth in series of 9–12, which arise behind the outer margin of the inner naris and extend inward and backward toward the mid-line, where they are separated by about the diameter of a naris. Parasphenoid teeth in 2 long patches narrowly separated from each other and from the vomerine by about twice the diameter of a naris.

COLOR. The general ground color above is black. Scattered over the dorsal surfaces of the head, trunk, legs, and basal part of the tail are many large, irregular, light gray, lichen-like patches which give these surfaces a frosted appearance. The light patches are somewhat concentrated along the sides of the head and in some individuals along the sides of the trunk. On the head, and at the base of the tail, the patches may be very lightly tinged with brassy. The throat is evenly and lightly

FIG. 64. *Plethodon clemsonae* Brimley. (1) Adult female, actual length 5⅜″ (136 mm.). (2) Same, dorsolateral view. (3) Same, ventral view. Jocassee, South Carolina.

pigmented so that the general effect is whitish. The belly and ventral surface of the tail black. The feet brownish. This species is easily distinguished from *P. glutinosus,* which is found in the same region, by the diffuse character of the light markings which are distributed abundantly over the dorsal surfaces, by the lighter throat, and by the fewer costal grooves.

BREEDING. Nothing is known of the breeding habits of this species and the eggs and larvae have not been described. In preservatives, the light markings largely disappear, and it is likely for this reason that the species was mistakenly relegated to the synonymy of *P. metcalfi.*

**DUNN'S SALAMANDER. WESTERN TAN-BACKED SALAMANDER.** *Plethodon dunni* Bishop. Fig. 65. Map 31.

TYPE LOCALITY. Near Portland, Oregon, in Clackamas County.

RANGE. West of the Cascade Mountains from Curry County, southwestern Oregon, to Wahkiakum and Cowlitz Counties, Washington.

HABITAT. On June 13, 1936, we collected at the type locality with Stanley G. Jewett, Jr. Here the salamanders were found beneath logs and bark and among fragments of rock on the sides of a deeply wooded, damp ravine. At Eagle Creek, near Estacada, Oregon, we took additional specimens from beneath fragments of shale on a clay bank bordering the stream. Fitch (1936, p. 637), collecting in a ravine on the north side of Rogue River, 11 miles above its mouth, dug 20 specimens from a rock slide.

SIZE. The adults attain a length of at least 131 mm. The proportions of the type specimen, a female, are as follows: total length, 131 mm., tail, 64.5 mm.; head length, 14 mm. An adult male 97 mm. long has the tail 56 mm. A female of the same length has the tail only 47 mm. long.

DESCRIPTION. *Plethodon dunni* is one of the larger species. The head in the female is moderately wide and with the sides converging only slightly behind. In the male, the head is distinctly widened above the angle of the jaws. The eyes are large and prominent and with the iris tinged with brassy. The trunk is roughly cylindrical, slightly flattened above and more strongly below. The tail is oval in cross section near the base but more strongly compressed distally. The costal grooves normally number 15 on each side, counting 1 each in the axilla and groin, but vary from 14 to 16. The legs are well developed and strong, the hind considerably larger than the fore. Toes 5–4, those of the hind feet 1–5–2–4–3 in order of length from the shortest; toes of the fore feet

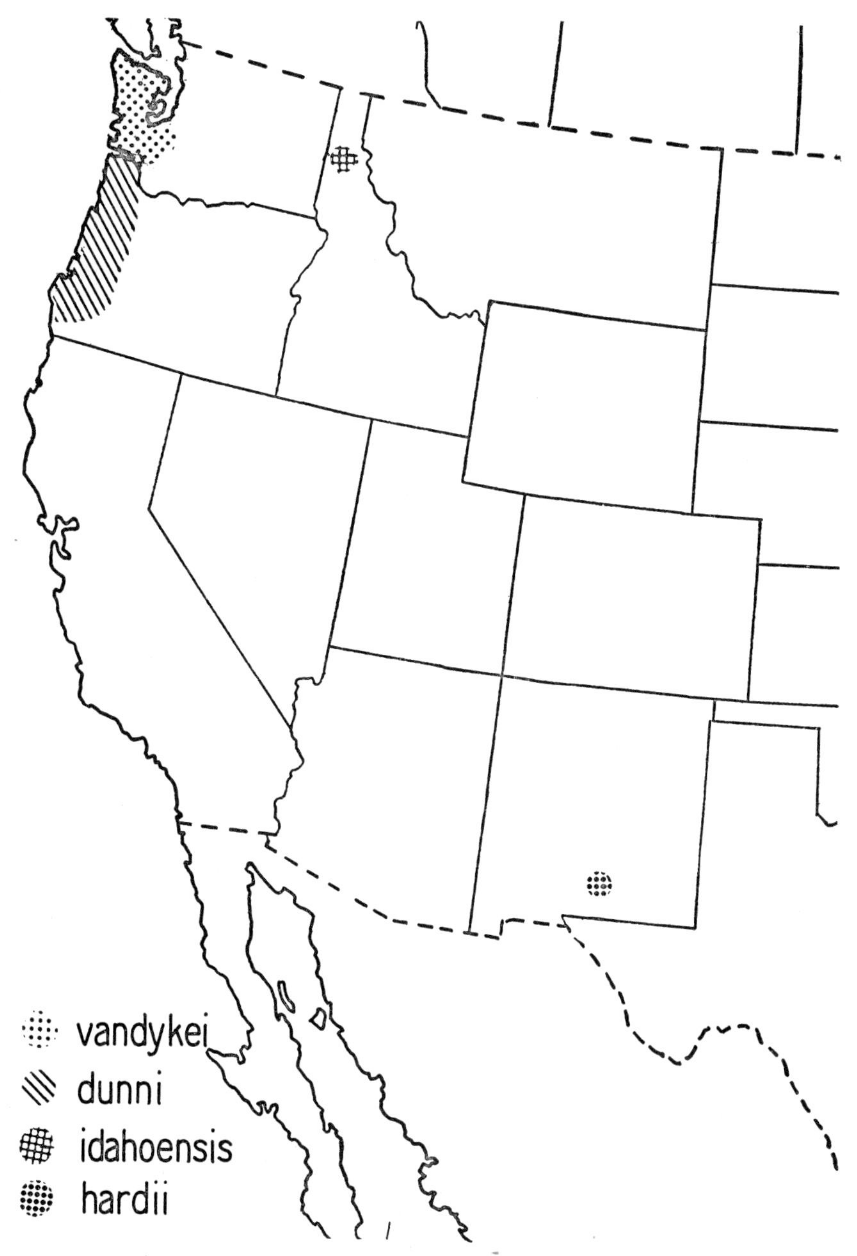


MAP 31.—Distribution of *Plethodon vandykei, P. dunni, P. idahoensis,* and *P. hardii.*

1-4-2-3. Tongue oval, moderately large, and with narrow plicae extending forward from the posterior field. Vomerine teeth in female in series of 8-13. The series originate behind the middle of the inner nares, curve inward and backward, separated posteriorly by about the diameter of an inner nares, and from the parasphenoids by a little more. Parasphenoid teeth in 2 elongate patches, wider behind than before, and narrowly separated throughout their length. In the male the vomerine teeth number 6-8 in each series.

COLOR. In life the ground color is best developed on the upper sides. Here the color is deep brown, fading to pale slate toward the belly. The broad dorsal light band extends from the end of the snout almost to the tip of the tail. It widens on the head back of the eyes, narrows on the neck, maintains its width to the base of the tail, then gradually tapers toward the tip. The dorsal band varies in color from dull tan to yellowish-green in different individuals and is flecked by irregular spots of brown or black. The dark flecks are fewer on the base of the tail above, so that here the band is brighter. The sides are mottled with irregular, dull, white and tan markings. The throat, belly, and lower surfaces of the tail and limbs are slate color, flecked with many small pale spots. In some individuals the pale fleckings are so numerous on the belly that this region appears more yellow than slate. The legs in adults are mottled above with tan and dark brown; in the young frequently yellowish-green, mottled with dusky.

SEXUAL DIFFERENCES. The sexual differences may be emphasized. The head of the male usually widens abruptly back of the eyes, the vomerine series are shorter, and the tail is proportionally longer. The tail in the female comprises 46.1-49.2 per cent of the total length, and in the male 49.1-57.7 per cent.

BREEDING. Little is known of the breeding habits. Slater (1939, p. 154) dissected a female 130 mm. long, taken May 1, 1937, at Portland, Oregon, and found 18 light-cream-colored eggs about 2 mm. in diameter and many smaller ones. Females taken June 1936, across the Willamette River from Portland, had ovarian eggs about 1.5 mm. in diameter. Six juvenile specimens taken at this time resemble the adults in color and pattern. They are nearly uniform in size, varying only from 40 to 42 mm. in total length.

There is a strong superficial resemblance between certain specimens of *Plethodon dunni* and tan-backed individuals of *Plethodon vehiculum*. The greater number of vomerine teeth, more strongly compressed tail, and more strongly mottled sides of *P. dunni* will distinguish it. The

FIG. 65. *Plethodon dunni* Bishop. (1) Adult female, actual length 5⅛″ (130 mm.). Eagle Creek, Clackamas County, Oregon. (2) Adult male, ventral view; actual length 4 7/16″ (113 mm.). Tidewater, Oregon.

mottling of the upper sides in *P. dunni* extends to the light dorsal band, the edges of which are somewhat irregular. In *P. vehiculum,* the upper sides next to the dorsal band are darker and less mottled, and the band itself has more regular edges. From *P. elongatus,* this species differs in its generally larger size, and in having fewer costal grooves, a brighter dorsal band, and the upper surface of the basal joints of the legs mottled with tan or yellowish-green like the dorsal stripe. In *P. elongatus,* the upper surface of the legs is dark like the upper sides.

Bishop, 1934, p. 169 (*Plethodon dunni*); Fitch, 1936, p. 637; Jewett, 1936, p. 71; Slater, 1939, p. 154.

**DEL NORTE SALAMANDER.** *Plethodon elongatus* Van Denburgh. Fig. 66. Map 32.

TYPE LOCALITY. Requa, Del Norte County, California.

RANGE. Known from Del Norte County, extreme northwestern California, and from Curry County, southwestern Oregon.

HABITAT. Wood (1934, p. 191) records specimens from a south-facing slope well covered by redwood, Sitka spruce, and madroño (Arbutus). Here the salamanders were found beneath rotting logs and slabs of bark. Slevin (1928, p. 57) reports this species from beneath decaying logs in damp woods. Fitch (1936, p. 638) found 2 specimens in a rock slide in company with *Plethodon dunni*. On June 6, 1936, Margaret Wright and I, collecting at Carpentersville, Oregon, found specimens beneath stones and heat-splintered rocks on a fairly dry, wooded hillside that had been burned over.

SIZE. Two large females are 115 and 113 mm. long respectively and in each the tail measures 54 mm. Nine of the 11 specimens captured had lost the tips of the tails and had partially regenerated them. The proportions of the largest female are as follows: total length 115 mm., tail 54 mm.; head length 14 mm., width 8 mm.; 8 intercostal folds between the appressed limbs.

DESCRIPTION. This is a slender species with a general resemblance to the eastern, *Plethodon cinereus*. The head is long, relatively narrow, and with the sides nearly parallel to the eyes. In front of the eyes the sides converge abruptly to the blunt snout. The eyes are of moderate size but prominent, with the iris above and below the pupil brassy to dark brown. The trunk is nearly cylindrical, the tail broadly oval in section on the basal half but becoming somewhat compressed distally. The costal grooves number 17 in the majority of specimens I have examined, with some individuals having 18, counting 1 each in the axilla and groin, and 6–7 intercostal folds between toes of appressed limbs. In specimens having 17 grooves, usually 2 run together in the groin. The legs are well developed but appear shorter than in most species of Plethodon because of the lengthening of the trunk. Toes 5–4, those of the hind feet with the innermost very short, others in order of length 5–2–3–4

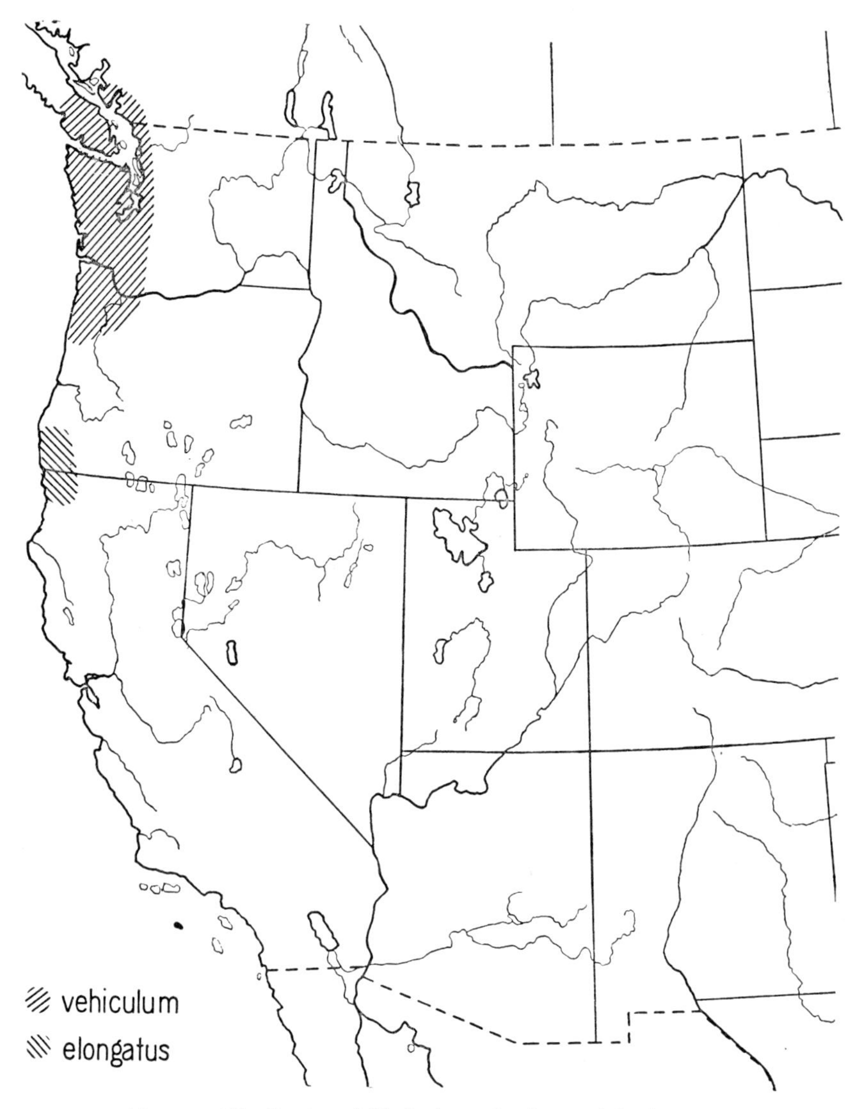


MAP 32.—Distribution of *Plethodon vehiculum* and *P. elongatus.*

or 2–5–(3–4); toes of the fore feet 1–4–2–3, the 2nd and 4th about equal. The tongue is oval in outline, fleshy, and fills the floor of the mouth. The vomerine teeth are in 2 short series of 4–6 teeth, widely diverging anteriorly, the outermost behind the inner nares, the innermost of each row separated by about half the distance between the inner nares. The

FIG. 66. *Plethodon elongatus* Van Denburgh. (1) Adult female, actual length 4½″ (115 mm.). (2) Same, ventral view. Carpentersville, Oregon.

parasphenoid teeth form a large wedge-shaped patch, widest behind, where a deep notch separates the right and left sides, confluent in front, separated from the vomerine by a distance nearly equal to that between the inner nares.

COLOR. The general color above is deep brown, almost black—black to casual inspection. There is a broad dorsal band which originates on the snout, passes between the eyes, broadens behind the eyes to the full width of the head, narrows to the neck, then continues the entire length of the trunk, gradually darkening toward the tip of the tail. In some individuals the dorsal band is scarcely evident, in others it may appear as a broad brown band strongly suffused with dusky, or a deep chestnut-brown. A half-grown individual has the lateral margins of the band tinged with orange, the center suffused with dusky. A small specimen 49 mm. long (tip of tail lost) has the margins of the band a bright reddish-orange, the center dusky. Storer (1926, p. 104), commenting on one of the original specimens described by Van Denburgh, remarks that the color of the dorsal band (in alcohol) is bright pink, clouded on the head and middorsal line with dark brown. The ground color on the lower sides fades to a dark slate. Venter slate. In some specimens the color on the sides and venter is quite uniform, but in others the sides of the trunk and head and the ventral surfaces of the trunk, throat, and legs are marked with many small, irregular, whitish flecks. The throat in front of the gular fold is dirty light brown. This species resembles *Plethodon vehiculum* in some of its color phases, but differs in eye color, in having the dorsal band more strongly diffused with dusky, and in its different costal-groove count. Furthermore, the upper surface of the basal joints of the legs in *P. vehiculum* are the same color as the dorsal band, while in *P. elongatus* the basal joints above are dark, like the upper sides.

BREEDING. Wood (1934, p. 191) collected in the region about Requa, Del Norte County, California, and on Novermber 4, 1933, he took an adult *P. elongatus* from beneath a piece of redwood bark, together with two small clusters of eggs. The eggs were aggregated by a small amount of viscous jelly-like material. The individual eggs were about 3 mm. in diameter, nearly spherical, but drawn out on one side to form a point. Some of the eggs contained embryos approximately 15 mm. long. The ovarian eggs in some of the females we collected June 6, 1936, had attained a diameter of 1.5 mm. and presumably would have been deposited sometime later in the fall or early winter.

Fitch, 1936, p. 638 (*Plethodon elongatus*); Grinnell and Camp, 1917, p. 134; Slevin, 1928, p. 55; 1934, pp. 46, 53; Storer, 1925, pp. 21, 104; Van Denburgh, 1916, p. 216; Wood, 1934, p. 191.

**SLIMY SALAMANDER.** *Plethodon glutinosus glutinosus* (Green). Fig. 67. Map 33.

TYPE LOCALITY. Probably in the vicinity of Princeton, New Jersey.

RANGE. New York south to northern half of Florida, the Gulf States

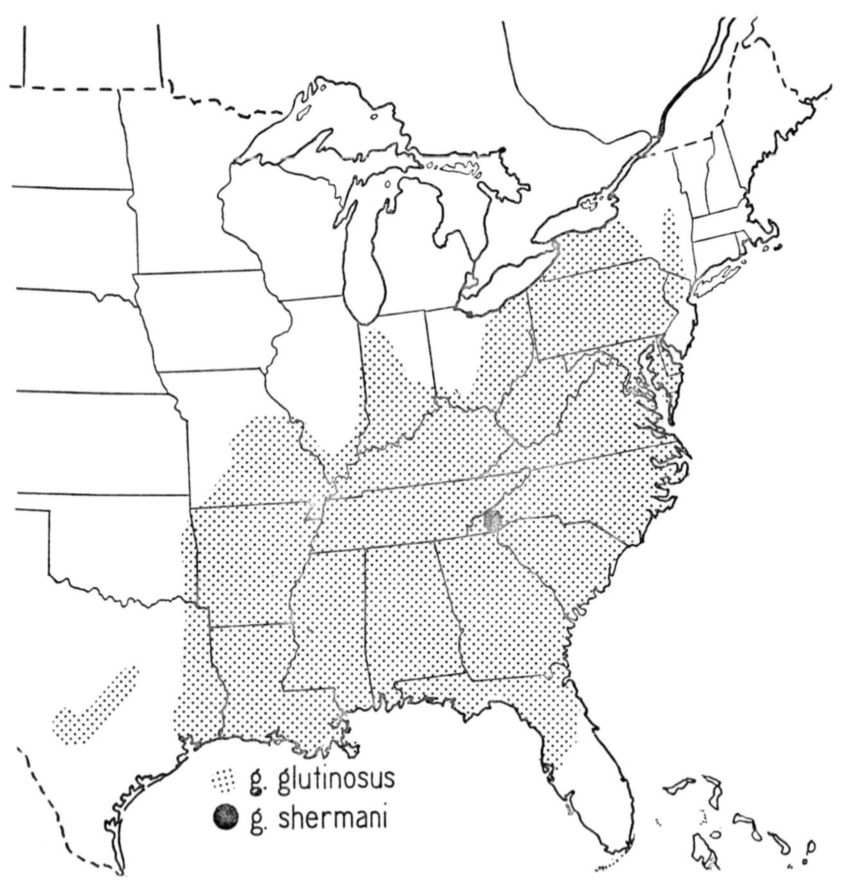


MAP 33.—Distribution of the subspecies of *Plethodon glutinosus.*

to Texas, northward through eastern Oklahoma, Missouri, southern Illinois to northern Indiana. This is a composite of several forms.

HABITAT. Usually found beneath logs and stones in woods, in crevices of shale banks, and along the sides of wooded gullies and ravines.

Frequently occurs under moist humus and is much more sensitive to dry air than the Red-backed salamander, burrowing deeply in dry seasons.

SIZE. This is one of the larger species of Plethodon, the males attaining a greater size than the females. Mature males vary from 4¾" to 7⁷⁄₁₆" (121–188 mm.) and average about 5⅞" (148 mm.). Adult females range in size from 4¼" to 6¹¹⁄₁₆" (105–168 mm.) and average 5¹¹⁄₁₆" (143.6 mm.).

DESCRIPTION. The Slimy salamander has the head only moderately broad, widest just behind the eyes, the sides nearly parallel or tapering slightly to the lateral extensions of the gular fold and more abruptly to the blunt snout. The trunk is rounded on the sides and slightly depressed above, the tail nearly circular in cross section and tapering to a slender tip. There are 16 costal grooves and the gular fold is developed as a transverse crease. The legs are strong, the toes 5–4, those of the fore feet 3–2–4–1 in order of length, rather short and slightly webbed at base; toes of the hind feet, 3–4 about equal, others 2–5–1. The vomerine teeth, 10–13 in each series, form somewhat irregular, backward-curving lines that are separated from each other and the parasphenoids, the latter usually in 2 patches. The tongue is broad and flat, nearly fills the floor of the mouth, and has the side margins free. The general ground color above is black or blue-black, with the belly light slate color and the throat and gular region very slightly paler and sometimes mottled. The distal third of the tail is dull brown in many adult specimens. Scattered over the sides of the body, and sometimes over the dorsal surface, except the tip of snout and tail, are many silvery-white flecks, variable in size and shape. In some individuals from the South, the light spots of the sides tend to run together and form incomplete bands; in others the spots may be nearly lacking. In the sexually mature males there is a well developed mental gland that appears as a rounded, slightly raised, light spot on the tip of the lower jaw, and the vent is lined with papillae.

BREEDING. Little is known of the breeding habits of this species. The eggs, accompanied in one instance by a female, were reported from Arkansas caves (Noble and Marshall, 1929). One lot containing 18, found August 17, was suspended from the vertical wall of a little pocket

FIG. 67. *Plethodon glutinosus glutinosus* (Green). (1) Adult female, actual length about 6⁵⁄₁₆″ (160 mm.). Woodville, Canandaigua Lake, New York. (2) Adult female, ventral view; actual length 5⅞″ (149 mm.). Allegany State Park, New York.

near the entrance, and the second, containing 10 well developed embryos, found Sept. 3, lay at the bottom of a little crevice about 200 feet from the entrance. The eggs were aggregated by a relatively thin, common envelope and were not stalked. The average egg has a diameter of 5.5 mm. and is creamy-white in color. In addition to the vitelline membrane there are 2 distinct envelopes which give a total diameter of 7.5 mm. An account of the discovery of a female with her eggs on Ice Mountain, Hampshire County, West Virginia, has recently been given by Fowler (1940, p. 133). On June 3, 1938, the female was discovered coiled about a group of about 15 eggs under a rotting stump on a wooded hillside. Nothing is known of the length of the period of incubation, the size at hatching, or the extent of the period the gills are retained. The well developed embryos resemble those of *Plethodon cinereus* in their possession of a dorsal stripe and broad, leaf-like gills arising from a common base. The dissection of females indicates that, farther north, the egg-laying season may be in the early spring months. Large ovarian eggs are found from October to December, and again, in some individuals, in May and June. Other females, taken as early as May and through the summer months, generally have very small eggs, indicating that the season's complement had been laid. Males taken in October have the vas deferens packed with sperm, suggesting an autumnal mating season, at least in the North. The Slimy salamander buries itself deeply at the approach of winter, and it is likely that the eggs will be found in the crevices of shale banks, or well below the surface of the ground, in situations which serve as hibernating retreats.

**RED-LEGGED SALAMANDER.** *Plethodon glutinosus shermani* Stejneger. Fig. 68. Map 33.

TYPE LOCALITY. Wayah Bald Mountain, North Carolina.

RANGE. Nantahala Mountains in southwestern North Carolina.

HABITAT. Wooded slopes and thickets of Rhododendron, hiding by day beneath old logs, planks, and bark, more rarely beneath stones and rocks, especially where moss-covered and somewhat imbedded. Sometimes associated with *Plethodon g. glutinosus.*

SIZE. My largest specimen, a female, has a total length of 6 1/16″ (153 mm.) and a tail length of 3 5/32″ (80 mm.). Five females average 5 1/8″ (130.2 mm.) and vary from 4 9/16″ to 5 15/16″ (115 to 147 mm.). The males I have measured average considerably smaller. The proportions of a typical female are as follows: total length 5 1/16" (128 mm.), tail 2 11/16" (68 mm.); head length 5/8" (15 mm.), width 13/32" (9.5 mm.). Male, total length 4 1/8" (115 mm.), tail 2 1/4" (57 mm.); head length 5/8" (15 mm.), width 3/8" (9 mm.).

DESCRIPTION. This is a slender, black salamander with red in irregular blotches on the legs. Viewed from above, the head is broadly oval in outline, widest immediately back of the eyes, tapering quite abruptly to the short snout and more gently to the lateral extensions of the gular fold. The eyes moderate but strongly protuberant, the posterior angle with a short vertical groove. Gular fold well developed and extending on the side of the neck, where it is met by a sinuous line from the posterior angle of the eye; a short vertical groove from this to the angle of the jaw. The trunk is rounded on the sides and above, slightly flattened ventrally. A median dorsal groove extends from the back of the head to the base of the tail. The tail is long, in some individuals comprising 56 per cent of the total length, nearly circular in section and gently tapering from base to tip. Costal grooves 15 or 16, counting 1 each in axilla and groin, where there is always space for one whether or not it is developed; 1–2 intercostal spaces between toes of appressed limbs. Legs moderately stout, toes 5–4, those of the hind feet 1–5–2–3–4 or 4–3, the innermost very short, all very slightly webbed at base; toes of the fore feet, 1–4–2–3. Tongue fairly large, nearly filling the floor of the mouth, the margins free, thin, and smooth. Vomerine teeth somewhat variable, in the females 9–9 to 9–11 or 11–12; in the males usually fewer, 7–8 to 8–9. The series begin outside the outer edge of the inner nares, curve gently inward and backward, and are separated at the mid-line by a distance about equal to the width of a naris. Parasphenoid teeth in 2 narrowly separated club-shaped patches and distant from the vomerine by about twice the diameter of an inner naris.

COLOR. In the living, typically colored individual the ground color above is a lustrous blue-black, with the tail a little lighter, tending toward slate. Lower sides light gray. In some individuals the ground color may be bluish-brown. The legs above are mottled with bright red in irregular blotches and varying much in extent; in some individuals reduced to a few scattered flecks. Feet generally lighter, brownish or tan. The throat is light, but very finely speckled with small black chromatophores, slightly darker on the sides. The belly and ventral

FIG. 68. *Plethodon glutinosus shermani* Stejneger. (1) Adult, actual length about 5⅛″ (131 mm.). (2) Same, lateral view. (3) Same, dorsolateral view. Nantahala Gap, North Carolina.

surface of the tail may be pale bluish-white or pale gray, sometimes blotched irregularly on the sides between the fore and hind legs. Some individuals may have the throat and feet flesh color, or the feet may be yellowish-gray.

SEXUAL DIFFERENCES. Sexual differences are not well marked. The sexually mature male has the tip of the lower jaw slightly pointed, and there is a well developed mental gland; the sides of the vent are raised, thickened, and crossed anteriorly by a number of narrow grooves. Posteriorly the sides of the vent are developed into a pair of flat, fleshy lobes which may have the hind margins free. The tip of the lower jaw in the female is evenly rounded and the sides of vent are not produced behind into fleshy lobes.

BREEDING. Nothing is known of the breeding habits. A large female, taken in October 1926, has ovarian eggs 2 mm. in diameter, while those collected in April 1938 have a considerable number of very small eggs.

**SACRAMENTO MOUNTAINS SALAMANDER.** *Plethodon hardii* Taylor. Fig. 69. Map 31.

TYPE LOCALITY. Sacramento Mountains, Cloudcroft, New Mexico.

RANGE. Known only from the vicinity of the type locality, altitude about 9000′, and from Aqua Chiquite, Otero County, New Mexico.

HABITAT. The type was found under the bark of a rotten pine log in heavy pine forest. On July 17, 1942, Dr. and Mrs. A. H. Wright collected, near Cloudcroft, a fine series of juveniles and adults in and beneath old logs and bark and under the bark at the base of stumps. At this locality the specimens were mostly in shady, moist situations beneath pines and spruces.

SIZE. The average length of 10 adults of both sexes from Aqua Chiquite, New Mexico, is $3\frac{1}{32}''$ (77 mm.), the extremes $2\frac{5}{8}''$ (67 mm.) and $3\frac{11}{16}''$ (94 mm.). The proportions of an adult male are as follows: total length $3\frac{11}{16}''$ (94 mm.), tail $1\frac{3}{4}''$ (45 mm.); head length $\frac{15}{32}''$ (12 mm.), width $\frac{5}{16}''$ (8 mm.). The measurements of a female are: total length $3\frac{1}{8}''$ (80 mm.), tail $1\frac{7}{16}''$ (37 mm.); head length $\frac{13}{32}''$ (11 mm.), width $\frac{9}{32}''$ (7 mm.).

DESCRIPTION. This is a fairly slender species of moderate size. The head is large and strongly depressed. Back of the eyes the sides are nearly parallel or slightly converging to the lateral extensions of the prominent gular fold, in front tapering rather abruptly to the truncated snout. The eye is large and strongly protuberant, the horizontal diameter about equal the snout, the iris with bright coppery tinges. A sinuous impressed line from the posterior angle of the eye to the lateral extension of the gular fold; a short vertical groove from this line to the

FIG. 69. *Plethodon hardii* Taylor. (1) Adult female, actual length 3 9/16" (91 mm.). (2) Same, dorsal view. (3) Same, ventral view. Cloudcroft, New Mexico (type locality). [Anna A. and A. H. Wright, collectors, July 17, 1942.]

angle of the jaw. Trunk slender, with 14–15 costal grooves and 3½ or more costal grooves between the toes of the appressed limbs. Tail slender, and comprising 42–53 per cent of the total length. Legs well developed. Toes 5-4, those of the hind feet 1–5–2–4–3 in order of length from the shortest, webbed at base, flattened and square-tipped; toes of the fore feet 1-4-2-3, webbed at base, first rudimentary. Tongue elongate oval, free at the sides and behind. Vomerine teeth in transverse series of 6-10 (the average about 8) which arise behind the middle of the inner nares and slant inward toward the mid-line, where they are separated by less than the width of a naris. Parasphenoid teeth in elongate patches, usually narrowly separated but occasionally contiguous anteriorly, and separated from the vomerine by about twice the diameter of a naris. The sexes may be distinguished by the form of the vent, that of the male having the margins crossed by transverse grooves and provided internally with many slender papillae; the vent of the female is a simple slit. An adult male has a circular mental gland.

COLOR. The ground color above is nearly black, mostly obscured by overlying pigment which varies in color from dull clay to grayish-tan and bronze. The head is usually more uniformly bronze, suffused with dusky, and the tail brighter than the trunk. The overlying pigment may be disposed in large blotches with irregular edges or form a nearly uniform, broad, dorsal stripe. The tail may be uniformly colored above or irregularly invaded by the dark gray of the sides. On some young individuals, the dorsal color is disposed in a fairly regular band, limited either side by a darker line. The upper sides are irregularly blotched and mottled between the conspicuous black costal grooves, the lower sides more sparsely spotted over a bluish-black background. The belly is slate, the throat very light but lightly mottled, the lower surface of the tail flesh to light brown. The legs are colored above like the back, below like the lower surface of the tail. The costal grooves are very conspicuous in life, with extensions dorsally and ventrally nearly or quite to the mid-line. The sides of the tail, on the basal part, are marked with impressed vertical grooves.

BREEDING. Nothing is known of the breeding habits, eggs, or recently hatched young.

**COEUR D'ALENE SALAMANDER.** *Plethodon idahoensis* Slater and Slipp. Fig. 70. Map 31.

TYPE LOCALITY. Coeur d'Alene Lake, Kootenai County, Idaho.

RANGE. Known only from the type locality.

HABITAT. "All five of the specimens were taken at the foot of a high cut bank above the road which follows the edge of Coeur d'Alene

FIG. 70. *Plethodon idahoensis* Slater and Slipp. (1) Adult, dorsal view; about natural size. (2) Same, lateral view. Coeur d'Alene Lake, Idaho. [Photographs by courtesy of J. R. Slater.]

Lake, the two adults from a rock and dirt talus and the three juveniles from the gravel floor at the entrance of a very wet mine .6 of a mile distant. . . . The forest in this vicinity has a humid aspect, containing much douglas fir and dwarf maple, while a few miles westward the yellow pine and open plains of the arid transition predominate" (Slater and Slipp, 1940, p. 43).

SIZE. This is a species of moderate size, the 2 adults having measurements as follows: Type, male (U.S.N.Mus. No. 110504), total length 4″ (101.5 mm.), tail 1 25/32″ (46 mm.); head length 15/32″ (12 mm.), width 5/16″ (8 mm.). Adult female, No. 2711 in the collection of the College of

Puget Sound, total length 3 1/16" (78 mm.), tail 1 11/32" (34 mm.); head length 3/8" (10 mm.), width 9/32" (6.4 mm.). The smallest specimen secured had a total length of 1 11/32" (34 mm.).

DESCRIPTION. The head is broad and depressed, the sides back of the eyes slightly converging to the lateral extensions of the gular fold, in front more abruptly to the squarely truncated snout, which is slightly swollen in the region of the nasolabial grooves. Eyes large and very strongly protuberant, the horizontal diameter once in the snout; iris brown, tinged with gold or brassy. Eye limited behind by a vertical fold; an impressed line from the posterior angle of the eye to the lateral extension of the gular fold, and a vertical groove from this line to the angle of the jaw. Trunk moderately stout, rounded on the sides and above, with an impressed median line. Costal grooves 13, or 14 counting 2 that run together in the groin, and about 3 intercostal folds between the toes of the appressed limbs. Tail subquadrate in section at base, long, slender, tapering, and slightly compressed distally. Legs stout. Toes 5–4, those of the hind feet 1–5–2–(4–3) in order of length from the shortest, first rudimentary, all slightly webbed at base. Toes of the fore feet 1–4–2–3, the 1st short, others slightly webbed at base. Tongue elongate oval, free at the sides and behind, thin at margins. Vomerine teeth in short series of 9–10 that arise behind the inner margin of the inner naris and slant obliquely backward and inward toward the midline, where they are separated by about the diameter of a naris. Parasphenoid teeth in 2 long slender patches, narrowly separated from one another and from the vomerine by about three times the diameter of a naris.

COLOR. The ground color above is black. There is a broad, middorsal, yellow band that arises on the snout and continues to the tip of the tail. The yellow band narrows to a blunt point on the snout, widens to include the upper eyelids, and narrows at the neck. It is strongly suffused with brownish on the head, becomes bright yellow on the tail base, and is again suffused with brownish toward the tip. The sides of the dorsal band are slightly irregular. The sides and venter are black, with lighter areas on the throat, sides of the head, and tips of the toes, and a general scattering of small gray flecks.

BREEDING. Nothing is known of the breeding habits, and the eggs and recently hatched young are unknown.

**RED-CHEEKED SALAMANDER. RED-NECKED SALAMANDER.** *Plethodon jordani* Blatchley. Figs. 49e, 71. Map 30.

TYPE LOCALITY. Mt. Collins and Indian Pass, Sevier County, Tennessee.

RANGE. The Great Smoky Mountains in western North Carolina and eastern Tennessee.

HABITAT. This species is abundant in and beneath old rotten, moss-covered logs and bark and, more rarely, beneath stones and slabs of rock, on the heavily forested slopes of the Smoky Mountains. On April 18, 1938, we collected at the type locality, Indian Pass, Sevier County, Tennessee, with Mr. Willis King, and found this species to be extremely abundant. We have also taken this species in numbers along the heavily shaded banks of Mill Creek on the slopes of Mt. Le Conte between 2500–5500′.

SIZE. A series of 10 adults of both sexes from Indian Gap, Tennessee, average $4\frac{7}{16}''$ (112.5 mm.). The largest specimen taken, a female, has a total length of $5\frac{5}{16}''$ (135 mm.), tail $2\frac{9}{16}''$ (65 mm.); head length $\frac{23}{32}''$ (18 mm.), width $\frac{3}{8}''$ (10 mm.). The proportions of a male are as follows: total length $4\frac{11}{16}''$ (119 mm.), tail $2\frac{3}{8}''$ (60.5 mm.); head length $\frac{9}{16}''$ (14 mm.), width $\frac{3}{8}''$ (9 mm.).

DESCRIPTION. A distinctively marked and easily recognized salamander, apparently limited to the Great Smoky Mountains. The head is widest immediately back of the eyes, where it appears slightly swollen. Behind this point the sides converge very slightly to the lateral extensions of the gular fold; in front of the eyes, the sides taper abruptly to the rather pointed snout. Eyes large and protuberant, posterior angle limited by a short vertical fold. Gular fold well developed; a deep sinuous groove from the posterior angle of the eye to the lateral extension of the gular fold; a short vertical groove from this to the angle of the jaw.

Trunk moderately stout, well rounded on the sides and above, slightly flattened beneath. Tail long, averaging about 50 per cent of the total length, nearly circular in cross section at base, evenly tapering to the slender tip. Costal grooves 15–16, counting 1 in the axilla, which is usually poorly developed, and 1 in the groin, where frequently 2 run together; 2–3 intercostal spaces between toes of appressed limbs. Limbs stout, larger than in most species of Plethodon. Toes 5–4, those of the hind feet 1–5–2–4–3 in order of length from the shortest, the innermost rudimentary; toes of fore feet 1–2–4–3 or 1–4–2–3, not webbed at base. Tongue fairly large, the margins thin and smooth. Vomerine teeth 7–10, the series arising behind or just outside the outer edge of the nares, curving gently inward and backward toward the mid-line, where they are separated by about twice the diameter of an inner naris. Parasphenoid teeth in a large patch narrower in front than behind, where it is broadly rounded, and with a narrow, elongate area free from teeth through the center.

COLOR. The general color above in life is a deep blue-black, with the legs and feet a little lighter, grayish-brown. The head in some individuals is a deep brown. Cheeks dull pink to bright red, the area limited behind by the lateral extensions of the gular fold, above by the impressed line running from the eye to the gular fold, in front by the eye, and below, anteriorly, by the lower margin of the upper jaw; posteriorly extending varying distances on the throat. The extreme lower sides and belly are dull bluish-gray; the throat, fading anteriorly to dull flesh color. Rarely the red may be lacking on the cheeks or greatly decreased in extent. In a few individuals, of the several hundred I have collected, the red completely encircled the throat behind the angle of the jaw. Other individuals have the red areas invaded by varying amounts of black pigment. The lower surfaces of the feet and legs are colored generally like the belly. Tail beneath, usually darker than belly, but lighter than dorsal surface. A single adult animal from Indian Gap, Tennessee, taken April 18, 1938, has the red encroaching broadly on each side of the throat, a few spots on the fore legs, and a series of dull red, paired

FIG. 71. *Plethodon jordani* Blatchley. (1) Adult, actual length about 4⁵⁄₁₆" (110 mm.). (2) Same, lateral view. (3) Same, dorsolateral view. Mount Le Conte, Tennessee.

spots on the dorsum of the trunk. My smallest specimen, 29 mm. long (tip of tail lost), is colored like the adults. Bailey (1937, p. 6) mentions two small specimens ("snout to vent, 22.3 and 24.0 mm.") having faint dorsal areas suggestive of juvenile dorsal spots. The sexes may be distinguished in adults by the presence, in the males, of a well developed mental gland, circular in outline, and by the slightly more pointed lower jaw.

BREEDING. Nothing has been published on the breeding habits of this species. Female taken at Indian Pass, Tennessee, in April 1938, had ovarian eggs about 1 mm. in diameter, and the males had large testes. The eggs and early stages have not been described.

**METCALF'S SALAMANDER.** *Plethodon metcalfi* Brimley. Fig. 72. Map 35.

TYPE LOCALITY. Sunburst, Haywood County, and Grandfather Mountain, North Carolina.

RANGE. Ranges of the southern Blue Ridge Mountains (except the Great Smokies and Nantahalas) from southwestern Virginia through western North Carolina, northern South Carolina, to northern Georgia; Roan Mountain, Tennessee, and possibly northeastern Alabama.

HABITAT. Common on wooded slopes, where they hide by day beneath old logs, bark, stones, and slabs of rock. On Big Pisgah, Big Bald, and Frying Pan Mountains, in western North Carolina, at localities where they were most abundant, the forests were mainly of oak and chestnut, with an undergrowth of *Cornus* and *Rhododendron*. Few specimens from above limits of large tree growth.

SIZE. Measurements given by Bailey (1937, p. 6) indicate that populations from high elevations, 4000–5800′, average considerably smaller than those from elevations of 2300–3600′. Fifteen males from Blackrock, North Carolina, averaged in total length 3⅞″ (98.7 mm.); 16 females, 4 3/16″ (106.4 mm.). Eleven males from Swannanoa, North Carolina, averaged 5¾″ (146.7 mm.); 12 females, 5⅝″ (142.8 mm.). The proportions of a male from the east side of Grandfather Mountain, North Carolina, are as follows: total length 5 3/16″ (132 mm.), tail 2⅞″ (73 mm.); head length 19/32″ (15 mm.), width ⅜″ (9 mm.). A female of the same total length from Mt. Pisgah, North Carolina, has a tail length of 2⅝″ (67 mm.); head length ⅝″ (16 mm.), width ⅜″ (9 mm.).

DESCRIPTION. This species is related to *P. jordani*, from which it differs in the absence of red cheek patches and in generally lighter color, to

*P. shermani,* from which it differs in the absence of red legs, and to *P. glutinosus.* The ground color is lighter than that of *P. glutinosus,* but occasionally specimens of *P. metcalfi* have scattered light spots quite suggestive of those so characteristic of the Slimy salamander. The head is rather broad and somewhat depressed, widest immediately back of the eyes, the sides gently converging behind; the snout short, bluntly

FIG. 72. *Plethodon metcalfi* Brimley. (1) Adult female, actual length 4¹³⁄₁₆″ (122 mm.). Frying Pan Gap near Mount Pisgah, North Carolina. (2) Adult male, actual length 4¹⁄₁₆″ (103 mm.). Grandfather Mountain, North Carolina.

pointed, and somewhat swollen in the region of the nasolabial grooves. Eyes large and protuberant, posterior angle limited by a vertical flap under which both eyelids fit. Gular fold well developed, the neck noticeably smaller than the head. A deep, slightly sinuous groove from the posterior angle of the eye to lateral extensions of the gular fold; a short curved, vertical groove from this to the angle of the jaw. Trunk well rounded on the sides and above, with a slight median impressed line, below flattened. Tail long, normally in adults longer than head and trunk, nearly circular in cross section at base, evenly tapering to the slender tip. Costal grooves 15–16, counting 1 each in axilla and groin, the usual number 16. About 2 intercostal spaces between the toes of appressed limbs in adults. Limbs stout; toes 5–4, slightly webbed at base, those of the hind feet 1–5–2–4–3 or 3–4 in order of length from the shortest; fore feet, 1–2–4–3. Tongue moderate, does not completely fill the floor of the mouth. Vomerine teeth 6–9, the series originating at the outer edge of the inner nares, curving inward and backward toward the mid-line, where they are separated by a little more than the diameter of a naris. Parasphenoid teeth in some forming 2 narrowly separated club-shaped patches; in other individuals the patches are united in front and narrowly separated behind.

COLOR. In life the general ground color varies from bluish-brown to purplish-gray above and is lighter than that of *P. jordani*, *P. g. shermani*, or *P. g. glutinosus*. The upper surface of the legs is lighter, dull grayish-white, and often there is a slightly lighter brownish area on the head between the eyes. The upper sides are colored like the back, the lower sides lighter, often bluish-white anteriorly and dull dirty white toward the hind legs. The belly is dull bluish or grayish. In some specimens there are tinges of lavender and flesh on the legs, the sides of the throat and just back of the gular fold. The throat is light, grayish-white with fleshy tinges around the margin of the lower jaw. The sexes may be distinguished by the presence, in the adult male, of a prominent mental gland and by the slightly more pointed lower jaw.

BREEDING. Nothing is known of the breeding habits of this species. The ovarian eggs are small in October and only slightly larger in April, so it is likely that egg-laying may take place in late summer.

**CHEAT MOUNTAIN SALAMANDER.** *Plethodon nettingi* Green. Fig. 73. Map 30.

TYPE LOCALITY. Barton Knob, Randolph County, West Virginia.

RANGE. Known only from the vicinity of Cheat Bridge and Barton Knob, Randolph County, West Virginia.

HABITAT. Mainly confined to cool, shady ravines, where it is often to be found hiding in moist, decaying logs and beneath rocks. In these situations it is often associated with *Plethodon cinereus, P. wehrlei* and *Desmognathus o. ochrophæus* (Green, 1938).

SIZE. This is one of the smaller species of Plethodon, the adults averag-

FIG. 73. *Plethodon nettingi* Green. (1) Adult female, actual length 3⅛″ (79 mm.). (2) Same, ventral view. Upper Cheat Bridge, Randolph County, West Virginia.

ing about 3″ (76 mm.) in total length. A mature female from Cheat Bridge, West Virginia, is proportioned as follows: total length 3 17/32″ (90 mm.), tail 1 11/16″ (43 mm.); head length 7/16″ (10.5 mm.), width ¼″ (6 mm.).

DESCRIPTION. A small Plethodon with a very dark ground color and, in life, with many small brassy flecks scattered over the dorsal surfaces.

The head is somewhat depressed, widest immediately back of the eyes, tapering behind very slightly to the lateral extensions of the gular fold, and more abruptly to the bluntly pointed snout. The eyes are small but protuberant, the iris tinged with gold. Gular fold well developed and without pigment in the crease. A deep groove from the posterior angle of the eye to the lateral extension of the gular fold; a short vertical groove from this to the angle of the mouth. Trunk slender, rounded above and on the sides, and with a distinct median dorsal impressed line; costal grooves 17–19, counting 1 each in the axilla and groin, the usual number 18; 5½–6 intercostal spaces between the toes of the appressed limbs. Tail circular in cross section near base, evenly tapering to the slender and sometimes slightly compressed tip. Legs of moderate size, the hind a little larger in proportion to the fore than is usual in *Plethodon;* toes 5–4, those of the hind feet 1–5–2–4–3 in order of length from the shortest; fore feet 1–4–2–3, the 2nd and 4th about equal, all slightly webbed at base. The tongue fills the floor of the mouth, its margins thin and smooth. Vomerine teeth few and rather widely separated, usually 5–7 in each series, which arise behind the middle of the inner nares and extend inward and sharply backward, and are separated mesally by about the diameter of an inner naris. Parasphenoid teeth few and rather large, in 2 elongate club-shaped patches narrowly separated.

COLOR. The general ground color is deep brown, almost black, the head sometimes slightly lighter. Scattered over the dorsal surface of the head, trunk, and tail are many fine pale-brassy flecks which are more highly concentrated on the head and fade out on the distal third of the tail. The brassy flecks are continued sparsely on the sides of the head and on the upper sides of the trunk to the level of the upper surface of the legs, but are mainly limited to the dorsal surface of the tail. Legs and toes are a little lighter, with a few scattered flecks. The belly and the ventral surface of the tail, dark slate; the throat definitely lighter, grayish. Under low magnification the belly, and particularly the throat, are seen to have many small pigment-free spots.

BREEDING. A cluster of 8 eggs was collected with 2 adults in a decayed hemlock log near Cheat Bridge on July 15, 1936, and reported by Green (1938, p. 297). Nine eggs with the attendant female were forwarded to me by Mr. M. Graham Netting from Upper Cheat Bridge, where they were found June 3, 1940. The eggs were held closely together by the adhesiveness of the outer envelopes and formed a compact mass. Individual eggs are about 4 mm. in diameter and are pale yellow, without pigment.

**OUACHITA SALAMANDER.** *Plethodon ouachitae* Dunn and Heinze. Fig. 74. Map 34.

TYPE LOCALITY. Rich Mountain, Ouachita National Forest, Polk County, Arkansas.

RANGE. Known from Rich Mountain, east of Page, Le Flore County, Oklahoma, and the Ouachita National Forest on Rich Mountain in Polk County, Arkansas. Doubtfully recorded by Burt (1935, p. 321) from 6 mi. northwest of the settlement of Rich Mountain.

HABITAT. The locality where most of the specimens have been secured is a moist, shaded ravine having a northern exposure. Rocks of boulder type are partially imbedded in the soil or piled one upon another, leaving deep crevices into which the salamanders may retreat.

SIZE. The series of specimens used by Dunn and Heinze (1933, p. 122) in their description of the species included males varying in length from $2\frac{15}{16}''$ (71 mm.) to $4\frac{3}{8}''$ (111 mm.) and females from $3\frac{9}{16}''$ (91 mm.) to $4\frac{11}{32}''$ (110 mm.). My largest specimen, a female, has the following proportions: total length $4\frac{3}{4}''$ (121 mm.), tail $2\frac{13}{32}''$ (61 mm.); head length $\frac{5}{8}''$ (15 mm.), width $\frac{3}{8}''$ (9 mm.).

DESCRIPTION. This species is closely allied to the group which includes *Plethodon glutinosus, P. yonahlossee,* and *P. wehrlei.* It is a slender species of moderate size. The somewhat depressed head has the sides back of the eyes nearly parallel or very slightly converging; in front the sides taper abruptly to the pointed snout. Eyes large and prominent, the lids limited behind by a short vertical fold of skin. From the posterior angle of the eye a sinuous line to the lateral extension of the gular fold;

a short vertical groove from this line to the angle of the jaw. The sides and dorsum of the trunk are rounded, but there is little evidence of a median dorsal impressed line; the belly is slightly flattened. The tail

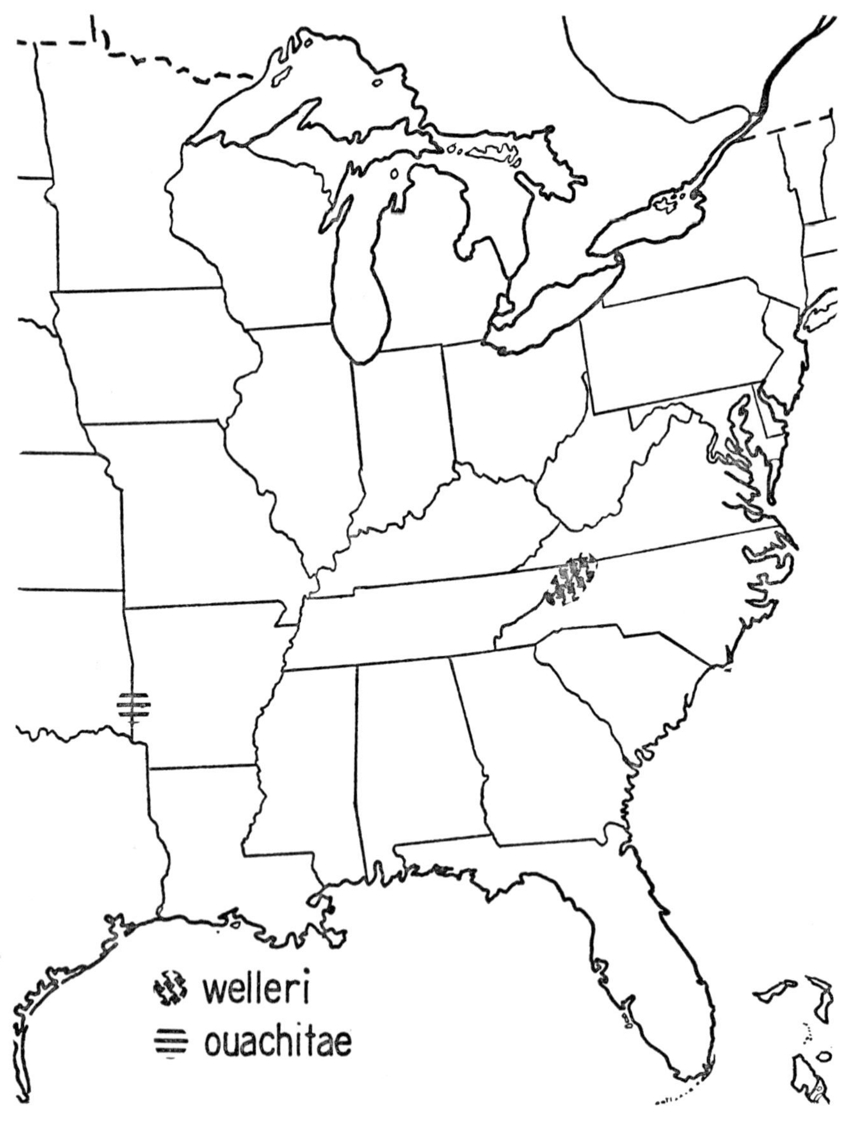


MAP 34.—Distribution of *Plethodon welleri* and *P. ouachitae.*

is nearly circular in cross section, slightly flattened, and strongly glandular at the base above. Costal grooves 15–17, the usual count being 16 when 1 each is counted in the axilla and groin. Usually 1–3 intercostal spaces between toes of appressed limbs. Legs moderate, toes short, thick, bluntly tipped, 5–4, those of the hind feet 1–2–5–3–4 in order of length

FIG. 74. *Plethodon ouachitae* Dunn and Heinze. (1) Adult female, actual length 4 11/16″ (120 mm.). (2) Same, dorsal view. Rich Mountain, Arkansas. Topotype.

from the shortest, the innermost rudimentary; toes of the fore feet 1–2–4–3. Tongue moderate, the sides thin, free, and smooth. Vomerine teeth 7–9 to 9–13 in each series, which arise just outside the outer margin of the inner naris and curve inward and backward and are separated at the mid-line by about the diameter of an inner naris and from the parasphenoid by a little more. Parasphenoid teeth usually in a broad patch only slightly notched anteriorly; in some the anterior halves separated by a narrow groove.

COLOR. The color, in life, of the head and trunk above is chestnut-brown overlying a generally darker ground color. In this dorsal area the numerous small and crowded light spots are bright brassy or golden. On the dorsal surface of the tail, where the ground color is generally darker, the light spots are silvery-white and are fewer and larger, the distal third usually nearly free. The upper sides are chestnut varied with black, the sides of the tail dull brown and with few light markings. On the lower half of the sides, between the fore and hind legs, there is a broad white band with an irregular upper edge. This light band is continued on the sides of the head, where it is more or less broken. The throat is light pink, nearly white, but mottled to a greater or less extent by aggregations of small dark pigment flecks. The lower surface of the fore limbs is usually mottled, that of the hind legs and tail black. The legs are spotted above with white on a brown or black ground.

Preserved specimens lose this striking color pattern and the general impression is that of a blackish salamander having a light throat and a few light spots along the lower sides, quite resembling some preserved specimens of *Plethodon g. glutinosus.* Some preserved specimens have the distal half of the tail a dull brown.

BREEDING. Nothing has been published on the breeding habits of this species, but a female sent me by Mr. Albert A. Heinze, taken May 1, 1933, at Rich Mountain, Polk County, Arkansas, has the ovarian eggs averaging about 3 mm. in diameter and apparently ready to be deposited.

**RAVINE SALAMANDER.** *Plethodon richmondi* Netting and Mittleman. Fig. 75. Map 30.

TYPE LOCALITY. Huntington, Cabell County, West Virginia.

RANGE. From Everett, Bedford, and Pittsburgh, Allegheny County, Pennsylvania, south to Wayne County, West Virginia, Carter and Fayette Counties, Kentucky, and southeastern Ohio west to Cincinnati.

HABITAT. "Exhibits a marked preference for the slopes of valleys and ravines, never occurring on hilltops, and only rarely on the valley floors" (Netting and Mittleman, 1938, p. 291).

SIZE. Netting and Mittleman (*ibid.*) give the maximum length as 130 mm., with the average of a series, selected at random, as 105 mm. A female sent to me by Mr. Neil D. Richmond has body proportions as

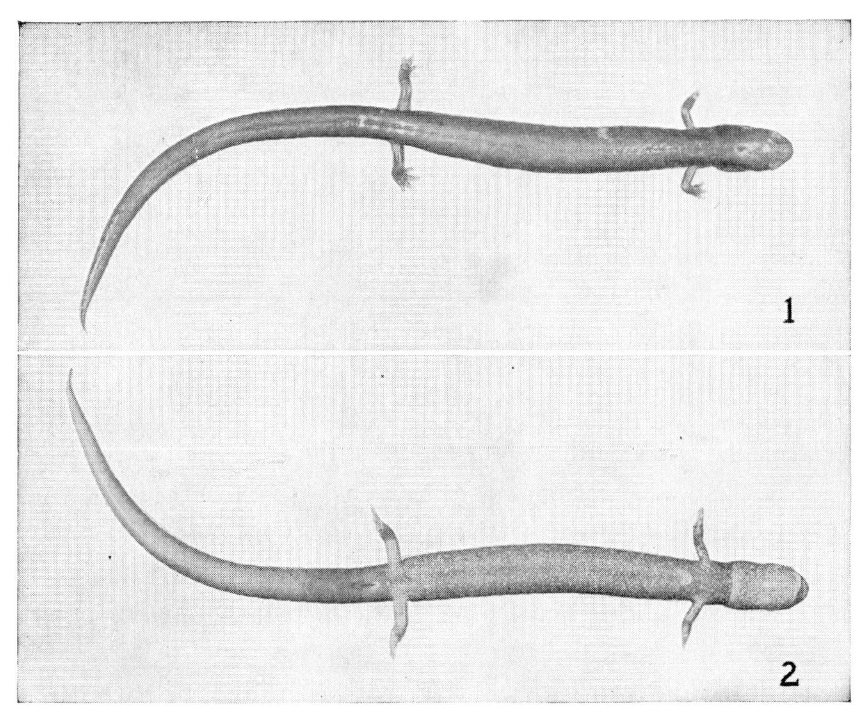


FIG. 75. *Plethodon richmondi* Netting and Mittleman. (1) Adult female, actual length 4 7/16″ (112 mm.). (2) Same, ventral view. Bristol, West Virginia. [Photographs from a preserved specimen.]

follows: total length 4¾″ (120 mm.), tail 2⅜″ (60 mm.); head length 15/32″ (12 mm.), width ¼″ (6 mm.).

DESCRIPTION. A slender Plethodon with the general appearance of a lead-colored *P. cinereus,* but with a higher costal groove count and other constant differences. To me it seems closely related to *P. cinereus* both in structural features and coloration. The head is long and slender, widest immediately back of the eyes, in front tapering to the bluntly pointed snout, behind with the sides slightly converging. Eyes with the iris and pupil black. Gular fold well developed, light in color and without pigment in the crease. A sinuous groove from the posterior angle of the eye to the lateral extensions of the gular fold, a short curved groove from this to the angle of the jaw. Trunk slender and rounded, costal grooves 20–21, rarely to 23, counting 1 each in axilla and groin; 9–10 costal folds between appressed limbs; tail circular in section, at base sometimes slightly flattened above, tip slender. Legs small, slender; toes 5–4, considerably webbed at base, those of the hind feet 1–5–2–4–3 in order of length from the shortest; fore feet 1–4–2–3, innermost rudimentary and entirely within the web. Tongue circular, moderate, the edges narrowly depressed. Vomerine teeth in 2 short series, 5–6 teeth in each, separated by about the width of an internal naris and extending at least to outer edge of naris; parasphenoid teeth in 2 narrowly separated, club-shaped patches.

COLOR. The ground color above is deep seal-brown in an indistinct band which arises on the snout, widens to the full width of the head back of the eyes, continues along the trunk and on the tail nearly to the tip. Minute silvery-white and bronze specks are scattered over the entire dorsal surface. The ground color of the sides of the trunk and tail immediately below dorsal band, dark brown to bluish-black. Lower sides with many small, irregular, light blotches. Sides of head with many dull white blotches, more abundant back of the eyes, these spots continuous with those of the throat, where they are larger and more numerous, giving a mottled effect. Belly dark with small, scattered, bluish-white spots. The legs above are lighter brown than the dorsal band, below with a few light flecks.

Preserved specimens become uniformly purplish-brown above on the trunk and bluish-brown on the tail, the light dorsal markings fade and finally disappear, the light spots of the lower sides and venter remain.

BREEDING. Nothing has been published on the breeding habits of this species and the larvae are unknown. A large female in the Cornell University collection from Buckhannon, West Virginia, had 7 ovarian eggs, each 3 mm. in diameter, but unfortunately the exact date of collection is lacking.

**VAN DYKE'S SALAMANDER. WASHINGTON SALAMANDER.** *Plethodon vandykei* Van Denburgh. Fig. 76. Map 31.

TYPE LOCALITY. Paradise Valley, Mt. Rainier Park, Washington.

RANGE. Western Washington; Coast, Olympic, and Cascade Mountains.

HABITAT. On June 21, 1936, we collected along the Calawah River, Washington, where small springs flowed from beneath overhanging banks covered by a tangle of vegetation. Here, among the small stones and gravel washed clean by the waters of the springs, we collected this salamander in company with *Plethodon vehiculum, Ascaphus truei,* and *Bufo boreas.* Slater (1933, p. 44) collected some specimens near Morton, Lewis County, Washington, from beneath moist slabs of bark at the base of a dead, giant Douglas fir, far from any stream. Other specimens collected by Slater were found under moist stones and moss near running water.

SIZE. The type specimen from Paradise Valley, Mt. Rainier Park, Washington, had a total length of 116 mm. and a tail length, measured from front of anus, of 56 mm. The proportions of an adult female from Forks, Washington, are as follows: total length 3 25/32" (96 mm.), tail (tip damaged) 1 3/8" (35 mm.); head length 9/16" (14 mm.), width 3/8" (9 mm.). A male has the following measurements: total length 3 3/8" (85 mm.), tail 1 7/16" (36 mm.); head length 1/2" (12.5 mm.), width 5/16" (7.5 mm.).

DESCRIPTION. This is one of the rarer species of Plethodon and is quite distinct in structural features from other members of the genus. Viewed from above, the head has the sides back of the eyes nearly parallel, the snout short and rapidly tapering. Parotoid glands rather prominently developed, in raised masses extending from the eyes along the dorsolateral angles of the head to the lateral extensions of the gular fold. The eyes are large and prominent, pupil horizontal, iris gilt and black. An impressed line extending from the posterior angle of the eye curves sharply down to the gular fold; a short curved line from this to the angle of the jaw and thence a short distance on either side of the throat.

FIG. 76. *Plethodon vandykei* Van Denburgh. (1) Adult male, actual length 3$\frac{5}{16}$″ (84 mm.). (2) Same, ventral view. Forks, Washington.

The trunk in cross section is wider below than above and with the sides slightly rounded to the somewhat flattened dorsum. There is a slight, impressed, median dorsal line extending from the back of the head to the base of the tail. The tail is roughly oval in section at the base, compressed and rather abruptly tapering to the tip. Costal grooves 13–14, the usual number 14 when 1 each is counted in axilla and groin; 2–3 intercostal spaces between toes of appressed limbs. Legs rather short and stout, the hind proportionally larger than in most species of Plethodon. Toes 5–4, those of the hind feet 1–5–2–4–3 in order of length from the shortest, strongly webbed at base; toes of fore feet 1–4–2–3, the innermost rudimentary and entirely within the web. Tongue moderate, bluntly pointed at tip, widest behind, the sides free, thin, and smooth. Vomerine teeth 10–10 in an adult female, the series arising behind the center of the inner nares and curving inward and sharply backward toward the middle line, where they are separated by about the width of an inner naris. In an adult male, the vomerine teeth 10–11, forming rather irregular series extending nearly to the mid-line and only narrowly separated from the parasphenoids. The parasphenoid teeth in 2 narrowly separated club-shaped patches in the female, less evidently separated in the male.

COLOR. This species is marked above by a broad median stripe which extends from the head to the tip of the tail. On the head and trunk the band is brownish-yellow, on the tail decidedly lighter and more yellowish. The sides below the dorsal band are a slightly darker brown, the color involving the sides of the head, the sides of the trunk to the level of a line passing through the legs, the dorsal surface of the legs, and the upper sides of the tail. Below the dark sides, the light color of the venter is visible up to the mid-line of the legs. The venter in some individuals is a very light flesh color, in others lightly mottled with pale brown. Throat and distal half of the tail beneath yellow. In younger specimens the dorsal stripe is sometimes brownish-yellow on the head, yellow tinged with greenish on the trunk, and bright orange-yellow on the tail. In these the sides of the trunk are Mummy Brown (Ridgway) flecked with white below, the sides of the head mottled with brown blotches and white flecks. Legs above brown, mottled with yellowish-white. The throat is light, yellowish, and with a few scattered blotches around the margins. The belly is brownish, lighter than the sides, and flecked with white.

BREEDING. Noble (1925, p. 6) remarks: "Mr. Phillips Putnam has sent me a cluster of eggs of this species which he found fastened together in a grape-like mass and attached to the stone by a string of elastic material. The stone was in a damp situation and covered by moss."

**WESTERN RED-BACKED SALAMANDER.** *Plethodon vehiculum* (Cooper). Fig. 77. Map 32.

TYPE LOCALITY. Astoria, Oregon.

RANGE. Western Oregon, Washington, and British Columbia including Vancouver Island.

HABITAT. Found beneath logs, bark, stones, and moss, in cool, damp, and shaded ravines. Common in the Douglas-fir forests near Portland, Oregon. In the Green River Gorge near Seattle, Washington, we collected many specimens from beneath stones and logs on the steep sides of the ravine and among the loose stones at the foot of the cliff bordering the river.

SIZE. The largest specimen, a female, measured by Dunn (1926, p. 155) had a total length of 105 mm. and a tail length of 43 mm. In a large series from Washington and Oregon the largest specimens measured were females. Measurements of a female from Skokomish River near Shelton, Oregon, are as follows: total length 4 5/32" (105 mm.), tail 1 31/32" (50 mm.); head length 15/32" (12 mm.), width 5/16" (8 mm.). An adult male from near Forks, Washington, measures as follows: total length 3 5/8" (92 mm.), tail 1 3/4" (45 mm.); head length 13/32" (10 mm.), width 9/32" (7 mm.). My smallest specimen is 32 mm. in total length, of which the tail comprises only 10.5 mm.

DESCRIPTION. This species resembles the eastern *Plethodon cinereus,* but is a more active species, attains a somewhat larger size, and has a stouter body and stronger limbs. The head has the sides nearly parallel back of the eyes, in front converging to the bluntly pointed snout. Eyes moderate but strongly protuberant; posterior corner of eye with a vertical flap under which both eyelids slip. A prominent gular fold; a deep sinuous groove from the posterior corner of eye to the lateral extensions of gular fold; a short deep groove from this to angle of jaw. Trunk moderately stout, well rounded, often slightly depressed above and with a slight median dorsal groove. Tail thick, fleshy, subquadrate in section near base, slightly tapering ¾ of its length, then quite abruptly

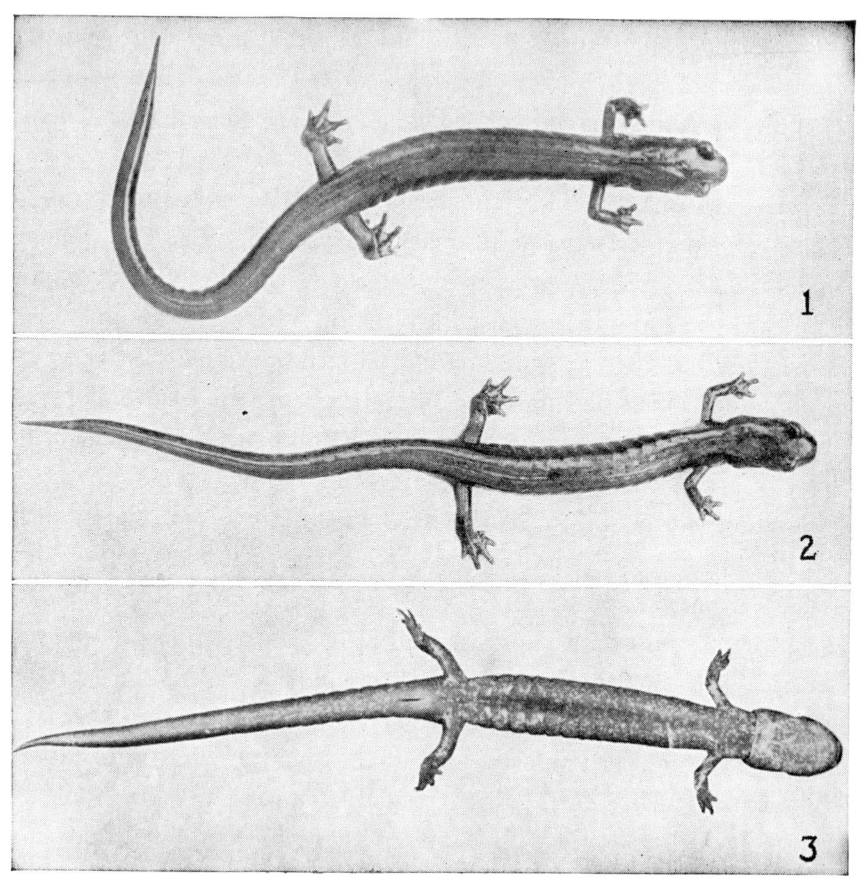


FIG. 77. *Plethodon vehiculum* (Cooper). (1) Adult female, dark phase; actual length 4⅛″ (105 mm.). (2) Adult male, light phase; actual length 3¹³⁄₁₆″ (97 mm.). (3) Adult female, ventral view; actual length 4¹⁄₁₈″ (103 mm.). Portland, Oregon.

to the pointed tip. Usually 15 costal grooves, counting 1 each in axilla and groin, rarely 16, or 15 on one side and 16 on the other; 5–6 intercostal spaces between toes of the appressed limbs. Legs fairly stout; toes 5–4, those of the hind feet 1–5–2–3–4 in order of length from the shortest, slightly webbed at base; toes of the fore feet, 1-4-2-3, the innermost rudimentary. Basal joints of legs above colored like the dorsal band. Lower legs and feet above usually colored like the side of trunk. Tongue fairly large and filling the floor of the mouth. Vomerine teeth 5-7, usually 6, the short series beginning behind inner edge of nares and curving inward and backward, separated by about the width of an inner naris and from the parasphenoid by a little more. Parasphenoids in 2 large, incompletely separated, club-shaped patches.

COLOR. The majority of specimens have a dorsal band which varies in color from dull, dirty white to light yellowish-red, dusky red, orange-red, bright red to deep brownish-red suffused with dusky. In some the band may be strongly flecked with black, in others almost obliterated, leaving only a very narrow lighter line on each side to indicate its limits. Young individuals may have the dorsal band a bright, clear orange-red without darker markings, or be uniformly dull pinkish-gray. The band may originate on the shoulders and continue to the tip of the tail, or the whole dorsal surface of the head may be involved. Upper sides darkest next to the band, fading gradually into the brownish sides, which are mottled with white. Venter, in life, definitely flecked with small white spots on a bluish ground color, throat lighter, dusky yellowish. The superficial white flecks on the sides and venter give a characteristic pepper-and-salt effect and apparently overlie the black chromatophores. The interspaces between them are sometimes quite yellow, particularly on the throat, sides of the vent, and lower surface of the basal third of the tail. In preserved specimens the red normally fades so that the dorsal band is usually tan or dull yellow and the sides brownish or bluish, with dull yellow interspaces. The superficial white fleckings disappear entirely on the sides and venter, the latter becoming grayish mottled with yellow.

SEXUAL DIFFERENCES. The sexes of adults may be distinguished by the shape of the lower jaw, which in the female is broadly rounded in front and in the male slightly pointed; or by the vent, which in the female has the sides thrown into narrow oblique folds, while in the male the sides are bordered by a depressed flange-like ridge broken by a few narrow cross lines and produced behind into free flaps.

BREEDING. Little is known of the breeding habits. Slevin (1928, p. 55) records that "a female collected near the Hoh River, Jefferson County, Washington, on September 20, 1919, contained eggs about two and one-half millimeters in diameter." In large females which we collected June 20, 1936, near Shelton, Washington, there were a few eggs 2 mm. in diameter and many small ones. Specimens taken in February at Portland, Oregon, had only small ovarian eggs. The larvae have not been described.

**WEHRLE'S SALAMANDER.** *Plethodon wehrlei* Fowler and Dunn. Fig. 78. Map. 35.

TYPE LOCALITY. Two Lick Hills, Indiana County, Pennsylvania.

RANGE. Southwestern New York, western Pennsylvania, West Virginia, and southeastern Ohio.

HABITAT. This species is commonly found on wooded hillsides, where it hides by day beneath stones or rocks; less frequently under or within old rotting logs. The majority of the specimens I have collected have been under flat stones in old second-growth, mixed deciduous and evergreen woods. Netting (1932, p. 6) found the species in West Virginia, in cave entrances, in deep rock crevices, and near "ice caves." In New York State the species is more commonly associated with *Plethodon cinereus* than with *P. glutinosus,* the latter apparently requiring more moisture.

SIZE. Large females reach a length of 6″ (152 mm.). My largest male, from Allegany State Park, New York, has a total length of $5\frac{5}{16}$″ (135 mm.), tail $2\frac{3}{4}$″ (70 mm.); head length $\frac{5}{8}$″ (15 mm.), width $\frac{3}{8}$″ (9 mm.). The proportions of an adult female from Port Allegany, Pennsylvania, are as follows: total length $4\frac{15}{16}$″ (122 mm.), tail $2\frac{9}{16}$″ (65 mm.); head length $\frac{9}{16}$″ (13.5 mm.), width $\frac{5}{16}$″ (7.5 mm.).

DESCRIPTION. This is a slender species related to and somewhat resembling the Slimy salamander, *P. glutinosus.* Differs from *P. glutinosus* in its more slender form, in having the light spots usually restricted to the sides, and in having a higher costal-groove count.

The head is broadly oval in outline, widest just back of the eyes, tapering slightly posteriorly, and abruptly to the bluntly pointed and de-

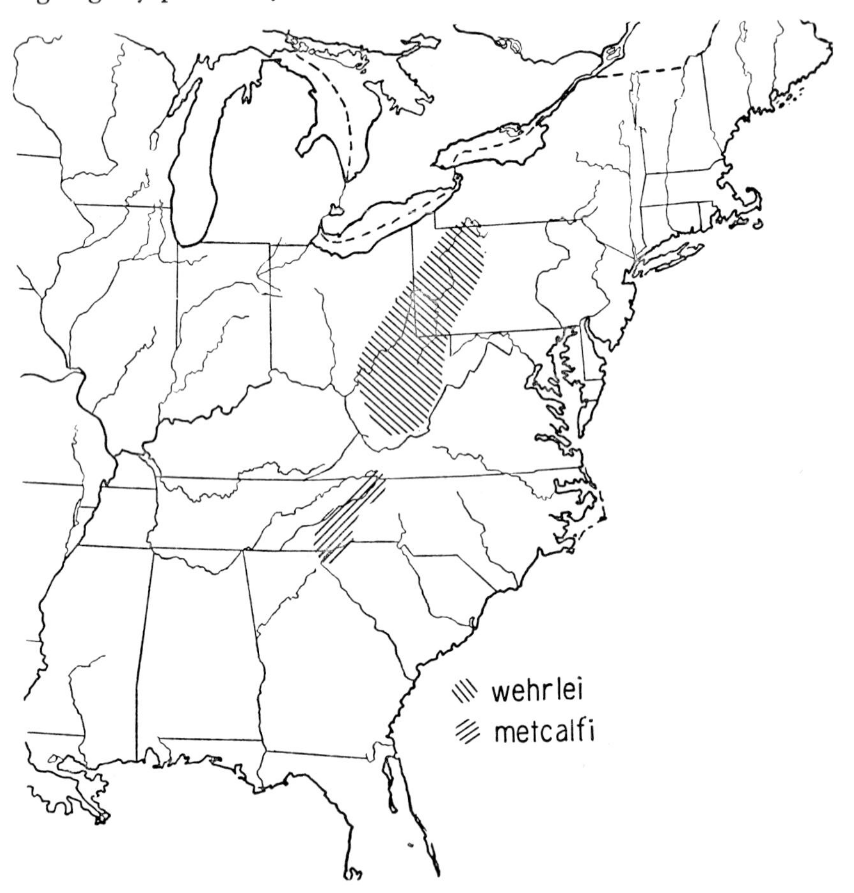


MAP 35.—Distribution of *Plethodon wehrlei* and *P. metcalfi*.

pressed snout. The eyes are prominent, with the pupil black and the iris deep brown. Eyelids limited behind by a short vertical groove. Gular fold prominent and extending well on the sides of the neck. A deep sinuous groove from the eye to the lateral extensions of the gular fold, and a short vertical groove from this line to the angle of the jaw. Trunk slender, rounded, and with a median impressed line from the back of the head to the base of the tail. Costal grooves 16–17, the usual number 16 when 1 each is counted in the axilla and groin; 2–3 intercostal spaces

FIG. 78. *Plethodon wehrlei* Fowler and Dunn. (1) Adult male, actual length 5$\frac{5}{16}$″ (135 mm.). (2) Same, lateral view. Allegany State Park, New York. (3) Adult female. Port Allegany, Pennsylvania.

between the toes of the appressed limbs. Tail slender, nearly circular in cross section at base, evenly tapering to the tip, which may be slightly compressed in some individuals. Legs well developed. Toes 5–4, those of the hind feet 1–5–2–3–4 or 4–3 in order of length from the shortest, short, blunt-tipped, 1st webbed nearly to the tip, others at base; toes of front feet, 1–4–2–3, the 1st rudimentary and webbed nearly to tip, others at base. Tongue broad and flat and nearly fills the floor of the mouth. Vomerine teeth 6–8 in each series in the male and 7–9 in the female. The series arise behind the outer margin of the inner nares and curve inward and gently backward toward the mid-line, where they are separated by about the diameter of a naris. Parasphenoid teeth in 2 well separated, slender, club-shaped patches, usually distant from the vomerine teeth by about the diameter of a naris, more rarely continuous with the vomerine series on either or both sides.

COLOR. The color in life is somewhat variable. The general impression is that of a black salamander with pale spots on the sides. In some individuals the ground color above is a deep brown, almost black, except on the head, where the brown may be a little lighter. The lower sides may be bluish-brown flecked with irregular whitish spots and dashes. The legs are usually colored above like the dorsal surface of the head and have a few light fleckings. The belly and lower surface of the tail uniformly light gray, the throat conspicuously lighter and often mottled. The lower surface of the legs and feet often with tinges of flesh. In some specimens the general color has bluish-black reflections, and the light markings of the sides are carried well on the back and sides of the head below the eyes. Dunn (1926, p. 135) writes that some young individuals have paired red spots on the back. His smallest specimen was 44 mm. in total length. I have collected a single specimen only 37 mm. long and several others only slightly longer, but in none has there been the slightest evidence of dorsal red spots.

SEXUAL DIFFERENCES. Sexual differences are not very strongly marked. Some adult males have a small mental gland, and the sides of the vent are thrown into narrow ridges. The vent of the female is a simple slit with smooth margins.

BREEDING. Little is known of the breeding habits. Ovarian eggs of females taken in June in New York State are very small, so it is likely that in this section egg-laying takes place later in the summer.

**WELLER'S SALAMANDER.** *Plethodon welleri* Walker. Fig. 79. Map 34.

TYPE LOCALITY. Grandfather Mountain, near Linville, North Carolina.

RANGE. Grandfather Mountain, North Carolina, above 5000′, and Whitetop Mountain, Virginia.

HABITAT. Generally found beneath logs, stones, and flakes of rock, in the spruce forests covering the higher slopes of the mountains. On April 19, 1938, U. B. Stone, Robert Van Auken and I collected along the trail leading to the summit of Grandfather Mountain and took many specimens within a few feet of the path.

SIZE. This is a small species, the adults attaining a length of a little more than 3″. The proportions of a female from Grandfather Mountain, North Carolina, taken April 19, 1938, are as follows: total length 3 1/16″ (77.5 mm.), tail (tip lost) 1 5/16″ (33 mm.); head length 13/32″ (10.5 mm.), width 1/4″ (6 mm.). A male of the same length from the same locality has the tail 1 1/4″ (32 mm.); head length 3/8″ (10 mm.), width 1/4″ (6 mm.).

DESCRIPTION. In the form of the body there is a general resemblance to a small *Plethodon cinereus.* Head widest at the angle of the jaws, converging slightly behind to the lateral extensions of the gular fold, the snout short and blunt, the tip truncated and slightly overhanging the lower jaws. The eyes are of moderate size, noticeably smaller than in *P. yonahlossee* of the same size, and with the iris only slightly flecked with brassy. A deep sinuous line from the posterior angle of the eye to the lateral extensions of the gular fold; a short curved groove from this line to the angle of the mouth. The trunk is rounded on the sides, flattened beneath, and with an impressed median dorsal line from the head to the base of the tail. The tail is slender and tapers evenly to the tip, nearly circular in cross section at the base, oval distally. Costal grooves 16, counting 1 each in axilla and groin; 3–4 intercostal spaces between toes of appressed limbs. Legs small, slender, toes 5–4, those of the hind feet 1–5–2–4–3 in order of length from the shortest, the innermost rudimentary and webbed nearly to the tip, the others webbed at base; toes of the fore feet 1–4–2–3. Tongue broad and flat, nearly filling the floor of the mouth. Vomerine series short, each with 4–6 teeth, the series arising behind the outer margin of the inner nares and extending obliquely inward and backward, separated at the mid-line by about

FIG. 79. *Plethodon welleri* Walker. (1) Adult female, actual length 3⁵⁄₁₆″ (75 mm.). (2) Same, ventral view. Grandfather Mountain, North Carolina.

twice the diameter of an inner naris. Parasphenoid teeth in 2 elongate club-shaped patches, widely separated from the vomerine.

COLOR. The ground color above is black, with the dorsal surfaces of the head, trunk, and tail with a broad irregular band formed by numerous dull golden blotches. In some individuals the blotches are aggregated along either side of the median line, in other more generally distributed. A few golden flecks extend down on the sides between the costal grooves, but the sides are mostly black, with a few small whitish flecks near the sides of the belly. The belly is slate with purplish reflections, sometimes with a few scattered light spots; the ventral surface of the tail darker than the belly and the throat lighter, a dull, dirty grayish-white. The lower surface of the legs is colored like the throat, with nearly white spots marking the joints. In preservative, the golden flecks disappear entirely and the ground color above becomes uniformly slate, the belly grayish-blue, the throat and lower surface of the legs grayish-white. A few pale spots may persist along the lower sides and on the sides of the belly.

SEXUAL DIFFERENCES. The sexes may be distinguished by the form of the vent, which is larger in the male and produced behind into 2 flat lobes; in the female it is a simple slit. Adult males have a poorly defined mental gland, the tip of the lower jaw is slightly more pointed, and the snout is swollen at the ends of the nasolabial grooves.

BREEDING. Nothing has been published concerning the breeding behavior.

**YONAHLOSSEE SALAMANDER.** *Plethodon yonahlossee* Dunn. Fig. 80. Map 30.

TYPE LOCALITY. Near the Yonahlossee Road, about 1½ miles from Linville, North Carolina.

RANGE. Blue Ridge Mountains in western North Carolina and in southwestern Virginia.

HABITAT. These salamanders have been taken between 3200′ and 5000′ elevation on the wooded slopes of the mountains, where they hide by day beneath logs and stones or occasionally beneath the bark of a rotted log. Dunn (1926, p. 133) writes, "They have long burrows in the floor of the forest, whose openings are usually under a fallen log or piece of bark." Bailey (1937, p. 3) found that this species could be most easily collected at night by means of a headlamp. *P. yonahlossee* is associated in nature with *Plethodon glutinosus, P. cinereus,* and *P. metcalfi,* but is rarer than any of these and much more restricted altitudinally.

SIZE. This is one of the larger species of Plethodon, adults attaining an extreme length of 6½″ (165 mm.). The measurements of the type, a moderate-sized individual, given by Dunn (1926, p. 132), are as follows: total length 5½″ (140 mm.), tail 2½″ (63 mm.); head length 11/16″ (17 mm.).

DESCRIPTION. This species is related to *Plethodon glutinosus, P. wehrlei,* and *P. ouachitae*. Of these only the last named has a chestnut-colored

FIG. 80. *Plethodon yonahlossee* Dunn. (1) Juvenile female, actual length 2 15/16″ (74 mm.). (2) Same, ventral view [photograph after preservation]. Near Linville, North Carolina.

band, all have whitish blotches on the sides or back, or on both areas. The head is widest at the eyes, converging slightly behind; the snout is short, the sides converging abruptly to the bluntly rounded tip. The eyes are large, strongly protuberant, and with the iris black; the eyelids limited behind by a crescentic fold. An impressed line from the eye to the lateral extension of the gular fold; a short curved groove from this line to the angle of the jaw. The trunk is well rounded on the sides and above, below flattened. The tail is subquadrate in section at the base, oval in section at about ½ its length, slightly compressed distally. Costal grooves 15, counting 1 each in the axilla and groin. In some individuals 2 grooves run together in the groin, and if both are counted the number is 16. In half-grown specimens the appressed toes may meet, in larger individuals they may be separated by an intercostal space. The legs are moderately stout, toes 5–4, those of the hind feet 1–5–2–3–4 or 4–3 in order of length from the shortest, the innermost webbed nearly to the tip, the others at base; toes of the fore feet 1–4–2–3, the 2nd and 4th nearly equal. Tongue rather broad and flat, the margins thin and smooth, the plicae radiating from the posterior field. Vomerine series long, originating outside and behind inner nares and curving inward and backward, separated by about the width of an inner naris. Teeth variable in number from 10 to 14 in each series in specimens I have examined. Dunn (1926, p. 130) records 17. Parasphenoid teeth in a single broad club-shaped patch.

COLOR. In life the ground color above is dark gray, almost black. A broad median chestnut-red band extends from the back of the head to the base of the tail; the band also with lateral extensions running down the sides between the costal grooves nearly to the level of the legs. Sides below the dorsal band gray, blotched strongly but irregularly with whitish spots. Sides of head and tail also with white blotches. Dorsal surface of the tail dark gray, almost black, and with only a few small light flecks. Fore legs light gray with a few small light spots, hind legs darker gray, lightly flecked; toes flesh color. The belly is bluish-gray, the ventral surface of the tail slightly darker, the throat lighter, pale dirty white. In young specimens, 2½″ long or less in total length, there is frequently a double series of dorsal spots brighter red than the adjacent band because of the absence in them of black pigment.

Specimens long in preservatives lose the chestnut color almost entirely; the back becomes uniformly dull brownish or bluish-black, the venter grayish or brownish mottled and blotched with lighter areas, and the legs brown or tan. The light blotches of the sides persist but are duller than in life.

BREEDING. Nothing is known of the breeding habits of this species.

## GENUS PLETHOPSIS

**WESTERN FOUR-TOED SALAMANDER.** *Plethopsis wrighti* Bishop. Fig. 81. Map 53.

TYPE LOCALITY. 8.7 miles southeast of Sandy, Clackamas County, Oregon.

RANGE. Known from the vicinity of the type locality, from near the mouth of Moose Creek and Middle Santiam River, Linn County, and from Cherryville, Clackamas County, Oregon.

HABITAT. Discovered on the wooded slopes bordering the Mount Hood highway, the forest consisting of mixed deciduous and evergreen trees, an undergrowth of shrubs, and a ground cover of moss and low herbs. Individuals were found beneath the bark of rotted logs, under chips and logs on the ground, and between pieces of bark piled in a heap at the base of a stump.

SIZE. In the type series, the adult males varied from 3 9/16" (91 mm.) to 3 13/16" (97 mm.), the females from 3 5/16" (84 mm.) to 3 3/4" (95 mm.). Other individuals of both sexes were mature at sizes less than those indicated above, but loss of portions of the tail made determination of total length impossible. The proportions of a male are as follows: total length 3 13/16" (97 mm.), tail 2" (51 mm.); head length 3/8" (9 mm.), width 9/32" (5.5 mm.). Female, total length 2 27/32" (72 mm.), tail 1 1/4" (32 mm.); head length 5/16" (8.5 mm.), width 3/16" (5 mm.).

DESCRIPTION. This genus is related to Plethodon but differs in several well marked structural characteristics. It is a slender species of moderate size, the head widest immediately behind the eyes, the sides converging slightly to the lateral extension of the gular fold and in front rather abruptly to the short blunt snout. The eyes are prominent, the lids limited posteriorly by an oblique fold. An impressed sinuous line from the posterior angle of the eye to the lateral extensions of the gular

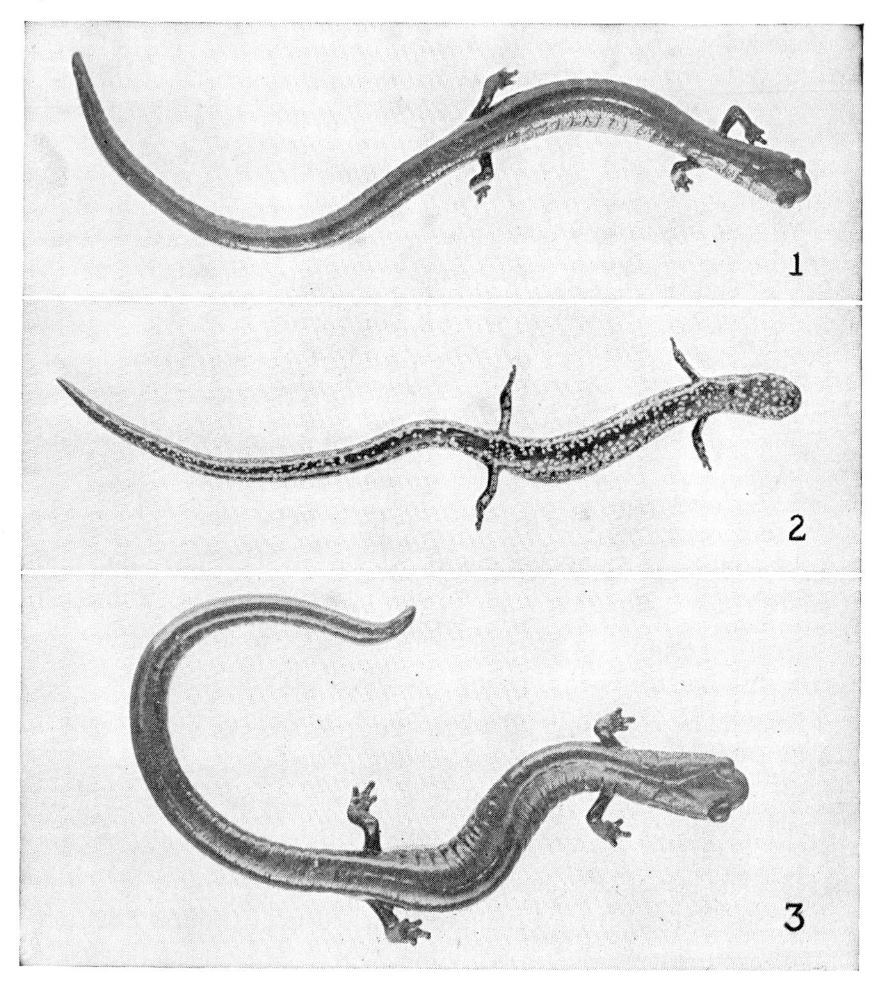


FIG. 81. *Plethopsis wrighti* Bishop. (1) Adult male, actual length 3¹³⁄₁₆″ (97 mm.). (2) Same, ventral view. (3) Another individual to show the four toes on both fore and hind feet. Paratypes from 8.7 miles southeast of Sandy, Oregon.

fold; a poorly defined vertical groove from this line passing back of the angle of the jaw. The trunk is well rounded on the sides and above, flattened ventrally, and with an impressed median line from the head to the base of the tail. Tail subquadrate in section near the base, slightly compressed distally. Costal grooves 16 or 17, counting 1 each in the axilla and groin; rarely with 16 on one side and 17 on the other; 6½–7 intercostal spaces between the toes of the appressed limbs. Legs rather small and slender, toes 4–4, 1–4–2–3 in order of length from the shortest, the innermost rudimentary and webbed to tip, the others webbed ½ their length. Tongue small, attached by a central pedicel and by a ligament in front. Vomerine series long, in the male each usually 12–15, the series originating behind the outer margin of the inner nares; in the female usually 10–14, the series originating behind the middle of the inner nares. Parasphenoid teeth in 2 long, narrowly separated, club-shaped patches. In 10 of a series of 20 specimens the vomerine and parasphenoid series are separate, in 3 males and 1 female the series are continuous on both sides, and in 1 male and 5 females the series are continuous on one side and slightly separated on the other.

COLOR. In life this species has a conspicuous dorsal band which varies in color from chestnut to reddish-brown. In a few individuals the band is reddish-brown along the margins and blotched with black through the center. Two of the largest individuals have the band dull brownish-yellow. In some the band is strongly suffused with dusky on the snout and trunk. The band is brightest on the basal half of the tail and is usually irregularly invaded with black on the posterior half. The band originates on the snout and involves the entire width of the head including the eyes, and extends along the trunk well above the level of the legs. The lateral margins of the band may be nearly straight, but in some individuals are irregular and broken. The upper sides, where they border the dorsal band, are deep brown or black to the level of the legs, below which the sides are heavily flecked with many small, irregular, bluish-white spots on a slate ground color; the belly is dark slate to almost black, conspicuously flecked with bluish-white, the middle line only being relatively free; upper surface of legs dark with small light spots. The throat, venter of the tail, and lower surface of the legs strongly flecked with lighter spots.

SEXUAL DIFFERENCES. The sexes may be distinguished by the form of the vent, which in the female is a simple slit; in the male it is bordered by a narrow depressed area and the lips are thrown into folds.

BREEDING. Nothing is known of the breeding habits of this species.

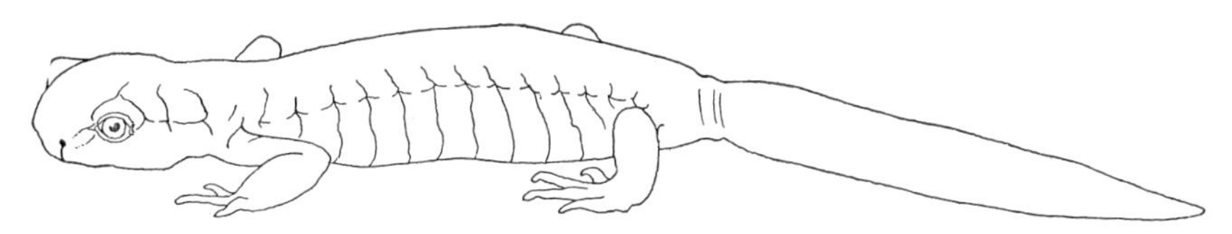

FIG. 82. Outline of *Ensatina e. eschscholtzii* to show the general body form and tail constriction. Tail shorter than normal. [Drawn by Hugh P. Chrisp.]

# GENUS ENSATINA

## KEY TO THE SPECIES AND SUBSPECIES OF ENSATINA

1. Dorsal color deep brown to black, with blotches, spots, or patches of orange or orange-yellow ........ 2
   Dorsal color not as above ........ 3
2. Dorsal blotches large, those of the trunk often confluent at mid-line above, sometimes alternating; tail from above broadly barred; legs mostly orange, dark near feet; length to 5⅞″ (150 mm.). From Kern and San Diego Counties, California; also reported from Lower California ........ *croceater* p. 295
   Dorsal blotches very irregular, lichen-like, on the trunk mostly confined to the sides; tail strongly blotched but seldom with appearance of being barred; basal segments of legs orange, distal segments blotched with gray and orange; length to 5 5/32″ (131 mm.). Sierra Nevadas from Butte to Tulare Counties, California ........ *sierra* p. 303
3. Uniformly yellowish-brown to light reddish-brown above; belly flesh color; length to 5 5/16″ (135 mm.). From British Columbia and Vancouver Island south to southern California, west of the Sierras and Cascades ........ *eschscholtzii eschscholtzii* p. 297
   Above light reddish-brown, with a dorsolateral series of irregular, deep brown or blackish marks, and usually with an interrupted middorsal black line; tail strongly blotched above; belly and ventral surface of legs yellowish; length to 4 3/32″ (104 mm.). From Humboldt County, California, northward through Del Norte County at least to Port Orford, Curry County, Oregon ........ *eschscholtzii picta* p. 300

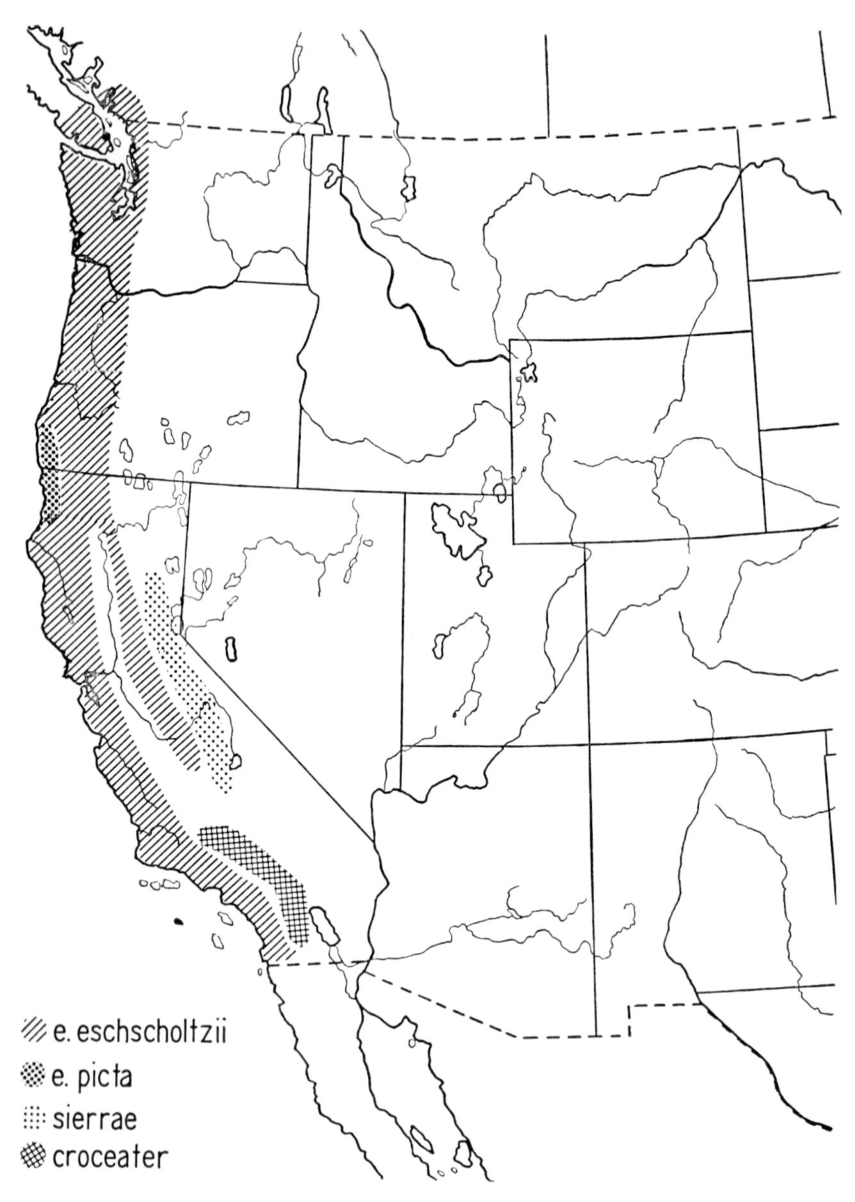


MAP 36.—Distribution of the subspecies of *Ensatina eschscholtzii*, *E. sierrae*, and *E. croceater*.

**YELLOW BLOTCHED SALAMANDER, YELLOW SPOTTED SALAMANDER.** *Ensatina croceater* (Cope). Fig. 83. Map 36.

TYPE LOCALITY. Fort Tejon, California.

RANGE. From Kern to San Diego Counties, southwestern California; also recorded from Lower California but probably erroneously.

HABITAT. This is a mountain species, having usually been taken at elevations above 4000′. Generally found in or beneath rotting logs or in leaf mold; occasionally abroad at night during rains.

FIG. 83. *Ensatina croceater* (Cope). Adult. San Diego County, California. [Photograph through the courtesy of L. M. Klauber.]

SIZE. This is the largest species of the genus, full-grown individuals reaching an extreme length of about $5\frac{7}{8}''$ (150 mm.). The average length of 18 adults of both sexes from San Diego County, California, is $4\frac{1}{4}''$ (108.5 mm.), the extremes $3\frac{5}{32}''$ (80 mm.) and $5\frac{15}{32}''$ (140 mm.). The proportions of an adult male from Pine Hills, San Diego County, California, are as follows: total length $4\frac{25}{32}''$ (122 mm.), tail $2\frac{13}{32}''$ (61 mm.); head length $\frac{11}{16}''$ (18 mm.), width $\frac{7}{16}''$ (11 mm.). A female from Cuyamaca, California, measures: total length $4\frac{13}{16}''$ (123 mm.), tail $2\frac{5}{32}''$ (55 mm.); head length $\frac{25}{32}''$ (20 mm.), width $\frac{15}{32}''$ (12 mm.).

DESCRIPTION. A vivid black-and-orange-blotched salamander with a definite tail constriction. The head is large, the sides behind the eyes nearly parallel or slightly narrowing to the gular fold; in front converging to the bluntly pointed snout. Eyes large and strongly protuberant, the pupil horizontally elliptic. A deep groove from the posterior angle of the eye to the lateral extension of the gular fold; a short vertical groove from this line passing behind the angle of the jaw. Trunk stout, above with an impressed median line. Usually 12 costal grooves, counting 1 each in the axilla and groin, occasionally 13; the toes of the appressed limbs overlapping 1 or 2 (in young) to 5½ intercostal folds. Tail stout and with a well marked constriction immediately behind the vent; broadly oval in section at base, widest and deepest at about the basal third, slightly compressed below, rounded above and tapering rather abruptly to a sharp point. The tail comprises 32.5–35.4 per cent of the total length in juveniles 43 and 62 mm. long respectively, 42.3–44.8 per cent in the females, and 45.1–52.2 per cent in the males, the length increasing with age. Legs stout, the hind noticeably larger than the fore. Toes 5–4, moderately slender and evenly tapering, those of the hind feet 1–5–2–3–4 in order of length from the shortest; toes of the fore feet 1–4–2–3. Tongue obovate, free behind and at the sides. Vomerine teeth in 2 very long, strongly arched series that arise well outside the outer margin of the inner nares, curve forward, then inward and backward, nearly meeting at the mid-line. In a series of 18 adults, the teeth average 26.58, the extremes 16 and 32. In juvenile specimens 43 and 45 mm. long, the teeth 8–8 and 10–8 respectively. Parasphenoid teeth in 2 narrowly separated, club-shaped patches distant from the vomerine by about 3 times the diameter of an inner naris.

COLOR. The color in life as described by Klauber (1927, p. 4) is as follows: "Dorsal color black, fading on the sides to Dusky Purplish Gray (Ridgway). The irregular series of dorsal blotches (one on the head engaging the eyelids, four on the body, four on the tail) are Orange Rufous; under surfaces transparent Light Vinaceous Lilac. The legs are of the same color as the dorsal blotches except the extremities, which are similar to the under surfaces." In a fine series of 23 specimens from San Diego County, California, loaned by Mr. L. M. Klauber, there is some variation in the number and disposition of the orange blotches.

Often there is a U-shaped mark on the dorsum of the head which involves the eyes and sends a broad branch down on each side of the head back of the angle of the mouth; frequently this blotch is broken at the mid-line above. The blotches on the trunk may be more or less rounded and separated, but usually form pairs in contact at the mid-line and set diagonally. Sometimes 2 or 3 pairs of blotches meet at the mid-line of the back nearly opposite one another, others are offset so that only the anterior corner of one is in contact with the posterior corner of another. The tail often has the appearance of being completely ringed when seen from above; or the blotches may be arranged alternately on either side of the mid-line.

SEXUAL DIFFERENCES. The sexes may usually be separated by the form of the vent, which in the male is larger and more strongly protuberant and has the margins grooved. The snout of the male is sometimes swollen at the nasolabial grooves, which are split distally (Y-shaped), the stem running to the nostril, one branch toward the mid-line of the upper lip, the other diagonally toward the side.

BREEDING. Nothing has been reported concerning the breeding habits. The eggs are known from a set of 14 collected by Mr. L. M. Klauber July 25, 1927, near Julian, California, and reported by Storer (1929, p. 447). The eggs were accompanied by an adult and adhered to one another, but were without evidence of a stalk. "Individually they were about five-sixteenths of an inch in diameter, yellowish in color, and opaque."

**RED SALAMANDER. OREGON SALAMANDER.** *Ensatina eschscholtzii eschscholtzii* Gray. Figs. 82, 84. Map 36.

TYPE LOCALITY. California.

RANGE. From British Columbia (mainland and Vancouver Island) south to southern California, west of the Sierra and Cascade Mountain systems.

HABITAT. Found beneath logs and bark and other suitable cover, usually in moist situations, but occasionally where it is fairly dry; often abundant in the Redwood and Douglas-fir forests and in mixed softwoods and hardwoods.

SIZE. Adults reach an extreme length of 5 5/16″ (135 mm.) and are thus intermediate in size between *E. sierra* and *E. croceater*. A series of 9 adults of both sexes varied in length from 3 7/32″ (82 mm.) to 5″ (117 mm.) and averaged 3 7/8″ (99 mm.). The proportions of an adult male from Alder Creek, Oregon, taken June 15, 1936, are as follows: total length 5″ (117 mm.), tail 2 6/16″ (60 mm.); head length 10/16″ (16 mm.), width 3/8″ (9.5 mm.). An adult female from Portland, Oregon, Dec. 23, 1933, has the following measurements: total length 3 9/16″ (91 mm.), tail 1 13/32″ (36 mm.); head length 19/32″ (15 mm.), width 6/16″ (10 mm.).

DESCRIPTION. Viewed from above, the head is roughly oval in outline, the sides behind the eyes curving gently to the lateral extensions of the gular fold and in front more abruptly to the truncated snout. The snout in side view obliquely truncate. The eyes are of moderate size and protuberant. An impressed sinuous line from the posterior angle of the eye to the lateral extension of the gular fold; a short vertical groove from this to the angle of the jaw. Gular fold well developed. Trunk well rounded and with a slight median impressed line above. There are 12 costal grooves, counting 1 each in the axilla and groin (that of the axilla scarcely developed), and the toes of the appressed limbs just meet or may overlap 3 intercostal spaces. Tail with a definite constriction at base immediately behind the vent. The tail is broadly oval in section at the base and, in the female, is widest at about the basal third, thence tapering evenly to the slightly compressed tip. In the male the tail is longer, slimmer, and only slightly swollen. In both sexes the tail is rounded above, where it is supplied with poison glands, and fairly sharp-edged below. Legs moderate in size but well developed. Toes 5–4, those of the hind feet 1–5–2–(3–4) in order of length from the shortest; toes of the fore feet 1–4–2–3. Tongue moderate in size, roughly oval in outline, and free at the sides and hind margins only, the surface finely papillose. Vomerine teeth in very long series which arise outside and behind the inner nares at a point about opposite the anterior angle of the eye and curve forward, then inward and sharply backward, to the mid-line, where they are nearly in contact. There are 23–30 teeth in each series in the male and 26–31 in the female. The parasphenoid teeth in 2 separate elongate club-shaped patches, separated from the vomerine by about 3 times the diameter of an inner naris.

COLOR. The general ground color above varies from yellowish-brown

FIG. 84. *Ensatina eschscholtzii eschscholtzii* Gray. (1) Adult female, actual length 3⁹⁄₃₂″ (84 mm.); (2) Same, dorsal view. Portland, Oregon. Tip of tail regenerated.

to light reddish-brown on the head, trunk, and tail; the basal segments of the legs are yellowish above, particularly at the points of insertion. The darker color of the upper sides fades slightly to a line immediately above the legs and is in rather strong contrast to the pale lower sides. In many specimens the sides are dusted with light gray. The belly is flesh color in life, the throat and ventral surface of the tail whitish; sides of tail frequently clouded with dusky. Usually, when the dorsal color is yellowish-brown, the legs at the insertion are yellowish-orange; in the reddish-brown specimens the legs at the base above are bright orange and there are scattered flecks of red-orange on the sides of the tail and fine reddish-orange specks on the lower sides of the trunk. In the adult male the tail is comparatively slender and comprises about 50 per cent of the total length, while in the female the tail is swollen at the basal third and comprises about 40 per cent of the length. The vent of the male is larger, longer, and more strongly protuberant, and the snout more abruptly truncated and swollen at the nasolabial region.

BREEDING. While little is known of the breeding habits, the eggs have been reported a number of times. Van Denburgh (1898, p. 140) mentions a lot found with a female at Mill Valley, Marin County, California, April 19, 1896, beneath a rotting log in the redwood forest. An adult with 34 eggs, taken Oct. 26, 1931, at Berkeley, California, is in the collection of the Museum of Vertebrate Zoology at Berkeley. These eggs are white and the size of a small pea. Eggs have also been found at Inverness, Marin County, California, June 4, 1913. This lot of 13, accompanied by 2 adults, was found in the tunnel system of the rodent *Aplodontia rufa phæa.* Individual eggs, after preservation in alcohol, measured 5.5–5.75 mm. in diameter, and, with their outer jelly coat, 5.9–7.5 mm. (Storer, 1925, p. 112). Another lot mentioned by Storer (*ibid.*) from Carlotta, Humboldt County, may belong to the subspecies *picta* recently described (see just below).

Four young from Portland, Oregon, collected Jan. 13, 1934, by Stanley G. Jewett, Jr., varied in total length from 29.5 to 45 mm. The color range is from deep brown to bluish-black above, very finely dotted with pale bluish-white points. The distal third of the tail is reddish brown and the legs above, at the insertion, bright red-orange. The belly is bluish, the lower surface of the tail and throat dull white.

**PAINTED SALAMANDER.** *Ensatina eschscholtzii picta* Wood. Fig. 85. Map 36.

TYPE LOCALITY. Klamath, Del Norte County, California.

RANGE. A narrow coastal area, from the vicinity of Weott, Humboldt

FIG. 85. *Ensatina eschscholtzii picta* Wood. (1) Adult female, actual length 3⅞₁₆" (88 mm.). (2) Same, dorsal view. (3) Same, ventral view. Del Norte County, California. [A. H. Wright, collector.]

County, California, northward through Del Norte County at least to Port Orford, Curry County, Oregon (Wood, 1940, p. 425).

HABITAT. This form has been taken in numbers in the dense redwood forests, in fir and spruce woods, and beneath debris in open areas (Wood, *ibid.*, p. 427). On June 2, 1936, we collected in the Alexander redwood grove, about 2 miles south of Miranda, California, and found 2 specimens beneath bark on the ground. On June 6, at Carpentersville, Oregon, adults were found beneath stones on a burned-over gully side.

SIZE. Wood (1940, p. 427) gives the average and extreme measurements of 40 specimens of both sexes from Del Norte County, California, as follows: average total length 3 13/32" (86.9 mm.), extremes 2 13/16" (72 mm.) and 4 3/32" (104 mm.). The proportions of an adult female from Carpentersville, Oregon, are: total length 2 27/32" (73 mm.), tail (tip regenerated) 1 7/32" (31 mm.); head length 1/2" (13 mm.), width 9/32" (7 mm.).

DESCRIPTION. This subspecies does not attain the size of the typical form and is mottled and blotched above. The head is large in proportion to the body, the sides behind the eyes nearly parallel to the lateral extensions of the gular fold, in front tapering abruptly to the short and bluntly pointed snout. The eyes are large and strongly protuberant, limited behind by a vertical fold. An impressed sinuous line from the posterior angle of the eye bends down posteriorly to meet the gular fold; a short vertical groove from this line extends to the angle of the jaw. The trunk is slightly depressed and with a slight middorsal impressed line. Costal grooves 12, counting 1 each in the axilla and groin, and 2 1/2–3 intercostal folds overlapped by the toes of the appressed limbs. Tail in the female short, flattened above at base, roughly spindle-shaped, and largest at about the middle of the length; tail of male longer and slimmer; a marked constriction immediately behind the vent. Legs rather short and stout. Toes 5–4, those of the hind feet 1–5–2–(3–4) in order of length from the shortest; toes of the fore feet 1–4–2–3. Tongue of moderate size, not completely filling the floor of the mouth, broadly oval in outline, free at the sides and behind. Vomerine teeth 19–21 in long series which arise outside the outer margin of the inner nares and arch broadly toward the mid-line, where they are separated by about 1/2 the diameter of a naris. Parasphenoid teeth in 2 club-shaped patches narrowly separated behind, more widely separated anteriorly, and distant from the vomerine teeth by about 3 times the long diameter of a naris.

COLOR. The general ground color of the trunk above is light reddish-brown. In many individuals, a dark line borders the upper edge of the groove running from the posterior angle of the eye to the gular fold. Extending from the back of the head to the base of the tail, a dorsolateral series of deep brown or blackish blotches; an interrupted middorsal line of small blotches, and another line on the lower sides between the fore and hind legs. The sides between the darker lines yellowish and lightly flecked. Sides and dorsum of tail strongly blotched with deep brown on a yellowish ground color. Belly and ventral surface of legs yellowish, venter of tail sometimes orange-yellow. Base of legs above yellowish, rest of legs mottled brown and yellow. Juvenile individuals often with the cheeks, tail, and basal segments of the legs above lighter, orange-red. The sides of the vent of the male somewhat swollen; the vent within lined with small papillae.

BREEDING. It is likely that the 2 adults, each found with a lot of 16 eggs at Carlotta, Humboldt County, California, on July 26, 1923, and reported by Storer (1925, p. 112), belong to this subspecies. The eggs and attendant adults were found under slabs of redwood on the ground. The eggs, in a relatively early stage of development, measured 5–5.5 mm. in diameter, and with the apparently single jelly envelope 7.6–7.8 mm. When found the eggs were nearly transparent; after preservation the yolk became opaque.

**SIERRA NEVADA SALAMANDER.** *Ensatina sierra* Storer. Fig. 86. Map 36.

TYPE LOCALITY. Yosemite Valley, Mariposa County, California.

RANGE. Transition zone of the Sierra Nevadas from Butte to Tulare Counties, California.

HABITAT. In and beneath logs and bark, usually in wooded areas. A specimen taken August 6, 1940, by Catherine Hemphill Brown, was found in a rotting log beside a trickling stream near Fern Spring, Yosemite Valley.

SIZE. Attains a length of 5 5/32" (131 mm.). A female of this size had a tail length of 2 3/16" (55 mm.) and a head length of 3/4" (19 mm.). (Dunn, 1926, p. 186.) The proportions of a small adult female from Fern Spring, Yosemite Valley, are as follows: total length 3 7/32" (81 mm.) tail 1 7/16" (36 mm.); head length 5/8" (16 mm.), width 11/32" (8 mm.).

DESCRIPTION. The strongly contrasting orange-and-black markings make this one of our most brightly colored salamanders. The head is large and long in proportion to the body, and convex above in the region back of the eyes. The sides of the head are nearly parallel but are constricted abruptly at the gular fold; in front tapering rather sharply to the bluntly pointed snout. The eyes are large and prominent, the iris dark brown slightly flecked with brassy. A deeply impressed line from the posterior angle of the eye extends on the side of the head to the lateral extension of the gular fold; a short vertical groove from this passes behind the angle of the jaw. The trunk is well rounded and provided above with a slightly impressed median line. There are 12 costal grooves, counting 1 each in the axilla and groin, and the toes of the appressed limbs may meet or overlap 2–3 intercostal spaces. Tail nearly oval in section at base and with a definite constriction immediately behind the vent. It is widest at about the basal third and tapers from that point uniformly to the tip. The tail is rounded above, sharp-edged below, and comprises 35.1–40 per cent of the total length. Legs well developed, the hind noticeably larger than the fore. Toes 5–4, those of the hind feet 1–5–2–3–4 (or 4–3) in order of length from the shortest, very slightly webbed at base; toes of the fore feet, 1–4–2–3. Tongue small, thin, bluntly pointed anteriorly, and with the sides and hind margins free. Vomerine teeth in long series of 17–24 or more. The series arise outside and behind the inner nares and curve inward toward the mid-line, where they are usually separated by about the width of an inner naris. Parasphenoid teeth in 2 elongate patches which may be separated or slightly in contact anteriorly and more divergent posteriorly. They are separated from the vomerine series by about twice the diameter of an internal naris.

COLOR. The general ground color is deep seal-brown to blackish, but is made to appear somewhat lighter by the presence of many small white

FIG. 86. *Ensatina sierra* Storer. (1) Adult male, actual length 2¹³⁄₁₆″ (76 mm.). (2) Same, ventral view. Fern Spring, Yosemite Valley, California. [Mrs. Stuart Brown, collector.]

pigment flecks scattered generally over the upper surfaces. The bright orange and orange-yellow blotches are irregular in size and shape and on the trunk confined largely to the sides. On the head they are largest on the sides back of the eyes, but are continued across the back of the head and between the eyes, leaving a median dorsal spot relatively free.

The snout is generally dark, relieved by a few white flecks. On the tail, the orange blotches tend to be brighter than elsewhere; they are greater in area than the dark brown interspaces and cross the dorsal surface. The basal joints of the legs are quite uniformly orange-yellow, the rest of the legs blotched with gray and orange; the feet mostly pale gray. The general color of the throat and belly is pale slate, the lower surface of the tail very light gray.

BREEDING. Nothing has been reported on the breeding habits of this species. Specimens only 36 mm. long have been found in June, and those having lengths of 39 and 42 mm. respectively have been taken in February and March (Storer, 1925, p. 107).

## GENUS HEMIDACTYLIUM

**EASTERN FOUR-TOED SALAMANDER.** *Hemidactylium scutatum* (Schlegel). Figs. 49a, 87. Map 37.

TYPE LOCALITY. Nashville, Tennessee.

RANGE. Southern Ontario and Maine westward to Wisconsin, south to Georgia and Alabama; also Arkansas, Missouri, and Illinois.

HABITAT. The adults are usually found in terrestrial situations, inhabiting wooded or more or less open areas adjacent to sphagnum bogs, swamps, quiet pools, and larch meadows.

SIZE. One of the smallest salamanders, adult males averaging only 2⁹⁄₁₆″ (65 mm.) in length and varying from 2 to 3″ (50–76 mm.). The females are a little larger and average nearly 3″ (75 mm.). They vary from 2⅞ to 3½″ (62–89 mm.).

DESCRIPTION. The body is nearly cylindrical, the trunk short, and provided with an impressed median line. The tail is broadly oval in section at the base, swollen at its mid-length, and only slightly compressed distally. At the base of the tail there is a definite constriction marking the point of detachment in autotomy. The head is somewhat flattened, the snout broadly rounded in the female and squarely truncate in the male. There are 13–14 costal grooves and the gular fold is well formed.

The legs are slender but relatively strong, toes 4–4, inner and outer rudimentary. Vomerine teeth in the male are in single backward-curving lines, usually separated from the parasphenoid patches. In the female the vomerine series may consist of 2 or more irregular rows. The tongue

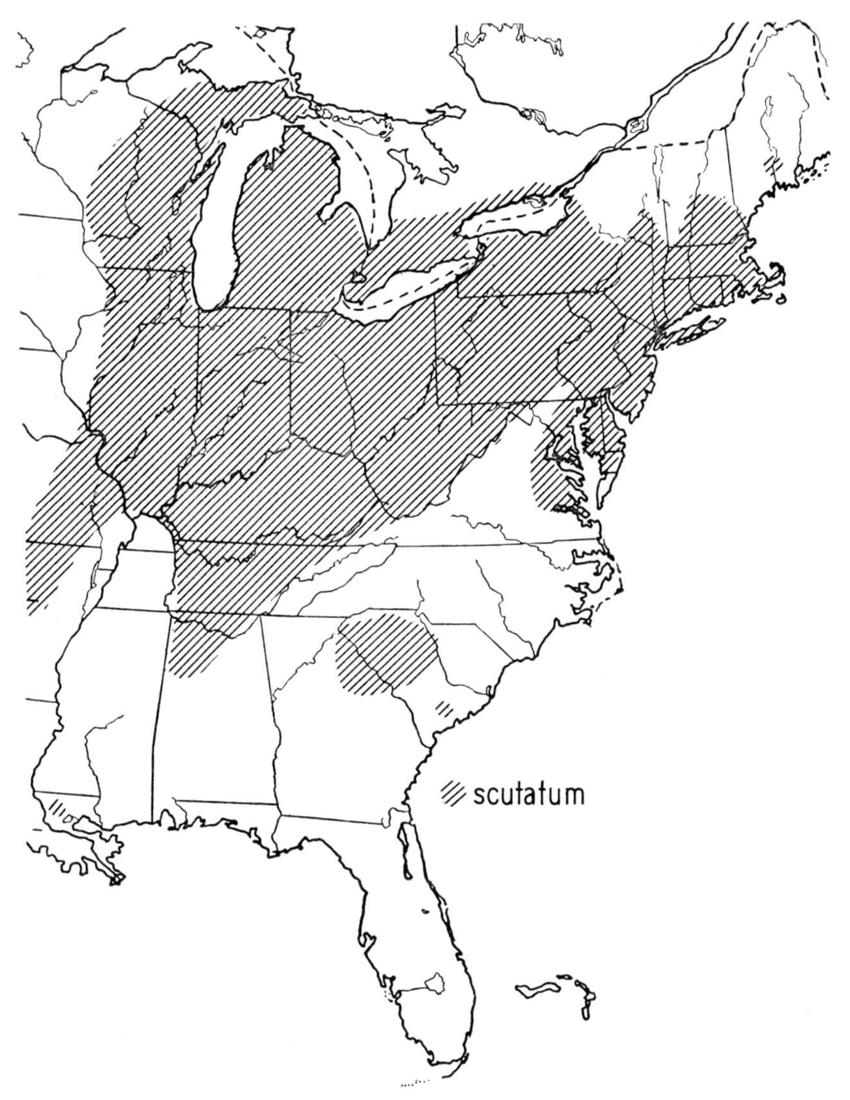


Map 37.—Distribution of *Hemidactylium scutatum*. (Also reported from Durham County, North Carolina.)

is small, narrow, and bluntly rounded behind and in front. The head and trunk above reddish-brown, fading slightly on the upper sides and

FIG. 87. *Hemidactylium scutatum* (Schlegel). (1) Adult male, actual length 2 7/16″ (62 mm.). (2) Same, ventral view. Mendon Ponds, Monroe County, New York. (3) Adult female with eggs. New Salem, New York, May 18, 1923.

becoming distinctly grayish on the lower sides between the costal grooves. The upper surface of the legs and tail has the ground color dull reddish-orange blotched with brown. The snout is bronzy and on each side above the shoulders there is a small, elongate, light yellowish-red patch. The venter is bluish-white, with small, irregular, inky-black flecks scattered over the surface. The mature males differ from the females in their smaller average size, more slender form, relatively longer tail, more squarely truncate snout, and in the disposition of the vomerine teeth. The 3 or 4 teeth on the premaxillary of the male are enlarged and sometimes perforate the lip.

BREEDING. There is an extended mating season through late summer and fall, and fertilization is accomplished by means of spermatophores. In the early spring, the females leave their winter quarters and migrate to the breeding grounds. The eggs are deposited in close proximity to water and usually in little cavities in sphagnum moss or among the roots of grass, rarely in rotting wood. The approximately 30 eggs are deposited singly and are not aggregated by a common envelope, but cling to one another and to the nest materials by the adhesive outer envelope. When depositing her eggs the female turns upon her back. Usually the female remains with the eggs until they hatch, and often many females will deposit their eggs in such close proximity to one another that it is impossible to determine the limits of individual masses. Under these circumstances only a few females remain in attendance. The individual egg has a diameter of 2.5–3 mm. and is surrounded by two envelopes, in addition to the vitelline membrane, which give a total diameter of about 5.3 mm. The incubation period may be limited to 38 days or extended to 60.

LARVAE. Larvae at hatching average about ½″ (12.4 mm.). The head is broad, the snout bluntly pointed. The trunk and tail are strongly compressed and wedge-shaped. The fore legs have 4 toes sometimes indicated, the hind legs short and directed backward. The gills are slender, tinged with orange at the base, and pigmented with black. The tail is keeled and continuous with the keel of the back, a condition unique among the larvae of the plethodontid salamanders. The head of the larva is tinged with orange, green, and yellow, and marked with dark brown or black. On the side of the head, a short, wide, dark bar extends from the eye to the gills. Along the mid-line of the trunk there is a light band with irregular edges bordered each side by dark worm-like markings. There is an aquatic larval stage of about 6 weeks and the transforming young attain a length of about ¾–1″ (18–25 mm.). Sexual maturity is probably attained at an age of about 2½ years.

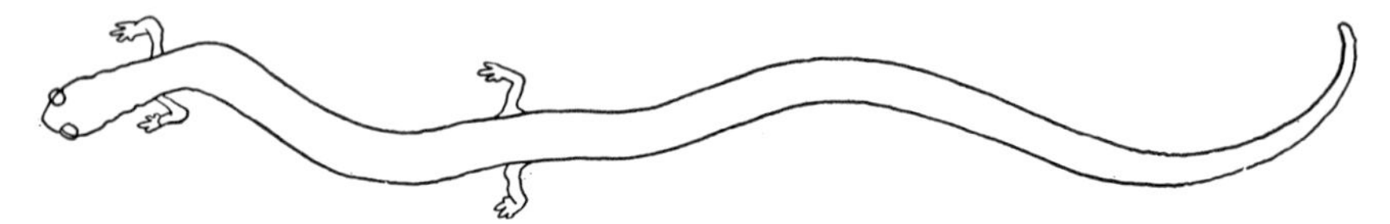

FIG. 88. Outline of *Batrachoseps attenuatus* to show the elongate form of the body and the four toes on both fore and hind feet. [H.P.C. *del.*]

## GENUS BATRACHOSEPS

### KEY TO THE SPECIES AND SUBSPECIES OF BATRACHOSEPS

1. Belly with at least some separate black flecks on a lighter ground color .......... 2
   Belly with a fine network of black, enclosing small, round, light spots .......... 4
2. Dark above, no dorsal band; hind legs when appressed to the sides overlapping 5½–6½ intercostal folds; vomerine teeth usually in a single series of 6–10, occasionally in irregular patches; parasphenoid teeth usually in 2 elongate patches narrowly separated, occasionally in a single broad patch as in *major;* length to 5⅟16″ (129 mm.). San Miguel, Santa Rosa, Santa Cruz, and Santa Barbara Islands off the west coast of California .......... *attenuatus pacificus* p. 323
   With at least a faint dorsal band .......... 3
3. Venter with separate black flecks only; hind legs when appressed to the sides overlapping 4–4½ intercostal folds; vomerine teeth in short patches of 1–3 rows; parasphenoid teeth in a single broad patch broadly rounded behind, pointed anteriorly, and narrowly separated from the vomerine; length to 6⅜″ (162 mm.). Los Angeles, Riverside, Orange, and San Bernardino Counties, California .......... *attenuatus major* p. 320
   Venter with black flecks confined to the central area, reticulated along sides; hind legs when appressed to the sides overlapping 4 intercostal folds; vomerine teeth in short patches, occasionally in rows of 6 or 7; parasphenoid teeth in a broad patch with a median groove incompletely separating the two sides or in 2 narrowly separated elongate patches; length to 5⁷⁄32″ (133 mm.). Santa Catalina Island, California .......... *attenuatus catalinae* p. 315
4. With a median dorsal stripe lighter than adjacent sides .......... 5

Without a conspicuous dorsal stripe; back scarcely lighter than sides . . 6

5. Dorsal light stripe well developed; venter fairly dark; head narrow; legs small, the hind when appressed to the sides overlapping 3–3½ intercostal folds; vomerine teeth in short patches of 2 or 3 rows; parasphenoid teeth in a single broad patch, broadly rounded behind, slightly narrowed and notched in front (southern part of range) or, patch more elongate, narrowed anteriorly and with a groove at mid-line extending ½ the length (northern); length to 5¾″ (136 mm.). Santa Cruz Island and Los Angeles County, California, northward to southwestern Oregon . . . . . . . . . . *attenuatus attenuatus* p. 311

6. Dark brown above, sometimes with a faint indication of a lighter dorsal band; sides dark, venter a little lighter; lower sides and margins of belly with many small grayish flecks; costal grooves 18–20; vomerine teeth in 2 imperfect rows of 5–8 or in short patches; parasphenoid teeth in 2 long club-shaped patches, united posteriorly, separated in front by a V-shaped groove, rarely in a single broad patch; length to 5″ (128 mm.). Southwestern California, Lower California, Los Coronados Island. Also recorded from Nevada de Colima, Colima, Mexico . . . . . . . . . . . . . . . . . . . . . . . . . . . . . . . . . . . . . *attenuatus leucopus* p. 317

   Dark brown above, slightly lighter below; no light gray flecks evident on lower sides and belly; costal grooves 21; vomerine teeth in short double series; length of type, 6²⁵⁄₃₂″ (172 mm.). Alaska . . . *caudatus* p. 325

**WORM-SALAMANDER.** *Batrachoseps attenuatus attenuatus* (Eschscholtz). Figs. 88–89. Map 38.

TYPE LOCALITY. Vicinity of the Bay of San Francisco, California.

RANGE. Southwestern Oregon, California to Los Angeles County, and lower western slopes of the Sierra Nevada (Stejneger and Barbour, 1939, p. 18). Also recorded from Santa Cruz Island (Campbell, 1931, p. 132).

HABITAT. Under logs, bark, boards, and stones on the ground, usually in moist situations, and in rotting logs, old stumps, and leaf mold. Burrows deeply during dry seasons.

SIZE. The average length of 15 adults of both sexes from various California localities is 3¹⁵⁄₃₂″ (89 mm.), the extremes 2⅞″ (73 mm.) and 4⁷⁄₁₆″ (114 mm.). Burke (1939, p. 211) indicates that an extreme length of 5¾″ (136 mm.) may be attained. The proportions of a male from Berkeley, California, are as follows: total length 4″ (102 mm.), tail

2 5/32″ (55 mm.); head length 11/32″ (9 mm.), width 7/32″ (4.5 mm.). A female from the same locality measures: total length 3 11/16″ (94 mm.), tail 1 31/32″ (50 mm.); head length 5/16″ (8 mm.), width 5/32″ (4 mm.).

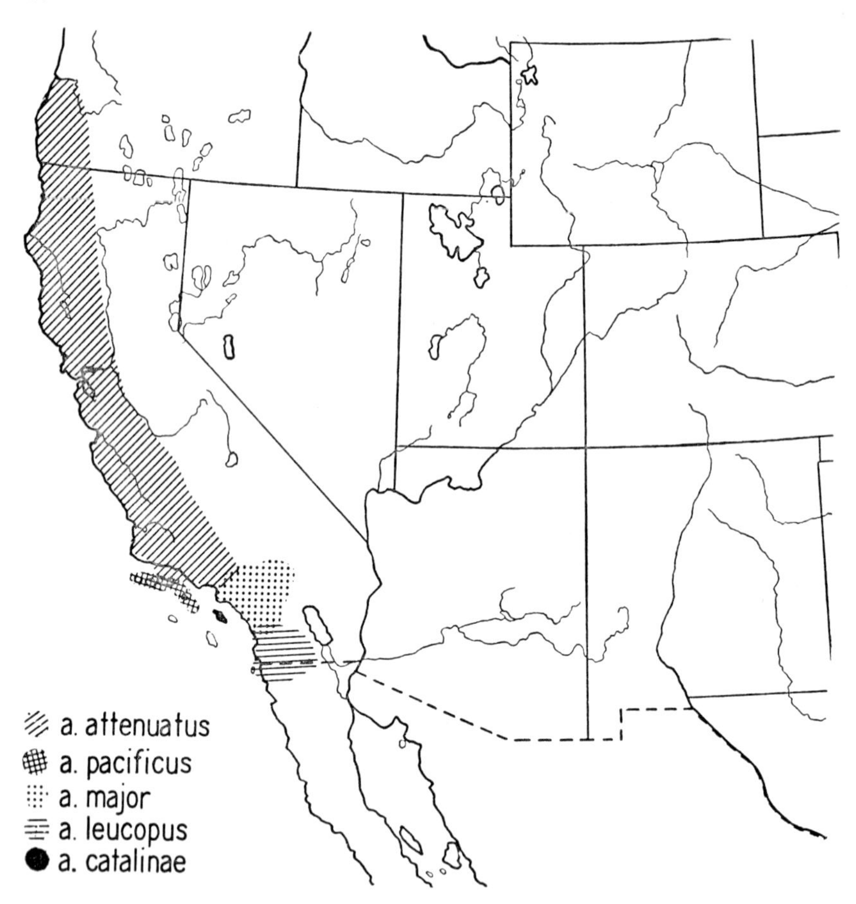


Map 38.—Distribution of the subspecies of *Batrachoseps attenuatus.*

Description. The light dorsal band, narrow head, and dark venter characterize this salamander. The head is widest immediately back of the eyes, the sides behind tapering gently to the lateral extensions of the gular fold and in front abruptly to the short, bluntly rounded snout. The eyes are of moderate size and limited behind by an oblique fold and groove. An impressed sinuous line from the posterior angle of the eye to the lateral extension of the gular fold; a short vertical groove from this line to a point behind the angle of the jaw and another about midway between the angle of the jaw and gular fold. Nasolabial grooves poorly developed. The trunk long, slender, and subcylindrical, slightly flattened below, and with an impressed median line above. The costal grooves usually 20, counting 1 each in the axilla and groin, but many specimens have only 19, occasionally 18 or 21, and there are 10–12½ intercostal spaces between the toes of the appressed limbs. The tail is nearly circular in section throughout its length, tapering only slightly to about the distal fourth, then more abruptly to the pointed tip. The tail is completely encircled by grooves that give a decidedly worm-like appearance. Legs very small and slender, the hind overlapping only 3–3½ intercostal folds when appressed to the sides. Toes 4–4, very short and ball-tipped, the innermost rudimentary, 1–4–2–3 in order of length from the shortest. Tongue small, roughly oval in outline, provided with a central pedicel and freely protrusible. Vomerine teeth in 2 short patches consisting of 2–3 rows, the patches usually separated at the midline by about the width of an inner naris and widely diverging anteriorly. The patches lie behind and wholly between the inner nares and are widely separated from the parasphenoids. Parasphenoid teeth may form a single large patch, broad behind and slightly narrowed anteriorly, or there may be 2 patches narrowly separated. Some adult males with enlarged premaxillary teeth and lower jaw narrowed anteriorly.

COLOR. The broad dorsal light band varies from light tan to brownish-red. The band may be almost uniform in color throughout, marked with a definite herringbone pattern, darker on the trunk and with a median dark line on the tail, or the median area of the band may be dark, the lateral margins light. The dorsal band is usually limited on either side by a narrow line somewhat darker, black or brown, and the sides below the line may appear bronze, chocolate-brown, or have a pepper-and-salt effect. The ventral pattern consists of a network of dark lines enclosing nearly circular light spots, so that the general color is dark, but some specimens have a few irregular lighter pigment flecks on sides and belly. The lower surface of the tail is sometimes greenish-yellow.

BREEDING. The egg-laying of this species was observed by Maslin (1939,

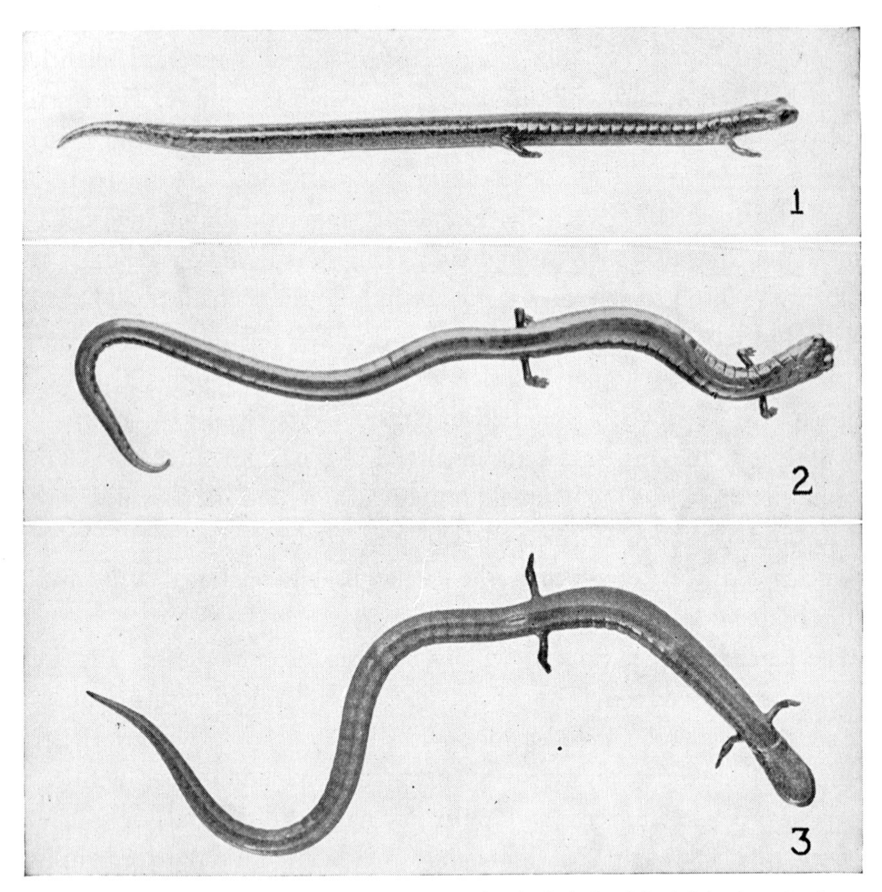


FIG. 89. *Batrachoseps attenuatus attenuatus* (Eschscholtz). (1) Adult, lateral view. Muir Woods, Marin County, California. (2) Adult female, actual length 4½″ (113 mm.). (3) Same, ventral view. Near Fort Ross, California.

p. 209) on Nov. 5, 1938. This is an early record for the vicinity of Berkeley, California, and the laying may have been initiated by the start of the rainy season. Other egg masses have been found in the field in January, near Palo Alto, California (Burke, 1911, p. 414), and at Snelling, Merced County, California (Grinnell and Storer, 1924, p. 654);

in March at Berkeley and Palo Alto (Storer, 1925, p. 24). Individual females deposit 12–15 eggs, but several may select the same retreat, beneath logs, rocks, or boards, or in pockets in the soil, in moist situations, and the total complements number as many as 74 (Burke, 1939, p. 211). Individual eggs when freshly deposited have a diameter of 4 mm. and are without pigment. They are surrounded by 3 envelopes, the inner of viscous jelly, the middle of semi-rigid jelly, and the outer thin and tough but not sticky. The eggs are attached to one another, sometimes in a bead-like string, and to surrounding objects by a gelatinous peduncle which may be 12 mm. long. The diameter of the outer envelope is 6 mm. Development under natural conditions is relatively slow and the time of hatching variable, but in the field probably occurs in the spring. (Burke, *ibid.*, pp. 209, 211.) A recently hatched young examined by Storer (1925, p. 96) measured as follows: total length 19 mm., tail 6 mm.; head width 2 mm. "The general coloration was black, with a slight brownish tinge; on the back were four parallel lines of spots silvery white and reddish gold in color; the ventral surface of the body was marked generally with scattered patches of silvery white."

**CATALINA ISLAND SALAMANDER.** *Batrachoseps attenuatus catalinae* Dunn. Fig. 90. Map 38.

TYPE LOCALITY. Santa Catalina Island, California.

RANGE. Santa Catalina Island, California.

HABITAT. During the rainy season not uncommon beneath surface materials and often discovered in numbers during the excavation of ditches, cellars, and sewers. During the dry season it burrows deeply.

SIZE. The average length of 9 adults of both sexes is 4³⁄₁₆" (105.7 mm.), the extremes 3³⁄₃₂" (79 mm.) and 5⁷⁄₃₂" (133 mm.). The proportions of a male, U. S. N. Mus. No. 89670, are as follows: total length 4¹⁷⁄₃₂" (116 mm.), tail 2¾" (70 mm.); head length ¹¹⁄₃₂" (9 mm.), width ⁷⁄₃₂" (6 mm.). An adult female, U. S. N. Mus. No. 89674, has the following measurements: total length 5⁷⁄₃₂" (133 mm.), tail 3¹⁄₁₆" (78 mm.); head length ⅜" (10 mm.), width ⁷⁄₃₂" (6 mm.).

DESCRIPTION. For living examples of this salamander I am indebted to Mr. Joseph R. Slevin, who collected a fine series at Avalon, Santa Catalina Island, in April 1941. The head is narrow, the sides behind the eyes converging slightly to the lateral extensions of the gular fold and

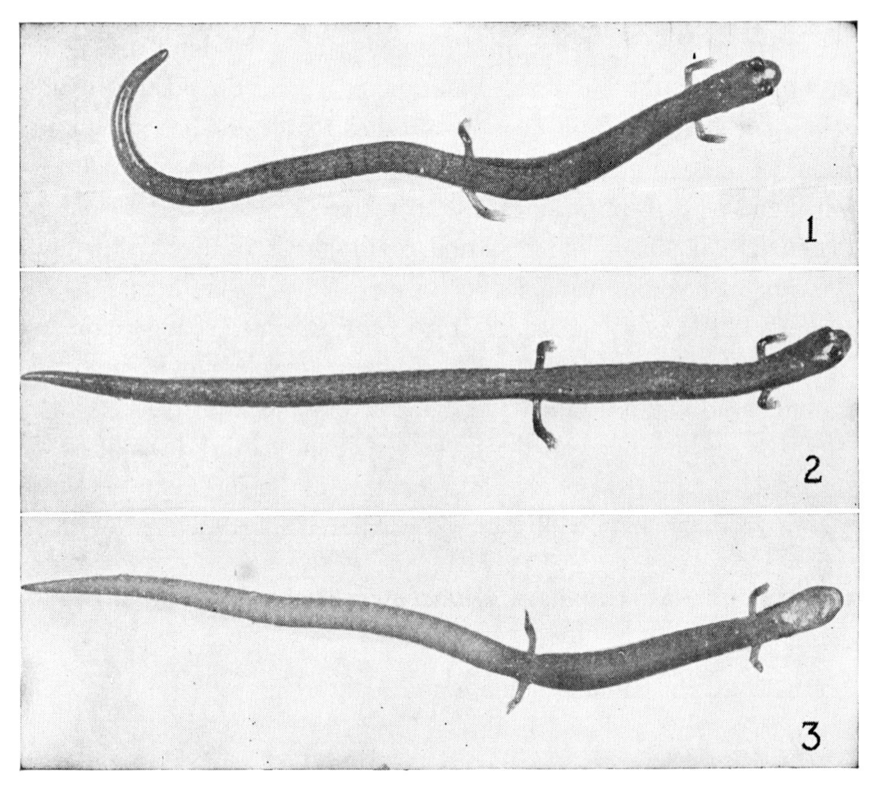


FIG. 90. *Batrachoseps attenuatus catalinae* Dunn. (1) Adult female, actual length 3½" (89 mm.). Santa Catalina Island, California. (2) Adult female, another individual; dorsal view. (3) Same, ventral view. [J. R. Slevin, collector.]

in front abruptly to the bluntly rounded snout. The eyes are large and strongly protuberant, the long diameter a little greater than the length of the snout; the skin above the eyes black with gold flecks; the iris flecked with brassy. An impressed line from the posterior angle of the eye to the lateral extension of the gular fold. The trunk is subquadrate in section and with a slightly impressed median line which forks at the back of the head and sends a branch to each eye. Costal grooves 20–21; and 10–13 costal folds between the toes of the appressed limbs. The tail is long, slender, and worm-like, and comprises from 53.2 per cent of the total length in the smaller individuals to 60.34 per cent in the large adults. The legs are slender and weak; toes 4–4, 1–4–2–3 in order of length from the shortest, the 1st rudimentary and partly fused to the 2nd. Tongue elongate oval, thin at margin, free at the sides and behind. Vomerine teeth in short irregular patches, or, rarely, in fairly regular rows of 6 or 7, which lie between the inner nares, narrowly separated at the mid-line, and about 3 times the diameter of a naris from the parasphenoids; the latter in 2 long, slender, club-shaped patches which may be narrowly separated or contiguous anteriorly.

COLOR. This salamander has a broad dorsal band which is scarcely lighter than the adjacent sides. The band is a dull bronze, darker in some individuals than in others, and extends from the snout the length of the trunk well onto the tail. The upper sides next to the dorsal band are brown, fading to grayish on the lower sides. The throat is tan or light gray, with a loose network of small black chromatophores and scattered white flecks. In most specimens, dark pigment of the belly and ventral surface of the tail forms a reticulated pattern over the purplish-gray ground color. In others the dark pigment is mostly in separate flecks along the central region of the belly and reticulated along the sides. The dorsal surface of the legs is colored like the back, the feet and toes grayish.

BREEDING. Nothing has been written on the breeding habits of this island form, and the eggs and recently hatched young have not been described.

**SOUTHERN SLENDER SALAMANDER.** *Batrachoseps attenuatus leucopus* Dunn. Fig. 91. Map 38.

TYPE LOCALITY. North Island, Los Coronados Islands, Baja California, Mexico.

RANGE. Los Coronados Islands; Baja California; San Diego and Imperial Counties, California.

HABITAT. Klauber (1927, p. 1) remarks concerning this species in San Diego County: "This salamander seems to be rather common in moist locations, especially shaded north slopes, throughout the western part of the County. It is frequently found in San Diego (city) gardens."

SIZE. The average length of 14 specimens from San Diego County, California, is $3\frac{13}{32}''$ (87 mm.), the extremes $2\frac{19}{32}''$ (66 mm.) and $4\frac{31}{32}''$ (127 mm.). The proportions of an adult female from San Diego, California, are as follows: total length $4\frac{15}{32}''$ (114 mm.), tail $2\frac{11}{16}''$ (68 mm.); head length $\frac{5}{16}''$ (8 mm.), width $\frac{7}{32}''$ (4.5 mm.). An adult male from the same locality has the following measurements: total length $3\frac{7}{32}''$ (82 mm.), tail 2″ (51 mm.); head length $\frac{9}{32}''$ (7 mm.), width $\frac{5}{32}''$ (4 mm.).

DESCRIPTION. The head is broadly oval in outline, widest at the angle of the jaws, the sides behind this point gently rounding to the lateral extension of the gular fold, the snout short and broadly rounded. The eyes large, the long diameter about equal the length of the snout, limited behind by an oblique fold; the iris flecked above the pupil with gold. A sinuous groove from the posterior angle of the eye to the lateral extension of the gular fold; a vertical groove from this passing behind the angle of the jaw. Trunk rounded above, with an impressed median line. Costal grooves usually 19, counting 1 each in the axilla and groin, occasionally 18 or 20, and 8–11 intercostal folds between the toes of the appressed limbs. Tail worm-like; long, slender, tapering, and encircled by grooves which give it a segmented appearance; in length comprising 50–62.2 per cent of the total length, the higher figures associated with males and larger individuals. Legs very small and slender. Toes 4–4, short, bluntly rounded at tip, the innermost rudimentary; those of the hind feet, 1–(4–2)–3 in order of length from the shortest; toes of the fore feet 1–4–2–3. Tongue broad behind, bluntly pointed in front, free at the sides and behind. Vomerine teeth in short irregular patches or in series consisting of 2 imperfect rows of 5–8 teeth which arise within or behind the inner margin of the inner naris and slant obliquely inward and backward toward the mid-line, where they may be in contact or narrowly separated. Parasphenoid teeth variable, in some in a single patch broad behind and narrowed anteriorly, in others in 2 imperfectly separated patches; in some females the patches in contact anteriorly, divergent posteriorly.

COLOR. The color is quite variable, but in general the sides are dark,

FIG. 91. *Batrachoseps attenuatus leucopus* Dunn. (1) Adult female, actual length 3 1/16" (78 mm.). (2) Same, ventral view. San Diego, California. [L. M. Klauber, collector.]

the dorsal surfaces somewhat lighter. In the majority of specimens I have examined, there is a dorsal band which arises on the head and continues the length of the trunk and tail. The dorsal band may be brown or dark gray, scarcely lighter than the upper sides, or the head and trunk deep red strongly suffused with dusky, the tail much brighter, almost orange-red. Occasional individuals will have the dorsal band coppery-red throughout its length from the back of the head to the tip of the tail; in others the band may be dark on the trunk, coppery-red on the tail, and with patches of red on the snout, above the eyes, on the shoulders, and in scattered flecks along the sides of the trunk. The ground color of the upper sides is deep brown, with minute spots of lighter brown and many flecks or lines of white. The legs above are generally colored like the sides and with a patch of red at the base. The belly has a finely reticulated black pattern over a lavender background, and along the sides many small white pigment flecks. The ventral surface of the legs, throat, and tail may be grayish or the venter of the tail yellowish.

BREEDING. Nothing is known of the breeding habits of this species, and the eggs have not been described. According to Klauber (1934, p. 5) it æstivates during the long, rainless summer.

**GARDEN SALAMANDER. CAMP'S SALAMANDER.** *Batrachoseps attenuatus major* Camp. Fig. 92. Map 38.

TYPE LOCALITY. Sierra Madre, Los Angeles County, California.

RANGE. Los Angeles, Riverside, Orange, and San Bernardino Counties in southwestern California.

HABITAT. Found beneath logs, boards, and rocks, or buried in loose gravel; occasionally in cellars. In May 1936 we collected a fine series from beneath the boards covering a cesspool in a yard at Claremont, California.

SIZE. A series of 6 sexually mature individuals measured by Camp (1915, p. 330) varied in length from 3¼" (82.7 mm.) to 6⅜" (162 mm.) and averaged 4⅝" (117 mm.). The proportions of a male from Claremont, California, are as follows: total length 4 13/32" (111 mm.), tail 2½" (63 mm.); head length 13/32" (9.5 mm.). A female from Sierra Madre, noted by Camp (*ibid.*) measured: total length 4 9/16" (115.5 mm.), tail 2 19/32" (66 mm.); head width 7/32" (5.9 mm.).

DESCRIPTION. The large size, light venter, and relatively wide head distinguish this species from *Batrachoseps a. attenuatus,* with which it is sometimes associated. The head is widest at the eyes, the width comprising as much as 63.1 per cent of the head length measured from the tip of the snout to the lateral extension of the gular fold. Behind the eyes, the sides of the head usually taper slightly to the gular fold, and

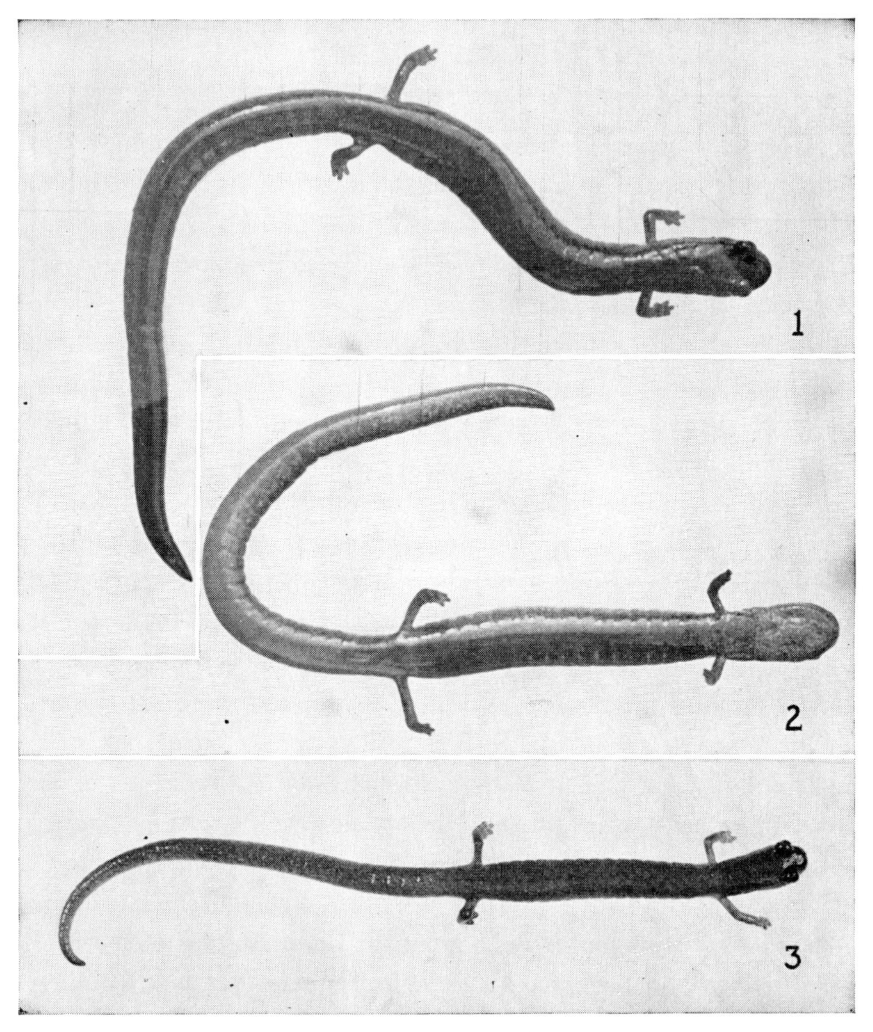


Fig. 92. *Batrachoseps attenuatus major* Camp. (1) Adult male, actual length 4½″ (113 mm.). (2) Same, ventral view. Claremont, California. (3) Another individual. Altadena, California.

in front abruptly to the short and bluntly rounded snout. The eyes large and strongly protuberant, limited behind by an oblique fold and groove; the iris brassy in narrow lines above the pupil. A poorly developed impressed line from the posterior angle of the eye to the lateral extension of the gular fold; and a short vertical groove from this line to behind the angle of the jaw. Nasolabial grooves scarcely evident. Trunk subcylindrical, slightly flattened below. There are usually 19 costal grooves, counting 1 each in the axilla and groin, but the number may vary from 18 to 20; 9–10 intercostal folds between the toes of the appressed limbs. Tail long, nearly circular in section throughout its length, and tapering to a slender tip. The tail length increases with size and may comprise 52–57 per cent of the total. Tail completely encircled by grooves. Legs short but not so slender as in *Batrachoseps a. attenuatus;* when appressed to the sides each overlapping 4–4½ intercostal folds. Toes 4–4, short, ball-tipped beneath, the innermost small but more nearly normal than in *B. a. attenuatus;* 1–4–2–3 in order of length from the shortest. Tongue moderate in size, roughly oval in outline, thin at the margins, and with a central pedicel; strongly protrusible. Vomerine teeth small, in short patches of 1–3 rows, narrowly separated at the mid-line and lying wholly between and slightly behind the inner nares. Parasphenoid teeth often forming a single patch, broad and rounded behind, slender and tapering anteriorly, or in 2 patches narrowly separated behind, united anteriorly; separated from the vomerine teeth by about 3 times the diameter of an inner naris. Premaxillary teeth of males not strikingly enlarged in the series examined, but point of lower jaw of the male somewhat narrower than in the female.

COLOR. Above somewhat variable, ranging from light grayish-brown to brownish-red; the dorsal surface of the tail often with brick-red spots overlaid by a stippling of light gray; the dorsal surface of the head often grayer than the adjacent trunk. Upper sides of trunk definitely brownish and sharply marked off from lower sides, which are bluish-gray. Sides of tail lighter than sides of trunk, silvery-gray with fleshy tinges. Legs brownish above, grayish on the sides. The throat is flesh color, the belly light bluish-gray, the lower surface of the tail flesh or dull yellowish-white. In some individuals there is a broad dorsal band extending from the snout to the tip of the tail, but it is never developed to the extent found in *Batrachoseps a. attenuatus.* The detailed pattern of the ventral surfaces consists of separated, small, black chromatophores distributed uniformly over the light ground color, and a few scattered white pigment flecks. In some young individuals the entire dorsal surface of the trunk and tail and the snout in front of the eyes may be tinged with brick-red, which is in strong contrast to the brownish sides. In others, the upper surface of the tail alone may be bright red in irregular blotches. In preservatives, the general color above fades to pinkish-brown, the ventral surfaces dull yellow, resembling the colors of an earthworm.

BREEDING. Nothing is known of the breeding habits of this species, and the eggs have not been described.

**PACIFIC WORM-SALAMANDER.** *Batrachoseps attenuatus pacificus* (Cope). Fig. 93. Map 38.

TYPE LOCALITY. Recorded from Santa Barbara, on the coast of Southern California, but probably from Santa Barbara Island.

RANGE. San Miguel, Santa Rosa, Santa Cruz, and Santa Barbara Islands off the west coast of California.

HABITAT. Usually found under sticks, stones, or lumps of earth in damp places; also found under loose bark of fallen trees (Slevin, 1928, p. 47).

SIZE. The average length of 13 adults of both sexes from Santa Cruz and San Miguel Islands, California, is $3\frac{25}{32}''$ (96.5 mm.), the extremes $2\frac{13}{16}''$ (72 mm.) and $5\frac{1}{16}''$ (129 mm.). The proportions of the largest female (from San Miguel) are as follows: total length $5\frac{1}{16}''$ (129 mm.), tail $2\frac{3}{4}''$ (70 mm.); head length $\frac{13}{32}''$ (11 mm.), width $\frac{7}{32}$ (6 mm.). An adult male from Santa Cruz has the following measurements: total length $3\frac{1}{32}''$ (77 mm.), tail $1\frac{19}{32}''$ (41 mm.); head length $\frac{3}{8}''$ (10 mm.), width $\frac{3}{16}''$ (5 mm.).

DESCRIPTION. The head is widest immediately behind the eyes, tapering slightly to the lateral extensions of the gular fold, in front more abruptly to the truncated and broadly rounded snout. Eyes large and protuberant, the horizontal diameter a little less than the length of the snout, limited behind by an oblique fold. An impressed line from the posterior angle of the eye to the lateral extension of the gular fold. Trunk slender, subcylindrical, marked above with a slight median impressed line. Costal grooves variable, 17–20, the usual number 19, counting 1 in the axilla and 2 that run together in the groin, and 9–11 intercostal folds between the toes of the appressed limbs. Tail long, slender, tapering, and marked with grooves that give it a segmented appearance. The tail may comprise 44–54.3 per cent of the total length in adults, the higher figures associated with the largest specimens. Legs slender, noticeably longer than in *B. p. major,* hind, when appressed to the sides overlapping 5½–6½ intercostal folds. Toes 4–4, ball-tipped below

FIG. 93. *Batrachoseps attenuatus pacificus* (Cope). Adult, about natural size. San Miguel Island, California. [Photograph through the courtesy of J. R. Slevin.]

but more nearly normal than in other species of the genus, 1–4–2–3 in order of length from the shortest, the 1st rudimentary, 2nd and 4th nearly equal; toes of the hind feet webbed at base. Nasolabial groove scarcely reaching the edge of the lip, sometimes branched at tip. Tongue small, oval, provided with a pedicel and membrane running to the anterior margin. Vomerine teeth often in a single irregular row of 7–10 teeth, sometimes in short double series or irregular patches which arise behind the inner margin of the inner naris and slant inward and backward toward the mid-line, where they are narrowly separated. Parasphenoid teeth sometimes in 2 slender elongate patches narrowly separated, or in a single broad patch narrowed anteriorly and separated from the the vomerine series by about 3 times the diameter of a naris.

COLOR. The general color resembles that of *B. p. major* in that the dorsal surfaces are dark and the ventral light; but there is considerable variation. The San Miguel specimens I have examined resemble *major* closely and have the ventral pigmentation in fine specks like that subspecies. In some Santa Cruz specimens the dark pigment of the venter forms a reticulated pattern like that of *B. a. attenuatus,* and there is an indication of a dorsal band, but the legs are longer and larger and the head wider and larger. Other individuals from Santa Cruz have the pigment of the venter in fine specks like *B. p. major.* In preserved specimens, the dorsal color becomes purplish-brown, with the legs, head, and tail somewhat lighter. The color fades slightly on the lower sides to the pale venter.

BREEDING. Nothing definite is known. A large female from San Miguel Island, No. 45226 in the collection of California Academy of Sciences, had ovarian eggs 2 mm. in diameter when captured May 20, 1919.

**ALASKA WORM-SALAMANDER.** *Batrachoseps caudatus* Cope. Fig. 94. Map 11.

TYPE LOCALITY. Hassler Harbor, Alaska.

RANGE. Known only from the type locality and doubtfully from Yukatat Bay.

HABITAT. Nothing has been published. Stejneger and Barbour (1939, p. 18) believe that the type may have been taken on Anette Island in southeastern Alaska.

SIZE. The type, U.S.N.Mus. No. 13561, is the only specimen in condition to be measured, and, with the passage of time, its dimensions seem to have changed somewhat. In Cope's original account the total length is given as 160 mm., tail 103 mm. (1889, p. 126). Dunn (1926, p. 234) has the total length 165 mm., tail 110 mm. I have recently examined the specimen and my measurements are as follows: total length 6 25/32″ (172 mm.), tail 4 3/16″ (107 mm.); head length 11/32″ (9 mm.), width 7/32″ (5.5 mm.).

DESCRIPTION. The head is narrow, the sides behind the eyes gently converging to the lateral extensions of the gular fold, in front converging more abruptly to the bluntly pointed snout. Eyes moderate, the horizontal diameter a little less than the snout. A slightly impressed line from the posterior angle of the eye to the lateral extensions of the gular fold, a vertical groove from this line to the angle of the mouth, and another groove to the angle of the jaw. The trunk is slender and well rounded. There are 21 costal grooves, and 12 intercostal folds between the toes of the appressed limbs. Tail long, slender, tapering, and comprising 62.2 per cent of the total length. Legs slender. Toes 4–4, those

FIG. 94. *Batrachoseps caudatus* Cope. (1) Type; adult female, dorsal view. (2) Same, ventral view. Hassler Harbor, Alaska. [Photograph from the U.S. National Museum.]

of the hind feet 1–4–2–3 in order of length from the shortest, 1st very short and webbed to tip. Toes of the fore feet 1–4–2–3, 1st within web. Tongue small, oval. Vomerine teeth in short double series which arise behind inner margin of inner naris and slant inward and backward toward the mid-line, where they are narrowly separated. Parasphenoid teeth in 2 elongate separate patches.

COLOR. The color at the time of Cope's description (1889, p. 126) was given as follows: "The general color is brown. It is deeper on the sides to a line on each side of the back and on the anterior half of the abdomen and on the superior surface of the distal part of the tail. Gular region and chin yellowish." At the present time (January 1941) the colors are about as follows: dark brown above, lighter below, the dorsal pattern consisting of small round spots which form a finely reticulated pattern.

Venter finely reticulated, the network of dark lines enclosing nearly circular spots.

BREEDING. Nothing is known.

## Genus ANEIDES

### KEY TO THE SPECIES AND SUBSPECIES OF ANEIDES

1. Ground color above black, or black with yellowish green blotches .... 2
   Ground color not as above .... 3
2. Ground color above black; belly black or dark slate; back and sides with small, scattered, whitish or yellowish specks (often fading in preservatives); soles of feet light; vomerine series short, 3–5; toe tips bluntly rounded and only slightly widened; 3½–5 intercostal folds between toes of appressed limbs; length to 5¹⁵⁄₁₆″ (152 mm.). Santa Cruz County, California, northward in the Coastal Range to the Klamath Mountains .... *flavipunctatus* p. 336
   Ground color above black (fades to brown with dull tan markings in preservatives), with yellowish-green lichen-like markings; base of legs above with a spot of bright orange; belly light, yellow to bluish-gray, with a few scattered light flecks; vomerine series 5–9; toe tips transversely widened and squarely truncate; toes of appressed limbs overlap 1–3 intercostal folds; length to 5″ (128 mm.). Virginia, West Virginia, Ohio, North and South Carolina, Georgia, Alabama, and eastern Tennessee .... *æneus* p. 328
3. Ground color above uniform light brown to yellowish- or chocolate-brown, with small whitish or yellowish spots; 15–16 costal grooves; mandibular teeth very large, flattened, sharp-pointed .... 4
   Ground color above deep brown in a reticulated pattern with a few small, whitish pigment flecks and many small brassy flecks most abundant on the snout and in a median dorsal band; base of tail above bronzy; venter bluish-gray, with a few scattered whitish flecks; mandibular teeth 5–8 each side, the posterior enlarged and flattened; 16–17 costal grooves; toe tips squarely truncate; length to 5¼″ (134 mm.). Southern Mendocino County, California, northward to Vancouver Island, British Columbia .... *ferreus* p. 332
4. Light spots of dorsum and sides few, small, and scattered; vomerine series 5–8; tail long, somewhat prehensile; toe tips slightly expanded; toes of appressed limbs meeting or overlapping 1 intercostal fold; length to 6⅜″ (162 mm.). Southwestern California from Humboldt and San Diego Counties; also from Mariposa and Madera Counties; Coronados Island, Lower California .... *lugubris lugubris* p. 340
   Light spots of dorsum abundant, varying in size from small flecks to elongate spots 3 or 4 mm. in length; vomerine series 5–11; toe tips

squarely truncate and slightly widened; toes of appressed limbs overlap 1½–3 intercostal folds; length to 5³⁄₃₂″ (130 mm.). South Farallon Island, California . . . . . . . . . . . . . . . . . . . . *lugubris farallonensis* p. 343

**BRONZED SALAMANDER. GREEN SALAMANDER.** *Aneides æneus* (Cope). Figs. 95a-b, 96. Map 39.

TYPE LOCALITY. Nickajack Cave, Tennessee.

RANGE. Virginia, West Virginia, Kentucky, Ohio, North and South Carolina, Georgia, Alabama, and eastern Tennessee.

HABITAT. Often abundant in crevices of sandstone cliffs, or beneath flakes of rock in fairly dry or damp situations. Sometimes found beneath the loose bark of standing or fallen trees, occasionally in or beneath logs on the ground.

SIZE. The average length of 19 adults of both sexes from Kentucky and West Virginia is 4″ (102 mm.), the extremes 3⁵⁄₃₂″ (81 mm.) and 5″ (128 mm.). Ten males average 4³⁄₃₂″ (104 mm.) and seven females 4¹⁄₃₂″ (103 mm.). The proportions of an adult male from Elkins, West Virginia, sent me by Mr. N. B. Green, are as follows: total length 5″ (128 mm.), tail 2²³⁄₃₂″ (69 mm.); head length ⅝″ (16 mm.), width ¹³⁄₃₂″ (11 mm.). An adult female from the same locality measures: total length 3⁵⁄₃₂″ (81 mm.), tail 1¼″ (32 mm.); head length ½″ (13 mm.), width ⁵⁄₁₆″ (8 mm.).

DESCRIPTION. This is a black salamander with yellowish-green, lichen-like blotches and expanded toe tips. The head is widest immediately back of the eyes, where, in the adult male, it is somewhat swollen; the sides behind the eyes converging slightly to the lateral extensions of the gular fold, in front tapering more abruptly to the truncated snout. Eyes large and strongly protuberant, the pupil horizontally elliptic, the iris golden. A sinuous impressed line from the posterior angle of the eye to the lateral extension of the gular fold; a short vertical groove from this extending behind the angle of the jaw. Gular fold well developed. Trunk rounded above, flattened below. The costal grooves usually 14; or 15 counting 1 in the axilla and 2 that run together in the groin. Toes of appressed legs overlapping 1–3 intercostal folds. Tail slightly flattened at base above, nearly circular in section beyond base, slender and tapering. Legs well developed, the animal capable of leaping 2 or 3 times its length. Toes 5–4, those of the hind feet 1–2–5–(3–4) in order of

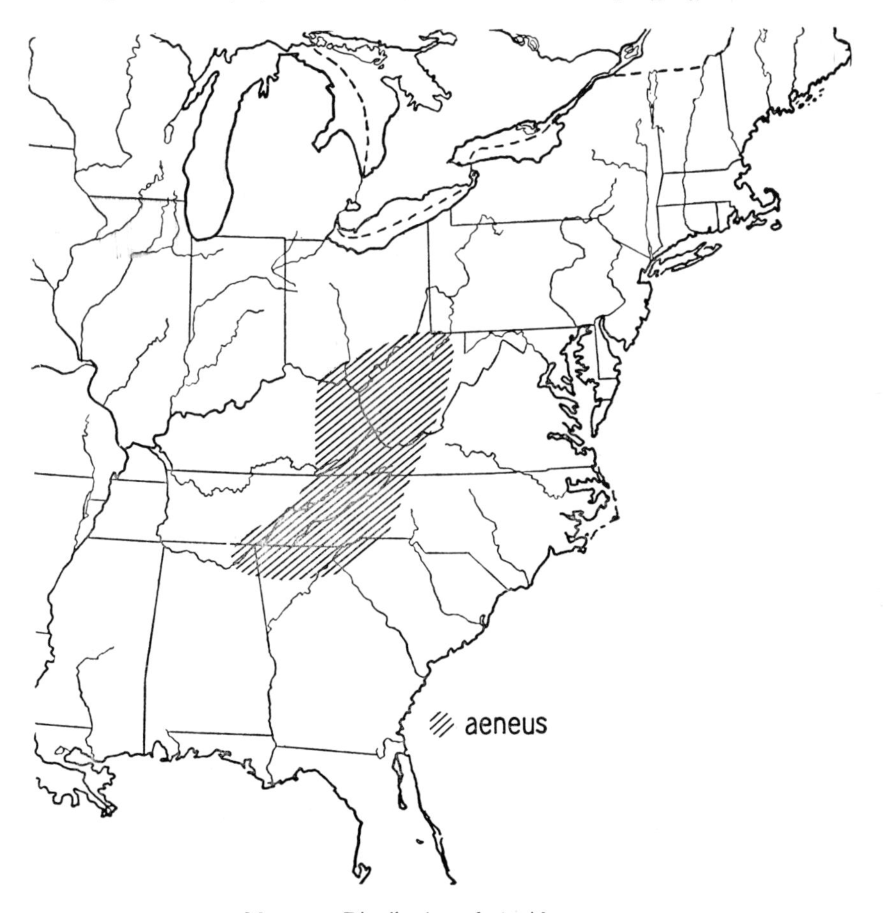


Map 39.—Distribution of *Aneides æneus.*

length from the shortest; 1st toe very short and nearly enclosed by web; others webbed ½ the length. Toes of the fore feet 1–2–4–3, the 1st enclosed in the web, others webbed at base. Toe tips transversely widened and squarely truncate. Tongue heart-shaped, free at sides only, the surface finely papillose. Vomerine teeth in series usually varying in number, in different individuals, from 5 to 9. The series arise behind the middle of the inner nares and curve gently inward and backward toward the

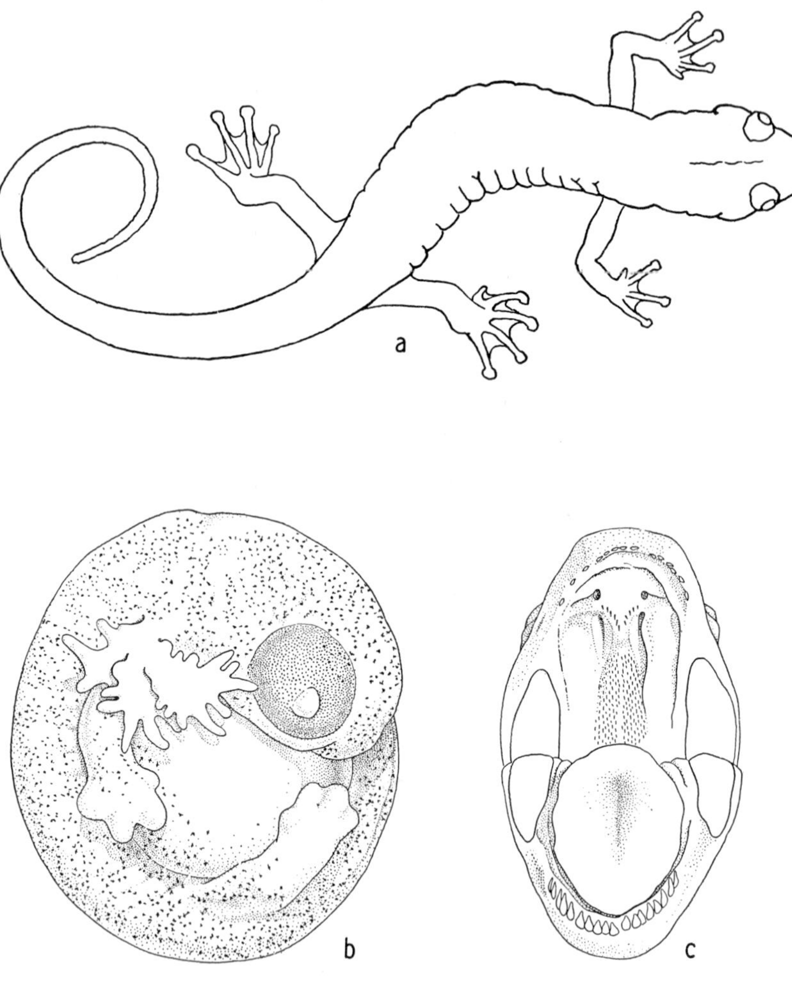


FIG. 95. (a) Outline of *Aneides æneus* to show the form of the body and the character of feet and toes. [H.P.C. *del.*] (b) Encapsuled larva of *Aneides æneus.* (c) Mouth parts of *Aneides lugubris lugubris* to show the character of tongue and teeth. [b, c, M.L.S. *del.*]

mid-line, where they are separated by about the width of an inner naris. Parasphenoid teeth may form a single patch, wide behind, narrowed in front, or 2 elongate groups narrowly separated; occasionally the patches may be separate behind and contiguous anteriorly. Teeth on maxillary extending to behind margin of inner naris, rest of maxillary sharp-edged. Teeth on mandible limited to anterior ⅔, those on the sides enlarged

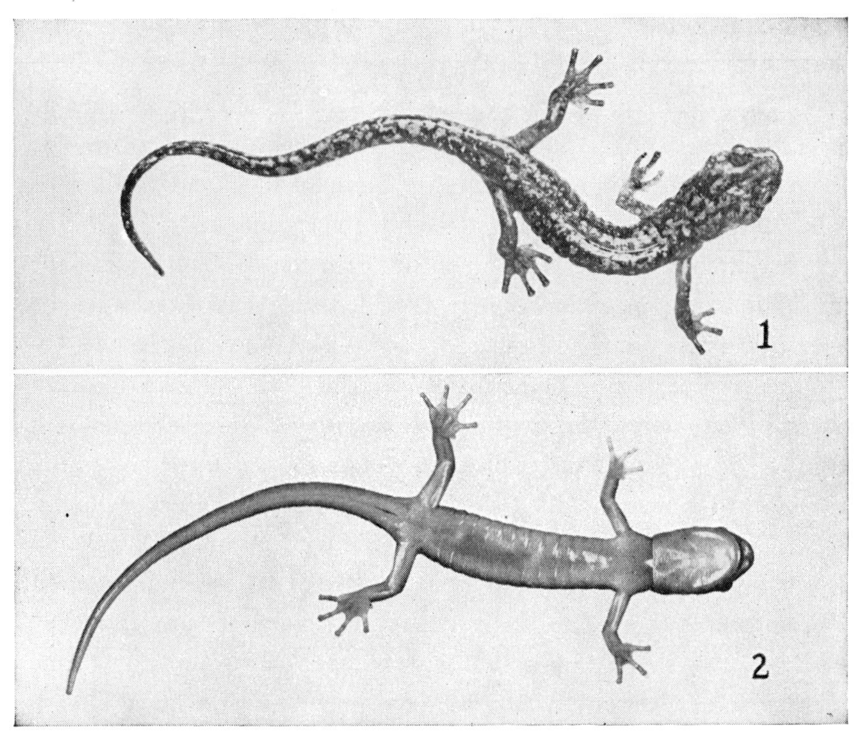


FIG. 96. *Aneides æneus* (Cope). (1) Adult male, actual length 5″ (127 mm.). (2) Adult male, another individual; actual length 4[13]/16″ (122 mm.). Near Elkins, West Virginia.

and flattened. Vent of male with fine papillae within lips anteriorly; vent of female smooth within.

COLOR. The general ground color is black, marked above with yellowish-green, lichen-like patches; on the sides the patches are fewer, smaller, and more widely scattered. At the base of the legs above, a bright orange patch; rest of legs blotched with greenish-yellow. Head generally darker than body and with a few small, whitish points among the larger patches. Belly light yellow to bluish-gray, with a few scattered light flecks. Lower surface of legs, throat, and tail pale yellowish-gray, the throat with many small irregular light spots.

BREEDING. Nothing is known of the mating habits, but eggs have been discovered a number of times and the encapsuled larvae described. Eggs were first reported by Pope (1928, p. 6). On July 23, 1937, an irregular cluster of 14 was discovered in the cavity of a prostrate water-oak limb near Pine Mountain, Kentucky. The eggs adhered to one another by the adhesiveness of the outer envelopes and were attached to the side of the cavity by 4 short mucous cables. In color they were dirt-brown with a tinge of yellow, perhaps stained by the rotting wood. Individual eggs with their envelopes have an outside diameter of about 5 mm. The egg itself is a light yellow sphere provided with two envelopes, the inner thin and resistant, the outer soft and thick (Pope, *ibid.*, p. 10). On Aug. 19, 1937, Mr. Leslie Hubricht collected a lot of 19 eggs on the north side of Pine Mountain, Kentucky. These eggs, with embryos well formed, had a long diameter of 5 mm. and a short diameter of 4 mm. The embryo was surrounded by 2 thin envelopes only. Gentry (1941, p. 329) reported the discovery of a female with a cluster of eggs on Sept. 5, and again on Sept. 6, in the Obey River drainage in Tennessee.

From evidence presented by Walker and Goodpaster (1941, p. 178) it is likely that the young remain with the guardian parent some time after hatching. Groups of young were twice found, each with an attending adult female. The first lot found numbered 13, the second 17, the size ranging from 21.5 to 24.8 mm. These are the smallest specimens so far reported. The young lacked gills and exhibited a color pattern similar to that of adults. The large size of the external nares is a striking characteristic of the recently hatched young.

**CLOUDED SALAMANDER. RUSTY SALAMANDER.** *Aneides ferreus* Cope. Fig. 97. Map 40.

TYPE LOCALITY. Fort Umpqua, Oregon.

RANGE. Southern Mendocino County, California, northward to Vancouver Island and smaller islands near by, British Columbia.

HABITAT. While not as arboreal in habit as *A. l. lugubris,* this species has sometimes been found living in the rotten wood of standing trees 20 feet above the ground. The majority of specimens have been taken from beneath logs, bark, or boards, or in rotting logs on the ground.

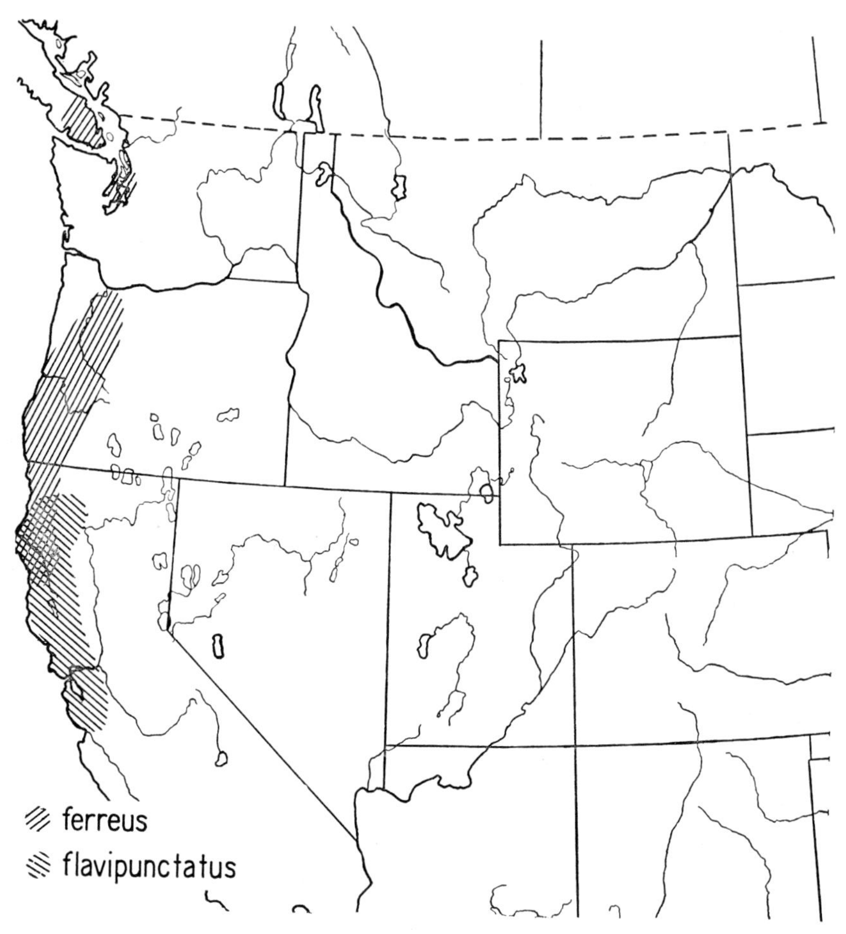


MAP 40.—Distribution of *Aneides ferreus* and *A. flavipunctatus.*

SIZE. Reaches an extreme length of about 5¼″ (134 mm.). Five adults of both sexes from Bayne Island, Union Bay, British Columbia, average 3 27/32″ (98 mm.) and vary from 3 1/32″ (77 mm.) to 4 15/32″ (114 mm.). The largest of this series, a male, has the following proportions: total length 4 15/32″ (114 mm.), tail 1 5/8″ (42 mm.); head length 21/32″ (17 mm.), width 3/8″ (10 mm.). A female from the same locality measures: total length 4 1/32″ (103 mm.), tail 1 11/16″ (43 mm.); head length 5/8″ (16 mm.), width 3/8″ (10 mm.).

DESCRIPTION. The head is widest midway between the eye and the lateral extension of the gular fold, the sides behind this point constricted to the neck, in front tapering to the bluntly pointed truncated snout. In side view, snout obliquely truncated. Eyes large, protuberant, the iris dark, tinged with brassy above the pupil, limited behind by a vertical groove. An impressed line from the posterior angle of the eye to the lateral extension of the gular fold. A short vertical groove from this line passing behind the angle of the mouth. Trunk with a median dorsal impressed line, rounded on the sides, flattened below; 16–17 costal grooves, counting 1 each in the axilla and groin, and 1–1½ intercostal folds between toes of appressed limbs. Tail broadly oval in section at base, nearly circular in section at mid-length, slightly compressed distally, and evenly tapering to the tip. Legs stout. Toes 5–4, those of the hind feet 1–(2–5)–3–4 in order of length from the shortest, the first small but free, others webbed at base; toes of the fore feet 1–4–2–3, scarcely webbed at base. Toes, except the 1st, all transversely widened at tip and squarely truncated. Tongue large, filling the floor of the mouth, thick and fleshy, wider behind than in front, free at the sides. Vomerine teeth in short series of 4–4 to 6–6 or 6–7, which arise behind the inner margin of the inner naris, curve obliquely inward and backward to the mid-line, where they may meet or be narrowly separated. Parasphenoid teeth in a single broad patch slightly narrowed anteriorly and separated from the vomerine by about the length of a vomerine series. Premaxillary teeth not greatly enlarged but sometimes perforating the lips in the male. Maxillary teeth few (usually 3 or 4), somewhat enlarged and flattened, and extending only to a point opposite anterior angle of eye. Mandibular teeth 5–8 on each side, the posterior enlarged, flattened.

COLOR. The general ground color above is dark brown, overlaid by large, irregular, grayish blotches which have a brassy tinge. There is considerable variation in the blotching, but in some specimens the head is mostly grayish with the exception of a dark central area, a dark bar

FIG. 97. *Aneides ferreus* Cope. (1) Adult, dorsolateral view; actual length about 4 11/16" (120 mm.). Comptche, Mendocino County, California. [Photograph by J. R. Slevin.] (2) Adult female, actual length 3 1/8" (80 mm.). (3) Same, ventral view. Mill Creek, near Crescent City, Del Norte County, California. [W. F. Wood, collector; from A. H. Wright.]

from the nostril to the eye, and dark pigment following the impressed line which extends from the posterior angle of the eye to the lateral extension of the gular fold. The legs are generally light, with a few brownish patches. The venter is light bluish-gray, with many very small, light yellow flecks scattered generally over the surface except, rarely, on the ventral surface of the tail. In a juvenile specimen, 46 mm. long, from near Tidewater, Oregon, the colors are essentially as follows: The general ground color is black, mottled and marbled with bronze, the bronze concentrated on the snout in a patch extending to the eyes, in an elongate spot on either side of the neck above the insertion of the arms, on the basal half of the proximal segments of the limbs, and along the dorsal surface of the tail. The sides are strongly mottled bronze and black, the belly uniformly bluish-gray with minute white flecks. The vent of the male is lined with papillae, anteriorly.

BREEDING. Nothing is known of the mating habits. Slevin (1928, p. 71) reported that two females collected on Bayne Island, British Columbia, May 16, 1906, had pigmentless, ovarian eggs 4 mm. in diameter, and Fitch (1936, p. 638) mentions a female taken May 22, 1935, in the Rogue River basin, Oregon, which contained 12 ovarian eggs, also about 4 mm. in diameter. The first account of eggs deposited under natural conditions is that of Dunn (1942, p. 52). A cluster of 9 eggs, accompanied by a female, was found Aug. 16, 1941, near Patrick Creek, Del Norte County, California, at an altitude of 2000′. The eggs were attached separately to a bit of bark of a fallen Douglas fir and measured approximately 6 mm. in diameter. A single egg opened by Dunn had 2 envelopes which enclosed a well developed young about 15 mm. long. The young salamander had a single, leaf-like, allantoic gill on each side and still retained considerable yolk.

**BLACK SALAMANDER. SHASTA SALAMANDER.** *Aneides flavipunctatus* (Strauch). Fig. 98. Map 40.

TYPE LOCALITY. New Albion, California (probably in Sonoma County).

RANGE. Santa Cruz County, California, northward in the Coastal Range to the Klamath Mountains.

HABITAT. Wood (1936, p. 171; 1939, p. 110) found this species most abundant in burnt-over areas in the southern redwood forests, under charred logs and slabs, and in the higher fir and pine forests of the outer Coast Ranges. It is not limited to such situations, for Van Denburgh (1895, p. 776) wrote that it had been "found under boards, decaying logs, and stones in the vicinity of running water, and in the drain from a spring"; and we have collected it near Fort Ross, California, from beneath loose stones and small rock slabs at the edge of a stream in a deeply shaded gully.

SIZE. Attains a length of $5\frac{15}{16}''$ (152 mm.). The average length of 16 adults of both sexes is $4\frac{1}{2}''$ (115 mm.), the extremes $3\frac{11}{32}''$ (85 mm.) and $5\frac{17}{32}''$ (141 mm.). An adult male from 9 miles south of Fort Ross, California, has the following measurements: total length $5\frac{3}{32}''$ (130 mm.), tail $2\frac{11}{32}''$ (60 mm.); head length $\frac{23}{32}''$ (19 mm.), width $\frac{13}{32}''$ (11 mm.). An adult female from Los Gatos has the following proportions: total length $4\frac{1}{2}''$ (115 mm.), trunk length $2\frac{27}{32}''$ (73 mm.) (tip of tail lost); head length $\frac{21}{32}''$ (17 mm.), width $\frac{3}{8}''$ (10 mm.).

DESCRIPTION. A black salamander with small whitish or yellowish spots. The head is widest immediately back of the eyes, where, in the male, it is noticeably swollen, the sides behind this point converging gently to the lateral extensions of the gular fold and in front abruptly to the bluntly pointed snout. Eyes only moderately large and protuberant, limited behind by a vertical fold. A sinuous fold or groove from the posterior angle of the eye to the lateral extension of the gular fold, a slightly impressed line from this to behind the angle of the jaw. Gular fold well developed. Trunk slightly depressed and with an impressed median dorsal line. Costal grooves strongly developed, variable in number but usually 14, sometimes 15 when 2 that run together in the groin are counted separately, rarely 16, and $3\frac{1}{2}$–5 intercostal folds between the toes of the appressed limbs. Tail stout, flattened above at base, beyond to distal third, nearly circular in section, slightly compressed distally. Legs stout, noticeably shorter than in other species of the genus, as indicated by the greater number of folds between the appressed toes. Toes 5–4, short, those of the hind feet 1–5–2–4–3 or 3–4 in order of length from the shortest, the innermost rudimentary but with the tip free, others webbed at base; toes of the fore feet 1–4–2–3. Toes only slightly widened at the tips and bluntly rounded. Tongue large, filling the floor of the mouth, broadly rounded behind, narrower and bluntly pointed in front, its surface with fine villae; free at the sides and slightly at the rear. Vomerine teeth in short series of 3–3 to 5–5. The series arise

FIG. 98. *Aneides flavipunctatus* (Straugh). (1) Adult male, actual length 5⅛″ (130 mm.). (2) Same, ventral view. Near Fort Ross, California. The apparent middorsal and midventral stripes are due to reflection from adjoining surfaces.

behind or just inside the inner margin of the inner naris and slant obliquely inward and backward toward the mid-line, where they are separated by about the width of a naris. The parasphenoid teeth usually form a single broad patch which is slightly narrowed anteriorally and separated from the vomerine by about the distance between the nares. Premaxillary teeth not enlarged, but with the tips perforating the edge of the lip; maxillary teeth enlarged, 2–4 in number, angular in section; mandibular teeth 3–4 on each side, enlarged, thickened anteriorally, knife-edged behind. The vent of the male is lined with papillae in front; that of the female with oblique folds.

COLOR. In the specimen from near Fort Ross, California, the general color above is black. Scattered over the dorsum and sides are a few small straw-yellow spots. A few light spots are present on the back of the head and on the upper surface of the limbs. The belly is deep slate with many minute points of light gray. The under surface of the head from the point of the jaw to the angle of the mouth, lighter gray. Gular fold whitish, soles of feet grayish, tinged with flesh. A young individual, 50 mm. long, in life had the dorsal surface of the head, trunk, and tail bronzy-green, the upper sides black. Scattered over back and sides were many small silvery-white points, and the upper surface of the basal segment of the legs was golden-yellow. The ventral surfaces were colored as in the adult. Adults south of San Francisco Bay are said to be unspotted (Myers, 1930, p. 60).

BREEDING. Nothing is known of the mating habits. Van Denburgh (1895, p. 776), the first to report the eggs of this species, received an adult female and eggs from Los Gatos, California, July 23, 1895. "Each egg was about 6 mm. in diameter, almost spherical, and inclosed in a thin, tough, gelatinous sheath. Each of these sheaths was drawn out, at one place, into a slender peduncle, which was attached to a basal mass of the same gelatinous substance. In this way, each egg was at the end of an individual stalk, and all were fastened to a common base. This base had evidently been anchored to a stone or lump of earth." The salamander and about 30 eggs had been found beneath a platform in dry earth at a depth of 15″ or more. A specimen killed July 30, 1895, was found to contain 25 large ovarian eggs and many minute ones. Storer (1925, p. 123) also reported the discovery of a female and seven pedunculated eggs found at Laytonville, California, July 1, 1913. This batch was found 9″ below the surface of the ground in a cellar. The eggs were deposited in a group and attached by their peduncles to the damp earth. After long preservation they measured 5.9–6.4 mm. in diameter, being elongated in the direction of the peduncle.

**ARBOREAL SALAMANDER.** *Aneides lugubris lugubris* (Hallowell). Figs. 95c, 99. Map 41.

TYPE LOCALITY. Monterey, California.

RANGE. Mainly confined to western California from Humboldt County

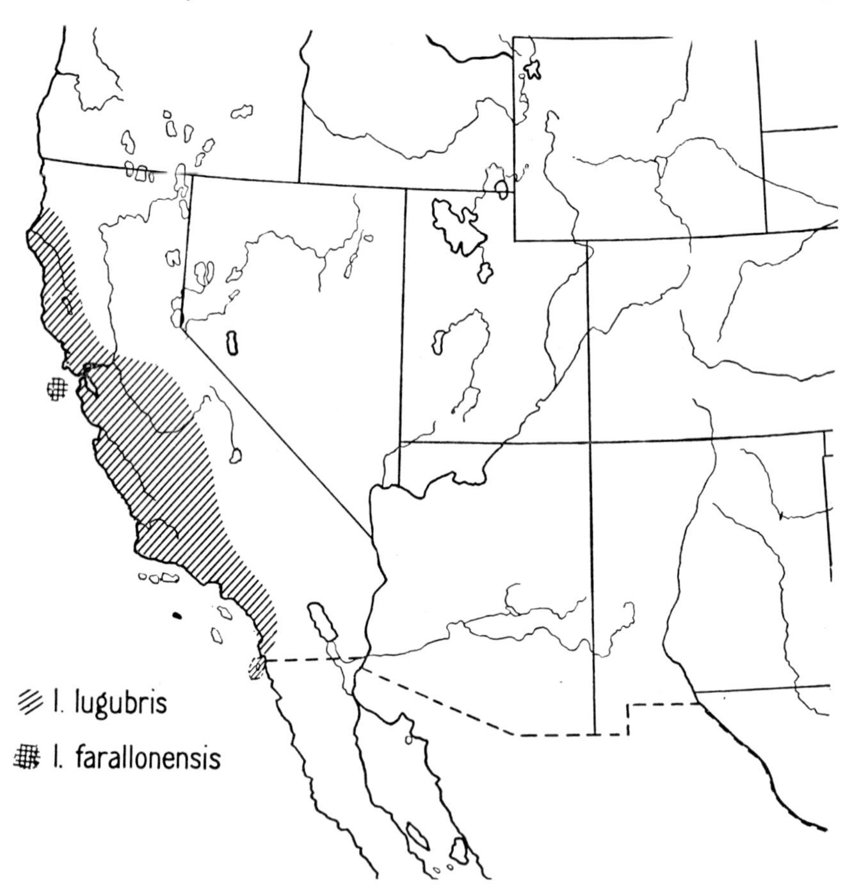


MAP 41.—Distribution of the subspecies of *Aneides lugubris.*

to San Diego County; also from Mariposa and Madera Counties; Anno Nuevo Island, Monterey Bay; Coronados Island, Lower California.

HABITAT. This species is found both on the ground and in the trees. When on the ground it often hides in logs and stumps or beneath surface materials such as stones, logs, bark, and boards. On the University of California campus at Berkeley, many individuals were found occupying the cavities in the "live oak" trees, some at a height of 30′ above the ground. The eggs have been discovered both in the soil and in the cavities of trees.

SIZE. Attains an extreme length of $6\frac{3}{8}''$ (162 mm.). The average length of 12 adults of both sexes is $4\frac{19}{32}''$ (118 mm.), the extremes $3\frac{13}{32}''$ (87 mm.) and $5\frac{19}{32}''$ (143 mm.). An adult male from Berkeley, California, has the following measurements: total length $5\frac{9}{16}''$ (142 mm.), tail $2\frac{17}{32}''$ (65 mm.); head length $\frac{26}{32}''$ (21 mm.), width $\frac{1}{2}''$ (13 mm.). An adult female measures: total length 5″ (128 mm.), tail $2\frac{15}{16}''$ (59 mm.); head length $\frac{23}{32}''$ (19 mm.), width $\frac{1}{2}''$ (13 mm.).

DESCRIPTION. The head is large, wider than the trunk, widest at the angle of the jaws, where it is swollen above, the sides behind this point converging to the gular fold and in front more abruptly to the bluntly rounded snout. The eyes are large and prominent, limited behind by a vertical groove. An impressed sinuous line from the posterior angle of the eye bends down posteriorly to join the gular fold. Gular fold strongly developed. Trunk subcylindrical. Usually 15 costal grooves, counting 1 each in the axilla and groin, rarely 14 or 16, the toes of the appressed legs just meeting or overlapping one intercostal fold. Tail long, somewhat prehensile, comprising 42–46 per cent of the total length, broadly oval in section at base, becoming slightly compressed distally; tail with impressed vertical lines on the sides. Legs large, stout, as befitting a climbing animal. Toes 5–4, those of the hind feet 1–5–2–(3–4) in order of length from the shortest; toes of the fore feet 1–4–2–3, the 2nd and 4th about equal; tips of toes only slightly expanded. Commissure of the mouth straight and horizontal to below the eye, then sharply deflected upward. Tongue large, filling the floor of the mouth, the margins thin, the surface spongy, free at the sides and behind. Vomerine teeth 5–8 in short series that arise behind the inner margin of the inner naris and bend sharply backward toward the mid-line, where they are separated by about ½ the diameter of a naris. Parasphenoid teeth usually in a single patch, wide behind, narrow in front, and separated from the vomerines by about 3 times the diameter of a naris; occasionally in 2 narrowly separated patches. Premaxillary teeth enlarged and somewhat flattened; maxillary teeth 6–8, the series extending to a point op-

FIG. 99. *Aneides lugubris lugubris* (Hallowell). (1) Adult male, actual length 5¼″ (134 mm.). (2) Same, dorsal view. Berkeley, California.

posite the anterior angle of the eye. Mandibular teeth 5–7 in each series, very large, flattened, sharp-pointed, and directed obliquely upward and backward.

COLOR. The general color above varies from a uniform light brown to light chocolate-brown, the upper surface of the tail sometimes brighter. Scattered over the entire dorsal surface of the head, trunk, tail, and limbs, especially in small individuals, are many small whitish or pale yellow spots. In some large specimens the pale spots or flecks are most abundant on the sides of the trunk, thinning out on the back, legs, and base of the tail. The venter is lighter, pale yellow, dull white, bluish-white, or grayish. The throat and legs are definitely flesh color in some individuals.

BREEDING. The Arboreal salamander is a nocturnal animal, unique among American species in its habit of climbing trees and making use of cavities as retreats and for egg-laying. Ritter (1903, p. 883) was the first to call attention to this interesting habit. The cavities of trees, often extensive, sometimes harbored as many as 35 individuals of various ages, including the young of the year, yearlings, and adults, but often only 1 or 2 salamanders were found. The egg-laying season is during the summer months, eggs having been found from July to September. Individual eggs have a diameter of about 5 mm. and, with their envelopes, from 6 or 7 to 9.5 mm. Whether in cavities in trees, buried in the soil, or hidden beneath some sheltering object on the ground, the eggs are usually attended by the female, sometimes by both parents. The eggs number 12–19 and are deposited singly, each attached by a slender stalk, 8–20 mm. long, to the roof of the cavity and twisted about one another. Young on hatching vary in length from 26.5 to 32 mm. in total length, and in general resemble the adult (Storer, 1925, p. 136).

**FARALLON SALAMANDER. FARALLON YELLOW-SPOTTED SALAMANDER.** *Aneides lugubris farallonensis* (Van Denburgh). Fig. 100. Map 41.

TYPE LOCALITY. South Farallon Island, San Francisco County, California.

RANGE. This subspecies is known only from the type locality.

HABITAT. The mainland form is more or less arboreal, but in the absence of any considerable number of trees on South Farallon Island, this subspecies is mainly terrestrial. It has been found beneath boards and piles of loose stones in moist situations.

SIZE. In a fine series from the California Academy of Sciences, loaned by Mr. Joseph R. Slevin, 8 males averaged $3\frac{31}{32}''$ (101 mm.) and varied from $3\frac{1}{4}''$ (83 mm.) to $4\frac{27}{32}''$ (124 mm.); 5 adult females averaged $4\frac{1}{8}''$ (105 mm.), the extremes $3\frac{21}{32}''$ (93 mm.) and $4\frac{5}{16}''$ (110 mm.). The proportions of an adult male are as follows: total length $4\frac{7}{32}''$ (108 mm.), tail $1\frac{25}{32}''$ (46 mm.); head length $\frac{25}{32}''$ (20 mm.), width $\frac{3}{8}''$ (10 mm.). A female has the following measurements: total length $4\frac{7}{32}''$ (108 mm.), tail $1\frac{3}{4}''$ (45 mm.); head length $\frac{11}{16}''$ (18 mm.), width $\frac{13}{32}''$ (11 mm.). Storer (1925, p. 141) records a specimen $5\frac{3}{32}''$ (130 mm.) in total length.

DESCRIPTION. The head is widest at the angle of the jaws, where it is somewhat swollen above, the sides behind this point converging to the lateral extension of the gular fold and in front abruptly to the bluntly pointed snout. The snout strongly depressed and in side view obliquely truncated, the commissure of the mouth sinuous, bent sharply upward from below the eye. The eyes are large and protuberant, the horizontal diameter about $1\frac{1}{4}$ in the snout, limited behind by a vertical fold. An impressed sinuous line from the posterior angle of the eye to the gular fold. The trunk is somewhat flattened and has a slightly impressed median line. There are usually 15 costal grooves, counting 1 in the axilla and 2 that run together in the groin. The tail is subcylindrical, becoming slightly compressed distally. In the males the tail comprises 40.4–45.16 per cent of the total length, and in the female 41.7–45.16 per cent. The legs are long and moderately stout, and, when appressed to the sides, the toes overlap $1\frac{1}{2}$–3 intercostal folds. The hind leg itself overlaps, on the average, 9 costal folds as compared with 8 in *Aneides l. lugubris.* Toes 5–4, those of the hind feet slender, square-tipped, 1–2–5–3–4 in order of length from the shortest, the 3rd and 4th about equal, all webbed at base; toes of the fore feet 1–4–2–3 or 1–2–4–3, the tips slightly widened or squarely truncate, webbed at base. Tongue broadly oval in outline, free at the sides and behind, the margins thin, the surface with fine plicae which diverge slightly from the posterior field. Mandibular teeth enlarged, flattened, and triangular. Vomerine teeth 5–11, average 7, in series that arise behind the inner margin of the inner nares and curve inward and backward toward the mid-line, where they may be nearly in contact or separated by about the width of a naris. Parasphenoid teeth in a single patch, broad at the middle of its length and pointed anteriorly, or in 2 narrow patches imperfectly separated.

COLOR. The color above varies from seal-brown to yellowish-brown (turning purplish-brown in preservative), darkest along the middle of the back, fading slightly on the lower sides, and lightest on the snout and legs. This subspecies is much more strongly spotted than the main-

FIG. 100. *Aneides lugubris farallonensis* (Van Denburgh). (1) Adult male, photograph from preserved specimen; actual length 4²⁷⁄₃₂″ (124 mm.). South Farallon Island, California. [J. R. Slevin, collector.] (2) Adult, from life. South Farallon Island, California. [Photograph through the courtesy of J. R. Slevin.]

land form, the spots of pale yellow varying in size, from mere specks to elongate blotches 3–4 mm. in length, and distributed generally over the dorsal surfaces but usually fewer on the head and tail. On the sides the light spots extend to the level of the legs and involve the sides of the neck and tail and the upper surface of the limbs. The ventral surfaces are pale yellowish or whitish, the tips of the toes reddish.

BREEDING. Nothing has been made known of the breeding habits, and the eggs and recently hatched young have not been described. Storer (1925, p. 142) gives some indication of the rate of growth, based on measurements of 11 individuals taken Oct. 26, 1922. "They range from 39 to 126 mm. in total length. Three which measure 39 and 40 mm. are undoubtedly of the 1922 brood. . . . One with a blunt (regenerated?) tail, 53 millimeters in total length, and a normal individual 62 millimeters long, are taken to represent the 'yearling' class. Five others 101 to 112 millimeters in length are either three or four years of age and one is 126 millimeters and hence four or five years old."

## Genus STEREOCHILUS

**MARGINED SALAMANDER.** *Stereochilus marginatus* (Hallowell). Figs. 101c-d, 102. Map 42.

TYPE LOCALITY. Liberty County, Georgia.

RANGE. Petersburg and the Dismal Swamp, Virginia, to Liberty County, Georgia, in the Coastal Plain (Stejneger and Barbour, 1939, p. 20).

HABITAT. Usually aquatic but occasionally found on land under logs in damp situations. They are perhaps most abundant in pools and slow streams in swampy woods, and may be collected by raking out the dead leaves and other bottom rubbish. Brady (1930, p. 58) regards this species as characteristic of the pine barrens.

SIZE. In a series of 14 adults measured by Brimley (1909, p. 132) the size range was from $2\frac{9}{16}''$ (65 mm.) to $3\frac{3}{4}''$ (95 mm.), the average length $3\frac{3}{16}''$ (81 mm.). Dunn (1926, p. 245) reports an adult only $2\frac{15}{32}''$ (63 mm.). The proportions of an adult female from Whitakers, North Carolina, taken Oct. 25, 1926, are as follows: total length $3\frac{3}{16}''$ (71 mm.), tail $1\frac{3}{8}''$ (35 mm.); head length $\frac{5}{16}''$ (7 mm.), width $\frac{7}{32}''$ (5.5 mm.). An adult male from Wilmington, North Carolina, July 25, 1907, C. S. Brimley, collector, measures as follows: total length $2\frac{29}{32}''$ (73 mm.), tail $1\frac{5}{16}''$ (33 mm.); head length $\frac{11}{32}''$ (8 mm.), width $\frac{3}{16}''$ (4.5 mm.).

DESCRIPTION. This is a lithe, active species characterized by the presence of alternating light and dark longitudinal lines on the sides. The head is small, narrow, pointed, and depressed, and with little indication

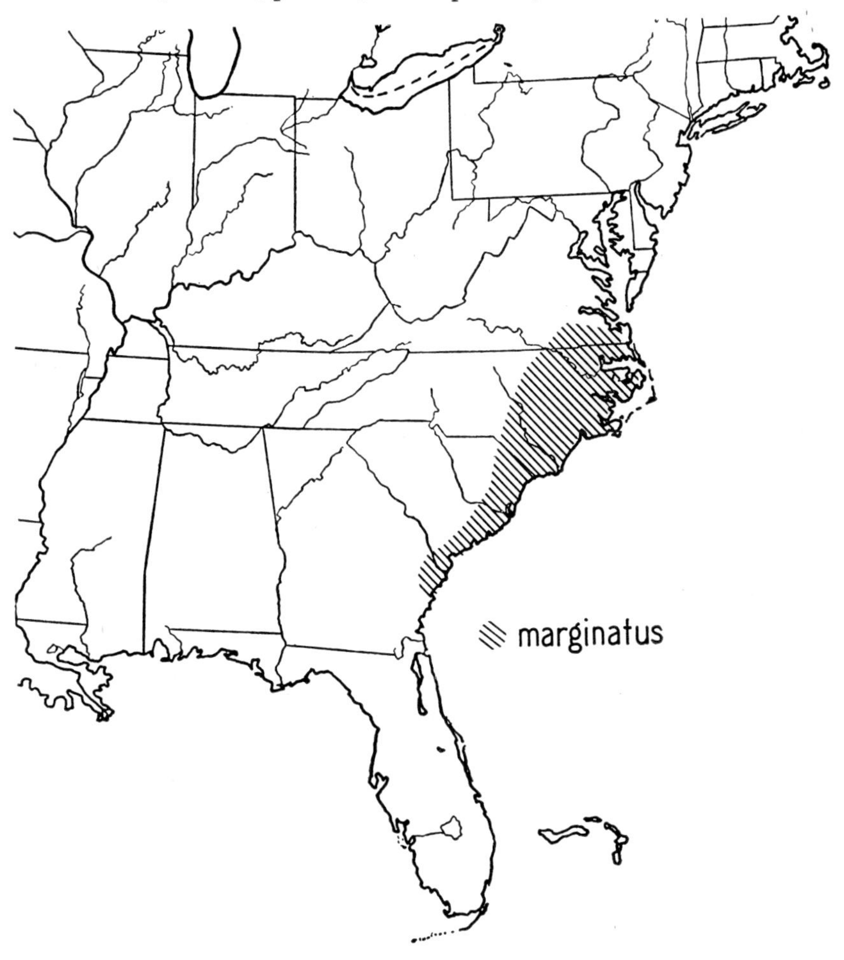


MAP 42.—Distribution of *Stereochilus marginatus.*

of a constriction at the neck. The sides behind the eyes are nearly parallel and in front converge to the sharp snout. There is a well defined system of pores which form a small group back of the eye and extend forward in a single line above the eye to a point about ½ the distance to the nostril. This series is overlapped at its anterior end by another which lies mesad to it and extends to the side of the snout. Another group on the side of the head behind the eye extends forward to the margin of the lip below the naris, and from this point sometimes turns backward in a

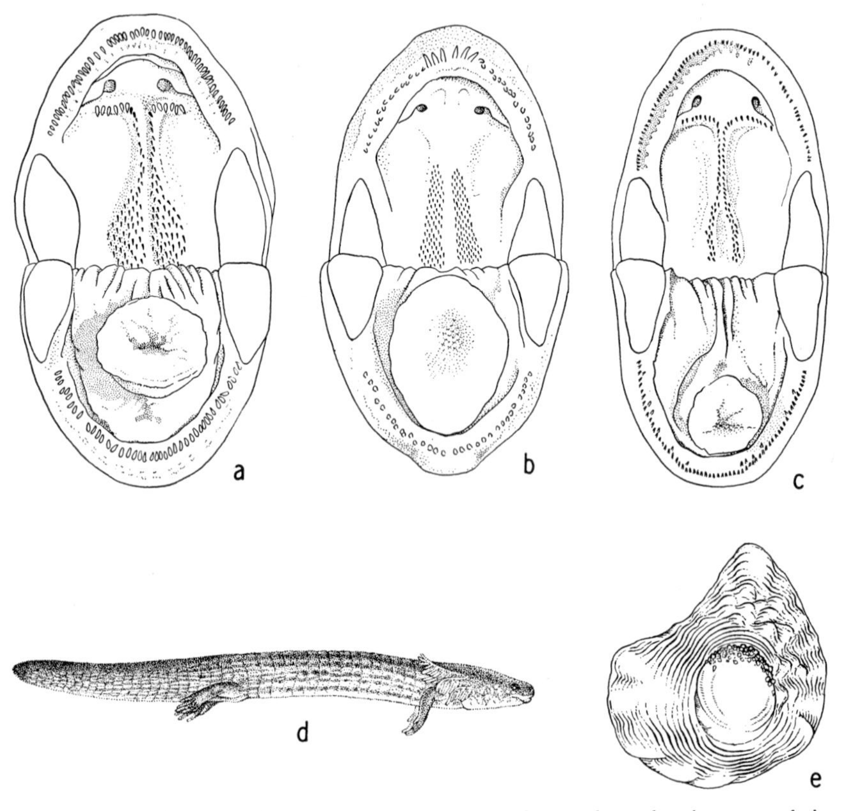


Fig. 101 (a) Open mouth of *Pseudotriton ruber ruber* to show the character of the tongue and teeth. (b) Open mouth of adult male *Desmognathus fuscus fuscus.* (c) Open mouth of *Stereochilus marginatus.* (d) Larva of *Stereochilus marginatus.* [H.R.B. *del.*] (e) Spermatophore of *Eurycea longicauda melanopleura.* [a–c, M.L.S. *del.;* e, H.M.Z. *del.*]

single line to the anterior angle of the eye. A deep, darkly pigmented groove extends from the posterior angle of the eye to the lateral extension of the gular fold, and from this a short vertical groove passes behind the angle of the jaw. The eyes are of moderate size and may be retracted to the contour of the head. Trunk cylindrical, somewhat flattened below, and with an impressed median dorsal line. Costal grooves 18, and about 8 or 9 intercostal spaces between the toes of the appressed limbs. Legs short, slender; toes long and slender without webs, in order of length 1–5–2–4–3 from the shortest. Toes of the fore feet 1–4–2–3. Tail oval in section at base, beyond rounded below, sharp-edged above, the tip compressed. Tongue small, broadly oval in outline, slightly de-

FIG. 102. *Stereochilus marginatus* (Hallowell). (1) Adult male, actual length about 3″ (77 mm.). (2) Same, lateral view. (3) Same, ventral view. Near Emporia, Virginia.

pressed into the floor of the mouth, and surrounded by a low fleshy ridge. It is rather broadly attached in front at the mid-line and this membrane continues to the pedicel near the center. Vomerine teeth continuous with parasphenoid. The vomerine series arise behind and just outside the outer margin of the inner naris, curve inward and gently forward toward the mid-line, where they are narrowly separated from the series of the opposite side, then sharply backward. Parasphenoid teeth in 2 slender patches which diverge slightly posteriorly.

COLOR. The general ground color above is a dull yellow, slightly lighter on the upper sides. The middorsal impressed line is darkly pigmented and in young individuals the entire dorsal surface is lightly reticulated with seal-brown. In old adults the dark brown pigment largely obscures the lighter ground color, which strikes through in small circular spots. The ground color of the lower sides is considerably lighter than that of the upper. The sides are narrowly, longitudinally lined with alternating brown and yellow, best developed in the younger specimens. The dark lines of the upper sides originate on the snout, continue through the eye and along the trunk to the tail, where they unite with the lower series to form a coarsely reticulate pattern. In many specimens, the narrow dark lines of the sides are arranged in threes, the upper group separated from the lower by a yellow interspace. The belly is dull yellow with a few dark specks scattered irregularly over the surface. Specimens from dark swamp waters are very dark-colored and with little indication of light lines on the sides.

BREEDING. The egg-laying habits of captive, stimulated specimens have been studied by Noble and Richards (1932, p. 5). As many as 121 eggs may be deposited by a single female, but the average of 19 was 57. Individual eggs with their external envelopes attain a diameter of 3-3.5 mm., the inner envelope 2.5-3 mm., the egg itself 2-2.5 mm. They are attached singly to a support in water, probably in nature to the lower surface of a stone, piece of bark, or log. The outer capsule is slightly elongated to form a stalk.

LARVAE. Larvae attain an extreme length of 3⁷⁄₁₆" (87 mm.), the older individuals resembling the adults in general form and color. The gills are long and extremely compressed, the filaments slender, flattened, and well pigmented. The younger larvae tend to be more mottled above than the adults, but the irregular streaking of the sides is noticeable.

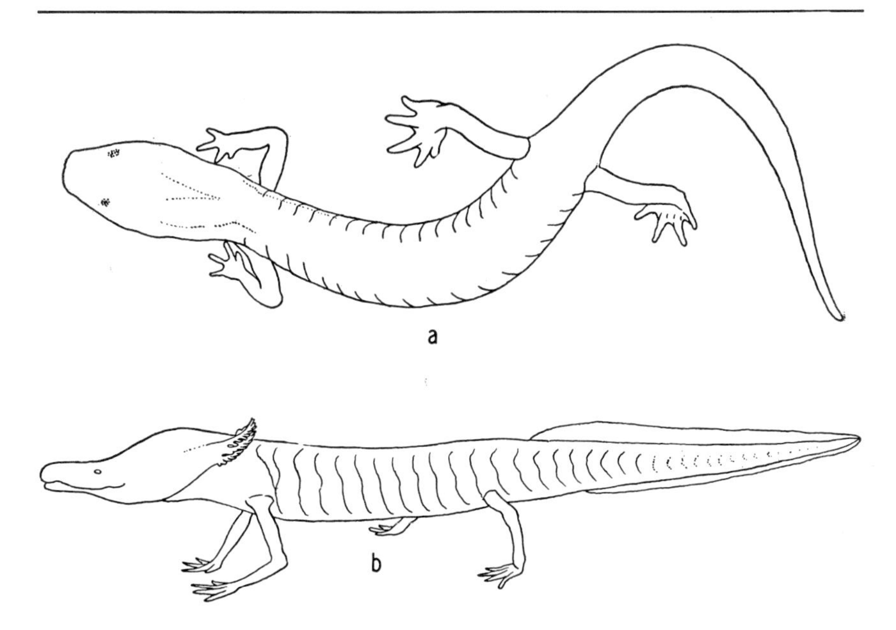


FIG. 103. (a) Outline of *Typhlotriton spelæus* to show the form of the body, reduced eyes, and character of legs and feet. (b) Outline of *Typhlomolge rathbuni* to show the form of the body, the slender and elongate legs, and the toes, gills, and vestigial eyes. [H.P.C. *del.*]

## GENUS TYPHLOTRITON

**OZARK BLIND SALAMANDER.** *Typhlotriton spelæus* Stejneger. Figs. 103a, 104. Map 43.

TYPE LOCALITY. Rock House Cave, Missouri.

RANGE. Ozark Plateau in Missouri, Oklahoma, Kansas, and Arkansas.

HABITAT. The adults are mainly limited to the caves and underground passages of the limestone regions in the Ozarks and, while freely entering water, are more terrestrial than other true cave salamanders. The larvae are found in springs and streams, both in the open and in caves.

SIZE. The species may attain a length of 5 5/16″ (135 mm.), but this is exceptional. The average of 7 adults which vary from 2 7/8″ (72 mm.) to 4 3/4″ (120 mm.) is 3 3/4″ (95 mm.). The proportions of an adult from Cave Springs Caverns, Carter County, Missouri, are as follows: total length 3$\frac{5}{16}$″ (84 mm.), tail 1$\frac{9}{16}$″ (40 mm.); head length $\frac{13}{32}$″ (10 mm.), width ¼″ (6 mm.).

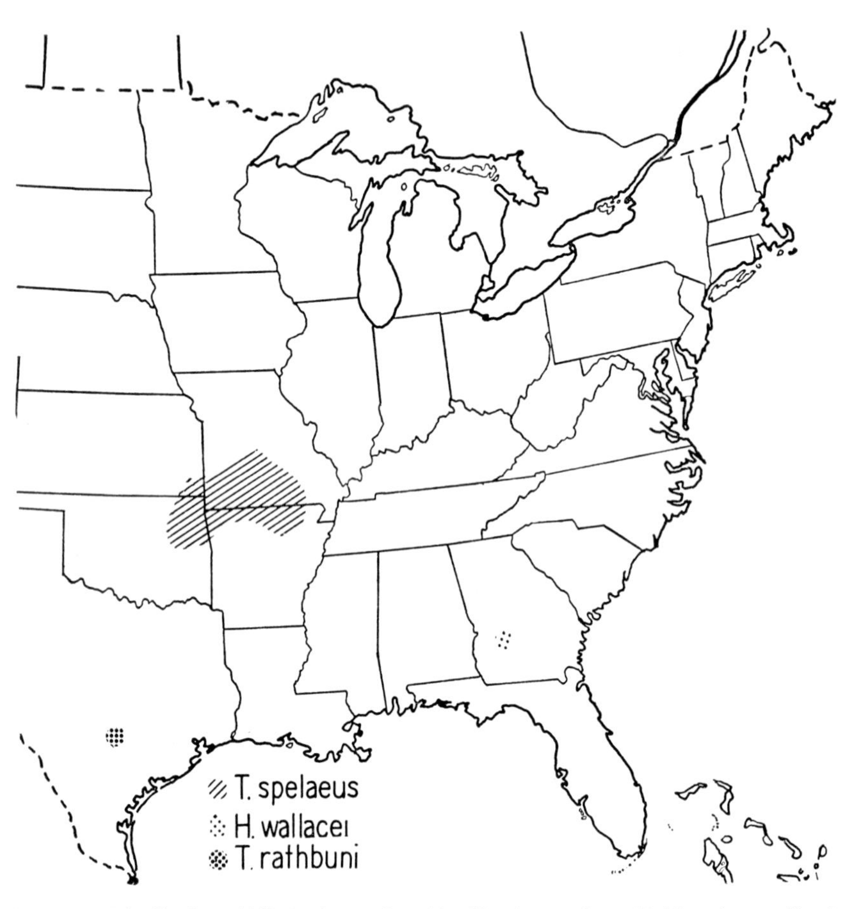


MAP 43.—Distribution of blind salamanders, *Typhlotriton spelæus, Haideotriton wallacei,* and *Typhlomolge rathbuni.*

DESCRIPTION. This pale denizen of the caves has a ghostly appearance when seen in its natural surroundings by the light of a torch or lamp. The strongly depressed head is long and slender, widest at the angle of the jaws, and gently tapering to the gular fold. In front of the eyes the sides converge more abruptly to the sharply truncated snout, which, in some individuals, is slightly convex between the nostrils. There is a slight swelling in the parotoid region, and a sinuous groove extending from below the eye to the gular fold, and from this a short oblique groove to the angle of the jaw. Viewed from the side, the commissure of the mouth is gently and evenly arched. The snout is swollen in the region of the nasolabial grooves and produced into blunt cirri below the nostrils, particularly in the males. The eyes are small and dark, partly or completely covered by the fused lids, and separated equally from each other and the nostril of the same side. The trunk is slender and rounded and with little evidence of an impressed median line except in some preserved material. There are 16 well developed costal grooves, or 17 if 2 that run together in the groin are counted, and about 4 intercostal spaces between the toes of the appressed limbs. The legs are rather slender, the hind somewhat larger than the fore. Toes 5–4, those of the hind feet 1–5–2–4–3 in order of length from the shortest; toes of the fore feet 1–4–2–3, all toes slightly webbed at base. The tail is long and slender, oval in section at base, rounded below, and provided with a low keel above. The tongue is small, roughly oval in outline, and with the margins thin. It is attached by a central pedicel, from which a thin membrane extends along the median line to the anterior margin and to the floor of the mouth. Vomerine teeth in series of 11–17. In a male, the series arise behind the middle of the inner naris, extend inward and forward, then abruptly backward parallel with the adjacent series, from which they are distant about ⅔ the diameter of an inner naris. In some the vomerine series are continuous with the parasphenoids, in others widely separated. Parasphenoid teeth in 2 elongate patches separated by about the diameter of an inner naris.

COLOR. The general color in life is pale flesh, with tinges of orange pigment in small round spots scattered over the dorsal surface of the tail and on the lower sides and feet, imparting a yellowish cast in these regions. A few individuals are sparsely pigmented over the dorsal surfaces generally and have a light gray appearance. The males may usually be recognized by the greater development of cirri and by the presence of papillae along the margins of the vent.

BREEDING. Nothing is known of the breeding habits and eggs.

LARVAE. Unlike the adults, which are only slightly pigmented and usually restricted to the caves, the larvae are highly pigmented and are

FIG. 104. *Typhlotriton spelæus* Stejneger. (1) Adult, actual length about 3¾″ (95 mm.). Marvel Cave, Stone County, Missouri. (2) Large larva from spring. Stone County, Missouri. (3) Large larvae. From River Cave, Camden County, Missouri. [Collected by James Kezer and photographed by Arthur L. Smith.]

most frequently taken in open streams. The smallest larva mentioned by Dunn had a total length of 24 mm., and the largest 93 mm. In my own collection the range is from 23 to 96 mm., the smaller individual taken April 30, 1927, at York Spring, Imboden, Arkansas, by Byron C. Marshall. Other specimens taken at the same time and place measure 28, 30, 35, and 35 mm. respectively. The fully grown larvae have a ground color of light grayish-lavender on the trunk and head, with small scattered light yellow flecks and an iridescent cast on the trunk above. The legs are pale, dirty white, the belly and throat dull white. The tail is broadly keeled above, from a point opposite the insertion of the hind legs to the bluntly rounded tip, and below about ½ the distance to the vent, thence to the vent as a low ridge. The general pigmentation of the sides of the trunk is continued on the tail, and the fins are lightly tinged with gray along the margins. Beginning at a point behind the gills, there is a dorsolateral series of small light spots which continues well above the insertion of the hind legs onto the sides of the tail. In some, a secondary row is developed at a lower level. The gills are short, strongly compressed, wide at the base but abruptly tapering, and with slender, darkly pigmented filaments. The head of the larva is decidedly different from that of the adult and the eyes are fully developed and functional. The head is widest in front of the gills, tapers gradually to just in front of the eyes, then slightly to the bluntly rounded snout. In some individuals, a narrow light line extends from the eye to the nostril.

**GEORGIA BLIND SALAMANDER.** *Haideotriton wallacei* Carr. Fig. 105. Map 43.

TYPE LOCALITY. From a 200′ well at Albany, Dougherty County, Georgia.

RANGE. Known only from the type locality.

HABITAT. Subterranean waters.

SIZE. The only known specimen, a sexually mature female, had the following measurements as given by Carr (1939, p. 335): total length 3″ (75.5 mm.), tail 1 11/32″ (33.5 mm.); head length 17/32″ (12.5 mm.), width 11/32″ (8 mm.).

DESCRIPTION. This subterranean species is blind, white, and semitransparent. The head is broad but not strongly depressed as in *Typhlomolge,* the sides slightly concave between the angle of the mouth and the gular fold, the snout very broadly rounded. The mouth is small and nearly terminal. Neither eyes nor eye spots developed. The gills

FIG. 105. *Haideotriton wallacei* Carr. (1) Adult female, type specimen; lateral view. (2) Same, ventral view. (3) Same, dorsal view. From a deep well at Albany, Georgia. [Photographs through the courtesy of A. F. Carr.]

are very long and slender, the 3rd longest, and when appressed to the sides reaching the 3rd costal groove. Neck markedly narrower than head and trunk. Trunk subcylindrical. Eleven costal grooves. Legs long and slender; toes 5–4, those of the hind feet 1–5–2–(3–4) in order of length from the shortest; toes of the fore feet 1–4–2–3. Tail moderately compressed, the dorsal fin arising above the vent and continuing without widening to the bluntly rounded tip. Lower tail fin narrower. Vent without papillae.

COLOR. "In life, body pale, pinkish white, vaguely opalescent; viscera and eggs plainly evident through body wall; limbs transparent, the larger blood vessels readily discernible; tail and fin very faintly suffused with yellow; gills blood red. After preservation, dull white, with widely scattered clusters of dark pigment cells" (Carr, 1939, p. 335).

BREEDING. Nothing is known.

Through the courtesy of Mr. A. F. Carr, the describer, I have been given the privilege of reproducing his excellent photographs of this remarkable salamander.

**TEXAS BLIND SALAMANDER. SAN MARCOS SALAMANDER. WHITE SALAMANDER.**

*Typhlomolge rathbuni* Stejneger. Figs. 103b, 106. Map 43.

TYPE LOCALITY. Artesian well, 188′ deep, at San Marcos, Hays County, Texas.

RANGE. Known from Hays, Kendall, and Comal Counties, and reported from Crockett County, Texas.

HABITAT. Known only from wells and underground streams in caves.

SIZE. The largest specimen seen by Dunn, a male, had a total length of $5\frac{13}{32}''$ (136 mm.), tail length, $2\frac{13}{32}''$ (60 mm.). Most specimens are considerably shorter, 4 sexually mature individuals varying from $3\frac{13}{32}''$ (85 mm.) to $4\frac{13}{32}''$ (110 mm.) and averaging $3\frac{13}{16}''$ (97 mm.). The proportions of an adult male from San Marcos, Texas, are as follows: total length $3\frac{15}{16}''$ (100 mm.), tail $1\frac{25}{32}''$ (45 mm.); head length $\frac{19}{32}''$ (15 mm.), width $\frac{17}{32}''$ (13 mm.).

DESCRIPTION. This is a blind, white salamander with long, slender legs. flattened snout, and permanent, pinkish gills. Viewed from above, the head is very broad, widest just in front of the gills, curving gently forward to the eyes, where there is a slight constriction, then evenly and gently to the broadly truncated snout. The snout is much flattened and in front of the eyes has a depth about $\frac{1}{3}$ that of the head at the gills. The eyes are small black dots wholly covered by the skin. A broad sinuous groove extends from the eye to the lateral extension of the gular fold. The commissure of the mouth is sinuous and the angle of the mouth lies immediately below the eye. The gills are slender, the filaments long and slender and without pigment. The trunk is deeper than wide and has a low median dorsal ridge. There are 12 costal

FIG. 106. *Typhlomolge rathbuni* Stejneger. (1) Juvenile, actual length 2⁹⁄₁₆″ (65 mm.). (2) Same, dorsal view. Ezell's Cave, San Marcos, Texas.

grooves, counting 1 in the axilla and 2 that run together at the groin, and about 6 intercostal spaces between the toes of the appressed limbs. The legs are extremely long and slender; toes 5–4, those of the hind feet 1–5–2–3–4 in order of length from the shortest; toes of the fore feet 1–4–2–3, all without webs. The tail is subquadrate in section at the base, strongly compressed toward the pointed tip. The dorsal tail fin arises at a point above the vent and is widest at about the distal third; the ventral fin extends about ½ the distance to the vent and is continued forward as a low, fleshy ridge. The tongue is crescent-shaped, free in front and for a short distance on the sides. The vomero-palatine teeth form an inverted U-shaped series, narrowly interrupted at the mid-line and at about halfway down the sides.

COLOR. The snout is dull white, very slightly suffused with darker except in a blotch below each eye, where all pigment is lacking; the remainder of the head, the trunk, and the tail are iridescent, with pastel shades of pink, lavender, and pale blue in a broad dorsal band; the band limited on each side by a faint row of elongate pale spots. The dorsal keel of the tail is lightly mottled with pale brownish blotches, and on the sides below the dorsal band there is just the slightest indication of pigment. The belly is uniformly dull white except for a midventral band, where it is iridescent. The heart in the region of the throat shows through as a pinkish spot. The legs appear opaque white, but in reality are lightly pigmented.

BREEDING. Nothing is known of the breeding habits under natural conditions. Eigenmann (1909, p. 31) reported that a specimen in captivity laid a few eggs about March 15, 1896, and Dunn (1926, p. 255) examined a female collected in the early fall of 1916 that had the spermatheca packed with sperm.

In this permanently larval form the sexually immature individuals resemble those that have attained full growth. Eigenmann (*ibid.*, p. 31) mentions a specimen only 30 mm. long.

On March 31, 1936, I visited Ezell's Cave, San Marcos, Texas, where many specimens have been collected, but the width of my shoulders did not conform to the diameter of the shaft leading to the underground waters and I failed to reach the proper level. A second attempt was made at Johnson's well, not the original which had been capped by concrete, but one recently opened a short distance away. Here, with the aid of a rope, I descended to the water level, and saw a fine large specimen resting quietly on a ledge a foot or two below the surface. My maneuvering disturbed the creature and it disappeared in the shadowy depths. The same or another individual was momentarily glimpsed in the deep recesses of a side channel, but it too darted away and was lost when an attempt was made to direct it into a dip net. Three living examples were secured for me by Mr. W. E. Ezell in Ezell's Cave, and I have seen other living specimens from New Braunfels, Comal County, and from Boerne, Kendall County.

## Genus GYRINOPHILUS

### KEY TO THE SPECIES AND SUBSPECIES OF GYRINOPHILUS

1. Back strongly clouded or mottled, light yellowish-brown, with reddish tinges to light salmon (purplish in preservatives); sides with darker reticulations enclosing pale spots; light line from eye to naris bordered below with a darker band; entire venter flesh in life (yellow in preserved specimens) sometimes with a few small scattered dark dots on the belly and more numerous spots on the throat and margin of the lower jaw; length to 7¹⁹⁄₃₂″ (193 mm.), rarely to 8⅝″ (219 mm.). Ontario, Canada opposite Buffalo, New York, through the eastern states to Virginia, Tennessee, and possibly Alabama . . . . . . . . . . . . . . . . . . . . . . . . . . . . . . . . . . . . . *porphyriticus porphyriticus* p. 367

   Back lightly, if at all, clouded or mottled; uniformly reddish suffused with dusky; or with dark dots, flecks, or spots on a lighter ground . . 2
2. Back uniformly reddish in life, with a faint chevron-like middorsal pattern of darker lines; or sometimes with a few small, scattered, dark flecks on the back and sides; internal nares large, nearly circular in outline; venter flesh color without darker markings; length to about 6¼″ (154 mm.). Unglaciated area of southern Ohio . . . . . . . . . . . . . . . . . . . . . . . . . . . . . . . . . . . . . . . . *porphyriticus inagnoscus* p. 373

   Back not uniformly reddish in life with chevron-like middorsal pattern; back with conspicuous dark dots, flecks, or spots on a lighter ground; or, back with few or no black dots but with dorsolateral series . . . . 3
3. Dark spots of back and sides few, small, and widely separated, usually forming a dorsolateral series, never extending below the level of the legs; venter immaculate; throat never blotched or reticulate, margin of lower jaw with a few small dark dots; light line from eye to naris indistinct, only lightly bordered with darker below; internal nares small, oval in outline; length to 6¹⁵⁄₃₂″ (164 mm.). Northeastern Kentucky, southern Ohio, northwestern West Virginia . . . . . . . . . . . . . . . . . . . . . . . . . . . . . . . . . . . . . . . . . . . . . . . . *porphyriticus duryi* p. 370

   Dark spots or flecks numerous, scattered generally on back and sides; margin of lower jaw and sometimes the throat with dark spots or strongly mottled black and white . . . . . . . . . . . . . . . . . . . . . . . . . . 4
4. Dark spots of black and sides dark brown to black, often fusing to form elongate blotches; venter of adults usually with scattered black dots; throat mottled or reticulated, black and white; length to 8½″ (204

mm.). Southwestern Virginia through eastern Tennessee, western North Carolina, and possibly to Georgia and South Carolina, above 3500′ ........................................ *danielsi danielsi* p. 361

Dark spots of back and sides small, brown, usually separate, occasionally forming indistinct chevrons; venter usually immaculate, occasionally with a few dark dots; margin of lower jaw with dark spots, dots, or reticulations; throat never heavily marked as in typical *danielsi;* light line from eye to naris heavily bordered with brown below, lightly margined above; length to 6⁵⁄₁₆″ (160 mm.). Northwestern South Carolina, western North Carolina, northeastern Georgia, and eastern Tennessee, below 3500′ ........................ *danielsi dunni* p. 365

**MOUNTAIN PURPLE SALAMANDER.** *Gyrinophilus danielsi danielsi* (Blatchley). Fig. 107. Map 44.

TYPE LOCALITY. Mt. Collins and Indian Pass, Sevier County, Tennessee.

RANGE. Mainly confined to the southern section of the Blue Ridge Mountains from near Cumberland Gap, Kentucky, through eastern Tennessee, western North Carolina, and possibly to Georgia and South Carolina.

HABITAT. Reaches its best development in high mountain streams, where it hides by day beneath stones or other sheltering object. Occasionally wanders some distance from the stream and may be found beneath logs or bark.

SIZE. Seventeen adults taken at an elevation of 6000′ and above on Mt. Mitchell, North Carolina, varied in length from 4¾″ (120 mm.) to 8⅙″ (204 mm.) and averaged 6⅛″ (155.5 mm.). The 10 largest averaged 6¹⁵⁄₁₆″ (172.7 mm.).

DESCRIPTION. Daniels' salamander is somewhat larger than its northern relative, *G. porphyriticus,* and is more strikingly marked. The head is broad and somewhat flattened, widest immediately behind the eyes, the sides converging slightly to the lateral extensions of the gular fold and more abruptly to the nostrils. The snout is depressed, the end broadly and evenly rounded. The eye is of moderate size, with the pupil small, black, and slightly elongate longitudinally; iris gold, brassy immediately surrounding the pupil, and crossed horizontally by a broad coppery-red band. A short vertical groove from the angle of the jaw to the long sinuous groove which extends from the eye to the lateral extension of the gular fold. The strong, muscular trunk is rounded on

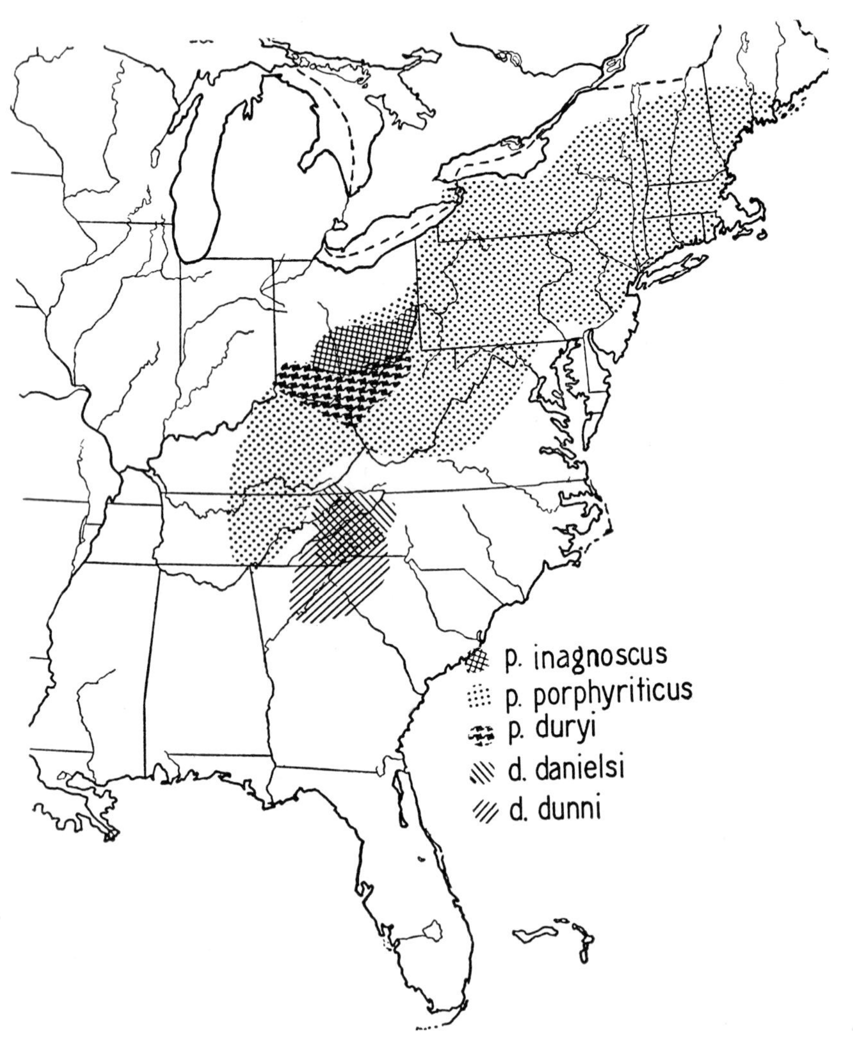


Map 44.—Distribution of the subspecies of *Gyrinophilus porphyriticus* and *G. danielsi*.

the sides and somewhat flattened above and below. Costal grooves usually 17 but occasionally 18. About 6 intercostal spaces between toes of appressed limbs. Tail rounded below; keeled above, beginning about opposite posterior end of vent and becoming knife-like above on the distal third. Legs strong, toes 5–4, those of the hind feet 1–5–2–3–4 or 4–3 in order of length from the shortest; fore feet 1–4–2–3; all slightly

Fig. 107. *Gyrinophilus danielsi danielsi* (Blatchley). (1) Adult female, actual length 8¹⁄₃₂″ (204 mm.). (2) Same, dorsal view. (3) Same, ventral view. Steprock Creek, Mount Mitchell, North Carolina; altitude, 6000′. [Photographs by E. J. Stein and S. C. Bishop.]

webbed at base. Tongue moderate in size, roughly circular in outline, the margins smooth, the center with fine papillae. Vomerine teeth continuous with the parasphenoid. The vomerine form strongly arched series which arise just outside and behind the inner nares, slant gently inward and forward, then turn sharply backward, their inner arms nearly parallel. The teeth form a single line, 9 or 10 to the bend, and 19–21 in each series. Parasphenoid teeth in 2 well separated, slender, club-shaped patches.

COLOR. This is a reddish or salmon-colored salamander, marked above with irregular flecks and spots of black. The ground color varies considerably, being brighter in recently transformed individuals and becoming tan or brownish in older ones. There are frequently tinges of orange-red on the legs and along the mid-line of the back. In water a bluish or purplish bloom or cast covers the dorsal surfaces. The sides are lighter than the back, and the deep brown or black flecks are smaller and become fewer and farther apart toward the belly and on the sides of the tail. The belly and under surface of the legs are flesh-colored in life, with scattered spots of tan and brown pigment, and sparsely scattered whitish flecks. Back of the hind legs the ground color may be slightly darker, light tan with darker markings. The entire throat and, in some old individuals, the region back of the gular fold, as well as the lower edge of the upper lip, may be finely reticulated with black and white in about equal proportions. In younger individuals these reticulations may be almost lacking. In most individuals there is a dark line extending from the anterior angle of the eye to the nostril which is bordered above by white. This dark line may be extended back of the eye along the groove which runs to the lateral extension of the gular fold. Feet reddish, toes tipped with black.

BREEDING. Nothing is known of the mating habits. A large female guarding her hatching young was found Oct. 22, 1923, in Steprock Creek, at an altitude of about 6000′, on Mt. Mitchell, North Carolina. The eggs had apparently been attached separately to the lower surface of the rock (Bishop, 1924, p. 90).

LARVAE. The recently hatched larvae are slender and delicate, and average about 1″ in length. The ground color is pale lavender, marked with small flecks of light yellow. A slightly darker median dorsal line extends from the head to the hind legs. A definite line separates the color of the back and upper sides from that of the venter, which is pale yellow, colored by the yolk. In their color both the larvae and the adults of *danielsi* differ from those of *G. porphyriticus* in being speckled or finely spotted rather than faintly reticulated.

**CAROLINA PURPLE SALAMANDER.** *Gyrinophilus danielsi dunni* Mittleman and Jopson. Fig. 108. Map 44.

TYPE LOCALITY. Clemson, Oconee County, South Carolina.

RANGE. Northwestern South Carolina, western North Carolina, northeastern Georgia, and eastern Tennessee.

HABITAT. The type specimen was collected in a small stream running through a shallow wooded ravine. In general this form is found at elevations less than 3500′.

SIZE. Adults attain an extreme length of $6\frac{5}{16}$″ (160 mm.) but the average is considerably less, the type series of 16 adults averaging $4\frac{29}{32}$″ (125 mm.) (Mittleman and Jopson, 1941, p. 2). The proportions of an adult male paratype from Sunburst, Haywood County, North Carolina, are as follows: total length $4\frac{7}{8}$″ (124 mm.), tail $1\frac{31}{32}$″ (50 mm.); head length $\frac{3}{4}$″ (19 mm.), width $\frac{3}{8}$″ (10 mm.).

DESCRIPTION. The head is rather narrow, the sides behind the eyes nearly parallel to the lateral extensions of the gular fold, and in front tapering to the bluntly rounded snout. The eye is small, its long diameter about twice in the snout, the iris narrowly ringed with brassy and crossed horizontally by a reddish-brown band. Above the pupil the iris may be brassy, below silvery-white. An impressed sinuous line from the posterior angle of the eye to the lateral extension of the gular fold; a short vertical groove from this line passing well back of the angle of the jaw. The trunk is rounded above and on the sides and with a slightly impressed median line. There are 18 costal grooves, counting 1 each in the axilla and groin, and 5–8 intercostal spaces between the toes of the appressed limbs. The tail is subquadrate in section at base, becoming trigonal at about the basal fourth and compressed and slightly keeled above distally. The legs are of moderate size. Toes 5–4, those of the hind feet 1–5–2–(4–3), slightly webbed at base; toes of the fore feet

1–4–2–3. The tongue is of moderate size, thin at the margin behind, the surface spongy. Vomerine teeth 12 or more in each series, which arise outside the outer margin of the inner nares, curve forward toward the mid-line, then angle sharply backward to meet the parasphenoids,

FIG. 108. *Gyrinophilus danielsi dunni* Mittleman and Jopson. (1) Adult female, actual length 5⁵⁄₁₆″ (136 mm.). (2) Same, ventral view. Type, from life. Clemson, Oconee County, South Carolina.

with which they are continuous. Parasphenoid teeth in 2 long, very slender, widely separated patches.

COLOR. In life the ground color above varies from orange-yellow to light reddish. Scattered generally over the dorsal and lateral surfaces are many small, separate, brown flecks. A slight concentration of this dark pigment is sometimes noticeable along either side of the dorsal ridge of the tail, and rarely the brown flecks form an indistinct herringbone pattern on the dorsum of the trunk. A light line extending from the anterior angle of the eye to the margin of the lip is bordered heavily below by dark brown or black and lightly above. The margins of the jaw, both upper and lower, are mottled with black and white. The legs are colored above like the back, the toes sometimes narrowly banded with black. On the lower sides the brown chromatophores are smaller and usually more widely separated, fading out entirely at about the lower level of the legs. Ventral surfaces salmon-pink in life, usually without darker markings. In preservatives the ground color above fades to dull tan with light brown markings, the ventral surfaces dull yellow. Sexual differences are not well marked externally, but the vent of the male is densely lined with short papillae, which makes identification possible.

BREEDING. Nothing is known of the breeding habits of this form.

LARVA. A larva collected by Arnold Grobman and M. B. Mittleman at a locality 2 miles east of Stockville, Buncombe County, North Carolina, is apparently this species. It is 63 mm. in total length and colored essentially like the adults, with the addition of conspicuous sense organs on the head and an indistinct series of small light spots dorsolaterally. The tail is provided with a broad fin above and below and tapers abruptly to a sharp point.

This is the form which I described but did not name (1924, p. 90) and which Stejneger listed as *Gyrinophilus porphyriticus dunni* (1937, p. 30).

**NORTHEASTERN PURPLE SALAMANDER.** *Gyrinophilus porphyriticus porphyriticus* (Green). Figs. 1c, 49d, 109. Map 44.

TYPE LOCALITY. French Creek, near Meadville, Crawford County, Pennsylvania.

RANGE. From New England states westward to Ontario, Canada, opposite Buffalo, southward through the mountains of the eastern states to Virginia, Tennessee, and possibly Alabama.

HABITAT. The purple salamander frequents cool springs and streams, or occupies little natural or excavated cavities beneath logs or stones at the margins. While the larvae are entirely aquatic, the adults are frequently found some distance from water but usually in damp situations.

SIZE. Adult males vary in length from 5½ to 7⅝″ (140–193 mm.) and the females from 4¾″ to 7⅛″ (120–180 mm.). Individuals may transform at a length of 3¹⁵⁄₁₆″ (100 mm.), but a well-fed giant living in the outlet from a fish-rearing station had attained a length of 8⅝″ (219 mm.).

DESCRIPTION. The Purple salamander is moderately stout and is strong and active. Viewed from above, the head is long and slender, with the sides behind the eyes nearly parallel, in front tapering slightly to the bluntly truncated snout. The eyes are of moderate size and are limited behind by a vertical crescentic fold. A sinuous horizontal groove from the posterior angle of the eye to the lateral extension of the gular fold, and a short vertical groove from the angle of the jaw to the horizontal groove. The trunk is rounded on the sides, somewhat flattened below, and marked above by a median impressed line. Costal grooves usually 17, rarely 18 if an imperfect groove in axilla is present and counted; 7–7½ intercostal spaces between toes of appressed limbs. Tail moderately long, rounded below, compressed above, and with a knife-like edge toward the tip. Legs stout, toes 5–4, those of the hind feet 1–5–2–3–4 or 4–3 in order of length from the shortest; fore feet 1–4–2–3; all slightly webbed at base. Tongue moderate, nearly circular in outline, boletoid. Vomerine teeth usually continuous with parasphenoid. They are in 2 strongly arched series which arise behind and slightly outside the inner nares, extend inward and gently forward, then bend sharply backward to form inverted L-shaped patches. The teeth form a single line 7 or 8 to the bend, 17–20 in each complete series. Parasphenoid teeth in 2 well separated club-shaped patches.

COLOR. The adult Purple salamander, belying its name, is light yellowish-brown with reddish tinges, or light salmon. Usually the back is darker than the sides and is clouded or mottled, and the sides are marked with darker reticulations enclosing pale spots. The venter is flesh-colored in life, and in old individuals there are often small, scattered, dark dots on the belly and many on the throat and along the margin of the lower jaw. The tips of the toes and joints often conspicuously darker than the rest of the legs. A conspicuous light line extending from the anterior angle of the eye to the nostril is bordered on the outside by gray. The color of the recently transformed individual is

FIG. 109. *Gyrinophilus porphyriticus porphyriticus* (Green). (1) Adult male, actual length 7⅛″ (181 mm.). Voorheesville, New York. (2) Adult male, lateral view; actual length 5²⁷⁄₃₂″ (150 mm.). (3) Same, ventral view. Allegany State Park, New York.

much brighter, salmon-red, and the darker reticulations are poorly developed.

BREEDING. Little is known of the mating habits under natural conditions. Preliminary activities were observed Oct. 8, 1924, when male and female grasped one another and wrestled vigorously. A spermatophore ready for deposition was recovered from the vent of a male taken Oct.

28, 1933, and sperm have been found in the spermathecae of females taken from June to November. The egg-laying season may be an extended one, since females with large eggs have been taken in April, May, and June, and in October and November (Bishop, 1941), and eggs supposedly of this species were found August 8 (Green, 1925, p. 32). The dissection of mature females has indicated that eggs may vary in number from 44 to 132. The individual egg is a light yellow sphere about 3½ mm. in diameter. In addition to the vitellus it is surrounded by 3 distinct envelopes which give it a total diameter of approximately 9 mm. The eggs are attached singly, by a flange-like enlargement of the outer envelope, to the lower surface of a support in water.

LARVAE. The recently hatched larvae have a length of about 1″. I have collected them in April and July in spring-fed runs in swamps, and from beneath stones in springs. They are light brown above, yellowish-white below, and with a narrow, dark, pigmented band from the back of the head to the base of the dorsal tail fin. Probably 3 years are spent as larvae, sexual maturity being attained at a length of about 5½″.

**KENTUCKY PURPLE SALAMANDER.** *Gyrinophilus porphyriticus duryi* (Weller). Fig. 110. Map 44.

TYPE LOCALITY. Cascade Caverns, near Grayson, Carter County, Kentucky.

RANGE. Northeastern Kentucky, south-central Ohio, and north-central West Virginia.

HABITAT. Although the type locality for this subspecies is Cascade Caverns near Grayson, Kentucky, the animal is not restricted to this kind of habitat. In the caverns it is to be searched for beneath flakes of rock, pieces of wood, or other debris, in damp situations. Outside of caves it has been found beneath stones, bark, and logs in the vicinity of streams or springs, and usually at fairly low elevations.

SIZE. The average length of 6 adults from Kentucky is 5 13/16″ (148 mm.) with extremes of 5¼″ (133 mm.) and 6 15/32″ (164 mm.). The proportions of a female which I collected at the type locality, Cascade Caverns near Grayson, Kentucky, are as follows: total length 5⅞″ (148.5 mm.), tail 2 7/16″ (61.5 mm.); head length 13/16″ (20 mm.), width ⅜″ (9.5 mm.).

DESCRIPTION. This is the smallest, slenderest, and palest in color of the five representatives of Gyrinophilus. The head is only moderately widened, and slenderer than in *G. d. danielsi.* It is widest just back of the eyes, the sides gently converging behind to the lateral extensions of the gular fold and more abruptly in front to the bluntly rounded snout. The head is slightly depressed. Eye moderate and less protuberant than in its near relatives; iris tinged with gold and brassy. A depressed horizontal line from the posterior angle of the eye to the lateral extension of the gular fold; a short vertical groove from this to the angle of the jaw, the angle of the mouth distant anteriorly from this point about the diameter of the eye. Trunk slender, rounded on the sides, flattened below, and with an impressed median dorsal line extending from a Y-shaped mark on the dorsum of the head to the base of the tail. There are 17 well developed costal grooves, or 18 if an imperfectly developed one in the groin is counted; 6½–7½ intercostal spaces between the toes of the appressed limbs. The tail is slender, rounded below, and keeled above to a point opposite the posterior end of the vent. Legs well developed but relatively short and stocky; toes 5–4, short, blunt, those of the hind feet 1–5–2–3–4 in order of length from the shortest; fore feet 1–4–2–3. Tongue roughly circular in outline and with the margins smooth. Vomerine teeth in strongly arched series which arise behind the outer margin of the inner nares, curve inward and forward to a point opposite the middle of the nares, and then sharply backward to join the parasphenoid about opposite the middle of the eye. There are 8 or 9 single teeth in each series to the bend and about an equal number to the point where they join the parasphenoid; parasphenoid teeth in 2 well separated, slender, club-shaped patches.

COLOR. The ground color in life varies from salmon-pink to light brownish-pink. In some the dorsal surface is very slightly darker than the sides. In some individuals from the type locality, there is a fairly well defined single row of small dark brown or blackish spots on either side, extending from the side of the head onto the basal half of the tail; in others there are numerous small scattered spots on the sides of the head, trunk and tail, and upper surface of the legs, but few or none on

FIG. 110. *Gyrinophilus porphyriticus duryi* (Weller). (1) Adult, actual length about 5″ (128 mm.). (2) Same, ventral view. Cascade Caverns, Carter County, Kentucky.

the dorsal surfaces. In a few individuals the small dark spots are scattered more or less generally over the dorsal as well as the lateral surfaces. The ventral surfaces, with the exception of the lower lip and occasionally the throat and gular region, are free from the dark spots. The throat and belly are flesh color, tinged with bluish where the color of the liver strikes through; ventral surface of the tail with tinges of orange. The light, dark-bordered line extending from the eye to the nostril, which in the other forms of Gyrinophilus is quite conspicuously developed, is here only faintly indicated.

BREEDING. Nothing is known of the breeding habits of this subspecies, but it is to be expected that the eggs will be attached to the lower surface of some support in water, as are those of its near relatives.

LARVAE. The larvae are very light and are marked dorsally with a dusky reticulate pattern, somewhat as in *G. porphyriticus,* but paler. In large larvae the adult pattern is suggested by the presence of a few poorly defined dark dots.

Supposed intergrades between *danielsi* and *duryi* were reported by King (1939, p. 554) from eastern Tennessee.

**OHIO PURPLE SALAMANDER.** *Gyrinophilus porphyriticus inagnoscus* Mittleman. Fig. 111. Map 44.

TYPE LOCALITY. Salt Creek, 4 miles southwest of Bloomingville, Good Hope Township, Hocking County, Ohio.

RANGE. The unglaciated parts of Ohio south to but not in the southern tier counties bordering the Ohio River. Apparently avoids the limestone regions.

HABITAT. Apparently less aquatic than typical *porphyriticus.* It has been collected from beneath logs, rocks, and debris at varying distances from water, but is perhaps found in greatest numbers in the vicinity of rocky, woodland streams.

SIZE. I have not seen enough material to be able to give average measurements of any significance, but the general impression is that of a slender salamander of moderate size. The proportions of an adult male from Salt Creek Township, Hocking County, Ohio, are as follows: total length $4\frac{13}{16}''$ (121 mm.), tail $2\frac{1}{32}''$ (51 mm.); head length $\frac{11}{16}''$ (18 mm.), width $\frac{3}{8}''$ (10 mm.). The measurements of an adult female from the same locality are: total length $5\frac{3}{4}''$ (146 mm.), tail $2\frac{15}{32}''$ (63 mm.); head length $\frac{25}{32}''$ (20 mm.), width $\frac{13}{32}''$ (11 mm.). It is evident that this salamander attains a size somewhat larger than is indicated by the measurements given above. The female measured has a trunk length of 82 mm. Other individuals of both sexes, with mutilated tails, have trunk lengths of 85 and 87 mm. respectively.

DESCRIPTION. This is a comparatively slender salamander which does not attain the size of the typical subspecies. The head is narrow, widest at the angle of the jaws, the sides behind this point converging gently to the lateral extensions of the gular fold, in front more abruptly to the short and bluntly rounded snout. The eye is of moderate size, the horizontal diameter about equal to the distance from the anterior angle to the naris of the same side; iris brassy, suffused with dusky, bright golden-yellow next to the pupil. Lips mottled, light and dark. An impressed line from the posterior angle of the eye to the lateral extension of the gular fold; a short vertical groove from this line to the angle of the jaw. Gular fold prominent. Trunk rounded on the sides and back, slightly flattened below. There are usually 18 costal grooves, counting 1 each in the axilla and groin, occasionally 17 or 19, and 6–7 intercostal folds between the toes of the appressed limbs. Tail broadly oval in section at base, becoming sharp-edged and keeled above immediately behind the vent; ventral tail fin confined to the distal fourth or third. Legs moderately stout, toes 5–4, those of the hind feet 1–5–2–(4–3) in order of length from the shortest; toes of the fore feet 1–4–2–3. Tongue small, broadly oval in outline, the surface spongy. Vomerine series continuous with parasphenoids. The series arise behind or just outside the outer margin of the inner naris and usually extend obliquely forward toward the mid-line, where they are separated by about the width of a naris, then turn abruptly backward, forming nearly parallel lines to the parasphenoids. The number of vomerine teeth is variable; in the few males examined they were fewer, larger, and more widely separated on the transverse limb of the series (7–9); smaller and more numerous (9–12) on the backward extension. Vomerine teeth of females small, 9–11 on either transverse limb, total 17–19. Parasphenoid teeth in long, slender, club-shaped patches, well separated. Internal nares large, nearly circular in outline.

COLOR. A recently killed specimen was Pompeiian Red (Ridgway, 1912) above, evenly and lightly suffused with dusky, slightly darker on the snout; lower sides of head and trunk pale salmon color. Upper surface of the legs and feet like the back but more strongly suffused with dusky. The light line from eye to nostril bordered above and below by dusky. Throat, belly, and venter of tail light flesh color. The dorsal extensions of the costal grooves turn forward and meet at the mid-line of the back at an acute angle. These grooves are more strongly marked with dusky than the adjacent surfaces and often form a fairly regular

FIG. 111. *Gyrinophilus porphyriticus inagnoscus* Mittleman. (1) Adult male, actual length 4⅛″ (104 mm.). (2) Same, ventral view. Tail tip regenerated. Near Cambridge, Guernsey County, Ohio. [Photographs of a preserved specimen.]

chevron-like pattern. On some specimens the sides of the head and trunk are irregularly marked with small dark flecks. Usually there is very little suggestion of the reticulated pattern of typical *porphyriticus*.

SEXUAL DIFFERENCES. The sexes may be distinguished by the form of the vent, which, in the male, is larger and lined anteriorly with short papillae; vent of female short, the sides at the opening thrown into folds.

BREEDING. Nothing is known of the breeding habits of this subspecies, and the larvae have not been described.

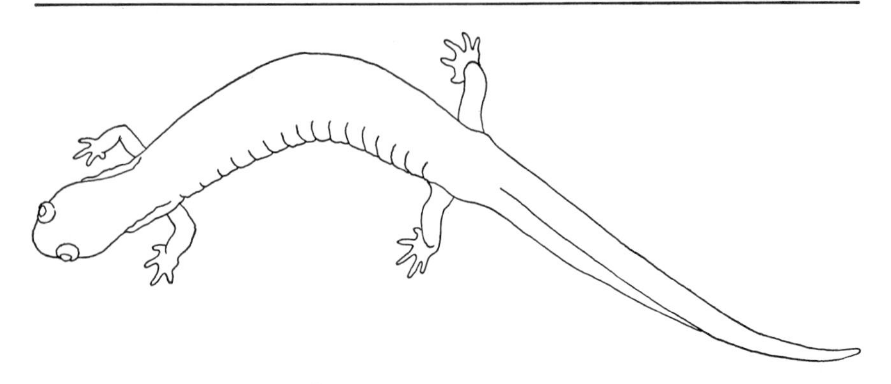

FIG. 112. Outline of *Pseudotriton montanus montanus* to show the form of the body, the short snout, and the character of the limbs. [H.P.C. *del.*]

# GENUS PSEUDOTRITON

## KEY TO THE SPECIES AND SUBSPECIES OF PSEUDOTRITON

1. Dorsal ground color clear, brilliant red to dark chocolate-brown; dorsal and lateral black spots small to large, round, well separated, never fused, but sometimes lost in dark ground color of old individuals; vomerine series broadly curved and usually obtusely angled at about the middle of the length, closely approximated at mid-line; snout short, usually strongly convex above; horizontal diameter of eye 1¼–1½ in snout; no light, dark-bordered line from eye to nostril (*flavissimus, montanus*) . . . . . . . . 2

   Dorsal ground color bright coral-red to dark purplish-brown; dorsal and lateral black spots small and well separated in young individuals, becoming enlarged and tending to fuse in old, dark specimens; vomerine series usually right-angled at about the middle of the length, occasionally curved, well separated at mid-line; snout longer, less convex above; horizontal diameter of eye 1½–2 in snout; at least a suggestion of a light, dark-bordered line from eye to nostril (*ruber*) . . . . . . . . 6

2. Costal grooves 18 . . . . . . . . 3

   Costal grooves 16 or 17 . . . . . . . . 4

3. Adults with ground color purplish-brown above, clouded with dull yellow; sides of trunk with short dull yellow dashes sometimes forming irregular lines; venter yellowish-white with small reddish-brown dots; length to 4⁵⁄₃₂″ (105 mm.). Peninsular Florida and possibly from Valdosta, Georgia . . . . . . . . *flavissimus floridanus* p. 381

4. Ground color above clear, light coral-pink, brownish salmon or bright red, but never purplish-brown clouded with yellow; dorsal and lateral dark spots small to large, but always present and well separated;

venter immaculate . . . . . . . . 5

Ground color above dull, light to dark chocolate-brown; dorsal and lateral dark spots usually distinct, sometimes obscure in old individuals, but when present extending well onto lower sides; venter with at least a few, often many, small, brown, widely separated dots; iris of eye dark brown; length to 7" (178 mm.). Carlisle, Pennsylvania, southward through Maryland to South Carolina and northwestern Georgia . . . . . . . . *montanus montanus* p. 383

5. Ground color above light brownish salmon; dorsal and lateral dark spots small; sides of head, trunk, and tail light salmon; usually 16 costal grooves; length to 3 29/32" (99 mm.). Georgia westward in the Coastal Plain of Alabama, Mississippi, and Louisiana . . . . . . . . *flavissimus flavissimus* p. 378

Ground color above bright coral-pink to brilliant red; dorsal dark spots usually large (occasionally small in juveniles), well separated, never reaching below dorsal insertion of legs; usually 17 costal grooves, occasionally 16; length to 6 1/8" (156 mm.). Unglaciated plateau region of southern Ohio, central and eastern Kentucky, southwestern West Virginia, western Virginia, and eastern Tennessee . . . . . . . . *montanus diastictus* p. 386

6. Adults coral-red (juvenile) to dark purplish-brown above (old individuals); dorsal and lateral dark spots irregular in size and shape and tending to fuse in old adults; ground color of back distinctly darker than ground color of venter in preserved specimens; venter often strongly spotted . . . . . . . . 7

Adults clear red; dark spots of back and sides well separated, larger dorsally; venter flesh in life, immaculate or with a few small, scattered fleckings on throat and belly; ground color of back and sides not greatly darker than that of belly in preserved specimens . . . . . . . . 8

7. Adults with dorsal dark spots tending to fuse; venter salmon-red in life, with small, round, brown or black spots; ventral surface of legs and tail usually immaculate; no whitish flecks around snout and on the head in adults; iris tinged with brassy; vomerine series broadly curved in some, right-angled in others; the section before the bend long and not widely separated from its fellow of the opposite side; length to 6 1/2" (165 mm.). Albany County, New York, southward to northern Georgia, westward to Alabama, Tennessee, Kentucky, and Ohio . . . . . . . . *ruber ruber* p. 389

Adults with dorsal spots usually distinct, occasionally somewhat fused and lost in general dark ground color; belly usually strongly spotted; venter of legs and tail not immaculate; many whitish flecks around snout and on the head in old individuals; section of vomerine series before the bend short and widely separated from its fellow of the opposite side; iris tinged with brassy, silver, and black; length to 5 25/32" (148 mm.). Gulf Coastal Plain of northwestern Florida,

Georgia, Alabama, Mississippi, and Louisiana ....... *ruber vioscai* p. 399

8. Margin of lower jaw black, usually in a solid bar, but occasionally broken; tail spotted above and on sides nearly or quite to the tip; throat and area between fore legs usually spotted or dotted; length to 4 29/32" (125 mm.). Southwestern North Carolina, northwestern South Carolina, eastern Tennessee, and northern Georgia ........ ........................................ *ruber schencki* p. 396

Margin of lower jaw flecked or spotted with brown or black but never heavily barred; distal half of tail usually unspotted; throat and area between fore limbs usually immaculate; length to 4 17/32" (116 mm.). From Abington and Whitetop Mountain, Virginia, to the valley of the French Broad in North Carolina, Roan Mountain in Tennessee ........................................ *ruber nitidus* p. 393

**GULF COAST RED SALAMANDER.** *Pseudotriton flavissimus flavissimus* Hallowell. Fig. 113. Map 45.

TYPE LOCALITY. Liberty County, Georgia.

RANGE. From Liberty and Lowndes Counties, Georgia, westward in the Coastal Plain of Alabama, Mississippi, and Louisiana.

HABITAT. Little is known of the habitat preferences. Allen (1932, p. 5) reported specimens found under the bark of a rotten log and ploughed up in a field. Mr. Henry Dietrich sent me 4 specimens which he collected under wood in sphagnum moss beside a small branch near Lucedale, Mississippi.

SIZE. The average length of 10 adults of both sexes from Mississippi and Alabama is 3 7/16" (87.8 mm.), the extremes 3 1/8" (80 mm.) and 3 29/32" (99 mm.). The proportions of an adult female from near Lucedale, Mississippi, are as follows: total length 3 13/32" (87 mm.), tail 1 11/32" (34 mm.); head length 17/32" (13 mm.), width 11/32" (8 mm.).

DESCRIPTION. This is a small, slender salamander. Viewed from above, the sides of the head back of the eyes are nearly parallel or slightly converging to the lateral extensions of the gular fold, in front converging abruptly to the bluntly pointed snout. The eye is moderate, the long diameter about 1½ in the snout. An impressed sinuous line from the posterior angle of the eye to the lateral extension of the gular fold; a short vertical groove from this line passing behind the angle of the jaw. Trunk slender, slightly compressed dorsally, rounded on the sides. There are usually 16 costal grooves, sometimes 17 counting 1 in the axilla and 2 that run together in the groin, and 6–7½ intercostal folds

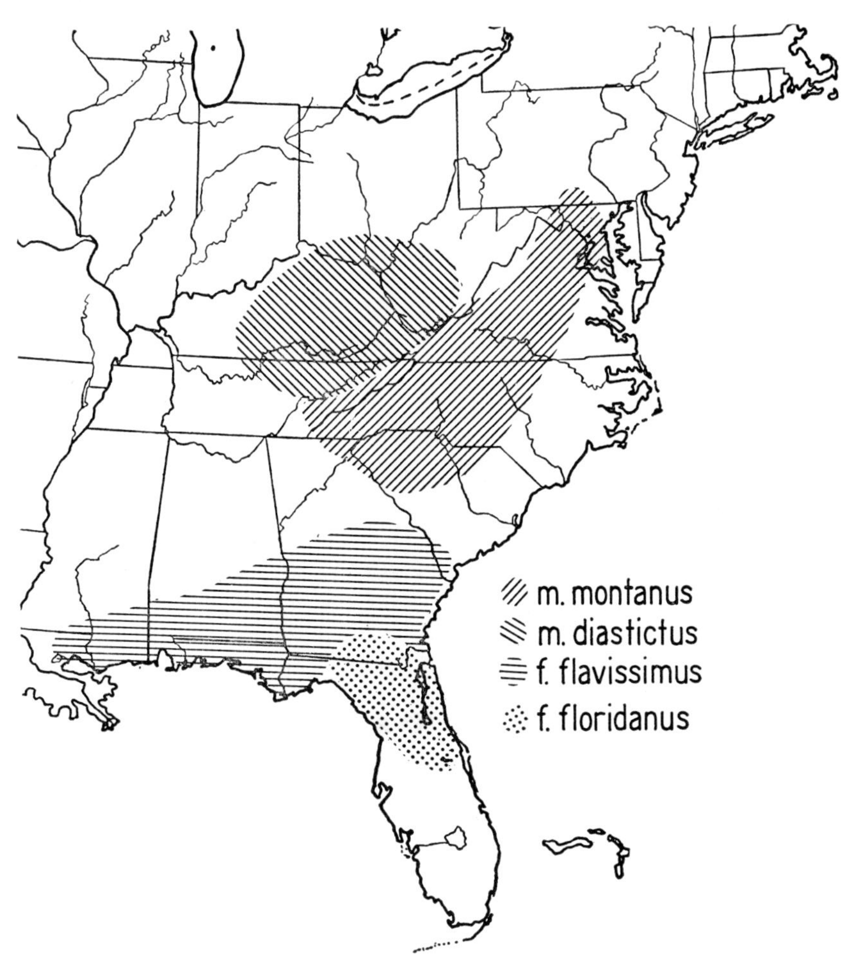


MAP 45.—Distribution of the subspecies of *Pseudotriton montanus* and *P. flavissimus.*

between the toes of the appressed limbs. Tail short, comprising 36–44 per cent of the total length, broadly oval in section at base, becoming compressed and slightly keeled above beginning a short distance behind the vent, and strongly compressed distally; below rounded on basal half and thin-edged distally. Legs short and only moderately stout. Toes 5-4, rather short and thick, those of the hind feet 1-5-2-(4-3), the 1st very short in some; toes of the fore feet 1-4-2-3. Tongue small, boletoid, elongate oval in outline. Vomero-parasphenoid teeth in continuous series. Vomerine teeth 12-15 in series that arise behind the middle or inner margin of the inner nares, extend toward the mid-line

FIG. 113. *Pseudotriton flavissimus flavissimus* Hallowell. (1) Adult female, actual length 3⅜″ (86 mm.). (2) Same, dorsal view. Near Lucedale, Mississippi. [Henry Dietrich, collector.]

in a nearly straight line or shallow curve before turning sharply backward to run nearly parallel to the parasphenoids. Parasphenoid patches long and narrow, divergent behind.

COLOR. The ground color in life is a clear, light, brownish-salmon above, from the tip of the snout to near the tip of the tail. The sides of the head, trunk, and tail, below a line which extends to the margin of the upper jaw on the head and just above the legs on the trunk, are lighter salmon, with tinges of yellow-orange on the tail. The dorsal and lateral surfaces of the head, trunk, and limbs, and the sides of the tail are sparsely marked with small, round, black dots. The ventral surface of the head and trunk is clear salmon-pink, which in some specimens gradually fades into the pale yellowish-orange color of the ventral surface of the tail. In preservatives, many specimens are pale yellowish-brown above and dull yellow below. Many specimens have the lower lip tinged with dusky.

This subspecies differs markedly from *P. m. floridanus* in color and pattern, having a relatively clear background color, as compared with the clouded condition in *floridanus,* and black spots on a lighter ground instead of light dashes on a dark ground. The venter of *flavissimus* may be entirely free from the dark specks which are characteristic of *floridanus.*

BREEDING. Nothing has been reported on the breeding habits, and eggs are unknown.

LARVAE. I have not seen the larvae.

**FLORIDA RED SALAMANDER.** *Pseudotriton flavissimus floridanus* Netting and Goin. Fig. 114. Map 45.

TYPE LOCALITY. Gainesville, Alachua County, Florida.

RANGE. Restricted to southern Georgia and the northern half of peninsular Florida.

HABITAT. Carr (1940, p. 49), under the name *P. m. flavissimus,* records this subspecies from springs, seepage areas, and small, sand-bottomed hammock streams.

SIZE. The few adults I have seen varied in length from $2\frac{27}{32}''$ (72 mm.) to $4\frac{5}{32}''$ (105 mm.) and averaged $3\frac{5}{8}''$ (92 mm.). The proportions of an adult male from Alachua County, Florida, are as follows: total length $3\frac{3}{4}''$ (95 mm.), tail $1\frac{3}{8}''$ (35 mm.); head length $\frac{1}{2}''$ (12.5 mm.), width $\frac{5}{16}''$ (8 mm.). A female from the same locality measures: total length $3\frac{9}{32}''$ (83 mm.), tail $1\frac{5}{16}''$ (33 mm.); head length $\frac{15}{32}''$ (12 mm.), width $\frac{9}{32}''$ (7.5 mm.).

DESCRIPTION. This is a dwarfed, peninsular-Florida derivative of *P. flavissimus.* The head is slightly convex above, the sides back of the eyes nearly parallel, in front converging to the bluntly pointed snout. Eye moderate, the horizontal diameter about 1½ in the snout. An impressed sinuous line from the posterior angle of the eye to the lateral extension of the gular fold, and a short vertical groove from this line to the angle of the jaw. Trunk slender, rounded; costal grooves 18, and 6–8 intercostal folds between the toes of the appressed limbs. Tail short, comprising 40–47 per cent of the total length, subquadrate in

FIG. 114. *Pseudotriton flavissimus floridanus* Netting and Goin. (1) Adult male, actual length 3²³⁄₃₂″ (95 mm.). (2) Same, dorsal view. Alachua County, Florida. [Photographs of a preserved specimen.]

section at base, becoming sharp-edged above a short distance behind the vent, and slightly keeled distally; rounded below to about the distal third and strongly compressed. Legs moderately short and stout. Toes short, those of the hind feet 1–5–2–4–3 in order of length from the shortest; toes of the fore feet 1–4–2–3. Tongue small, nearly circular in outline, boletoid. Vomero-parasphenoid series continuous. The vomerine series arise behind the middle or outer margin of the inner nares, curve inward toward the mid-line, then posteriorly, where they are usually narrowly separated but occasionally meet; 11–17 teeth in each series. Parasphenoid teeth in long, narrow patches, nearly contiguous at the anterior end but evenly separated by about ½ the width of a patch behind this point.

COLOR. In preserved specimens the dorsal surface of the head, trunk, and tail is purplish-brown clouded with dull yellow. The sides of the trunk to the level of the legs are marked with short, dull yellow dashes which are in contrast with the dark ground color. In some the light marks tend to form irregular longitudinal lines. The sides of the head and tail are mottled, blotched, or clouded with dull yellow. The venter is yellowish-white, with separate, small, reddish-brown dots scattered generally over the surface and sometimes concentrated in series along the sides of the trunk at the line which separates the darker sides from the pale venter. In an occasional old individual these dark brown dots are present in limited numbers on the sides. The margin of the lower jaw is usually marked with brown; legs above colored like the back.

SEXUAL DIFFERENCES. The sexes may be distinguished by the form of the vent, which in the male is lined anteriorly with short papillae forming pigmented patches. The vent of the female is margined by narrow vertical folds.

BREEDING. Nothing has been published.

LARVAE. Carr (1940, p. 50) records a small larva only 22 mm. long, found in the sandy bottom of a rill from a seepage area near Gainesville, Florida.

**BAIRD'S RED SALAMANDER.** *Pseudotriton montanus montanus* Baird. Figs. 112, 115. Map 45.

TYPE LOCALITY. South Mountain, near Carlisle, Pennsylvania.

RANGE. From near Carlisle, Pennsylvania, and Maryland southward to South Carolina and northeastern Georgia, mainly in the Piedmont but extending into the Coastal Plain in the northern parts of the range.

HABITAT. This species is usually more aquatic than *P. ruber* and often lives in or about muddy springs. Frequently it occurs in the wet, low grounds bordering streams, where it may be found hiding beneath logs, bark, and other objects on the ground. Occasionally a specimen may be found in a somewhat drier situation, but this is unusual.

SIZE. This species occasionally attains a length of 7″ (178 mm.) and is thus larger than its relative *P. ruber,* but this is exceptional and the average is considerably less. Twelve adults of both sexes from eastern Virginia and North Carolina average $4\frac{15}{16}$″ (126 mm.) with extremes of 6″ (153 mm.) and $3\frac{21}{32}$″ (93 mm.). The proportions of a sexually mature male from Raleigh, North Carolina, are as follows: total length $4\frac{27}{32}$″ (124 mm.), tail $2\frac{1}{32}$″ (52 mm.); head length $\frac{5}{8}$″ (16 mm.), width $\frac{11}{32}$″ (9 mm.). An adult female from the same locality measures: total length $5\frac{5}{32}$″ (132 mm.), tail $1\frac{3}{4}$″ (45 mm.); head length $\frac{19}{32}$″ (15 mm.), width $\frac{3}{8}$″ (10 mm.).

DESCRIPTION. The head is oval when viewed from above, and strongly convex, widest at the angle of the jaws, the snout short and bluntly rounded. The eye is of moderate size, the long diameter about $1\frac{1}{4}$ in the snout; iris black or dark brown, in this respect differing from *P. ruber,* which has the iris flecked with silvery, brassy, or gold. The trunk is stout and well rounded on the sides and above. There are usually 17 costal grooves, counting 1 in the axilla and 2 that are crowded in the groin, occasionally 16 or 18, and 5–7 intercostal spaces between the toes of the appressed limbs. The tail is nearly circular in section at base, becoming sharp-edged above opposite the posterior end of the vent, and compressed and keeled distally; venter of tail rounded at base; thin-edged at the distal third. Legs short and stout. Toes 5–4, those of the hind feet 1–5–2–(4–3) in order of length from the shortest; toes of the fore feet 1–4–2–3. Tongue small, circular in outline, boletoid. Vomero-parasphenoid teeth in continuous series. The vomerine teeth usually 11–13 in series that arise behind the middle or outer margin of the inner nares and curve inward and backward toward the mid-line, not sharply angled as usually found in *P. ruber.* Parasphenoid teeth in club-shaped patches, slender, and nearly or quite in contact anteriorly and divergent posteriorly.

COLOR. The color and pattern vary with age, young individuals having a clearer ground color and fewer and smaller black spots. In adults of large size the ground color above varies from light reddish-brown or purplish-brown to chocolate, often lighter on the tail and with the sides of the head, trunk, and tail brighter, salmon color. The belly varies

FIG. 115. *Pseudotriton montanus montanus* Baird. (1) Adult female, old dark individual; actual length 4¹⁵⁄₁₆″ (122 mm.). Raleigh, North Carolina. (2) Another individual, ventral view. Mount Vernon, Virginia.

from salmon-red to a dusky orange-red. Scattered over the dorsal surfaces are rounded black spots variable in size but usually smaller and more widely separated than in *P. ruber*. The spots are continued on the sides of the head, on the trunk to the lower level of the legs, and on the tail often to the ventral surface. In young individuals the venter may be almost immaculate, in older ones the throat, belly, and ventral surface of the tail are often spotted or flecked with brown or black. In old individuals the ground color may become so dark as to nearly obscure the rounded black spots, and in animals of this type the margins of the jaws are frequently entirely black.

BREEDING. For knowledge of the breeding habits we are mainly indebted to C. S. Brimley. The eggs have been found on a number of occasions, in November, December, and January, in the leafy trickle from a little spring near Raleigh, North Carolina. The unpigmented white eggs were attached to dead leaves or other objects lying in the bed of the trickle and on one occasion were attended by a large female (Brimley, 1939, p. 19). A few eggs sent by Dr. Brimley to E. R. Dunn measured 6 mm. in diameter and were provided with an attachment stalk 4 mm. long (Dunn, 1926, p. 290). These measurements probably include the envelopes. The breeding season may be even more extended than indicated above. In a collection from Raleigh, made Feb. 24, 1923, is a female distended by eggs which have a diameter of approximately 3½ mm.

LARVAE. Larvae attain a length of at least $2\frac{13}{16}''$ (72 mm.) but may transform when considerably smaller. A juvenile from Raleigh, North Carolina, had completed transformation at a length of only $2\frac{5}{8}''$ (67 mm.), and others were still sexually immature at a length of $3\frac{21}{32}''$ (93 mm.). The larvae I have examined are light brownish above, with darker pigment, in small irregular flecks, scattered over the dorsal and lateral surfaces but entirely lacking on the ventral.

**CENTRAL RED SALAMANDER.** *Pseudotriton montanus diastictus* Bishop. Fig. 116. Map 45.

TYPE LOCALITY. Cascade Caverns, Carter County, Kentucky.

RANGE. So far as now known, limited to the unglaciated plateau region of southern Ohio, central and eastern Kentucky, southwestern West Virginia, western Virginia, and eastern Tennessee.

HABITAT. Perhaps somewhat less aquatic than *P. m. montanus*. Of the two living examples I have seen, one was taken in a cave near Mill Creek. near Greenbrier, Kentucky, and the other from beneath a large block of wood in a slightly damp situation near the entrance to Cascade Caverns, Carter County, Kentucky.

SIZE. The average length of 13 adults of both sexes from West Virginia and Kentucky is $5\frac{5}{16}''$ (135 mm.), the extremes $3\frac{13}{16}''$ (97 mm.) and $6\frac{1}{8}''$ (156 mm.). The proportions of an adult male are as follows: total length $5\frac{9}{16}''$ (142 mm.), tail $2\frac{7}{16}''$ (62 mm.); head length $\frac{11}{16}''$ (18

FIG. 116. *Pseudotriton montanus diasticticus* Bishop. (1) Adult male, actual length $5\frac{9}{16}''$ (142 mm.). (2) Same, dorsal view. (3) Same, ventral view. Type. Cascade Caverns, Carter County, Kentucky. [Photographs of a preserved specimen.]

mm.), width $\frac{11}{32}''$ (11 mm.). The measurements of an adult female are: total length 6″ (153 mm.), tail $2\frac{17}{32}''$ (65 mm.); head length $\frac{25}{32}''$ (20 mm.), width $\frac{15}{32}''$ (12 mm.).

DESCRIPTION. The head is small and comparatively slender, the sides back of the eyes nearly parallel, in front tapering to the truncated snout. Back of head between eyes slightly depressed or broadly rounded over, not highly convex as in typical *montanus*. The eye is small, the horizontal diameter about 1½ in the snout; an impressed line from the posterior angle of the eye to the lateral extension of the gular fold; a short vertical groove from this line to the angle of the jaw. Trunk not so stout as in typical *montanus;* usually 17 costal grooves, counting 1 in the axilla and 2 that run together in the groin, occasionally 16, and 5½–6½ intercostal folds between the toes of the appressed limbs in adults. Tail subquadrate in section at base, becoming broadly oval a short distance behind the vent, and compressed and sharp-edged above distally; no ventral tail keel. Legs stout, toes 5–4, those of the hind feet 1–5–2–(3–4) in order of length from the shortest; toes of fore feet 1–4–2–3. Vomero-parasphenoid teeth in series that arise behind or just outside the outer margin of the inner naris and curve inward toward the mid-line, then backward for a short distance before joining parasphenoid. Parasphenoid patches long and club-shaped, separated by more than the diameter of a naris.

COLOR. The clearness of the ground color is in strong contrast to the clouded condition which is usually characteristic of *P. m. montanus.* In life, younger individuals may be light coral-pink, the older adults clear brilliant red to clear brown above, the lower sides and belly lighter, flesh to salmon. Scattered over the dorsal surface of the head, trunk, tail and legs, and sides of trunk and tail to the upper level of the legs, are rounded black spots, fewer, larger, and more uniform in size than in typical *montanus.* The ventral surfaces are entirely without darker spots or markings except for an occasional dark line along the margin of the lower jaw. In many specimens the black spots are somewhat concentrated dorsolaterally, the mid-line of the back being relatively free. The dark spots on the sides of typical *montanus* reach a lower level and, in old individuals, frequently cover the entire venter.

BREEDING. Nothing has been published on the breeding habits of this race as distinguished from *P. m. montanus.*

LARVAE. A larva at the point of transformation, from Boyd County, Kentucky, has a total length of 3²¹⁄₃₂″ (93 mm.), tail 1¹⁵⁄₃₂″ (38 mm.). It is sparsely spotted above, immaculate below.

**NORTHERN RED SALAMANDER.** *Pseudotriton ruber ruber* (Latreille). Figs. 101a, 117. Map 46.

TYPE LOCALITY. The United States, probably near Philadelphia.

RANGE. From Albany County, New York, southward to northern Georgia, westward to Alabama, Tennessee, Kentucky, and Ohio. In the Coastal Plain only from northern Virginia northward.

HABITAT. Found in and about clear, cold springs and small streams of wooded ravines, swamps, open fields, and meadows. The adults are often terrestrial during the summer months and may be found hiding beneath logs, bark, and stones, some distance from the water.

SIZE. This salamander attains an extreme length of about 6½" (165 mm.) but this is much beyond the average. Twenty adults of both sexes from various northeastern localities average 5" (128 mm.), the extremes 5²⁹⁄₃₂" (151 mm.) and 4³⁄₃₂" (103 mm.). The females in this series are a little longer than the males. The proportions of an adult male from Columbia, South Carolina, are as follows: total length 4¹⁷⁄₃₂" (116 mm.), tail 1²³⁄₃₂" (44 mm.); head length ²¹⁄₃₂" (17 mm.), width ¹³⁄₃₂" (11 mm.). A female from the same locality measures: total length 4¹⁹⁄₃₂" (118 mm.), tail 1¹¹⁄₁₆" (43 mm.); head length ¹¹⁄₁₆" (18 mm.), width ⁶⁄₁₆" (10 mm.).

DESCRIPTION. The head is widest immediately behind the eyes, the sides behind this point nearly parallel or slightly converging to the lateral extension of the gular fold, in front tapering rather abruptly to the bluntly rounded snout. The eye is rather small, the long diameter about twice in the snout, the iris tinged with brassy. An ill-defined impressed line from the posterior angle of the eye to the lateral extension of the gular fold, and a short vertical line crossing this at the angle of the jaw. The trunk is stout and in old individuals well rounded, with a slightly impressed median dorsal line. There are usually 16 costal grooves, occasionally 17, and 5–7 intercostal folds between the toes of the appressed limbs. The legs are rather short and stout. Toes 5–4, short, those of the hind feet 1–5–2–(3–4) in order of length from the shortest;

toes of fore feet 1–4–2–3. Tail subquadrate in section at base, becoming broadly oval in section a short distance behind the vent, and sharp-edged

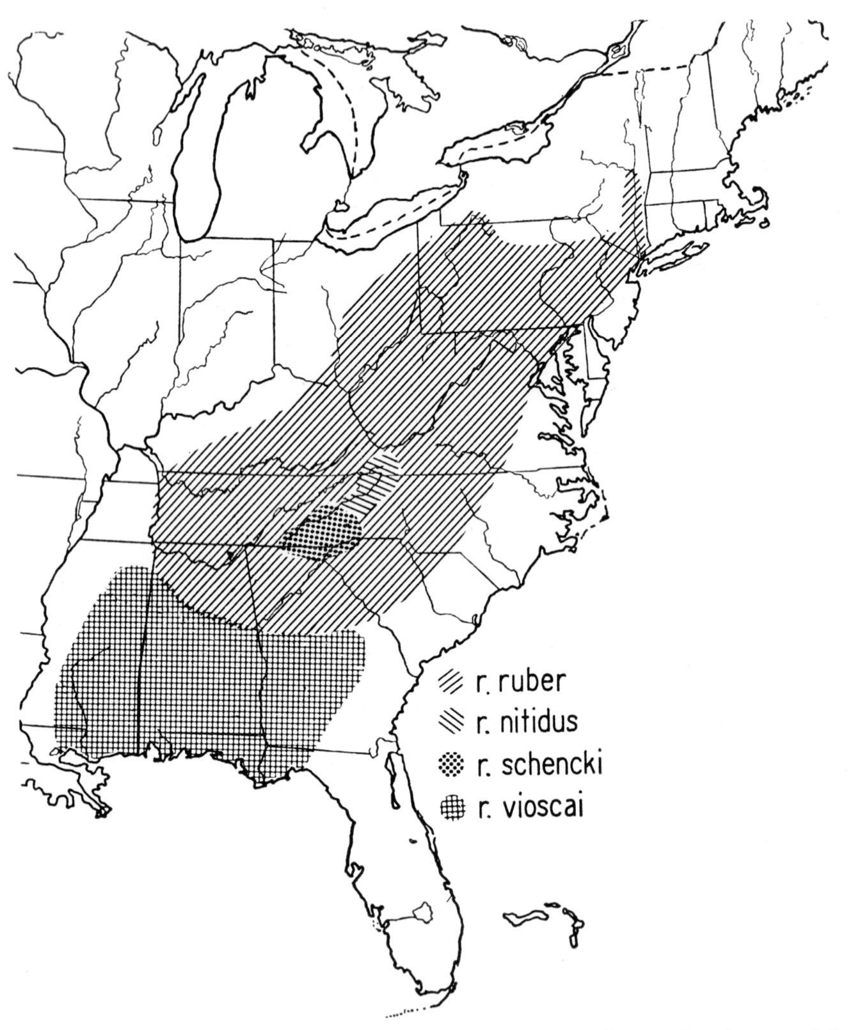


MAP 46.—Distribution of the subspecies of *Pseudotriton ruber*. (*P. r. ruber* also recorded from Burlington, Vermont.)

and keeled above at about the basal third. Tip of tail strongly compressed, but no true ventral keel. Tongue boletoid, roughly circular in outline. Vomero-parasphenoid teeth usually forming continuous series, occasionally a break on one side or the other. The vomerine series arise outside the outer margin of the inner naris, extend inward toward the mid-line, then turn sharply backward, where they continue, narrowly separated, until they join the parasphenoids. The parasphenoids some-

FIG. 117. *Pseudotriton ruber ruber* (Latreille). (1) Adult, dorsal view; actual length about 5″ (122 mm.). (2) Adult female, ventral view; actual length 5 13/16″ (147 mm.). (3) Portion of egg cluster. (4) Larva. Voorheesville, New York.

times in club-shaped patches; often they extend anteriorly as a narrow line for some distance before joining the vomerine.

COLOR. At transformation, and sometimes a little before, the ground color above is a beautiful coral-red, the lower sides and venter lighter, flesh color. Scattered over the dorsal surface of the head, trunk, and basal half of the tail are many small rounded black spots in strong contrast with the bright ground color. At this stage, the dark spots are usually absent on the lower sides, the belly, and the distal half of the tail. With increase in age and size, there is a gradual extension and darkening of the ground color and fusion of the dark spots, old individuals becoming purplish-brown above and well speckled with small brown or black spots on a salmon-red ventral color. Often the margin of the chin is flecked with black. Sexual differences are not well marked externally. The vent of the male is lined anteriorly with short papillae which are normally absent in the female.

BREEDING. In the northern parts of its range at least, the egg-laying season is apparently initiated by falling temperatures. The first lot of eggs found were attached to the lower surface of a flat stone which was well imbedded below the surface of the water in the bank of a spring near Voorheesville, New York, October 3, 1924. Additional lots were taken in November 1930 by Dr. V. C. Twitty and Dr. Crawford Hutchinson near Woodstock, Ulster County, New York. The eggs are without pigment and are attached separately in little clusters, the entire complement numbering 72 or more. Often several females select the same stone for attachment, but the groups belonging to different individuals may usually be recognized by differences in color and stage of development. Individual eggs are approximately 4 mm. in diameter and are surrounded by envelopes as follows: 1st, a thin tenuous vitelline membrane closely appressed to the yolk; 2nd, a layer enclosing a thin, clear jelly; and, 3rd, an outer tougher layer which is drawn out to form the attachment stalk. The egg with its envelopes measures about 6 mm. in diameter.

LARVAE. Larvae at hatching have a length of 13–15 mm. They are lightly pigmented above on the head to the lower level of the eyes, on the trunk to the level of the legs, on the tail to the ventral keel. The gills, buds of the legs, ventral surface of the head and trunk, dull yellow without pigment. Larvae may attain a length of 4⁵⁄₁₆″ (110 mm.), but usually transform when considerably smaller, 3″ (77 mm.) to 3¾″ (96 mm.). Large larvae have a ground color light reddish-brown, with tinges of orange at the base of the dorsal tail fin and sometimes on the snout. The dark pigment spots are usually larger on the trunk than on the head, legs, and distal part of the tail. On several occasions I have found half-grown larvae which had acquired the bright red characteristic of recently transformed individuals.

**DUNN'S RED SALAMANDER.** *Pseudotriton ruber nitidus* Dunn. Fig. 118. Map 46.

TYPE LOCALITY. Whitetop Mountain, Virginia.

RANGE. The northern part of the Southern Blue Ridge at elevations to 5000′. From Abington and Whitetop Mountain, Virginia, to the valley of the French Broad in North Carolina and Roan Mountain in Tennessee.

HABITAT. This subspecies has been found under logs in wooded areas and open pastures, and in and about springs and streams.

SIZE. The average length of 12 adults of both sexes from western North Carolina is $3\frac{15}{32}''$ (88 mm.), the extremes $2\frac{29}{32}''$ (74 mm.) and $4\frac{17}{32}''$ (116 mm.). The six females average $3\frac{9}{16}''$ (91 mm.), the males $3\frac{11}{32}''$ (85 mm.). The proportions of a male from Swannanoa, North Carolina, are as follows: total length $4\frac{17}{32}''$ (116 mm.), tail $1\frac{5}{8}''$ (42 mm.); head length $\frac{5}{8}''$ (16 mm.), width $\frac{3}{8}''$ (10 mm.). A female from Cave River, North Carolina, has the following measurements: total length $4\frac{15}{32}''$ (114 mm.), tail $1\frac{11}{16}''$ (43 mm.); head length $\frac{17}{32}''$ (14 mm.), width $\frac{11}{32}''$ (9 mm.).

DESCRIPTION. This subspecies is smaller than typical *ruber* and differs otherwise in shape, proportions, color, and pattern. The head is broadly oval in outline, the sides behind the eyes curving to the lateral extensions of the gular fold and in front to the broadly rounded snout. The eye is small, the horizontal diameter about twice in the snout. Eye limited behind by a fold which arises from the lower posterior margin of the lower lid and extends upward behind upper lid. A slightly impressed line from the posterior angle of the eye to the lateral extension of the gular fold; a short vertical groove from this line to the angle of the jaw. Trunk relatively slender, rounded, and with only a slight mid-dorsal impressed line. There are 16 costal grooves, counting 1 each in the axilla and groin and 3–6, usually 5, intercostal folds between the

FIG. 118. *Pseudotriton ruber nitidus* Dunn. (1) Adult male, actual length 3¼″ (83 mm.). (2) Same, dorsal view. (3) Same, ventral view. Weaversville, Buncombe County, North Carolina. [Photographs of a preserved specimen.]

toes of the appressed limbs. Legs stout, rather short. Toes 5–4, short and blunt-pointed, those of the hind feet 1–5–2–(3–4) in order of length from the shortest; toes of the fore feet 1–4–2–3. Tongue small, boletoid, nearly circular in outline. The vomero-parasphenoid series continuous. Vomerine series arise behind outer margin of inner naris, extend inward toward the mid-line, then turn sharply backward, where they are separated by about the diameter of an inner naris. Fewer teeth on vomerine than in *P. r. ruber*. Parasphenoid teeth usually in 2 long, slender, club-shaped patches somewhat divergent posteriorly.

COLOR. The general ground color above is red, below lighter, flesh color. The ground color, unlike that of *P. r. ruber,* does not become greatly darkened with age. Scattered over the dorsal surface of the head, trunk, and basal half of the tail are many separate, small, black spots, usually a little larger on the trunk than on the head or tail. On the sides of the head the dark spots extend to the margins of the jaws, on the trunk and tail to the upper level of the limbs. On some individuals there is a single row of small black dots along the margin of the lower jaw. The ventral surfaces are without black spots, except as noted on the lower lip. The dark spots remain separate through life and do not tend to fuse as in old individuals of *P. r. ruber*. There are no well marked sexual differences, but the males may usually be recognized by the form of the vent, which is larger than that of the female and provided within, anteriorly, with vertical imbricated folds. Adult *nitidus* resembles some juvenile *P. r. ruber,* but differs conspicuously from adult *ruber* in its smaller size, clear spotting, lack of black spots on the distal half of tail, and immaculate venter.

BREEDING. Nothing is known of the breeding habits of this salamander. A female (Univ. Mich. No. 76323) from Banners Elk, North Carolina, taken Sept. 4, 1934, has large eggs apparently ready to be deposited.

LARVAE. Dunn (1926, p. 283) records a larva 68 mm. long as the largest he had encountered, and the smallest transformed individual as 69.5 mm. in total length. Dunn (*ibid.*) writes: "A larva A.N.S. 4389, Roan Mt., Tenn., 4,000 feet, very near transformation is chiefly noticeable for having a narrow dark mid-dorsal line; the absence of spots on the tail is marked."

**BRIMLEY'S RED SALAMANDER.** *Pseudotriton ruber schencki* (Brimley). Fig. 119. Map 46.

TYPE LOCALITY. Sunburst, North Carolina.

RANGE. The southern part of the Southern Blue Ridge, ranging to above 5000′ altitude. Southwestern North Carolina, northwestern South Carolina, eastern Tennessee, and northern Georgia. Also recorded from Ohio but probably erroneously.

HABITAT. I have collected this salamander under the moss covering the gravelly bed of a small trickle. Brimley (1939, p. 22) reported it mostly from beneath logs in pastures near running water, but occasionally from streams, and Dunn (1926, p. 286) found it in swampy springs and one specimen beneath a log some distance from water.

SIZE. This subspecies attains an extreme length of 4 29/32″ (125 mm.). The average length of 25 adults of both sexes, mainly from western North Carolina, is 3 7/8″ (99 mm.), the extremes 4 25/32″ (122 mm.) and 2 21/32″(68 mm.). Eleven adult females average 4 1/8″ (105 mm.) and 13 adult males average 3 21/32″ (93 mm.). The proportions of an adult male from Cullosaja, North Carolina, are as follows: total length 4″ (102 mm.), tail 1 9/16″ (40 mm.); head length 5/8″ (16 mm.), width 3/8″ (10 mm.). A female from the same locality measures: total length 4 5/16″ (110 mm.), tail 1 11/16″ (43 mm.); head length 21/32″ (17 mm.), width 3/8″ (10 mm.).

DESCRIPTION. The head is widest at the angle of the jaws, the sides behind the eyes converging slightly to the lateral extensions of the gular fold and in front curving more abruptly to the bluntly pointed snout. The eye is small, the long diameter fully twice in the snout. Eye limited behind by a fold which arises from the lower, posterior margin of the lower lid and extends upward behind posterior margin of upper lid. An impressed sinuous line from the posterior angle of the eye to the gular fold; a short vertical groove from this line passes behind the angle of the jaw. Trunk moderately stout, rounded above and on the sides, flattened beneath, the middorsal line slightly impressed. There are 16 costal grooves, counting 1 each in the axilla and groin, and 4–5½ intercostal folds between the toes of the appressed limbs. Legs short

FIG. 119. *Pseudotriton ruber schencki* (Brimley). (1) Adult male, actual length 3¹⁹⁄₃₂″ (92 mm.). (2) Same, ventral view. Jocassee, South Carolina. (3) Male, actual length 3⁹⁄₃₂″ (84 mm.). Elkmont, Tennessee.

and stout. Toes short, blunt-pointed, 5–4, those of the hind feet 1–5–2–(3–4) in order of length from the shortest; toes of the fore feet 1–4–2–3. Tongue small, broadly oval in outline. Vomero-parasphenoid teeth in continuous series. Vomerine teeth 10–15 in each series, the usual number about 12. The vomerine series arise behind or just outside the outer margin of the inner naris and usually extend inward and slightly forward before turning abruptly backward to join the parasphenoids. Parasphenoid teeth usually in long, club-shaped patches, narrow and line-like anteriorly in many specimens and somewhat divergent posteriorly. In sexually mature males the vent is sometimes lined anteriorly with short papillae, that of the female thrown into folds.

COLOR. This subspecies attains a somewhat larger size than *P. r. nitidus* and differs otherwise mainly in having the chin black and the tail spotted nearly to the tip. The general ground color is red, and this form, like *nitidus,* does not darken greatly with age. Scattered over the surface of the head, trunk, and tail are many separate, rounded, black spots which are larger on the trunk and back of the head than on the lower sides, snout, and legs. The black spots extend on the sides of the head to involve both jaws, on the trunk to the level of the legs, and on the tail to the ventral third. The ventral surfaces are lighter than the dorsal, flesh color in life, and may be quite free from black spots, except for a black bar which follows the margin of the lower jaw, and scattered fleckings on the throat. In some specimens the entire venter, with the exception of the tail, is minutely dotted with black, often more highly concentrated along the mid-line of the belly.

BREEDING. Very little is known. Adult females in the University of Michigan collection from Pisgah National Forest No. 52516–7, taken July 26, 1926, and No. 81051 from Dillsboro, North Carolina, taken August 6, 1936, have ovarian eggs which appear ready to be laid, and males taken at the same time and place have the testes enlarged.

LARVAE. The larvae may attain a length of nearly 3″ before losing the gills, or transform at a length of 2⁹⁄₃₂″ (58 mm.) (Univ. Mich. No. 76321). A larva 51 mm. long from near Chatsworth, Georgia, taken with a transformed individual and presumably of this subspecies, is finely specked with black over the same surfaces as the adults. The gills of this specimen are small, pigmented on the rachises, and with long, slender, unpigmented filaments.

**VIOSCA'S RED SALAMANDER.** *Pseudotriton ruber vioscai* Bishop. Fig. 120. Map 46.

TYPE LOCALITY. Ten miles west of Bogalusa, Louisiana.

RANGE. Mainly limited to the Gulf Coastal Plain of northwestern Florida, Georgia, Alabama, Mississippi, and Louisiana.

HABITAT. Specimens taken at the type locality, Bogalusa, Louisiana, were found in and about springs and spring-fed streams or under logs on nearby hillsides. At Marianna, Florida, we took several large adults under logs and bark near springs, and Coleman J. Goin (Copeia, No. 4, 1939, p. 231) reports it in northwestern Florida from the cool deep ravines near the Apalachicola River.

SIZE. The average length of 19 sexually mature individuals of both sexes is 5″ (128 mm.), the extremes 5 25/32″ (148 mm.) and 3 7/8″ (99 mm.). There appears to be no difference in the average length of males and females. The proportions of an adult male from Bogalusa, Louisiana, are as follows: total length 4½″ (115 mm.), tail 1 9/16″ (40 mm.); head length 5/8″ (16 mm.), width 3/8″ (10 mm.). An adult female from Marianna, Florida, measures: total length 5 11/32″ (137 mm.), tail 2⅛″ (54 mm.); head length 23/32″ (19 mm.), width ½″ (13 mm.).

DESCRIPTION. This subspecies attains an average size equal that of the typical race, *P. r. ruber*. The head is widest at the angle of the jaws, and has the sides back of this point nearly parallel or slightly converging to the lateral extension of the gular fold; in front of the eyes the sides converge abruptly to the bluntly pointed snout. The eye is of moderate size, the long diameter about 1½ in the snout. The iris is mottled with brassy, silver, and black. There is a sinuous impressed line from the posterior angle of the eye to the gular fold, and a short vertical groove from this line to the angle of the jaw. The trunk is stout, well rounded, and with an impressed middorsal line. There are usually 16 costal grooves, counting 1 in the groin and 1 often imperfectly developed in the axilla, and 5–6½ intercostal spaces between the toes of the appressed limbs. The legs are short and stout; toes 5–4, rather more slender in this subspecies than others, those of the hind feet usually 1–5–2–4–3 in order of length from the shortest; toes of the fore feet 1–4–2–3. The tail is broadly oval or nearly circular in section at base, becoming com-

FIG. 120. *Pseudotriton ruber vioscai* Bishop. (1) Adult male, actual length 5¾″ (147 mm.). (2) Adult female, actual length 4½″ (115 mm.). Bogalusa, Louisiana. (3) Adult female, ventral view; actual length 5 7/16″ (139 mm.). Marianna, Florida.

pressed and keeled above a short distance behind the vent and strongly compressed distally. The tail is rounded below at base, becoming thin-edged at about the distal third. Tongue small, boletoid. Vomero-parasphenoid teeth in continuous series. The vomerine teeth 12–15 in each series, the usual number about 13. The vomerine series arise behind or just outside the outer margin of the inner naris, slant inward or obliquely backward toward the mid-line, then sharply backward, where they are separated by about the diameter of a naris. The parasphenoid patches are narrow anteriorly, sometimes reduced to a single line of teeth, posteriorly wider and slightly divergent. Sexual differences are not well marked externally, but the mature males may usually be distinguished by the form of the vent, which is lined internally at the anterior end with short papillae.

COLOR. In life, the ground color is dark salmon on the dorsal surface of the trunk and tail; the lower sides and venter much lighter. Scattered over the dorsal surface of the head, trunk, and tail are many fairly large, separate, blue-black blotches. Around the snout, on the sides of the head, and between the eyes there are many very small white flecks; the lips of both jaws are dark. On the sides of the trunk and tail the dark spots are smaller. Legs and feet with large spots above, smaller ones below. The ventral surfaces are spotted, those of the throat smaller and closer together than elsewhere. The ground color darkens with age and the black spots become less distinct but rarely fuse as in *P. r. ruber*. Preserved specimens become dark purplish-brown, the dark spots fading to deep brown, the ventral surfaces dull yellow. Occasionally, the dark dorsal spots suggest a herringbone arrangement. This subspecies differs from *P. r. nitidus* and *P. r. schencki* in its larger size, strongly spotted venter, tail spotted to tip, and darker ground color.

BREEDING. Nothing is known of the breeding habits. An adult female, (Univ. of Michigan No. 81182), taken August 31, 1935, at Pensacola, Florida, has very large ovarian eggs.

LARVAE. Larvae, some apparently recently hatched, were collected at the type locality, Bogalusa, Louisiana, March 19, 1936. The smallest, 25 mm. in total length and still retaining considerable yolk, was marked above with a very fine reticulated pattern of purplish-brown over a dull yellow background. On each side of the back, and extending onto the tail, there was a fairly regular row of faint yellow spots. The legs were light gray spotted with yellow. The gills were well developed, long and slender, the filaments without pigment. Belly unpigmented, dirty white. A 62-mm. larva had the ground color dull tan with flecks of brown pigment of irregular size and shape scattered over the dorsal surfaces and forming a mottled pattern on the sides of the tail. The belly was bluish-white.

## Genus EURYCEA

### KEY TO THE SPECIES AND SUBSPECIES OF EURYCEA

1. Costal grooves 19–20 ....... 2
   Costal grooves 17 or less ....... 4
2. Neotenic, adults with gills; usually dull yellow above, lightly suffused with dusky in a broad band lighter than adjacent sides, sometimes nearly uniformly dusky above or with remnants of paired larval spots; venter of trunk and head immaculate, yellowish; gills short, pigmented; lateral-line sense organs abundant along sides of trunk and on head; length to 2¹³⁄₃₂″ (61 mm.). Known only from vicinity of Proctor, Oklahoma ....... *tynerensis* p. 444
   Not neotenic, adults without gills ....... 3
3. Grayish or brownish above in a broad band limited either side by a poorly defined black line, or, with upper sides darker than back; venter evenly pigmented, gray over dull yellow; vomerine teeth 9–11; length to 3⁵⁄₃₂″ (81 mm.). Known only from the vicinity of Gore, Oklahoma ....... *griseogaster* p. 418
   Yellowish above, suffused with brownish in a broad dorsal band slightly lighter than adjacent sides; rarely with a series of dark flecks dorsolaterally; belly and venter of tail bright lemon-yellow, throat flesh; vomerine teeth 6–10; length to 3¹⁷⁄₃₂″ (90 mm.). Missouri, Arkansas, Kansas, and (?) Jemez Mountains, New Mexico ....... *multiplicata* p. 435
4. Toes of appressed limbs never separated by more than 2 intercostal folds, often meeting or overlapping; 14 costal grooves; long, slender, mostly terrestrial species ....... 5
   Toes of appressed limbs separated by 3–7 intercostal folds; body and tail not particularly long and slender; semi-aquatic or aquatic species; 15–17 costal grooves (except in *cirrigera,* which normally has 14) ....... 8
5. With a narrow, black, middorsal line within a tan or yellow dorsal band; upper sides black; venter mottled, gray and dull yellow; sides of tail with imperfect vertical black bars and narrow yellow interspaces; length to 7³⁄₁₆″ (183 mm.). Virginia to northwest Florida west-

ward through Alabama and Mississippi to Louisiana; western Tennessee ........ *longicauda guttolineata* p. 425

Without a narrow, black, middorsal line within a broad tan or yellow dorsal band ........ 6

6. Venter more or less spotted or mottled; back and tail with a median tan or yellow band within which are many small brownish or blackish spots; sides below light dorsal band dark brown with scattered light flecks or spots; length to $5^{25}/_{32}''$ (148 mm.). Southern half of Missouri, Arkansas, eastern Oklahoma, and southeastern Kansas ........ *longicauda melanopleura* p. 428

Venter immaculate; ground color of back and sides yellow to orange or reddish-orange ........ 7

7. Without a definite broad dorsal band; back and sides of head, trunk, and tail with many small, irregular or rounded, separate black spots, rarely a dorsolateral linear series; venter light yellow; length to $6^{11}/_{32}''$ (161 mm.). Illinois, Indiana, Ohio, West Virginia, Kentucky, and possibly Alabama, westward to Arkansas, Oklahoma, and Missouri ........ *lucifuga* p. 431

With a definite broad dorsal band limited either side by elongate dots or dashes and enclosing small, irregular, separate black spots; sides of head and trunk with separate black spots; sides of tail with vertical, black, crescentic or dumbbell-shaped bars; venter yellow; length to $7^{5}/_{32}''$ (182 mm.). New York south to Georgia, west to Arkansas, and north to Missouri and Illinois ........ *longicauda longicauda* p. 421

8. With a definite dorsolateral black stripe on either side of a light dorsal band; not neotenic ........ 9

Without a definite dorsolateral black stripe on either side of a light dorsal band; neotenic ........ 12

9. With 14 costal grooves; dorsolateral dark stripe usually continued on the tail to its tip; adult males with prominent cirri; sides below dark stripe often brown and white mottled or with a series of white spots in a line above the legs; belly yellowish; length to $3^{27}/_{32}''$ (98 mm.). Coastal Plain and Piedmont, North Carolina, south through western Florida, west through Alabama and southern Louisiana; western Tennessee ........ *bislineata cirrigera* p. 408

With 15 or more costal grooves; dorsolateral dark stripe not continuous on tail to its tip as a solid line; cirri on males small ........ 10

10. Dorsolateral dark stripe heavy, black, with straight upper edge, continuous on basal half or two-thirds of tail, broken into a series of spots on distal half; sides below dorsolateral band marked with definite black spots of irregular size and shape; length to $4^{11}/_{16}''$ (120 mm.). Whitetop Mountain, Virginia, southward through North Carolina and South Carolina to northern Georgia ........ *bislineata wilderae* p. 415

Dorsolateral dark stripe more or less irregular and invaded by small light areas, not reaching tip of tail; sides below dorsolateral stripe mottled ........ 11

11. Dorsal band dull yellowish ochre, suffused with dusky, and marked with small separate black dots; sides below dorsolateral stripe dark, strongly mottled to lower level of the legs; base of tail of sexually active male often with a well developed glandular hump above and slightly behind vent; length to 4⅜″ (112 mm.). South central Quebec on both sides of the St. Lawrence River ........ *bislineata major* p. 411

Dorsal band bright greenish-yellow to orange-yellow or brownish, not strongly suffused with dusky, but with small black flecks often arranged in a linear series along the mid-line; sides below dorsolateral dark strip uniformly grayish or mottled but lighter than in *major* and extending only to upper level of the legs; glandular protuberance on base of tail above in adult males when present, small; length 3 25/32″ (97 mm.). Eastern Quebec and New Brunswick to Illinois, North Carolina, and Tennessee. Also reported from South Carolina ........ *bislineata bislineata* p. 404

12. Uniformly light brown above except for a dorsolateral series of small, separate, yellowish flecks; belly and throat white, venter of tail often with chromatophores to mid-line; eyes normal, partly surrounded by black; gills with separate dark brown chromatophores; size small to 2″ (51 mm.). Known from a small lake at the head of San Marcos River, San Marcos, Texas ........ *nana* p. 439

Light yellowish above in a mottled or reticulated pattern; throat, belly, and mid-line of tail below without pigment; eyes normal in individuals from open streams, much reduced in cave specimens; gills pigmented in open streams, white or only lightly pigmented in cave specimens; size larger, to 3 11/16″ (94 mm.). Known from the type locality, 5 miles north of Helotes, from Helotes Creek and from a cave near Boerne, all in Texas ........ *neotenes* p. 441

**NORTHERN TWO-LINED SALAMANDER.** *Eurycea bislineata bislineata* (Green). Fig. 121. Map 47.

TYPE LOCALITY. Probably the vicinity of Princeton, New Jersey.

RANGE. Quebec, New Brunswick, to Illinois, North Carolina, and Tennessee. Also reported from South Carolina.

HABITAT. This is essentially a species of the brooksides, where it hides beneath stones and logs in the well saturated soil. It is at home in the water, where it resorts for the egg-laying season, and swims freely, assuming the habits of the larvae. Occasionally found in drier situations some distance from water.

SIZE. Twenty adult males from New York localities vary in length from $2^{21}/_{32}''$ (67 mm.) to $3^{5}/_{8}''$ (92 mm.) and average $3''$ (78.8 mm.). A similar series of females vary from $2^{13}/_{32}''$ (61 mm.) to $3^{13}/_{16}''$ (97 mm.). The proportions of an adult male from New York are as

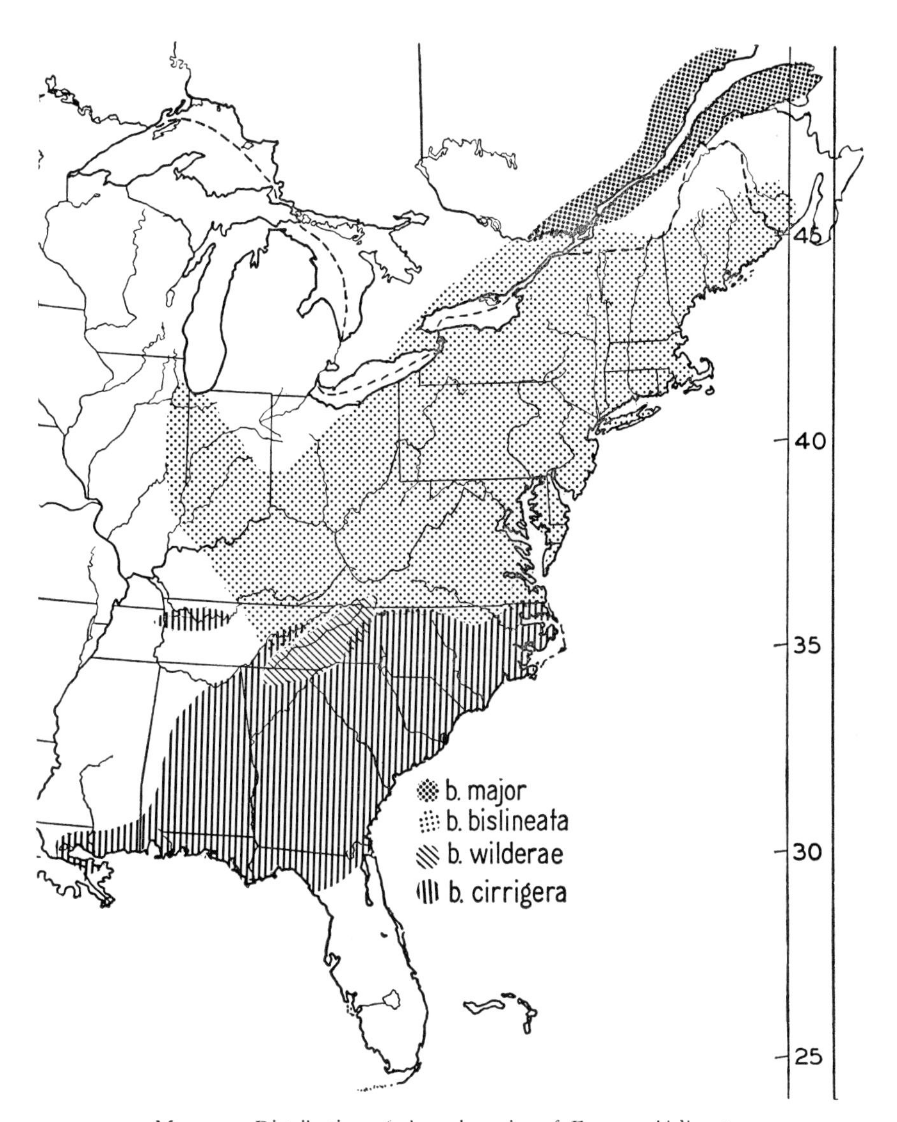


MAP 47.—Distribution of the subspecies of *Eurycea bislineata.*

follows: total length $3\frac{15}{32}''$ (88 mm.), tail $1\frac{31}{32}''$ (50 mm.); head length $\frac{11}{32}''$ (9 mm.), width $\frac{3}{16}''$ (5 mm.). A female measures: total length $3\frac{7}{32}''$ (82 mm.), tail $1\frac{9}{16}''$ (40 mm.); head length $\frac{5}{16}''$ (8 mm.), width $\frac{3}{16}''$ (5 mm.).

DESCRIPTION. The body is slender and subcylindrical. In the female, the sides of the head are nearly parallel back of the eyes and converge in front to the bluntly rounded snout. The head of the male may be widened back of the eyes, the snout somewhat swollen in the region of the nasolabial grooves, and, in some, the margin of the upper lip produced into small blunt cirri below the nostrils. There is an impressed line extending from the posterior angle of the eye to the lateral extension of the gular fold, and a short vertical groove passing behind the posterior angle of the eye to the angle of the mouth. The eyes are small, the long diameter equal to the length of the snout. The trunk is rounded above and slightly flattened below. There are 15 costal grooves in the majority of individuals, or 16 counting 1 in the axilla and 2 that run together in the groin, and about 3-4 intercostal spaces between the toes of the appressed limbs. Tail subquadrate in section at base, compressed and tapering to the slender tip; it is thinner above than below and often keeled. Legs well developed, toes 5-4, those of the hind feet 1-5-2-4-3 in order of length from the shortest; toes of fore feet 1-4-2-3. Tongue small and oval in outline and with a central pedicel. Vomerine teeth in series of 7-12, in the male forming an irregular line which arises behind the inner margin of the inner naris, arches gently toward the mid-line, then turns sharply backward, where it is narrowly separated from its fellow of the opposite side. The parasphenoid teeth are in elongate separate patches, widely separated from the vomerine.

COLOR. The ground color is variable and may be dull greenish-yellow to bright orange-yellow and brown. There is a broad dorsal light band which originates on the snout, widens back of the eyes to nearly the full width of the head, and continues along the trunk and the tail to its tip. Within the dorsal light band are many small, irregular, black spots and, in some individuals, a fairly regular median line. The dorsal band is limited either side by a dorsolateral black stripe which originates back of the eye and continues along the trunk and onto the tail. In most, the black stripe breaks up into short dashes on the tail and may be

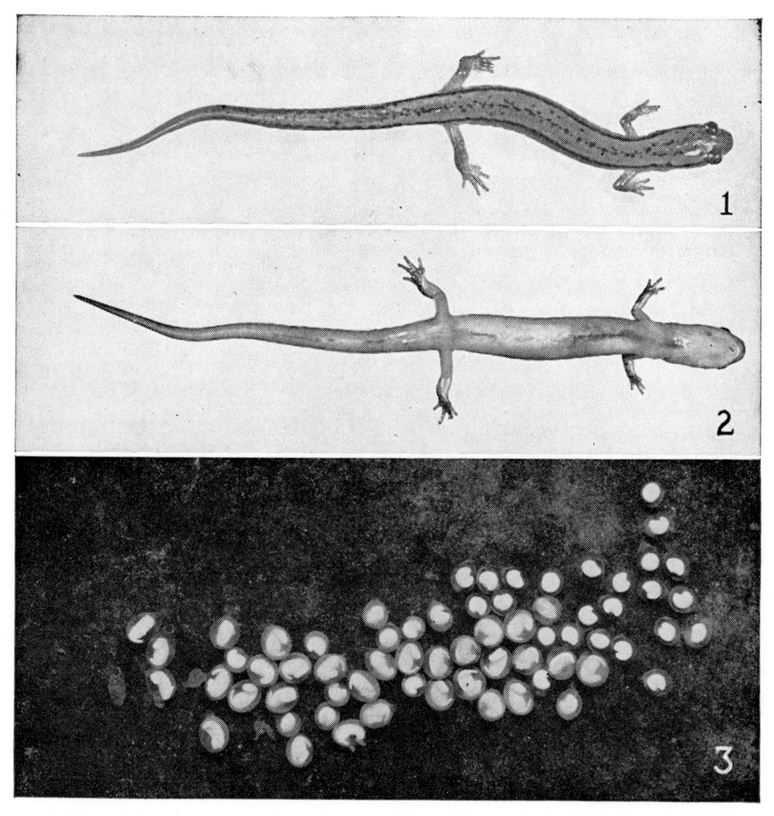


FIG. 121. *Eurycea bislineata bislineata* (Green). (1) Adult male, actual length 2 15/16″ (75 mm.). (2) Same, ventral view. Tufa Glen, Rochester, New York. (3) Eggs, slightly reduced. Stamford, New York, June 25.

limited to the basal half. Below the black stripes the sides are mottled, darkest next to the stripes, and fading on the lower sides to the bright yellow venter. Legs and feet lightly mottled above. In some individuals a row of small, round, light spots extends along the upper side below the black stripe and represents the remnant of the larval pattern. The sexes at the breeding season may usually be distinguished by the form of the head, as indicated above. The lower jaw of the male is usually more pointed and the snout more swollen, and often there are short cirri.

BREEDING. The breeding season usually extends from January to April (Noble, 1929, p. 2). The males nose one another and the females until the latter are stimulated to follow the males and pick up the spermatophores. The complete spermatophore consists of a gelatinous base capped by an elongate head (*ibid.,* p. 3). The eggs are attached singly to the lower surface of a support in running water. They may be found as early as April and with well developed embryos in August. The average number deposited is about 30, and often the complements of several females are deposited on the same stone. The eggs are without pigment, white to pale yellow. The egg has a diameter of 2.5–3 mm. and is surrounded by 2 definite envelopes which give a total diameter of about 4.5–5 mm., the outer envelope drawn out to form an attachment disk.

LARVAE. The newly hatched larva has a length of about 12 mm. There is a broad dorsal light band, below which a dark band encloses a series of small round light spots. Larvae may attain a length of 2 15/16″ (68 mm.) before transforming at an age of 2–3 years. The gills are pigmented.

**SOUTHERN TWO-LINED SALAMANDER.** *Eurycea bislineata cirrigera* (Green). Figs. 49b-c, 122. Map 47.

TYPE LOCALITY. New Orleans, Louisiana.

RANGE. Lowlands of eastern North Carolina, south through northwestern Florida, west through Alabama and southern Louisiana; western Tennessee (Stejneger and Barbour, 1939, p. 24).

HABITAT. This is mainly a Coastal-Plains form, and is found under rocks, logs, and leaves in the river and creek swamps and the outlets of springs. Mainly terrestrial, but enters the water for the egg-laying season.

SIZE. Attains a total length of 3 27/32″ (98 mm.), but the average is nearer 3 3/8″ (86 mm.). In a female from Marianna, Florida, 98 mm long, the tail comprises 58.1 per cent of the total (57 mm.); head length 11/16″ (9 mm.), width 7/16″ (6 mm.). A fully mature male from the same locality had the following proportions: total length 3⅝″ (94

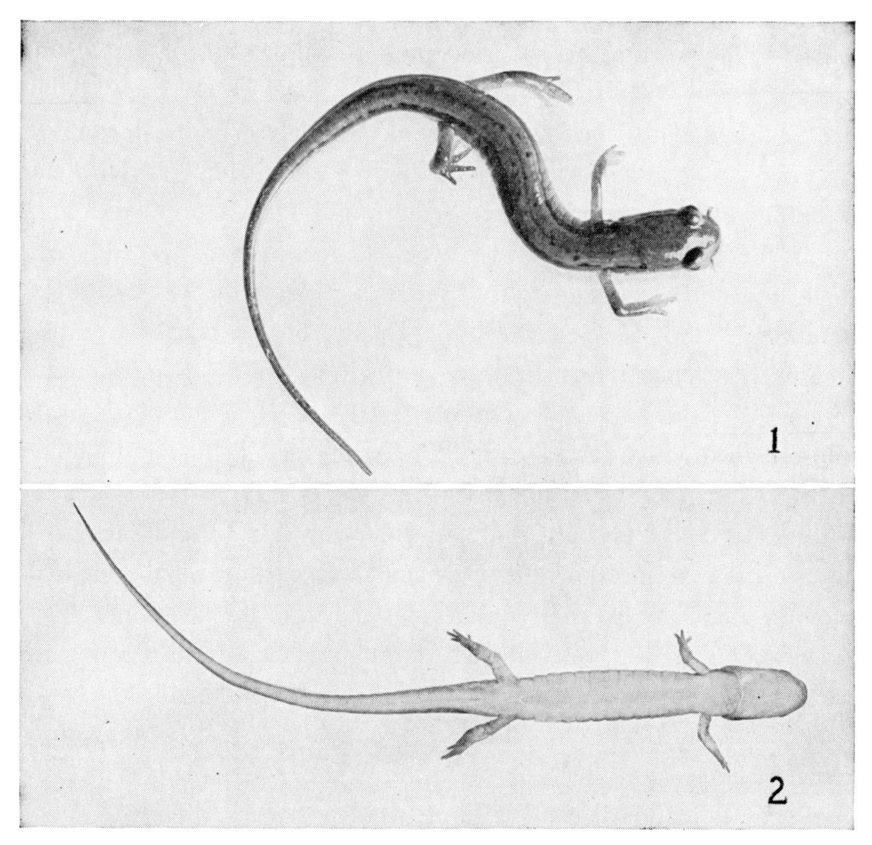


FIG. 122. *Eurycea bislineata cirrigera* (Green). (1) Adult male, actual length about 3¼″ (83 mm.). Pensacola, Florida. (2) Adult female, ventral view; actual length 3 21/32″ (93 mm.). Marianna, Florida. [Photograph of a preserved specimen.]

mm.), tail 2⅛″ (54 mm.); head length 11/16″ (9 mm.), width 7/16″ (6 mm.).

DESCRIPTION. Very closely related to *Eurycea b. bislineata,* but differs in its lower costal groove count, in the greater development of cirri, and to some extent in color and pattern. The head is moderately long and with the sides back of the eyes parallel or slightly widened just in front of the gular fold. In front of the eyes, the snout of the male is broadly truncated, swollen at the nostrils, the nasolabial grooves continued to the tip of the well developed cirri; in the female the snout is bluntly pointed. The eyes are of moderate size and protuberant. A sinuous groove from the posterior angle of the eye to the lateral extension of the gular fold, but no definite vertical groove back of the eye to the angle of the jaw. The trunk is nearly cylindrical, slightly flattened below, and provided dorsally with a slightly impressed median line. There are 14 costal grooves, counting 1 each in the axilla and groin, and 3–4 intercostal spaces between the toes of the appressed limbs. Tail long and slender, roughly oval in section at the base, strongly compressed and sharp-edged above distally, but without a free fin. Legs well developed, toes 5–4, those of the hind feet 1–5–2–4–3 in order of length for the shortest; toes of fore feet 1–4–2–3. Tongue small, oval, and provided with a central pedicel; vomerine teeth behind and wholly between the inner nares; the series are short in most, comprising 5 or 6 teeth, in others to 10 or more, and curving inward and backward toward the mid-line, where they are narrowly separated. Parasphenoid teeth in 2 elongate patches widely separated from the vomerine.

COLOR. There is a broad median dorsal light band which arises on the snout and continues along the trunk to the tip of the tail. The band varies in color from light yellowish to russet, and encloses numerous small black specks, which on some individuals form a single median line. The light dorsal band is limited either side by a narrow dark-brown-to-black stripe extending from the hind angle of the eye, along the trunk, and onto the tail. In most specimens this dark stripe continues to the tip of the tail, where it becomes lost in the general darkening of the sides; in a few it breaks up into a series of dashes. Usually the tail base above has a spot of color brighter than the surrounding areas. Sides below lateral dark stripe usually dark brown mottled with yellow, and often with a row of small circular light spots above the legs and an imperfect row at the junction of the lower sides and belly. Legs brown, mottled above. Belly yellowish, under sides of throat and legs flesh color, venter of tail greenish-yellow. The sexes may be distinguished by the greater development of cirri and more squarely truncated snout in the male, and by the presence in some males of enlarged premaxillary teeth.

BREEDING. Carr (1940, p. 49) reported gravid females taken in Liberty and Jackson Counties, Florida, Nov. 17. The males at this time were with swollen cirri. A large female sent me by Professor William M. Barrows from Pensacola, Florida, was distended with eggs Jan. 20, 1925. Carr (*ibid.*) reported the discovery of a lot of 50 eggs, "cemented in one layer to the under surface of a stone in running water," but did not give the date. Dunn (1926, p. 310) reported some larvae hatched April 12 from eggs found April 4. At Raleigh, North Carolina, Brimley (1939, p. 18) found eggs in March and April attached to the lower surfaces of stones in rapid water. The eggs were white and securely attached.

LARVAE. The larvae hatch in about a month and may attain a length of 75 mm. before transforming a year or two later. The larvae are characterized by a light dorsal stripe and a double row of light spots on the sides, the upper ones larger than the lower. The tail is keeled above to the base.

**GREATER TWO-LINED SALAMANDER.** *Eurycea bislineata major* Trapido and Clausen. Fig. 123. Map 47.

TYPE LOCALITY. Ouiatchouan River, Lake St. John County, Quebec, Canada.

RANGE. South central Quebec, Canada, on both sides of the St. Lawrence River.

HABITAT. Along the Ouiatchouan River, Lake St. John County, Quebec, Trapido and Clausen took many specimens from beneath limestone slabs during September.

SIZE. This subspecies averages conspicuously larger than the typical form. Trapido and Clausen (1938, p. 119) found 16 specimens from the Ouiatchouan River to average $3^{13}/_{16}''$ (96.06 mm.) in total length. Of this the tail comprised from 53–60 per cent. Fully adult females seem to average a little larger than the males. Seven females, each with fully developed eggs in the ovaries, varied in length from $3\frac{9}{16}''$ to $4\frac{7}{16}''$ (90–112 mm.); 7 males with prominent, glandular protuberances at the base of the tail varied in length from $3\frac{1}{2}''$ to $4\frac{5}{16}''$ (89–109 mm.). The proportions of a male from Val Jalbert are as follows: total length $4\frac{5}{16}''$ (109 mm.), tail $2\frac{9}{16}''$ (65 mm.); head length $\frac{1}{2}''$ (12.5 mm.), width $\frac{5}{16}''$ (7.5 mm.). A female from the same locality measures: total length $4\frac{9}{32}$ (109 mm.), tail $2\frac{11}{32}''$ (60 mm.); head length $\frac{14}{32}''$ (11 mm.), width $\frac{9}{32}''$ (6.5 mm.).

DESCRIPTION. Quite similar to *Eurycea b. bislineata,* but larger and with a proportionally longer tail. The fully adult males have a glandular enlargement at the base of the tail above, like that found in some males of the typical subspecies but more conspicuous. The head of the female is rather slender, widest immediately back of the eyes, the sides nearly parallel. The snout is short and bluntly rounded. In the male the head is actually and proportionally larger and longer. Back of the eyes the head widens conspicuously, the sides converging behind to the lateral extensions of the gular fold; the snout is a little longer and slightly more pointed. The eyes are of moderate size, protuberant, and with a dark horizontal bar, the iris above and below the bar brassy. A sinuous groove from the posterior angle of the eye to the lateral extension of the gular fold, and a short groove immediately behind the eye to the angle of the mouth. The trunk is moderately stout, nearly cylindrical, slightly flattened below, and with a median impressed line above. Tail subquadrate in section at the base, compressed distally and uniformly tapering to the slender tip. Costal grooves usually 16, occasionally 17 or 15. About 4–5 intercostal spaces between the toes of the appressed limbs. Legs well developed, apparently stouter than in typical *E. b. bislineata.* Toes 5–4, those of the hind feet 1–5–2–4–3 in order of length from the shortest; fore legs, 1–4–2–3. Tongue small, oval in outline. Vomerine series short, 6–7 in the female, 5–6 in the male, the series arising between and slightly behind the inner nares and curving in-

FIG. 123. *Eurycea bislineata major* Trapido and Clausen. (1) Adult female, actual length 4⁵⁄₁₆″ (109 mm.). (2) Same, ventral view. (3) Adult male, actual length 3¾″ (95 mm.). Val Jalbert, Lake St. John County, Quebec. [Harold Trapido, collector.]

ward, then sharply backward. Parasphenoid teeth in 2 slender, club-shaped patches which are separated the entire length.

COLOR. The ground color of the dorsal band in life is quite uniformly dull yellowish ochre somewhat obscured by a finely reticulate pattern of gray, except on the basal half of the tail, which is usually much brighter. Small, irregular, dark brown or black spots are scattered irregularly over the surface within the band on the trunk, but are fewer or absent on the tail. The dorsal band is limited on either side by a narrow black stripe which originates at the eye and extends along the upper side of the trunk and basal half of the tail. The sides below the dorsolateral dark stripe are usually dark brown somewhat mottled and blotched with lighter. In some individuals there is a well defined line of small light spots quite conspicuous against the darker brown of the sides and probably marking the position of the lateral-line sense organs. The legs above are usually mottled like the sides but are slightly lighter. The throat and ventral surface of the trunk are dull flesh color, lightly pigmented with gray, but becoming yellowish toward the hind legs, and often quite bright yellow-orange on the basal half of the tail. The tip of the tail beneath becomes strongly suffused with dusky.

SEXUAL DIFFERENCES. Sexual differences are well marked. As indicated above, the sexually mature males have a glandular protuberance at the base of the tail above. The vent of the male has fleshy lips and the margins are grooved, while in the female the opening of the vent lies in a groove and the sides of the vent are smooth along the posterior half. The tip of the lower jaw of the male frequently has a small depressed area, and the chin is a little more pointed than in the female.

BREEDING. Nothing has been published on the breeding habits of this subspecies. Adult females collected for me at the type locality, Val Jalbert, Quebec, Sept. 28, 1939, by Mr. Harold Trapido, have the ovarian eggs well developed, individual eggs having a diameter of 2–3 mm.

LARVAE. Larvae of all sizes from 15 to 70 mm. in length have been found by Trapido and Clausen (1940, p. 244), and the smaller figure probably represents, approximately, the length at hatching. The general color is darker than in typical *bislineata,* and the dorsal pattern consists of a median, longitudinal, light band with irregular edges. Dark pigment encroaches on the sides of the throat and, in large specimens, frequently on the belly. The dorsal fin of the tail arises abruptly at a point opposite the vent, and the tail tip is bluntly rounded.

**BLUE-RIDGE TWO-LINED SALAMANDER.** *Eurycea bislineata wilderae* Dunn. Fig. 124. Map 47.

TYPE LOCALITY. Whitetop Mountain, Virginia.

RANGE. The Southern Blue Ridge region from Whitetop Mountain, Virginia, south through North Carolina, South Carolina, and Tennessee to Rabun and Gilmer Counties, Georgia (Stejneger and Barbour, 1939, p. 24).

HABITAT. In and along the margins of mountain springs and streams, where they hide beneath sheltering objects such as stones, logs, bark, and other debris. Occasionally found far from open water under logs, bark, moss, and leaves in damp situations. Most abundant above 2000′ and extending to the summits of the highest mountains.

SIZE. Adults commonly vary from 2⅜″ (60 mm.) to 3 17/32″ (90 mm.). The average length of 17 adult males from Mt. Le Conte, Tennessee, is 3¾″ (94.4 mm.), with extremes of 3⅛″ (80 mm.) and 4 5/16″ (110 mm.). Eight adult females from North Carolina and Tennessee average 3 17/32″ (90 mm.) and vary from 2 25/32″ (71 mm.) to 4 5/32″ (106 mm.). King (1939, p. 559) records a specimen 4¾″ (120 mm.). In the male the tail comprises 52.3–61.3 per cent of the total length, the higher percentage from the largest specimens. In the females measured the range was 52.1–58.2 per cent. The proportions of an adult male from Mt. Le Conte are as follows: total length 4⅛″ (105 mm.), tail 2 11/32″ (60 mm.); head length 7/16″ (11 mm.), width ¼″ (6 mm.).

DESCRIPTION. Closely related to typical *E. b. bislineata,* but averages larger and with colors and pattern more distinct. The head of the female is slender, the sides behind the eyes nearly parallel, and the snout bluntly pointed. The male has the head conspicuously swollen back of the eyes, the snout broader and slightly swollen in the region of the nasolabial grooves; in some males the cirri are developed, in others lacking. A sinuous groove from the posterior angle of the eye to the lateral extension of the gular fold, and a vertical fold immediately behind the eye to the angle of the mouth. Eyes of moderate size and protuberant. The trunk is rounded above and on the sides, flattened below. There are 15 costal grooves, counting 1 each in the axilla and groin, and 3–5 intercostal spaces between the toes of the appressed limbs. Tail subquadrate in section at base, rounded below, and becoming sharp-edged and compressed distally. No free tail fin developed in the adults. The legs are well developed and stouter than in typical *E. b. bislineata.* Toes 5–4, those of the hind feet 1–5–2–4–3 in order of length from the shortest, the innermost rudimentary, webbed at base; toes of fore feet 1–4–2–3. Tongue broadly oval and with a central pedicel. Vomerine series short, usually in imperfect series, 4–7, which arise, usually between the inner nares but occasionally back of the inner margin of the inner nares, and curve inward and backward, separated by about ½ the width of an inner naris and from the parasphenoid by about 3 times that distance. Parasphenoid teeth in long and slender patches separated by about the diameter of an inner naris.

COLOR. Generally brighter than in *E. b. bislineata,* the ground color varying from yellow to light brown. The broad dorsal light band arises on the snout and continues along the trunk and the tail to its tip. The dorsal band may be almost immaculate, finely specked with black or with larger black spots of irregular size and shape. In a few individuals a single broken middorsal line is present. The dorsal band is limited either side by a definite black stripe which originates back of the eye, passes along the side above the legs, and continues along the side of the tail, where it usually is broken or lacking on the distal half. The sides below the dorsolateral stripes may be yellow without darker markings, with small scattered flecks, or with larger irregular black spots so numerous that a strongly mottled pattern is developed. When present and well developed, the darker markings of the sides extend to the lower level of the legs, and the legs themselves are well spotted. The entire ventral surface is yellow, without darker markings. The dark markings of the lower sides may be continued on the tail as a series of dots or small blotches, and when present are mainly limited to the basal half. The markings of the sides of the trunk and tail, below the dark dorsolateral stripes, are much more diffuse in *E. b. bislineata* than in *E. b. wilderae,* and this difference provides one of the best recognition marks. Sexual differences are well developed. The head of the male is swollen back of the eyes, the snout is broader, and there is a glandular hump at

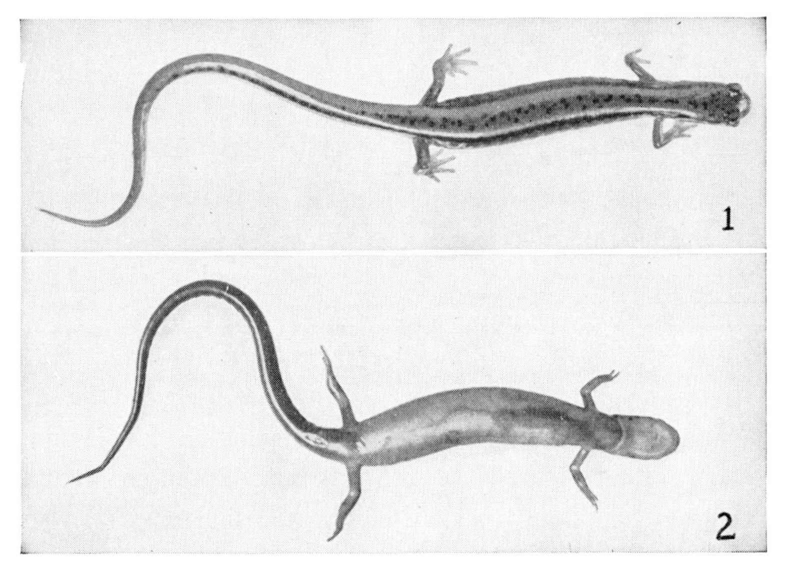


FIG. 124. *Eurycea bislineata wilderae* Dunn. (1) Adult female, actual size. (2) Same, ventral view. Mount Le Conte, Tennessee.

the base of the tail above, quite conspicuous in large specimens. The margins of the vent of the male are slightly raised and crossed by narrow grooves; the vent of the female is a simple slit.

BREEDING. Not much has been published. Dunn (1920, p. 135) writes: "At Linville a batch of eggs was found hatching on July 19. They were attached to the under side of a stone in a brook just as are the eggs of *bislineata.*"

LARVAE. Larvae reported by Dunn (1926, p. 312) ranged in size from 17 to 51 mm. In my own series the range is 30–55 mm. Small larvae have a narrow middorsal light line, bordered either side by a darker stripe, within which is a series of small round white spots. A second row of light spots extends on the sides above the level of the legs, and sometimes a third poorly developed row at the lower edge of the pigment of the sides. Dunn (*ibid.*) records a transformed individual 36 mm. in total length. My smallest fully transformed specimen is 40 mm. long.

**GRAY-BELLIED SALAMANDER.** *Eurycea griseogaster* Moore and Hughes. Fig. 125. Map 48.

TYPE LOCALITY. Swimmer's Creek, 1 mile above junction with Illinois River and about 10 miles northeast of Gore, Oklahoma.

RANGE. Known only from the type locality.

HABITAT. Found beneath stones in pools of relatively quiet water (Moore and Hughes, Copeia No. 3, p. 139, 1941).

SIZE. The average length of 19 adults of both sexes is 2¾″ (69.7 mm.), the extremes 1²⁹⁄₃₂″ (48 mm.) and 3³⁄₁₆″ (81 mm.). (*Ibid.*) The proportions of an adult male are: total length 2¾″ (70 mm.), tail 1⅛″ (29 mm.); head length ¹⁵⁄₃₂″ (12 mm.), width ³⁄₁₆″ (5 mm.). An adult female has the following measurements: total length 2¹³⁄₁₆″ (72 mm.), tail 1⁷⁄₃₂″ (31 mm.); head length ½″ (12.5 mm.), width ⁵⁄₃₂″ (4 mm.).

DESCRIPTION. The head is widest opposite the angle of the jaws, the sides behind this point converging slightly to the lateral extensions of the gular fold, in front more abruptly to the truncated snout. The trunk is slender, somewhat flattened above, and with a median dorsal groove, the sides rounded. There are usually 20 costal grooves, occasionally 19, and 7–9 intercostal folds between the toes of the appressed limbs. Tail subquadrate in section at base, becoming compressed a short distance behind the vent, the dorsal keel narrow and extending from a point above the vent to the tip. Ventral fin confined to the distal half. Legs small and slender. Toes 5–4, slightly webbed at base, flattened and blunt-pointed; those of hind feet 1–5–2–4–3 in order of length from the shortest; toes of fore feet 1–4–2–3. Tongue small, nearly circular in outline. Vomerine teeth 9–11, in series which arise behind the middle of the inner nares, curve inward, then sharply backward toward the mid-line, where they are narrowly separated. Parasphenoid teeth in 2 elongate patches separated the entire length by about the diameter of a naris and from the vomerine by a little more.

COLOR. The general color above is gray, with tinges of tan on some

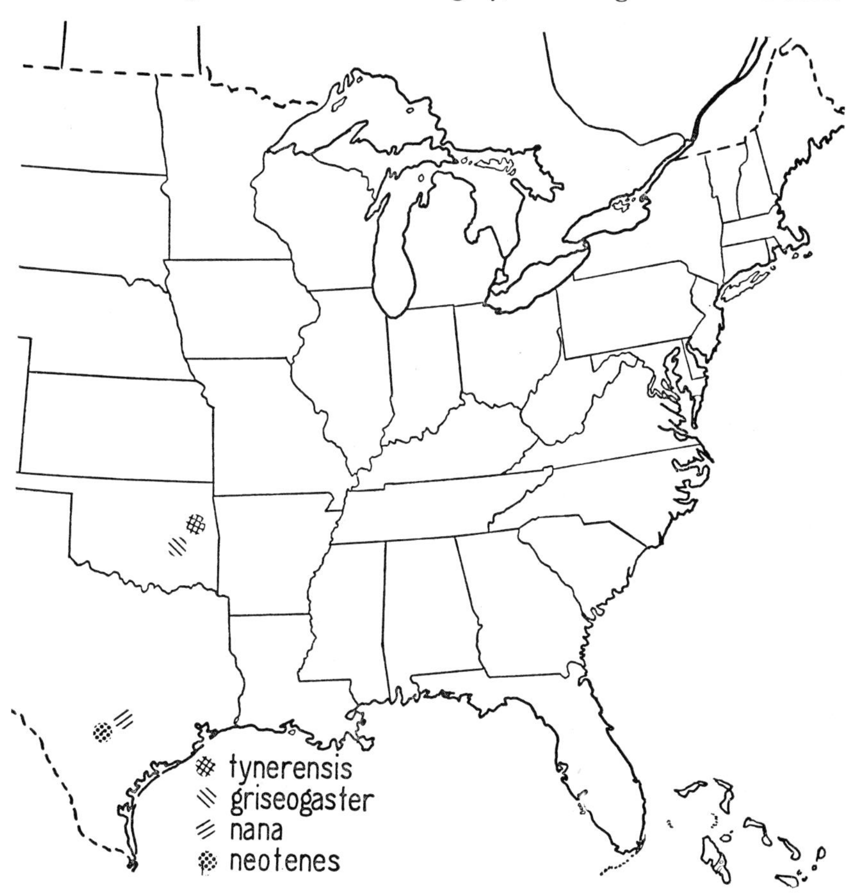


MAP 48.—Distribution of *Eurycea tynerensis, E. griseogaster, E. nana,* and *E. neotenes.*

preserved specimens. A broad median dorsal band is present in most individuals which is lighter than the adjacent sides and limited either side by an irregular and faintly developed dark line extending from the side of the head onto the base of the tail. The dark pigment of the sides fades somewhat toward the lighter venter, which is dotted with separate black chromatophores. In life this salamander is said to be "blotched with silvery patches particularly noticeable about the lateral line organs" (*ibid.*, p. 140).

SEXUAL DIFFERENCES. The sexes may be distinguished by the form of

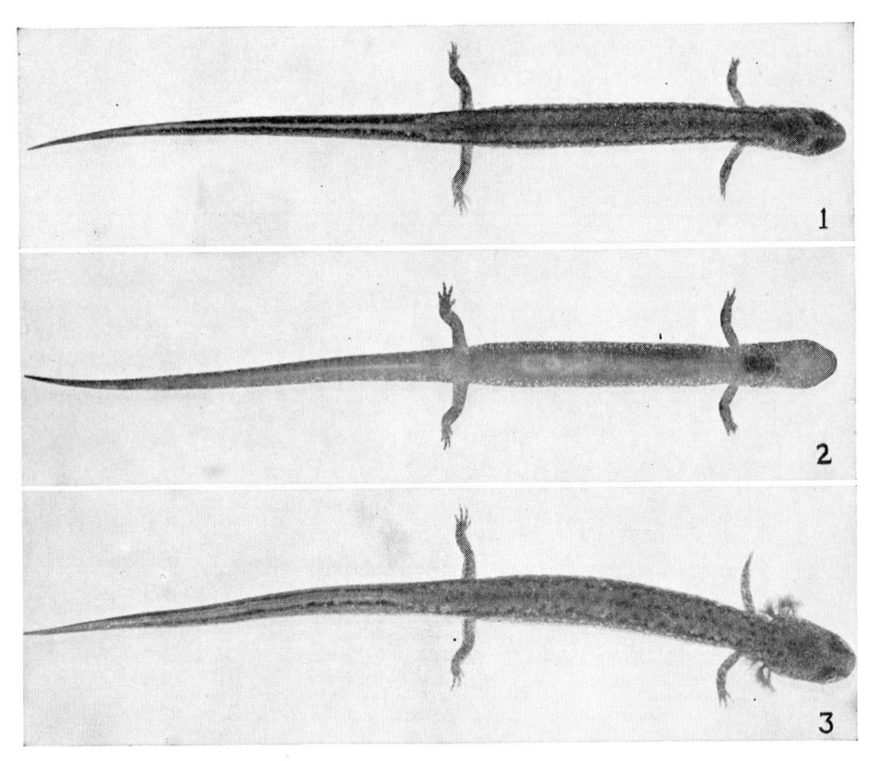


FIG. 125. *Eurycea griseogaster* Moore and Hughes. (1) Adult, actual length about 2¾" (70 mm.). (2) Same, ventral view. (3) Larva. Swimmer's Creek, near Gore, Sequoyah County, Oklahoma. [Photographs by Arthur L. Smith.]

the vent, which is larger in the male, pigmented within, and has the sides thrown into narrow ridges. In the female the vent is bordered by a narrow, smooth, flange-like area, and is relatively smooth or shallowly, obliquely ridged within.

BREEDING. Nothing has been reported on the breeding habits.

LARVAE. The larvae in general resemble the adults but are more uniformly colored and usually have an immaculate venter. The known specimens vary in length from 13 to 72 mm. The gills are well developed and highly pigmented, the filaments long and slender. The dorsal tail keel is wider than in the adult, highest toward the tip; ventral fin confined to the distal half.

This species is related to *E. multiplicata* and *E. tynerensis,* both of which are found in Oklahoma, but neither at the locality where *E. griseogaster* was taken. *E. griseogaster* attains a larger size than *E. tynerensis,* has a more generally pigmented venter, and is not neotenic. From *E. multiplicata* it differs in its general darker coloration, evenly pigmented venter, relatively shorter tail, and in lacking yellow on the tail venter and posterior part of belly.

I am greatly indebted to Professor George A. Moore for a series of larvae and adults.

**LONG-TAILED SALAMANDER.** *Eurycea longicauda longicauda* (Green). Fig. 126. Map 49.

TYPE LOCALITY. Probably from the vicinity of Princeton, New Jersey.

RANGE. Southern-tier counties in New York, south to Georgia, west to Arkansas, and north to Missouri and Illinois.

HABITAT. The adults are mainly terrestrial and are found in and beneath old rotting logs and under stones. Often they abound in crevices of shale banks and beneath stones and rock fragments near the margins of streams. Like some other species of the genus, they enter caves. The larvae are aquatic, and adults sometimes enter the water and swim freely.

SIZE. The average length of 18 mature individuals of both sexes is $4\frac{13}{16}''$ (123 mm.) with extremes of $3\frac{13}{32}''$ (87 mm.) and $7\frac{5}{32}''$ (182 mm.). The 182-mm. specimen, a female from Pine Ridge, Kentucky, has a tail length of $4\frac{3}{4}''$ (121 mm.); head length $\frac{5}{8}''$ (16 mm.), width $\frac{5}{16}''$ (8 mm.). An adult male from the same locality has a total length of $5\frac{15}{16}''$ (152 mm.), tail $3\frac{11}{16}''$ (94 mm.); head length $\frac{9}{16}''$ (14 mm.), width $\frac{5}{16}''$ (8 mm.). In the female the tail comprises 67 per cent of the total length, in the male 61.8 per cent, and the tail in both sexes is proportionally longer in the larger specimens than in the smaller ones. The smallest fully transformed individual I have examined was 1¾″ (45 mm.) long.

DESCRIPTION. The long tail and vertical dark bars on the sides of the

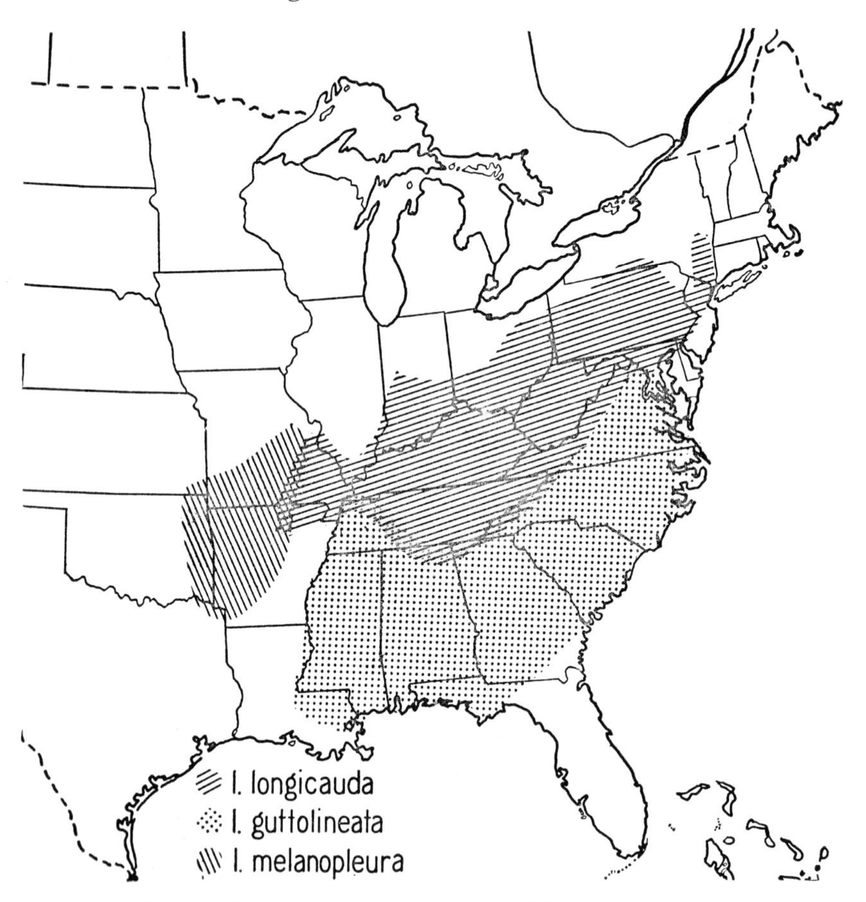


MAP 49.—Distribution of the subspecies of *Eurycea longicauda.*

tail serve to distinguish this species from its near relatives. The head is long and somewhat flattened above, widest immediately back of the eyes, the sides behind the eyes converging gently to the lateral extensions of the gular fold and in front abruptly to the rounded snout. The snout is swollen in the region of the nasolabial grooves, and some males and a very few females develop short cirri, more slender and pointed in the male. The eyes are large and strongly protuberant, the iris brassy. An impressed sinuous line from the posterior angle of the eye to the lateral extension of the gular fold, and a short groove from the eye to a point just in front of the angle of the mouth. The slender trunk is well rounded above and on the sides, flattened below, and without much evidence of an impressed median line above in most specimens. There are 14 costal grooves, counting 1 each in the axilla and groin, and 1–2 intercostal spaces between the toes of the appressed limbs. The legs are rather long and slender. Toes 5–4, those of the hind feet 1–5–2–4–3 in order of length from the shortest; toes of the fore feet 1–4–2–3. The tail is very long and slender; rounded below, sharp-edged above; broadly oval in section at base. The tongue is small, oval in outline, thin at the margin, and attached near the front by a pedicel. Vomerine teeth in series of 7–12; the series arise behind the inner margin of the inner naris and curve inward and backward toward the mid-line, where they are separated by about the width of a naris, and from the parasphenoid by about twice that distance. Parasphenoid teeth in 2 elongate patches narrowly separated.

COLOR. The ground color varies from yellow to bright orange; specimens from the North are usually brighter than those from further south. In most specimens, there is a median dorsal band which is limited either side by a series of elongate black spots. Between these dorsolateral lines the dorsal surface has many irregular black spots of various sizes, which in some form a regular median series and in others are scattered generally over the surface. The sides below the dorsolateral lines have the spots smaller approaching the venter. The spots are few on the dorsum of the head and may be entirely lacking on the snout. The base of the tail above is marked as on the trunk, but distally, along the thin edge, the black spots fail to cross the middle line. The sides of the tail are marked with short, vertical, crescent or dumbbell-shaped, black bars, the lighter interspaces slightly wider than the bars. Sides of the head with a few scattered black spots. The venter and lower surface of the limbs usually immaculate; the belly yellowish along the sides, bluish along the midventral line, and with greenish tinges below the liver. Legs and toes well spotted above and on the sides. The sexes may be distinguished by the form of the vent, which in the male has a flange-like area around it and a few papillae along the margins. Some males develop short pointed cirri, in contrast to the very blunt ones on a few females. The covering of the testis is gray or black; the ovaries are white.

BREEDING. Little is known. Morse (1901, p. 115) reported the species

FIG. 126. *Eurycea longicauda longicauda* (Green). (1) Adult, actual length about 4½″ (115 mm.). (2) Same, lateral view. Port Allegany, Pennsylvania.

as found "in May under stones at the edge of the water together with its eggs; the eggs are attached to the under side of a hollow stone."

LARVAE. The very young larvae, 19 mm. long, have been taken in February in Arkansas. These were uniformly pigmented above, with small black chromatophores. In larger specimens, 25 mm. long, a row of small, round, pigment-free spots appears on the upper sides, and a second row on the lower sides between the legs. Larvae attain a length of at least 60 mm. In a specimen of this size the light dorsal band had a few black spots and many fine ones scattered over the surface. The band is limited on either side by elongate black spots which are continued along the tail to its tip. The sides of the trunk and tail have many small, irregular, pigment-free spots which are arranged more or less in lines. On the sides of the tail the light areas form vertical series which probably mark the light interspaces of the adult.

**THREE-LINED SALAMANDER.** *Eurycea longicauda guttolineata* (Holbrook). Fig. 127. Map 49.

TYPE LOCALITY. "Carolina in the middle country."

RANGE. From Fairfax County, Virginia, to northwest Florida, west through Alabama and Mississippi to Louisiana; western Tennessee. Includes the East Gulf and Atlantic Coastal Plain and Piedmont, southern Blue Ridge, and extreme southern part of Appalachian Valley.

HABITAT. Most abundant in river-bottom swamps and the vicinity of springs and streams where seepage keeps the ground moist. Occasionally found some distance from water. It is nocturnal and mainly terrestrial as an adult, but is perfectly at home in the water.

SIZE. Attains a length of $7\frac{7}{32}''$ (183 mm.). The average length of 15 sexually mature adults of both sexes from various localities is $6\frac{5}{32}''$ (156.4 mm.), the extremes $5\frac{3}{32}''$ (129 mm.) and $7\frac{7}{32}''$ (183 mm.). The females average a little longer than the males. The tail of the female may comprise 58.1–64.4 per cent of the total length, and in the male the range is 59.8–64 per cent. The proportions of an adult female are as follows: total length $6\frac{23}{32}''$ (170 mm.), tail $4\frac{1}{8}''$ (105 mm.); head length $\frac{11}{16}''$ (17.5 mm.), width $\frac{11}{32}''$ (9 mm.). An adult male has the following measurements: total length $6\frac{15}{32}''$ (164 mm.), tail $4\frac{1}{8}''$ (105 mm.); head length $\frac{21}{32}''$ (17 mm.), width $\frac{5}{16}''$ (8 mm.).

DESCRIPTION. The head is widest immediately behind the eyes, the sides behind gently converging to the lateral extensions of the gular fold and in front abruptly to the bluntly rounded snout. The eyes are large and strongly protuberant, the iris veined and lined. An impressed line from the posterior angle of the eye to the lateral extension of the gular fold, and, in some specimens, a short vertical groove from this line to the angle of the jaw. The trunk is subcylindrical, flattened below. There are 14 costal grooves, counting 1 each in the axilla and groin, and ½–2 intercostal spaces between the toes of the appressed limbs. The tail is broadly oval in section at base, soon becoming rounded below and sharp-edged above, and distally much compressed and sharp-pointed. Legs and feet well developed, toes 5–4, those of the hind feet 1–5–2–4–3 in order of length from the shortest, webbed at base; toes of the fore feet 1–4–2–3. Tongue broadly oval in outline, the pedicel attached near the center. Vomerine series moderately long; in the female usually with 11–14 teeth; in the male 8–12. The series may arise behind the middle or inner margin of the inner nares or lie wholly between the nares. They curve inward and backward toward the mid-line, where they are separated by about the width of a naris. The parasphenoid teeth in 2 elongate patches, usually separated from the vomerine by about twice the diameter of a naris, but occasionally continuous. The male has the snout somewhat more strongly swollen in the region of the nasolabial grooves, and often there are short cirri. The vent of the male has a flange-like margin, that of the female is a simple slit.

COLOR. This is one of our most graceful and beautiful salamanders. It is marked above by a broad tan or yellow median band which arises on the snout and extends the length of the trunk and the tail to its tip. Within the light band there is a narrow, middorsal, deep brown or black stripe or series of fused spots, which arises on the head opposite the angle of the jaw and continues along the trunk to the base of the tail. The head, in front of the stripe and including the snout, usually has a few scattered black spots, but may be almost immaculate. The light dorsal band is limited either side by a broad black stripe which arises back of the eye and continues along the trunk and the tail to its tip. The lateral dark band is usually invaded by a series of small, rounded, yellow spots on the trunk and basal part of the tail, or narrow light lines extend along the costal grooves to give a segmented appearance. Below the lateral dark band there is a narrower light stripe extending from the base of the gills along the side of the trunk, passing above the legs and onto the tail, where it frequently breaks up into spots. The lower sides

FIG. 127. *Eurycea longicauda guttolineata* (Holbrook). (1) Adult male, actual length 5 7/16" (138 mm.). (2) Same, ventral view. Neuse River, Raleigh, North Carolina. (3) Adult male, actual length 5" (127 mm.). Davis Lake, Mobile, Alabama.

and ventral surfaces are usually strongly mottled with dull greenish-gray on a dull yellow ground. The upper surface of the legs is mottled with yellow and black.

BREEDING. Brimley (1939, p. 18) has found gravid females in November, and eggs, presumably of this form, have been reported by Dunn (1927, p. 105).

LARVAE. The larvae attain a length of at least 50 mm. A series taken June 18, 1940, on the Cherokee Indian Reservation, North Carolina (Univ. Mich. No. 86764), vary in length from 22 to 39 mm. The pattern of the larva is essentially that of the adult, except that in the smaller individuals there is a double row of small, round, pigment-free spots within the dorsal light band, and the light stripe of the lower sides is narrow and line-like. In larger larvae, there is an indication of the mid-dorsal dark line. In other words, the dorsal surface is light with a dark mid-line, the lateral surfaces are dark with a light longitudinal line. A dark bar from the nostril passes through the eye and widens on the side of the head. The legs and feet are strongly mottled, and usually there is a pigment-free band across the bases of the toes. The dorsal tail fin arises above the vent and is strongly mottled; the ventral fin is narrow and lightly mottled.

**COPE'S CAVE SALAMANDER.** *Eurycea longicauda melanopleura* (Cope). Figs. 101e, 128. Map 49.

TYPE LOCALITY. Raley's Creek, White River, Missouri.

RANGE. Mainly limited to the Interior Highlands except the southeastern portion. Southern half of Missouri, northern and southwestern Arkansas, eastern Oklahoma, southeastern Kansas. Also recorded from Texas.

HABITAT. Found under rocks at the margins of streams and springs and in the twilight regions of caves; usually on land, but occasionally in the water.

SIZE. Twelve sexually mature individuals of both sexes averaged $4\frac{17}{32}''$ (116 mm.) and varied from $3\frac{5}{8}''$ (92 mm.) to $5\frac{25}{32}''$ (148 mm.). Dunn (1926, p. 319) records a transformed individual only $1\frac{25}{32}''$ (44 mm.) in total length. The proportions of an adult female from Imboden, Arkansas, are as follows: total length $5\frac{9}{32}''$ (135 mm.), tail $3\frac{5}{32}''$ (80 mm.); head length $9/16''$ (14.5 mm.), width $9/32''$ (7.5 mm.). An adult male from the same locality measures: total length $4\,15/16''$ (126 mm.),

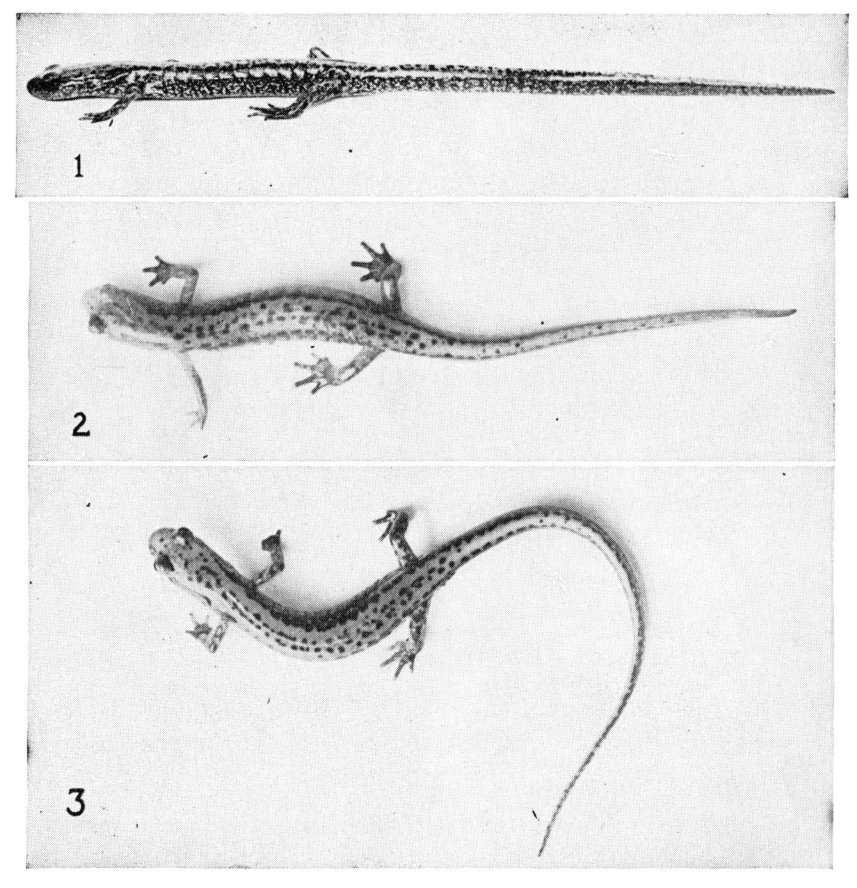


FIG. 128. *Eurycea longicauda melanopleura* (Cope). (1) Adult, actual length about $5\,1/8''$ (131 mm.). Spout Spring, Ravenden, Arkansas. (2) Same, dorsal view. (3) Another individual, dorsal view. [Photograph by R. Redman.]

tail $3\,1/16''$ (78 mm.); head length $17/32''$ (13.5 mm.), width $9/32''$ (7.5 mm.).

DESCRIPTION. Closely related to the Long-tailed salamander, which it resembles in general body form. The sides of the head back of the eyes nearly parallel, in front of the eyes converging to the bluntly pointed and truncated snout. The eyes are large and prominent, the iris brassy above. A slightly impressed line from the posterior angle of the eye to the lateral extension of the gular fold; a short vertical groove from this line to the angle of the jaw in some individuals. The trunk is cylindrical, flattened below. There are 14 costal grooves, counting 1 in the axilla and 2 that run together in the groin, and the toes of the appressed limbs may meet or be separated by 1 or 2 costal interspaces. The tail is broadly oval in section at the base, compressed and sharp-edged above; distally, slender and pointed. In the fully adult male the tail may comprise 62 per cent of the total length. Legs and feet well developed, the hind noticeably larger and stouter than the fore. Toes 5–4, those of the hind feet 1–5–2–(3–4) in order of length from the shortest, slightly webbed at base; toes of the fore feet 1–4–2–3. Tongue broadly oval in outline, the margins thin and free, the pedicel attached near the center. The vomerine teeth in rather long series of 12–16 which arise behind the middle or inner margin of the inner naris, curve inward and forward, then sharply backward, separated by about the width of a naris. Parasphenoid teeth in 2 separate elongate patches, widely separated from the vomerine.

COLOR. This species is marked above with a broad dorsal band much lighter than the adjacent sides. The ground color of the dorsal band varies from dull brownish-yellow in old individuals, through shades of greenish-yellow, to bright yellow in the juveniles. The band originates on the snout, passes between the eyes, back of which it widens to the full width of the head, then, narrowing in the neck region, continues the length of the trunk and on the tail sometimes to the tip. Within the dorsal band are many small, irregular, dark brown or black spots. The spots are few and scattered on the snout, more abundant on the head back of the eyes, and rather uniformly distributed along the trunk. In some individuals the spots from a fairly regular median series, in others a double row either side of the middorsal line; in most, however, there is no particular arrangement apparent. On the basal half of the tail above, the spots may be few and scattered, but in others continued to the tip. The upper sides bordering the dorsal band are dark, varying from grayish in the young to deep reddish-brown in mature individuals. The dark sides are marked with light gray or yellow flecks and spots, which are generally absent next the dorsal band and most abundant on the lower sides. On the tail the dark ground color of the sides may be mottled and blotched, or the sides may be almost uniformly reddish-brown with only the lighter yellow of the dorsal stripe showing as a narrow line along the top. The venter is usually free from dark spots, but occasionally an individual will be lightly blotched with brown. The throat is flesh color, the belly dull white, and the ventral surface of the tail yellowish. Rarely the yellow of the ventral surface of the tail extends forward some distance on the belly. The male may usually be recognized by the swollen snout and greater development of blunt cirri.

BREEDING. Direct observations are lacking. Females collected at Ravenden, Arkansas, in November, and sent me by B. C. Marshall, had well developed ovarian eggs. On the evening of Nov. 18, I confined males and females together and the following morning a spermatophore was found attached to a bit of damp moss (text fig. 12e.).

LARVAE. A series of small larvae sent from Imboden, Arkansas, by B. C. Marshall, was collected Feb. 10, 1928. These varied in length from 17 to 31 mm. and were uniformly pigmented on the dorsal and lateral surfaces except for a dorsolateral series of small, round, light spots. Chin and anterior part of throat pigmented. Larvae attain a length of at least 1 13/16″ (46 mm.).

**CAVE SALAMANDER. SPOTTED TAIL SALAMANDER.** *Eurycea lucifuga* (Rafinesque). Fig. 129. Map 50.

TYPE LOCALITY. Caves near Lexington, Kentucky.

RANGE. Illinois, Indiana, Ohio, Virginia, West Virginia, Kentucky, and possibly Alabama east of the Mississippi River, and Arkansas, Oklahoma, and Missouri west of the river.

HABITAT. Most abundant in the twilight regions of caves, where they climb about over the walls and ledges. Frequently found under logs, stones, and rubbish, in damp situations outside of caves. The adults are essentially terrestrial.

SIZE. The average length of 30 individuals of both sexes is $4\frac{25}{32}''$

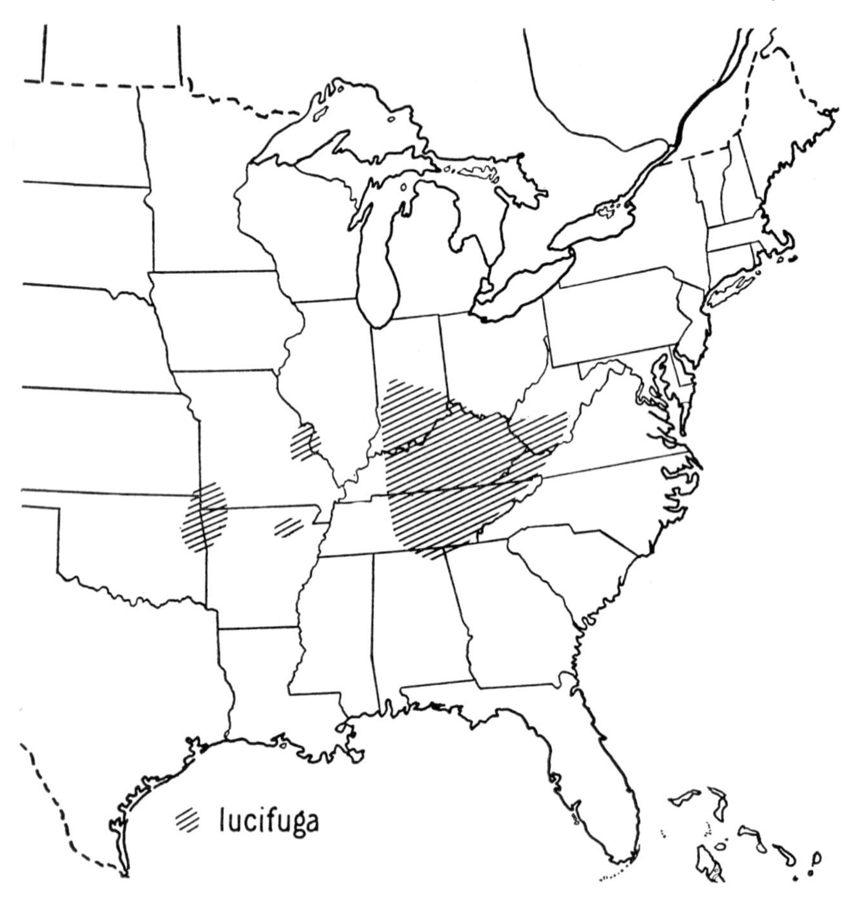


MAP 50.—Distribution of *Eurycea lucifuga.*

(122.2 mm.), with extremes of $2\frac{3}{4}''$ (70 mm.) and $6\frac{13}{32}''$ (161 mm.). In my collection the 19 males average $5\frac{3}{16}''$ (132.3 mm.), the 11 females $4\frac{1}{8}''$ (104.7 mm.). Individual females, however, exceed in length the largest males, Dunn (1926, p. 340) recording one individual $6\frac{19}{32}''$ (167 mm.) in total length, and largest male $6\frac{15}{32}''$ (164 mm.). The proportions of a male from Mammoth Cave, Kentucky, are as follows:

FIG. 129. *Eurycea lucifuga* Rafinesque. (1) Adult male, lateral view; actual length 6″ (153 mm.). (2) Same, dorsal view. [Photograph by E. J. Stein.] (3) Adult female, actual length 5⅛″ (131 mm.). Imboden, Arkansas.

total length $6\frac{7}{32}''$ (157 mm.), tail $3\frac{25}{32}''$ (95 mm.); head length $\frac{5}{8}''$ (16 mm.), width $^{13}/_{16}''$ (10 mm.). A female from near Lexington, Kentucky, has the following proportions: total length $6\frac{1}{8}''$ (155 mm.), tail $3\frac{19}{32}''$ (92 mm.); head length $\frac{5}{8}''$ (16 mm.); width $\frac{3}{8}''$ (9.5 mm.).

DESCRIPTION. A slender species with reddish-orange ground color and many black spots. The head is widest immediately back of the eyes, the sides behind the eyes gently converging to the lateral extensions of the gular fold and in front rather abruptly to the bluntly truncated snout. The males have the snout more strongly swollen in the region of the nasolabial grooves, and the short, blunt cirri developed to a greater extent than in the females. An impressed line extending from the posterior angle of the eye to the lateral extension of the gular fold is usually present, and a short vertical groove from the posterior angle of the eye to the angle of the mouth. The eyes are large, the horizontal diameter about equal to the length of the snout. The trunk is subcylindrical, flattened below. There are 14 costal grooves, counting 1 each in the axilla and groin, and the toes of the appressed limbs may just meet or overlap 2 intercostal spaces. Tail subquadrate in section at the base, becoming compressed into a flat oval at about the mid-length, and more strongly flattened and sharp pointed distally. The legs are well developed and moderately long. Toes 5–4, those of the hind feet short and webbed at base, 1–5–2–(3–4) in order of length from the shortest; toes of the fore feet 1–4–2–3. The boletoid tongue is broadly oval in outline and fills the floor of the mouth; vomerine teeth in long, sharply angled series of 11–20. The series arise behind the middle or inner margin of the inner nares, slant forward and inward, then strongly backward toward the center line, where they are separated by about the diameter of a naris. The parasphenoid teeth in 2 long club-shaped patches, separated from each other by about the diameter of a naris, and twice that distance from the vomerine. In this species the cirri are usually longer in the females than in the males, and longer in adults than in the recently transformed individuals.

COLOR. Ground color variable, dull yellow through orange to bright orange-red. The younger individuals usually yellowish, the orange-red developing with age. Scattered generally over the dorsal and lateral surfaces are many irregular, or rounded, or elongate black spots of variable size, usually 1–3 mm. in diameter. These spots sometimes form a dorsolateral series on either side of the trunk, enclosing a broad dorsal band within which there may be a single median series or many irregularly scattered spots. The lower surfaces are generally light yellow and without spots. In old and large individuals the spots of the lower sides may become enlarged and somewhat fused. The sexes may be distinguished by the form of the vent, which, in the male, has the margins raised and finely papillose, and by the greater development of cirri.

BREEDING. The mating habits have not been reported, nor have the eggs been described. Small larvae 17.5 mm. long have been found early in February and as late as March 20 in Mayfield's Cave, near Bloomington, Indiana, and have been described and figured by McAtee (1906, p. 74). The young larvae are nearly uniformly pigmented above and on the sides, except for 3 longitudinal series of small rounded or elongate spots on each side. The upper series of spots is near the middorsal line, the 2nd near the middle of the sides, and the 3rd between the limbs. As the larvae increase in size, the pigment tends to become concentrated in the characteristic black spots of the adult. Larvae attain a length of about 2¼″ (58 mm.). Larvae have a broad tail fin which arises above the insertion of the hind legs and continues to the broadly pointed tip and ventrally to the vent.

**MANY-RIBBED SALAMANDER.** *Eurycea multiplicata* (Cope). Fig. 130. Map 51.

TYPE LOCALITY. Red River, eastern Oklahoma.

RANGE. Stone County, Missouri, to Pulaski County, Arkansas, and Kansas; also reported from the Jemez Mountains, New Mexico, but probably erroneously.

HABITAT. Found under stones, logs, and other debris, in streams and springs, both in the open and in caves. Essentially an aquatic species but occasionally on land. In Arkansas found by Strecker (1908, p. 88) to be associated with *Desmognathus f. brimleyorum,* and in Missouri with *Typhlotriton* (Noble, 1927, p. 418).

SIZE. The largest specimen measured by Dunn (1926, p. 315) was a

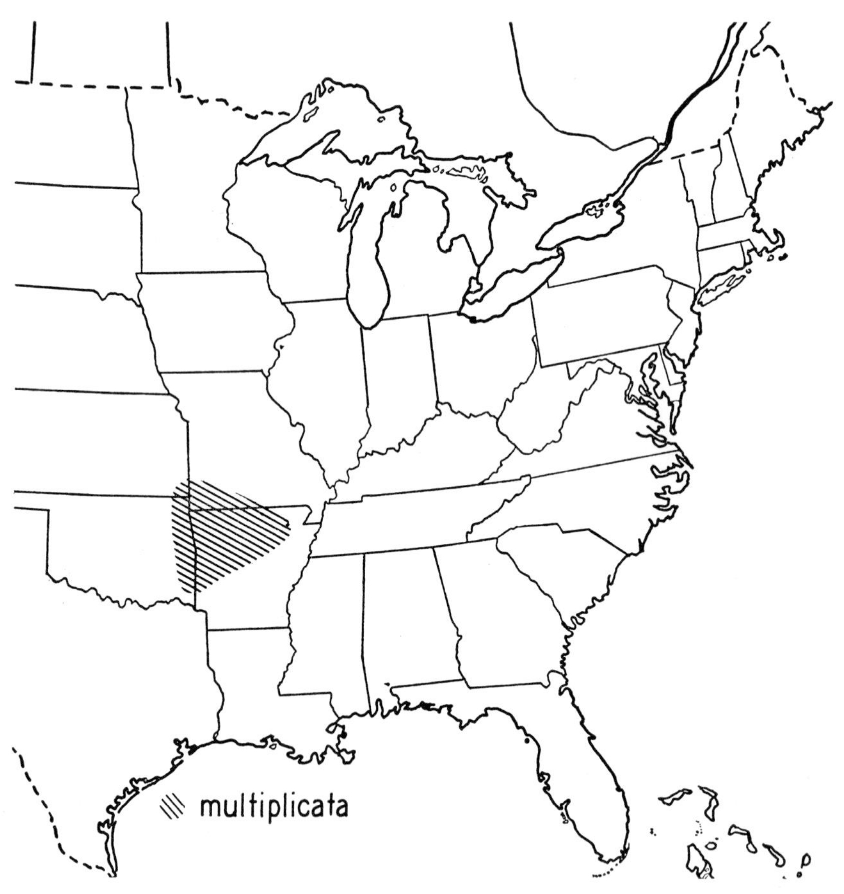


MAP 51.—Distribution of *Eurycea multiplicata.* (Also recorded from Jemez Mountains, New Mexico, but doubtless in error.)

male having a total length of $3^{17}/_{32}''$ (90 mm.). My largest specimen is a female from near Bono, Arkansas, which measures: total length $3^{15}/_{32}''$ (87 mm.), tail $1^{11}/_{16}''$ (43 mm.); head length $^{13}/_{32}''$ (10 mm.), width $^{7}/_{32}''$ (6 mm.). The average length of 6 adults from Arkansas and Missouri is $3\frac{1}{4}''$ (82.7 mm.), with extremes of $2\frac{29}{32}''$ (74 mm.) and $3\frac{15}{32}''$ (87 mm.). Dunn (*ibid.*) records a transformed specimen only $1\frac{19}{32}''$ (41 mm.) in total length.

DESCRIPTION. The head is slender, with the sides back of the eyes nearly parallel and in front tapering abruptly to the pointed snout. There is an impressed line extending from the posterior angle of the eye to the lateral extension of the gular fold, and a short vertical groove from this

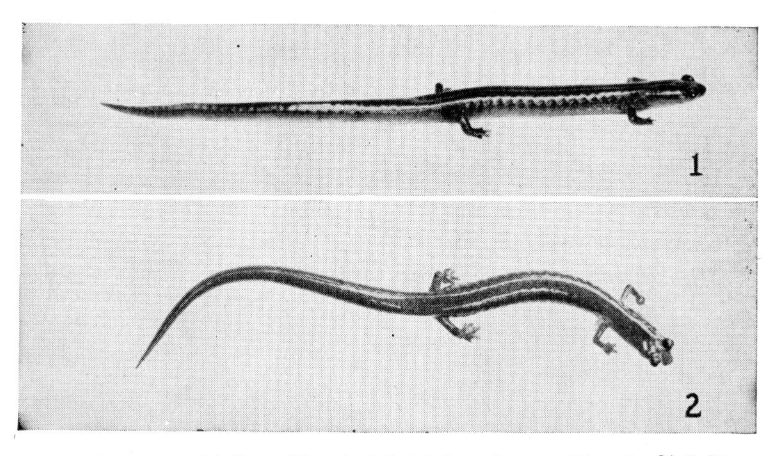


FIG. 130. *Eurycea multiplicata* (Cope). (1) Adult male, actual length $3\frac{9}{32}''$ (84 mm.). (2) Same, dorsal view. Rich Mountain, Polk County, Arkansas.

passing behind the angle of the jaw. The eyes are of moderate size and bordered behind by a vertical fold; the iris is brassy mixed with reddish-brown. The trunk is subcylindrical and provided above with a slight impressed median line. Costal grooves usually 19, counting 1 each in the axilla and groin, occasionally 20, and 8–10 intercostal spaces between the toes of the appressed limbs. The tail is subquadrate in section at the base, rounded below, sharp-edged above, and becoming strongly compressed and slightly keeled above, distally. Legs rather small, toes 5–4, those of the hind feet 1–5–2–4–3 in order of length from the shortest, the inner rudimentary, all rather short and blunt; toes of the fore feet 1–4–2–3, the inner very short. Tongue small, oval in outline, rather thick, and provided with a central pedicel. Vomerine teeth in short irregular series of 6–10, the series arising behind the inner margin of the inner naris and curving inward and backward toward the mid-line, where it is separated from its fellow of the opposite side by about ½ the diameter of an inner naris. The parasphenoid teeth in 2 elongate patches, separated from one another by nearly the width of a patch and from the vomerine by about 3 times that distance.

COLOR. The general ground color is yellow, resembling *Manculus,* but without the darker longitudinal stripes. The entire dorsal surface is usually evenly pigmented with dilute chocolate-brown, over which are scattered small, stellate, pigment-free spots, through which the ground color strikes and imparts a yellowish tinge. There is in some individuals an indication of a broad, middorsal, light stripe, limited on either side by an interrupted line of small dark spots. In some a series of small dark flecks follows the median impressed line to the base of the tail. The entire ventral surface is a bright lemon-yellow except the throat and lower surface of the limbs, which are flesh. The yellow of the venter extends on the sides of the tail and trunk to the level of the lower sides of the limbs. In some individuals the sides of the head and trunk are finely flecked with white pigment dots, and a few are often present on the dorsal surface of the head and trunk and on the sides of the tail. In alcohol, specimens may be dull yellow above, grayish on the sides, and with an indication of larval spots in a line below the edges of the dorsal band.

BREEDING. Little is known of the breeding habits, and the eggs have not been reported.

LARVAE. I have measured larvae from Scott County, Arkansas, collected April 30, 1934, by A. A. Heinze, which were only 23 mm. long, and others from Reed Spring, Missouri, sent me by B. C. Marshall, which had attained a total length of 85 mm. The small larvae are yellowish above and on the sides, finely flecked with brown pigment, and with a dorsolateral series of rounded pigment-free spots; a second series is sometimes developed near the lower edge of the pigmentation of the sides. The tail is keeled above to a point above the hind legs and below to the vent. The large larvae are colored essentially as the adults, but perhaps somewhat lighter. The dorsal fin of the tail is broad and lightly mottled, and extends to a point above the vent; the ventral tail fin is mainly confined to the distal half.

**DWARF EURYCEA.** *Eurycea nana* Bishop. Fig. 131. Map 48.

TYPE LOCALITY. Lake at head of San Marcos River, at San Marcos, Hays County, Texas.

RANGE. Known only from the type locality.

HABITAT. Collected among aquatic plants close to the surface of the water by C. E. Mohr.

SIZE. Smallest among the species of *Eurycea,* 6 adults average only 1 13/16″ (46 mm.), the extremes 1 5/8″ (41 mm.) and 1 31/32″ (50 mm.). The proportions of an adult male are as follows: total length 1 15/16″ (49.5 mm.), tail 7/8″ (22 mm.); head length 3/16″ (4.5 mm.), width 5/32″ (3.5 mm.). An adult female measures: total length 1 31/32″ (50 mm.), tail 25/32″ (20 mm.); head length 7/32″ (5 mm.), width 3/16″ (4.5 mm.).

DESCRIPTION. The known specimens are neotenic. The head is narrow, with the sides back of the eyes parallel, in front converging to the bluntly rounded snout. The eyes are of moderate size, partly or completely surrounded by a dark ring, the horizontal diameter about 1 1/4 in the snout; the iris is dark, with only a few light flecks evident in preserved specimens. Gills well developed and highly pigmented, the rachises flattened and increasing in length posteriorly, the filaments slender and pigmented nearly to the tips. Trunk slender, somewhat compressed, flattened above, and with a slightly impressed median line. There are 16–17 costal grooves and 6–7 intercostal folds between the toes of the appressed limbs. The tail is subquadrate in section at base, slender and compressed a short distance behind the vent, and with a dorsal keel that arises behind the posterior end of vent; ventral tail keel limited to the distal third. The legs are small and slender; toes 5–4, long and slender, those of the hind feet 1–5–2–4–3 in order of length from the shortest; toes of the fore feet 1–4–2–3. Tongue and teeth larval in character. The teeth on the premaxilla usually 10–13 and average about 11. On a single sexually mature female these teeth are enlarged and reduced in number to 7; vomerine teeth 9–14, average 11.7; pterygoid teeth 4–6, average 4.8. In two adult males, the pterygoid teeth are reduced to 2 on each side, suggesting incipient metamorphosis which

FIG. 131. *Eurycea nana* Bishop. (1) Adult male, actual length 1 17/32″ (39 mm.). (2) Same, ventral view. San Marcos, Hays County, Texas. [Photographs from a preserved specimen.]

may or may not be completed. Both sexes may be mature at a length of only 41 mm.

COLOR. The general ground color above is uniform light brown, relieved by a dorsolateral series of small, separate, yellowish flecks of irregular size and shape on each side of the mid-line. The chromatophores giving the general color above are in little clusters separated by inconspicuous, very narrow, light lines. The pigment extends on the side of the head to the level of the base of the first gills and involves the upper jaw and posterior part of the lower jaw, on the sides of the trunk to involve the upper half of the legs, and on the tail nearly to the ventral keel. The ventral surfaces are white, tinged with yellowish on the tail.

The dorsolateral series of light spots vary in number from 7 to 9, and there is sometimes a second incomplete line on the sides above the insertion of the legs. A few small light spots also invade the middorsal region in most specimens.

SEXUAL DIFFERENCES. The sexes may be distinguished by the form of the vent, which is larger in the male and has the opening lined with short papillae. In the female, the vent is a simple slit, with the side anteriorly thrown into narrow folds. The testes are strongly pigmented with black, and the peritoneum of both sexes is specked with scattered black chromatophores.

BREEDING. Nothing is known of the breeding habits in nature, but a female 50 mm. long had well developed eggs when collected June 22, 1938.

LARVAE. A series of 12 larvae, which have not attained sexual maturity, vary in length from 20 to 41 mm. These differ in no essential respect from the sexually adult individuals.

**TEXAS NEOTENIC SALAMANDER.** *Eurycea neotenes* Bishop and Wright. Fig. 132. Map 48.

TYPE LOCALITY. A small stream 5 miles north of Helotes, Texas.

RANGE. Known from the type locality, a small stream 5 miles north of Helotes; from Helotes Creek, and from a cave near Boerne, all in Texas.

HABITAT. The type specimens were found among dead leaves in pools varying from 12″ to 18″ in depth. C. E. Mohr, collecting at Cascade Cave, 3 miles east of Boerne, found several in a shallow pool deep within the recesses of the cave. Wright and Wright (1938, p. 31) report having seen specimens in the leafy mats along the edges of Helotes Creek.

SIZE. Measurements of the largest specimen in the type series, a male, are as follows: total length $2\frac{27}{32}$″ (72 mm.), tail $1\frac{5}{16}$″ (33 mm.); head length $\frac{5}{16}$″ (8 mm.), width $\frac{1}{4}$″ (6 mm.). Female: total length $2\frac{3}{4}$″ (69 mm.), tail $1\frac{7}{32}$″ (31 mm.); head length $\frac{5}{16}$″ (8 mm.), width $\frac{1}{4}$″

(6 mm.). The largest among the specimens from Cascade Cave has a total length of 3 11/16″ (94 mm.).

DESCRIPTION. The head is only moderately broad, widest immediately in front of the gills, the sides gently converging to the eyes, then more

FIG. 132. *Eurycea neotenes* Bishop and Wright. (1) Adult, dorsal view. West Fork of Cibola Creek, near Bracken, Bexar County, Texas. [A. H. Wright and A. A. Wright, collectors.] (2) Same, another individual. (3) Same, another individual. [Photographs by Arthur L. Smith.]

abruptly to the bluntly pointed snout. Eyes moderate, nearly circular in outline, and without lids, apparently normal in individuals found living in open streams but considerably reduced in the cave specimens from near Boerne. The gill filaments are long, slender, and bright red in life. The trunk is slightly compressed, the tail oval in section near the base, somewhat flattened below, compressed toward the tip. Dorsal tail fin narrow, and extending from a point above the posterior end of vent to the tip, where it ends in a sharp point; ventral tail fin narrow and extending anteriorly about ½ the tail length and continued a short distance as a low ridge. Costal grooves 15–17, the usual number 16 when 1 each is counted in the axilla and groin; 5–7 intercostal spaces between toes of appressed limbs. Legs small and slender. Toes 5–4, those of the hind feet 1–5–2–4–3 in order of length from the shortest; fore feet, 1–4–2–3. Tongue larval in character, moderately large, fleshy. Teeth larval, the vomero-pterygoid series following the lines of the upper jaw and extending posteriorly to a point about opposite the hind angle of the eye.

COLOR. The general ground color above in life is quite uniformly light yellowish, with light brown chromatophores aggregated on the back and sides to give a mottled appearance. The light dorsal band characteristic of *Eurycea b. bislineata* is here poorly developed, and the dorsolateral series of small light spots are but faintly defined. The secondary row of light larval areas on the sides is evident only on the smaller specimens. The sides of the head and chin are lightly pigmented, and there is a fairly prominent dark bar extending from the eye to the nostril. The lower sides and belly are normally without pigment, light yellow.

SEXUAL DIFFERENCES. The sexes may be distinguished by the form of the vent, which in the female is a simple slit with a few low tubercles along the margin; in the male the margins of the vent are thrown into low folds.

BREEDING. In the original lot of specimens taken April 1, 1936, were several females having unpigmented ovarian eggs about 2 mm. in diameter and apparently nearly ready to be deposited. On August 16–17, 1938, Dr. Leo Murray collected at the type locality and secured a fine series of 47 larvae varying in length from 14 to 33.5 mm., the smallest specimens quite recently hatched, the others apparently the young of the year but at least some weeks older. The variation in size of these small larvae suggests that the egg-laying season may extend over several weeks, perhaps from April to July.

LARVAE. The recently hatched larvae have the dorsal surfaces light brown, with the dorsolateral series of small, circular, light spots 10–12 in number, extending from the back of the head onto the basal half of the tail. A secondary row of light larval spots is inconspicuously developed on the sides of some individuals between the hind and fore legs. Small, irregular, pigment-free spots and lines are scattered generally over the back and sides to give a reticulated effect. The dorsal surfaces of the legs are colored like the back. The larval tail fins are much more strongly developed than in the sexually mature individuals, the dorsal fin arising above the insertion of the hind legs and continuing to the bluntly pointed tip, the ventral fin extending to the vent. Sexual maturity may be reached in the males at a length of about 50 mm.; in the females the smallest mature individual collected was 69 mm. in total length.

**OKLAHOMA NEOTENIC SALAMANDER.** *Eurycea tynerensis* Moore and Hughes. Fig. 133. Map 48.

TYPE LOCALITY. Tyner Creek near Proctor, Adair County, Oklahoma.

RANGE. Known only from the vicinity of the type locality.

HABITAT. Sexually mature individuals and larvae were raked out of loose gravel shallowly covered by the cold swift waters of Tyner Creek and Marvin's spring; associated, in this area, with *Eurycea l. melanopleura.*

SIZE. Thirteen specimens measured by Moore and Hughes (1939, p. 697) averaged 2 1/36″ (55.5 mm.) in total length and varied from 1 23/32″ (44 mm.) to 2 13/32″ (61 mm.). In this same series the tail averaged 7/8″ (22.4 mm.) and varied from 19/32″ (16 mm.) to 1 3/32″ (28 mm.).

DESCRIPTION. This is a small, neotenic species apparently related to *Eurycea neotenes,* but generally darker and with a higher costal-groove count. The head is somewhat depressed, the sides back of the eyes nearly parallel, the snout broadly rounded. Eye rather small, its horizontal diameter about twice in snout; no eyelids. Gills progressively longer from in front, the filaments well pigmented. Gular fold well developed. Trunk subcylindrical. Costal grooves 19–20, usually 20 counting 1 each in the axilla and groin, and about 10 intercostal spaces between the toes of the appressed limbs. Tail subcylindrical at base, broadly oval in section at mid-length, and becoming compressed distally. The tail fin arises as a low ridge at a point above the vent and becomes thin-

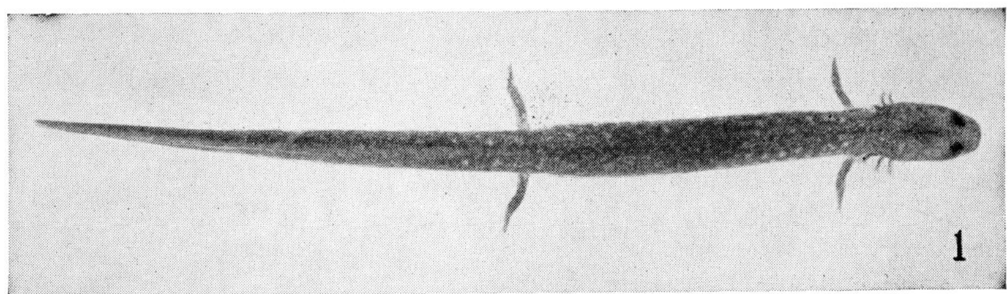

FIG. 133. *Eurycea tynerensis* Moore and Hughes. (1) Adult, actual length about 2″ (51 mm.). (2) Enlarged view of dorsum of head. Tyner Creek, Adair County, Oklahoma. [Photographs by Arthur L. Smith.]

ner and wider toward the tip; on the ventral side confined to the distal half. Legs slender, toes 5–4, those of the hind feet 1–5–2–4–3 in order of length from the shortest; toes of the fore feet, 1–4–2–3.

COLOR. I have not seen living animals, but in the specimens kindly sent me by the describers the pattern is essentially as follows: The general ground color above is dull yellow (cream in the original description), which is concentrated to form a middorsal band, and within which brown pigment is aggregated in the form of irregular lines and blotches. The dark pigment is concentrated in a narrow band along the upper sides, extending from the back of the head onto the basal third of the tail. The lower sides have the dark pigment more diffuse and surrounding a series of rounded pigment-free spots extending between the legs and on the basal part of the tail. The legs are mottled and blotched above with yellow and brown. The dorsal surface of the tail lighter than adjacent trunk, and with a row of small light spots on either side of the middorsal line. Ventral surfaces dull white, without pigment except on chin and sides of throat in front of gular fold.

BREEDING. Nothing is known of the breeding habits or egg-laying. Some females taken in April 1939 had ovarian eggs 1.8 x 1.5 mm. in diameter (Moore and Hughes, *ibid.*, p. 697). Larvae resemble the adults in general form and color. Six specimens varied in length from 17.5 to 31 mm. in total length and averaged 23.5 mm. (*ibid.*, p. 698). A larva 27.5 mm. long has the dorsal light band sharply cut off from the darker sides at a line which extends from the side of the head onto the basal third of the tail. The dorsal tail fin arises above the vent, reaches its greatest width at mid-length, and tapers to a blunt point. The narrow ventral fin extends to the vent.

## GENUS MANCULUS

**DWARF FOUR-TOED SALAMANDER.** *Manculus quadridigitatus* (Holbrook). Fig. 134. Map 52.

TYPE LOCALITY. Georgia, South Carolina, and Florida.

RANGE. North Carolina south to middle Florida, west through Gulf States to Texas east of the Trinity River, northward to Arkansas and Oklahoma.

HABITAT. I have found it most abundant in low, swampy places, beneath logs, bark, and other surface rubbish, in leaf-filled trickles from springs, and in the debris along the margins of pools in river swamps and marshes in the Coastal Plain. Mainly terrestrial in summer and fall, but enters the water for the egg-laying season, usually in December.

SIZE. Attains an extreme length of 3⁵⁄₁₆″ (84 mm.), but this is exceptional. A series of 10 adult females from Raleigh, North Carolina, taken June 12, 1924, and sent me by Mr. C. S. Brimley, varied from 2⅛″

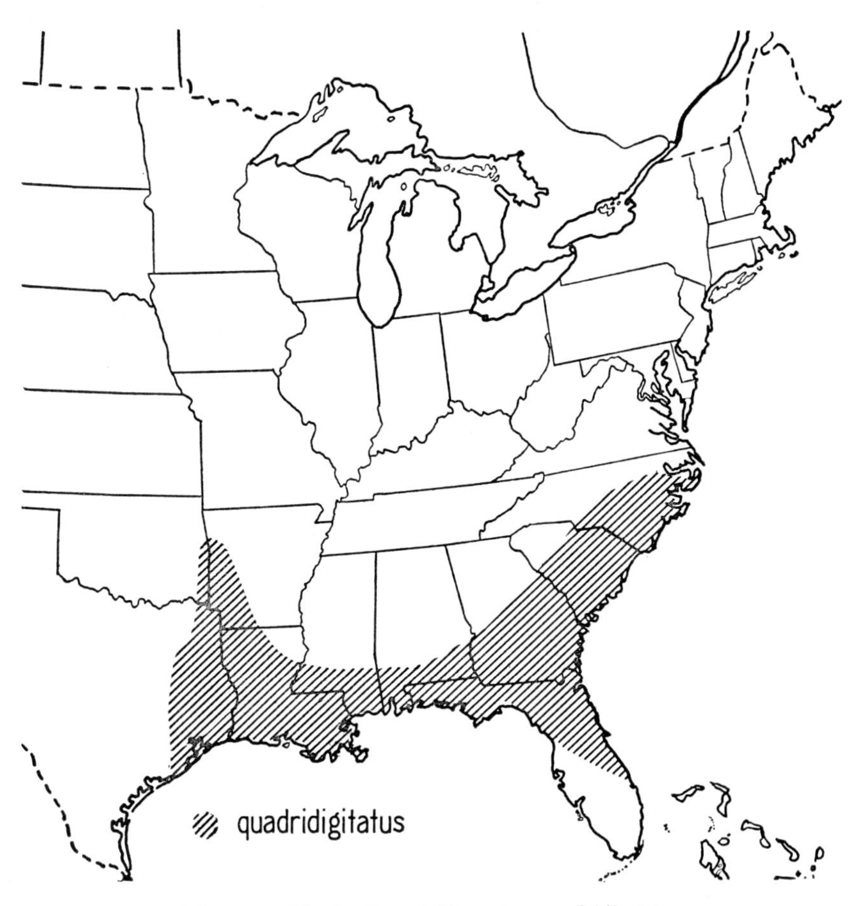


MAP 52.—Distribution of *Manculus quadridigitatus.*

(53 mm.) to 3⁵⁄₁₆″ (84 mm.) and averaged 2¹¹⁄₁₆″ (67.5 mm.). The males average a little shorter. The proportions of a male from near Spanish Creek, 3 miles west of Folkston, Georgia, are as follows: total length 3¹⁄₁₆″ (74 mm.), tail 1¾″ (44 mm.); head length ⁵⁄₁₆″ (7 mm.), width ³⁄₁₆″ (4 mm.).

DESCRIPTION. The Dwarf salamander is a small, slender, four-toed species which superficially resembles the Two-lined salamander, *Eurycea b. bislineata*. The head is long and slender, widest at a point about midway between the eyes and the lateral extensions of the gular fold; the sides in front of the eye converging rather abruptly to the very short and bluntly rounded snout. The snout of the male is more strongly truncated than that of the female, slightly swollen in the region of the nasolabial grooves, and the upper lip at the lower end of the grooves is produced into short cirri. Commissure of mouth slanting upward toward and extending back to posterior angle of the eye. The eyes are large and strongly protuberant, the pupil horizontally elliptic, the iris golden flecked with black. There is an impressed line extending from the posterior angle of the eye to the lateral extension of the gular fold, but no well defined vertical groove extending from this back of the jaw. The trunk is slender, well rounded above and on the sides, slightly flattened below. In some specimens there is a slightly impressed median dorsal line which splits at the back of the head and sends a branch to each eye. There are usually 16 costal grooves, counting 1 each in the axilla and groin, but occasionally the number is 15, and about 5 intercostal spaces between the toes of the appressed limbs. Legs small, toes 4–4, those of the hind feet 1–4–2–3 or 1–2–4–3 in order of length from the shortest, the 1st rudimentary, the 2nd and 4th about equal; toes of the fore feet 1–4–2–3, both inner and outer short. The tail is long and slender, subquadrate in section at the base, and becoming compressed and oval in section distally and slightly keeled above. The tongue is of moderate size, nearly circular in outline, and with a central pedicel. Vomerine teeth in short series of 4–8, the series beginning behind the inner margin of the inner naris and extending inward and backward to the center line, where they are scarcely separated. Parasphenoid teeth in narrowly separated elongate patches and widely separated from the vomerine.

COLOR. (Light form.) Marked above with a broad, bronzy band, with irregular edges, extending from the tip of the snout to the end of the tail. Near the base of the tail the band is usually brighter and with tinges of gold. Many individuals have a median dorsal line of small, irregular, dark brown spots; in others the spots may be lacking or may

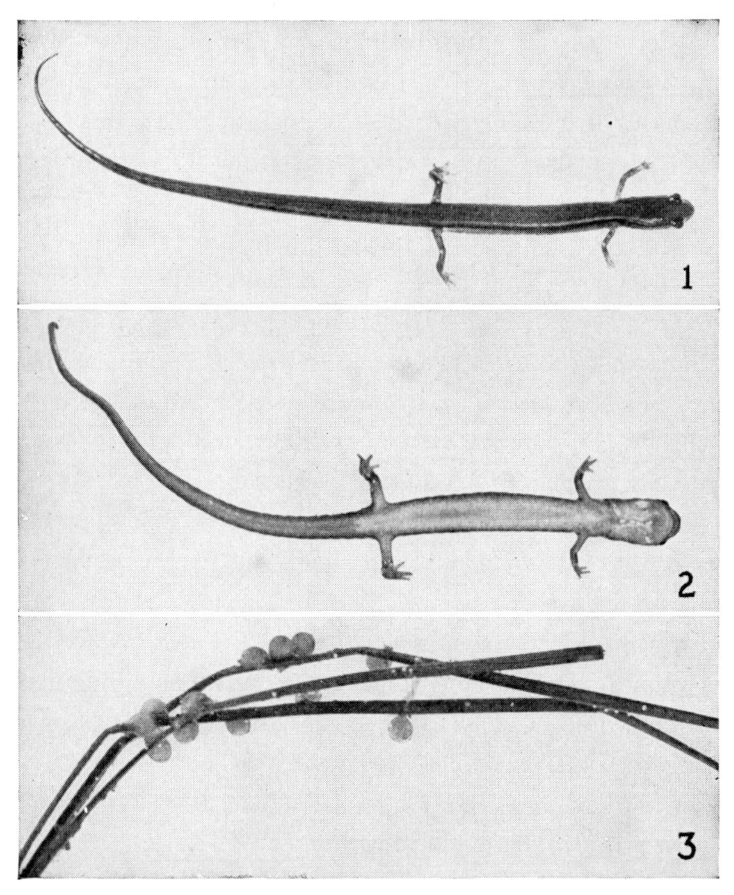


FIG. 134. *Manculus quadridigitatus* (Holbrook). (1) Adult female, actual length 2¾″ (70 mm.). Florida. (2) Adult male, actual length 2½″ (63 mm.). Amite, Louisiana. (3) Eggs, slightly enlarged. Raleigh, North Carolina, Jan. 30.

appear only at the base of the tail or back of the head. The dorsal band is limited on either side by a narrow brown stripe which originates on the side of the snout and continues along the trunk onto the basal half of the tail. The upper sides below the dark stripe are of lighter brown, gradually fading on the lower sides to permit the golden-yellow ground color to strike through. The venter is bright yellow except on the distal third of the tail, which is dusky yellow, and on the throat and lower sides of the legs and feet, which are flesh color.

(Dark form.) Many individuals exhibit a much darker pigmentation and somewhat different pattern. In these the sides are very dark brown, with narrow longitudinal or oblique lines of yellow or white. The light lateral lines are best developed at the level of the legs, and often with additional imperfect lines above and below. The ventral surfaces are quite uniformly pigmented, with small dark brown flecks and a few scattered light spots.

BREEDING. The egg-laying season at Raleigh, North Carolina, extends from late December to early February (Brimley, 1923, p. 81). When the eggs are laid in leafy trickles, they are usually attached, singly or in small clusters of 3–6, to the leaves and debris. In a pond at Gainesville, Florida, Carr (1940, p. 48) found three sets Jan. 15, 1935, each containing about 20 eggs attached to the lower surface of a log in shallow water. A female collected Jan. 12, 1924, and sent me by Mr. C. S. Brimley, deposited 48 eggs on the night of Jan. 30. The container was provided with damp leaves and the eggs were attached to the lower surfaces in small groups of 5–8. The eggs are creamy white, without pigment, and are provided with 2 envelopes. The egg itself is about 2 mm. in diameter and, with the outer envelopes, about 3 mm. They are attached by a short and thin tubular extension of the outer envelope, which, at the point of attachment, flares out to form a flange.

LARVAE. At Raleigh larvae are found mainly in March, and transform 2 or 3 months later. Dunn records a transformed individual only 32 mm. long, but the larvae frequently attain a much larger size before transformation.

Fig. 135. Outline of *Hydromantes platycephalus* to show the form of the body and the character of the legs and feet. [H.P.C. *del.*]

## Genus HYDROMANTES

**MOUNT LYELL SALAMANDER.** *Hydromantes platycephalus* (Camp). Figs. 135–136. Map 53.

TYPE LOCALITY. Head of Lyell Canyon, Yosemite National Park, California.

RANGE. The high Sierras from Alpine County south to northern Tulare County, California.

HABITAT. This is an Alpine species, known only from high elevations in the Sierras, where it has been found beneath stones and rock slabs and, infrequently, crawling in the open. The first specimens discovered were taken at an elevation of 10,800′ at the head of Lyell Canyon, Yosemite National Park. The 2 specimens then found were caught in a mousetrap set in a patch of heather in front of a small hole running into moist soil beneath some rocks. A fine series of 13 specimens was taken August 1, 1940, by Miss Catherine Hemphill and Miss Orlie Anderson, at Triple Divide Peak, Alpine County, California, at an elevation of 10,700′. The specimens, all collected in the course of about one hour, were found in slightly damp situations under rocks in a meadow about 15–20 yards below a snowslide, but neither in nor very near trickling streams of melting snow water. Two specimens exposed by the turning of a rock were occupying a small burrow some 10″ in length and ½″–⅝″ in diameter.

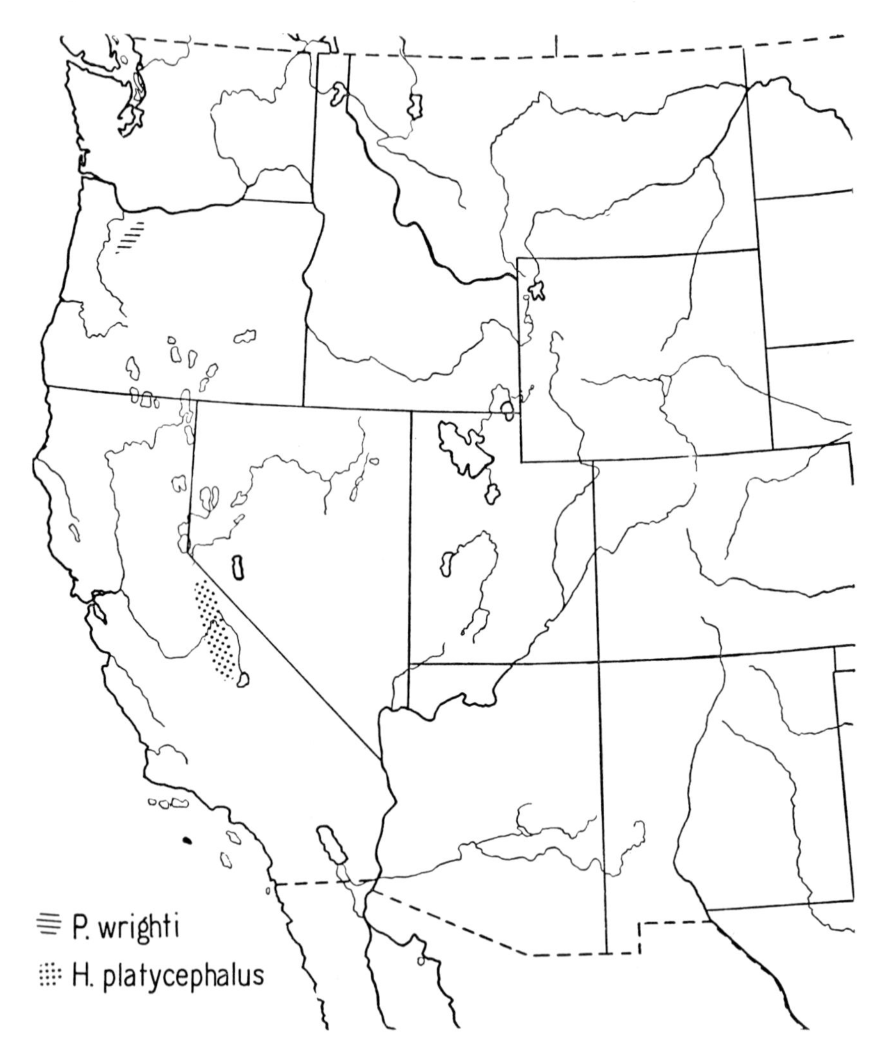


MAP 53.—Distribution of *Plethopsis wrighti* and *Hydromantes platycephalus.*

SIZE. The 29 individuals I have measured varied in length from 1 11/32″ (34 mm.) to 4⅛″ (104 mm.). The proportions of an adult male are: total length 3 11/16″ (94 mm.), tail 1 11/32″ (34 mm.); head length 9/16″

(14.5 mm.), width 7/16" (11 mm.). An adult female has the following measurements: total length 4 1/8" (104 mm.); tail 1 15/32" (37 mm.); head length 9/16" (14.5 mm.), width 7/16" (11 mm.).

DESCRIPTION. This salamander has a deep brown or blackish ground color, thickly overlaid with many irregular, light gray, lichen-like patches. The head and trunk are strongly depressed, perhaps an adaptation to the mode of life in crevices or beneath flakes of rock. The broad head is widest just back of the eyes, the sides tapering to the bluntly rounded snout, and behind gently to the lateral extensions of the gular fold. The snout in side view is obliquely truncate. The eyes are large and protuberant, the iris flecked with tan and brassy. An impressed sinuous line from the posterior angle of the eye to the lateral extension of the gular fold, and from this a vertical groove on the side of the head behind the angle of the jaw which is continued across the throat to connect with its fellow of the opposite side. The dorsally flattened trunk has the sides rounded and the belly flat. There are 13 costal grooves, counting 1 each in the axilla and groin and 1–2 intercostal spaces between the toes of the appressed limbs. The tail is flattened above at the base, broadly oval in section distally in the adults, and slightly compressed in the young. Legs short and stout; toes 5–4, those of the hind feet 1–5–2–3–4 in order of length from the shortest, the 2nd, 3rd, and 4th nearly equal in length and all strongly webbed; soles of feet very broad; tips of toes swollen ventrally; toes of the fore feet 1–2–4–3, webbed at base. The tongue is small, oval in outline, thin at the margins, and with a central area finely papillose. It is freely protrusible, the pedicel rather broadly attached at the anterior third. Vomerine teeth 8–13 in each series. In an adult male, there are 12 and 13 teeth respectively in the vomerine series, which arise behind and outside the inner nares and curve inward and backward toward the mid-line, where they are separated by about the diameter of an inner naris, and from the parasphenoid patches by about 3 times the diameter. The parasphenoid teeth in 2 elongate club-shaped patches narrowly separated anteriorly, divergent posteriorly. In the female the maxillary teeth are small, short, and mainly limited to the anterior part of the jaw; in the adult male these teeth are fewer in number, slender, longer, and directed obliquely backward and continued to the angle of the jaw.

COLOR. The general ground color is deep brown to black, marked above and on the sides with many light spots and speckings, which in

FIG. 136. *Hydromantes platycephalus* (Camp). (1) Adult male, actual length 3¼″ (82 mm.). (2) Same, ventral view. Sonora Pass, California, elevation 9000′. [Laura Henry, collector.]

some individuals run together to form lichen-like blotches of considerable extent. The light markings vary from light buff or gray, tinged with greenish yellow in the smaller individuals, to clay color in the adults. In the large specimens the light markings are so extensive that much of the ground color is obscured. In most small specimens the light markings are brighter on the dorsal surface of the tail than elsewhere, and extend on the sides to below the level of the legs. Legs and toes strongly mottled and blotched. The lower surface of the head is blotched and a few light spots encroach on the belly. The belly is light slate, the lower surface of the feet and tail a little lighter, dilute chocolate. In some individuals the light markings are more or less concentrated in a band along each side of the back.

BREEDING. Nothing is known of the breeding habits of this species. Females taken August 1 had well developed ovarian eggs which showed as elongate light patches on either side of the belly.

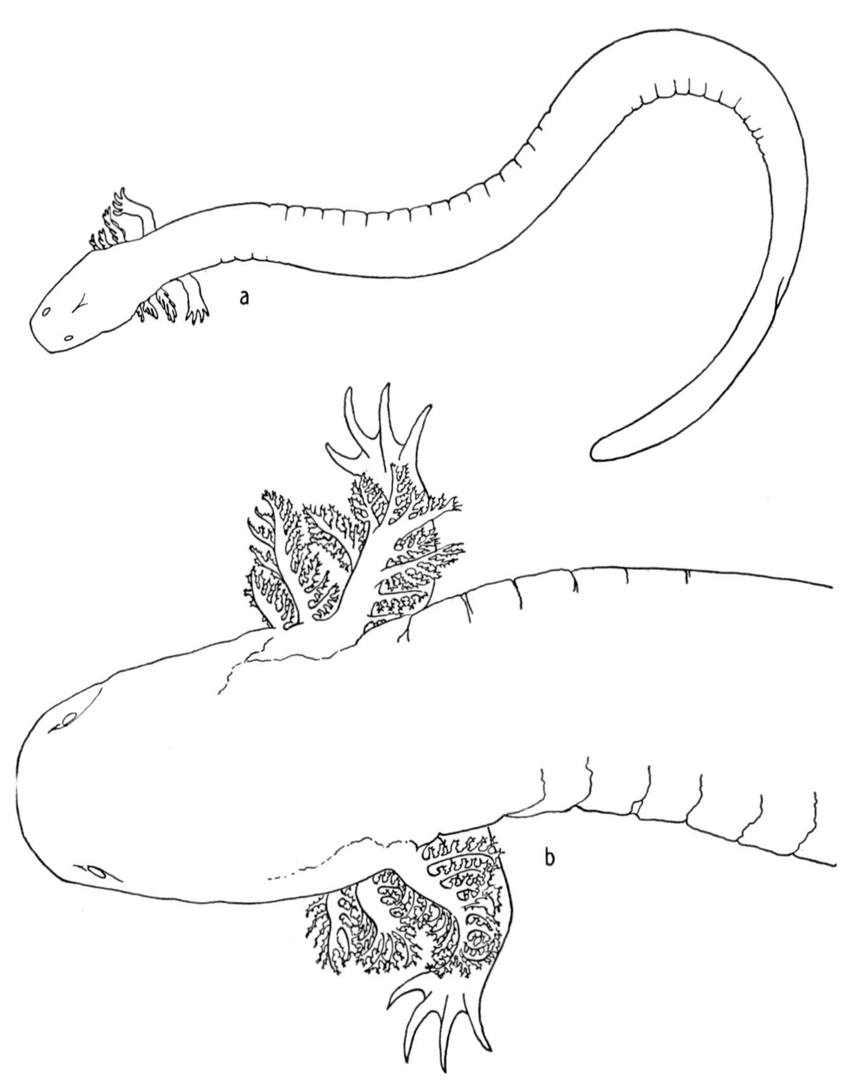


Fig. 137. (a) Outline of *Siren intermedia nettingi* to show the form of the body, the gills, and the anterior limbs. (b) Head and fore part of trunk of *Siren lacertina* to show the character of gills and legs. [H.P.C. *del.*]

# Family SIRENIDAE

Aquatic; body elongate, eel-like; size small, 8¼″ (210 mm.) in *Pseudobranchus,* to large, 36″ (915 mm.) in *Siren;* anterior legs, only, present; 3 pairs of gills; eyelids, maxillae, and cloacal glands absent; margins of jaws with horny sheaths; dentition larval; fertilization probably external. Two genera in the family.

### KEY TO THE GENERA OF SIRENIDAE

With 4 toes; 3 pairs of gill slits; body without longitudinal light lines or stripes; size to 36″ (915 mm.). Southeastern Atlantic states, northward in the Mississippi Valley to Illinois and Indiana . . . . . . . . . . . . . . . . . . . . SIREN p. 457
With 3 toes; a single pair of gill slits; body with longitudinal light lines or stripes; size to 8¼″ (210 mm.). Vicinity of Charleston, South Carolina, southward to Dade County, Florida . . . . . . . . . . . . . . . . . . PSEUDOBRANCHUS p. 468

## GENUS SIREN

### KEY TO THE SPECIES AND SUBSPECIES OF SIREN

1. Size large, to 36″ (915 mm.); costal grooves 36–39, the usual number 37; color in life light gray, the sides lighter than the back; venter bluish, with many small, dull yellow flecks (in preservatives, slate color above, dull gray below). District of Columbia south to southern Florida, west into Leon County, Florida . . . . . . . . . . . . . . . . *lacertina* p. 464
   Size small, from 15⁹⁄₁₆″ (396 mm.) to 26″ (660 mm.) in southeasern Texas; costal grooves 31–36, rarely to 38 . . . . . . . . . . . . . . . . . . . . . 2
2. With 31–34 costal grooves, the usual number 33; venter usually without light spots; length to about 13⅝″ (347 mm.). Virginia to Pasco County, Florida, west to the Florida Parishes of Louisiana . . . . . . . . . . . . . . . . . . . . . . . . . . . . . . . . . . . . . . . . . . . . . . *intermedia intermedia* p. 458
   With 34–36 costal grooves, the usual number 35 (rarely to 37 or 38

in southern Texas and northern Tamaulipas); length to 15 9/16" (396 mm.); sides and venter often with small light spots. Southern Louisiana northward to southern Illinois and Indiana, west and south to Maverick County, Texas, and northern Tamaulipas, Mexico ........................................... *intermedia nettingi* p. 461

**EASTERN DWARF SIREN.** *Siren intermedia intermedia* Le Conte. Fig. 138. Map 54.

TYPE LOCALITY. Restricted type locality, Riceborough, Liberty County, Georgia (Harper, 1935, p. 279).

RANGE. From Georgetown County, South Carolina, south to Pasco County, Florida, westward to the Florida Parishes of Louisiana. Also reported from Guiney Station, Virginia.

HABITAT. Common in cypress and pinewoods ponds and ditches, where they hide by day beneath bottom vegetation and stranded logs and bark.

SIZE. The average of 12 adults of both sexes from South Carolina, Georgia, and Florida is 8 5/8" (219 mm.), the extremes 5 9/32" (135 mm.) and 13 5/8" (347 mm.). The proportions of an adult male from Gainesville, Florida, are: total length 9 1/16" (230 mm.), tail 3 1/16" (78 mm.); head length 1 1/32" (26 mm.), width 17/32" (13 mm.). An adult female from the same locality measures: total length 7 1/2" (190 mm.), tail 2 9/32" (58 mm.); head length 26/32" (21 mm.), width 15/32" (12 mm.).

DESCRIPTION. The body is long, slender, and eel-like. In life it is slightly flattened below and above and provided with a weak middorsal impressed line and a more strongly impressed midventral line. The head is long, widest at a point 1/3 the distance from the base of the anterior gills to the end of the snout, the sides behind this point converging very slightly to the gills and in front gently to the broadly rounded snout. The eyes are small and without lids, the interorbital distance about 3 times in the length of the head from the end of the snout to the base of the first gills. Mouth small, crescentic, subterminal, and slightly overhung by the upper jaw. Nostrils small and slit-like and placed at the ventrolateral angles of the snout. Gills variable, with rachises and filaments short and knobby to comparatively long and brushy. Costal grooves 31–34, the usual number 33. The tail is broadly oval in section at the base, becoming more strongly compressed progressively toward

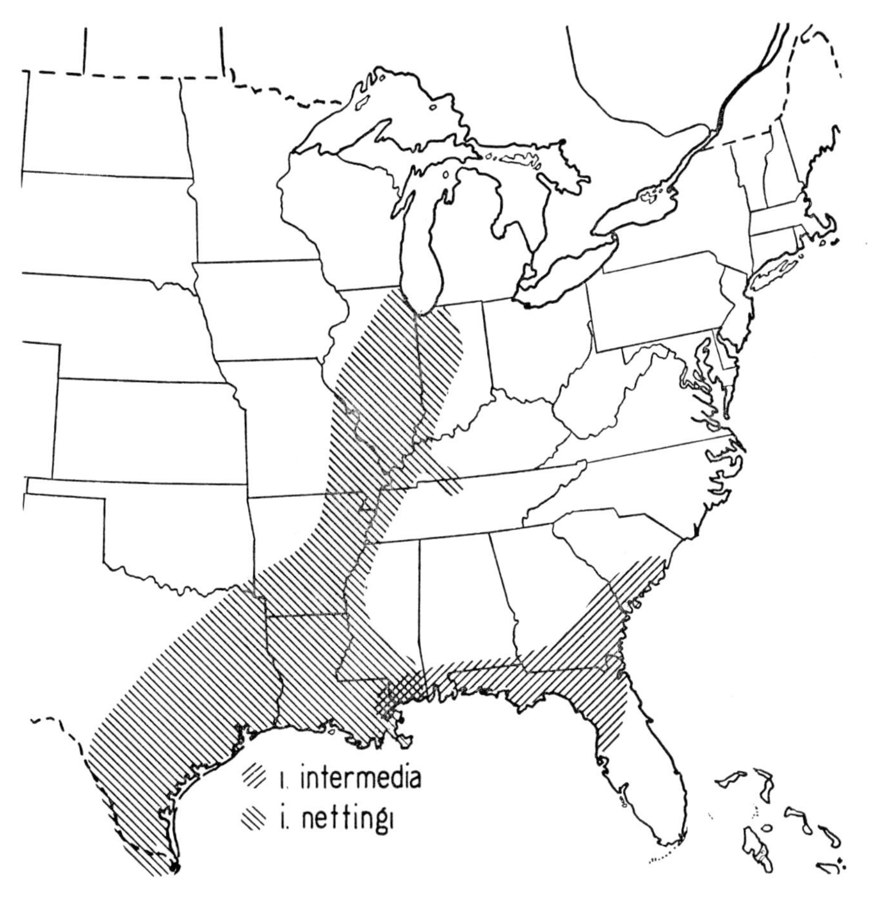


MAP 54.—Distribution of the subspecies of *Siren intermedia*. (*S. i. intermedia* also reported from Guiney Station, Virginia.)

the tip. The dorsal tail fin arises as a low ridge above the vent, but immediately thins and continues to the tip. The ventral fin is narrow and confined to the distal ½ or ⅓, continuing to the vent as a low ridge. The fore legs, only, present. Toes 4, in order of length from the shortest 4-3-1-2 or 4-1-3-2, the tips sharp-pointed, horny, often black beneath.

Tongue moderate, bluntly rounded, and free anteriorly. Palatine teeth in a broad V-shaped series slightly separated in front, the apex directed

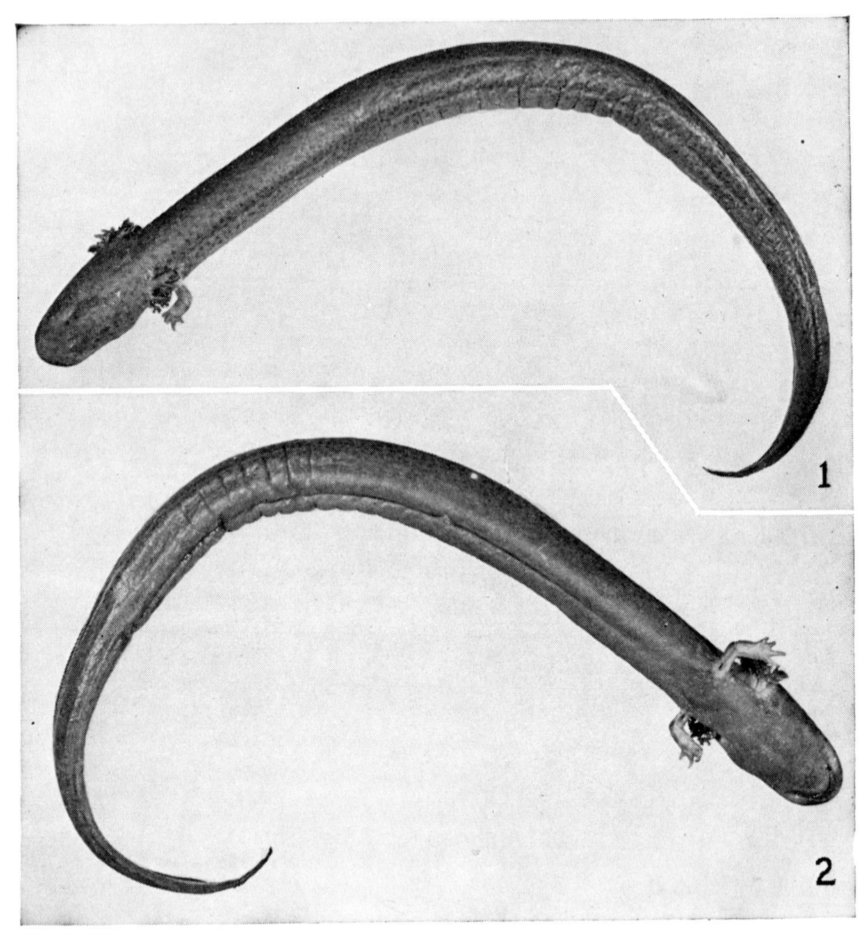


FIG. 138. *Siren intermedia intermedia* Le Conte. (1) Adult male, actual length 9″ (228 mm.). (2) Same, ventrolateral view. Dedge Pond, two miles east of Chesser's Island, Okefinokee Swamp, Georgia. [Photographs of a preserved specimen.]

forward, each lateral group in 2 patches slightly separated, the posterior patch small. A short, thin, black, horny sheath covering the premaxillary, a longer, sharp-edged sheath covering the tip of lower jaw; behind this sheath, mandibular teeth in 2 short patches forming divergent series.

COLOR. The color in life of several specimens taken in Dedge Pond near Chesser's Island, Okefinokee Swamp, Georgia, is as follows: above, varying from deep brown to olive-green. Scattered irregularly over the back and sides of the head, trunk, and tail are many small, rounded black spots, decreasing in size on the lower sides and fading out toward the tip of the tail in some specimens. The ventral surface is dull slate color, slightly lighter than the back. Lower surface of the legs and snout lighter. The ventral surfaces lack the light spots frequently well developed in *nettingi*. Some juvenile specimens are olive-green with minute brown spots.

In a series of preserved specimens taken in the vicinity of Gainesville, Florida, and sent me by Mr. Coleman J. Goin, the general color above and on the sides is uniformly bluish-black, the ventral surfaces slightly lighter.

The typical subspecies differs from *nettingi* in its smaller size, in having fewer costal grooves, and usually in its generally darker color.

BREEDING. Nothing has been reported concerning the breeding habits of this race, and the eggs and larvae have not been described.

**TEXAS DWARF SIREN.** *Siren intermedia nettingi* Goin. Figs. 137a, 139. Map 54.

TYPE LOCALITY. Imboden, Lawrence County, Arkansas.

RANGE. Southern Louisiana northward to southern Illinois and Indiana, west and south to Maverick County, Texas, and northern Tamaulipas, Mexico (Goin, 1942, p. 217).

HABITAT. Found in mucky and muddy ditches, sloughs, and flatlands ponds.

SIZE. Attains a maximum length of 15 9/16″ (396 mm.). The average length of 10 adults of both sexes from Mississippi and Arkansas is 9 1/8″ (232 mm.). The proportions of an adult female from Imboden,

Arkansas, are as follows: total length $8\frac{7}{32}''$ (209 mm.), tail $2\frac{7}{32}''$ (57 mm.); head length $\frac{27}{32}''$ (22 mm.), width $\frac{13}{32}''$ (11 mm.). An adult male from the same locality measures: total length $7\frac{3}{32}''$ (180 mm.), tail $2\frac{3}{16}''$ (56 mm.); head length $\frac{23}{32}''$ (19 mm.), width $\frac{3}{8}''$ (10 mm.).

DESCRIPTION. Body slender, slightly compressed; a lightly impressed middorsal line and a stronger midventral line. Head elongate, widest immediately in front of the gills, from this point gently rounding to the angle of the jaws, then tapering to the bluntly rounded snout. Eyes small, without lids, the interorbital distance $2\frac{1}{2}$–$2\frac{3}{4}$ in the length of the head from tip of snout to base of first gills. Mouth small, crescentic, partly overhung by tip of snout. Nostrils small, slit-like, ventral in position at the lateral angles of the snout. Gill slits normally 3. Gills variable, perhaps depending on environmental conditions; they may be short and stubby or fairly long and with bushy filaments. Costal grooves 33–37, the usual number 35. Tail broadly oval in section at base, becoming compressed distally. The dorsal tail fin arises above the vent and continues to the tip; ventral tail fin scarcely developed in juveniles, in adults narrow and mainly confined to distal $\frac{2}{3}$, sometimes reduced to a low ridge. Fore feet only present. Toes 4–4, in order of length from the shortest usually 4–1–3–2, occasionally 4–3–1–2 or 4–1–2–3; toes flattened, wide at base, tapering to the black-pointed horny tips. Tongue small, pointed in front, the anterior third free. Premaxillary with a short, flat, black, horny sheath; a long, narrow, curved, black sheath at tip of lower jaw in front of mandibular teeth, which are in 2 short, strongly divergent patches. Palatine teeth in strongly divergent patches narrowly separated at the mid-line in front, each lateral patch consisting of 2 groups of teeth, the posterior smaller.

COLOR. In some living specimens from the type locality, the color in life is uniform deep seal-brown. Scattered irregularly over the dorsal surface of head and trunk, and to a lesser extent on the tail, are many small, round, black spots. The ground color of the ventral surface is dull slate color, with here and there a scattering of small light markings most abundant on the throat and on the area between and immediately behind the legs. The lower surface of the legs is lighter than the belly, and there is a light area on the side of the neck between the gills and the legs. Gills pale bluish. On some individuals, the lateral-line sense

FIG. 139. *Siren intermedia nettingi* Goin. Adult female, actual length 8¼″ (210 mm.). Imboden, Arkansas (type locality).

organs are well developed as short white dashes in 2 well separated longitudinal lines extending along the sides from the gills to the base of the tail. Preserved specimens may be light olive to dark gray above and lighter below.

BREEDING. Eggs far advanced in development were found April 8 and April 10, 1931, in the vicinity of Imboden, Arkansas, and reported by Noble and Marshall (1932, p. 9). Eggs to the number of 555 were found in one lot disposed in a mass in a hollow in the mud at the bottom of a shallow pond. The individual egg averages about 3 mm. in diameter and is provided with 3 envelopes, the innermost thin, about 4 mm. in diameter, the middle one 4.2 mm., and the outer 4.4 mm. and slightly opaque.

LARVAE. Larvae soon after hatching are about 13 mm. in length, the tail comprising only 2.5 mm., or 19.2 per cent. The gills are well developed, the 3rd about as long as the head. A well developed dorsal fin extends from the back of the head to the tip of the tail, and a ventral fin from the tip to the vent. The legs are represented by short buds. The larvae are marked as follows: A dark bar through the eye to the base of the 3rd gills; a bar on the snout between the eyes widens on the dorsum of the head and sends a narrow branch to join the bar from the eye. On the trunk a dark stripe on either side of the dorsal keel extends to the tip of the tail; below this a narrow light line from the base of the 3rd gills to a point above the vent; below the light line, a dark stripe from the side of the head to the base of the tail, below which the sides are usually yellow, with small, scattered, round, brown spots. In larger larvae the dorsal fin is restricted to the tail, and the light markings are lost except on the snout, a narrow transverse bar on the dorsum of the head, a bar from the angle of the mouth to the base of the third gills, and a broken line of small dashes on the side from the gills to the base of the tail. The size at which sexual maturity is attained has not been determined.

**GREAT SIREN.** *Siren lacertina* Linné. Figs. 137b, 140. Map 55.

TYPE LOCALITY. "Habitat in Carolinae paludosis."

RANGE. District of Columbia south to southern Florida, westward to Leon County, Florida, and lower Alabama, all in the Coastal Plain.

HABITAT. We have collected this species in shallow roadside ditches, from beneath rocks in the bed of a swift-running stream, from weedy ponds and pools and muddy swamps. It is particularly abundant in Lake Miccosukee, Florida. The lake is weed-choked and mud-bottomed, its outlet crossed by a low dam, perhaps 18″ high, which offered little obstruction to large Siren which were observed to surmount it with ease.

SIZE. Attains an extreme length of about 36″ (915 mm.). A fine adult

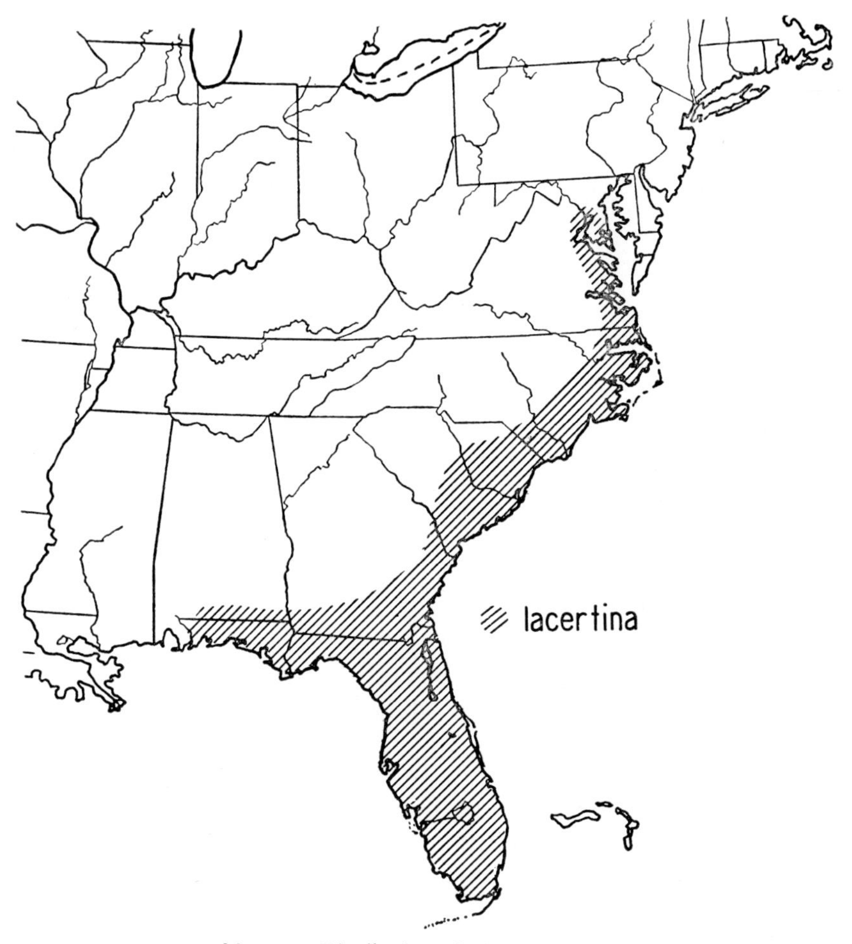


Map 55.—Distribution of *Siren lacertina.*

31⅜″ (798 mm.) from the outlet of Lake Miccosukee, Florida, is proportioned as follows: total length 31⅜″ (798 mm.), tail 9 7/16″ (240 mm.); head length (snout to posterior margin of third gills) 2 13/16″ (72 mm.), width 1¾″ (45 mm.); girth 6 15/16″ (170 mm.).

DESCRIPTION. The body is stout, considerably deeper than wide, and with a slightly impressed median dorsal line. The head is widest a short distance in front of the gills, the sides in front of this point somewhat convex and converging to the broadly rounded snout. Eyes small, with-

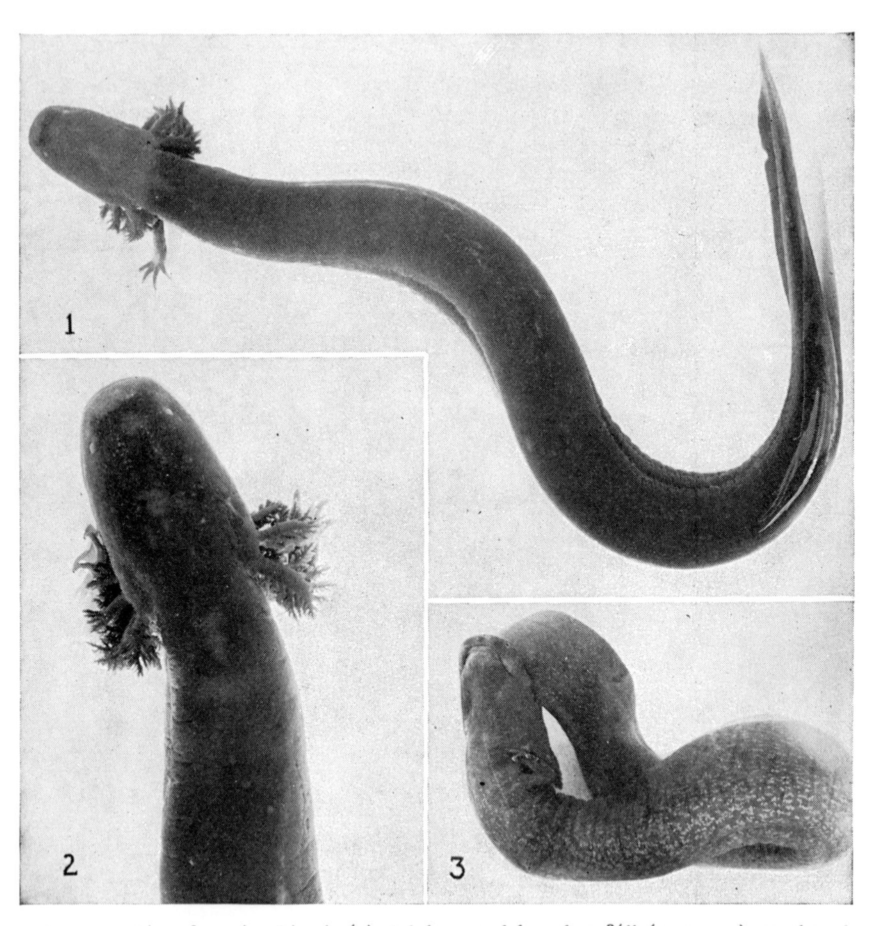


Fig. 140. *Siren lacertina* Linné. (1) Adult, actual length 31⅜" (915 mm.). Outlet of Lake Miccosukee, Florida. (2) Same, head somewhat enlarged. (3) Adult, actual length 19" (482 mm.). Near St. Augustine, Florida.

out lids, the interorbital distance about twice in the distance from the nostril to the base of the anterior gills. Mouth small, crescentic, subterminal. Nostrils small, round or oval in outline, and overhung by the anteriolateral angles of the snout. Gills in adults with compressed rami which are often branched or provided with a secondary series bearing slender filaments. Gills frequently reduced and sometimes non-functional. Gill slits, 3. The muscular trunk is rather strongly compressed, slightly flattened below. Costal grooves 36–39, the usual number 37. The tail is broadly oval in section at base, becoming strongly compressed beyond. The dorsal tail fin may arise above the vent or a short distance before it. The ventral tail fin narrower, and extends ½–⅔ the distance from the tail tip to the vent. Fore legs only developed. Toes 4, variable, usually 4–1–3–2 in order of length from the shortest, sometimes 4–3–1–2. Toe tips blunt, horny. Tongue large, filling the floor of the mouth, and free anteriorly and slightly at the sides. Palatine teeth in 2 contiguous patches on either side, the anterior patch longer and wider than the posterior, the teeth in oblique rows extending from the outer to the inner side of the patch. The 2 groups of teeth form a V-shaped patch with the apex directed forward. A short, black, horny sheath over the premaxillary, and a longer, sharp-edged sheath covering the dentary.

COLOR. The adult in life is light gray, the sides lighter than the back, and with inconspicuous yellow dashes and blotches. The venter has the ground color bluish, marked with many small, dirty, yellow chromatophores. The gills have a decided greenish cast; toes yellowish, tipped with black. The snout is mottled with yellow and light brown. In preservative, the general color becomes slate above and dull gray below. Young usually darker than the adults.

BREEDING. Nothing is known of the breeding habits of Siren in nature, and the very young have not been described. Eggs of *Siren lacertina* deposited by specimens in captivity have been reported by Noble and Richards (1932, p. 15, Fig. 5B). As in *Pseudobranchus striatus,* the egg is surrounded by 3 definite envelopes, the innermost filled with thin fluid. The upper ⅓ or ⅖ of the egg is heavily pigmented with brown. The average diameter of the fixed egg is 4 mm., of the inner envelope

5.3 mm., of the middle envelope 6.2 mm., and of the outermost 9 mm. (Noble and Marshall, 1932, p. 12). The captive specimens deposited the eggs singly or in small clusters.

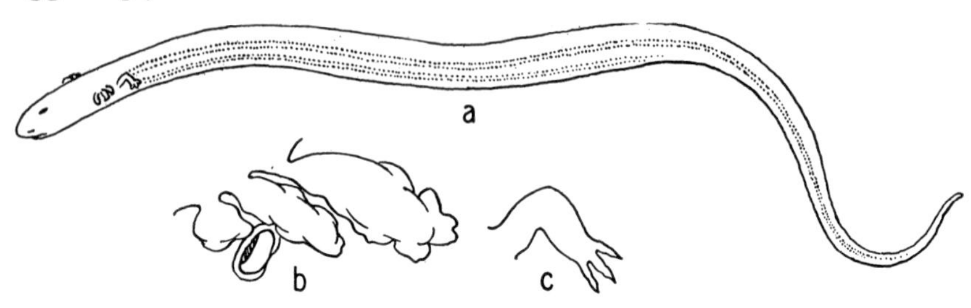


FIG. 141. (a) Outline of *Pseudobranchus striatus axanthus* to show the eel-like body, the anterior limbs, and the gills. (b) Gills and spiracle enlarged. (c) Fore leg enlarged. [H.P.C. *del.*]

## GENUS PSEUDOBRANCHUS

### KEY TO THE SUBSPECIES OF PSEUDOBRANCHUS STRIATUS

Ground color brownish; lateral light stripes broad, yellowish; head stripes distinct; venter with some yellow markings; length at least to $5^{27}/_{32}''$ (150 mm.), probably larger. Atlantic Coastal Plain from near Charleston, South Carolina, south to the Okefinokee Swamp, Georgia .............................................................. *striatus striatus* p. 468

Ground color grayish; lateral light stripes narrow and often incomplete; head stripes indistinct; venter uniformly gray without yellow spots; length to $8\frac{1}{4}''$ (210 mm.). Peninsular Florida from region of intergradation with typical *striatus* in the Okefinokee Swamp, Georgia, south to Dade County .............................................................. *striatus axanthus* p. 471

**BROAD-STRIPED MUD-SIREN.** *Pseudobranchus striatus striatus* (Le Conte). Fig. 142. Map 56.

TYPE LOCALITY. Riceborough, Liberty County, Georgia.

RANGE. Vicinity of Charleston, South Carolina, southward in the Coastal Plain to the Okefinokee Swamp, Georgia, where it intergrades with *P. s. axanthus.*

HABITAT. Common in cypress ponds, hiding among plants or burrowing in the mud or muck of the bottom.

SIZE. Apparently a smaller race than *P. striatus axanthus,* the few specimens I have measured varying in total length from 5¼″ (134 mm.) to 5$^{27}$/$_{32}$″ (150 mm.). The proportions of an adult female from Georgia

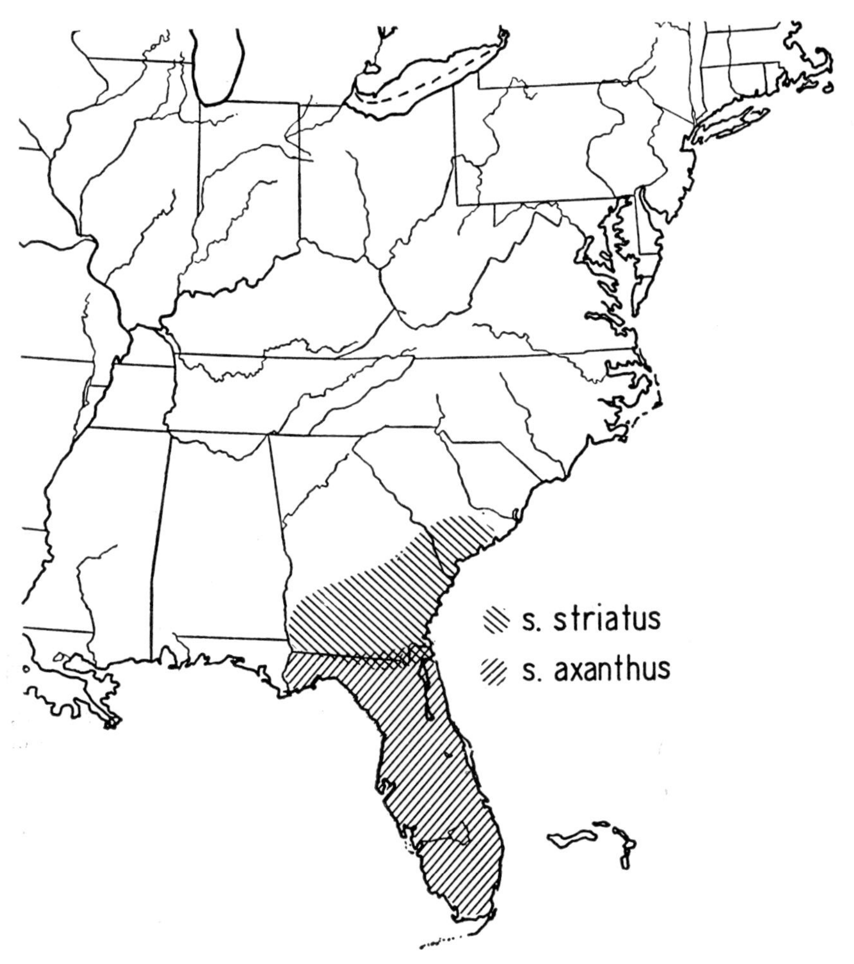


MAP 56.—Distribution of the subspecies of *Pseudobranchus striatus.*

are as follows: total length 5$^{27}$/$_{32}$″ (150 mm.), tail 2$^{7}$/$_{16}$″ (62 mm.); head length $^{17}$/$_{32}$″ (13 mm.), width $^{7}$/$_{32}$″ (6 mm.).

DESCRIPTION. The head is widest immediately in front of the gills, the sides in front of this point gently converging to the eyes, then rather abruptly to the pointed snout. Head in side view wedge-shaped, the upper slope the greater. Eyes small, without lids, the interorbital distance $2\frac{1}{4}$–$2\frac{3}{4}$ in the length of the head from the nostril to the base of the first gills. Nostrils small, slit-like, located in a ventrolateral position near the tip of the snout. Mouth subterminal, small, crescentic. Gills usually shorter and stubbier than in *P. s. axanthus,* sometimes com-

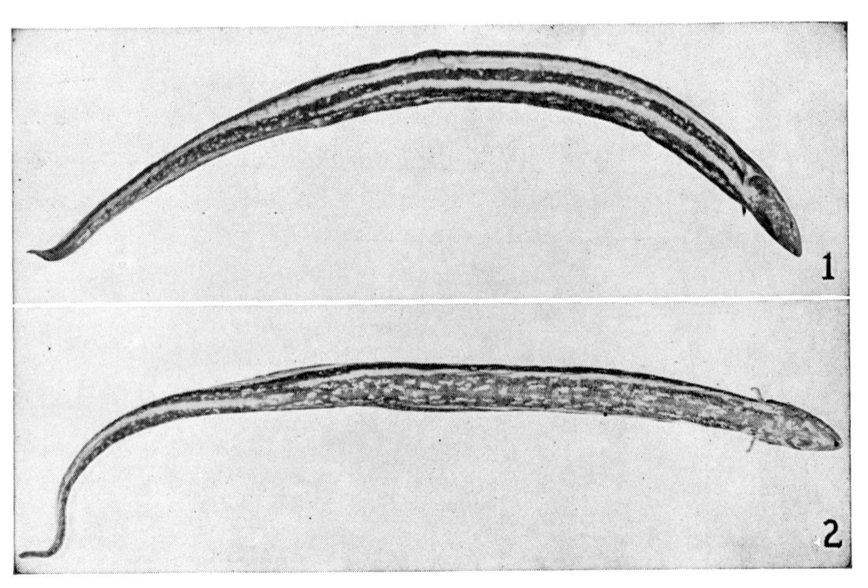


FIG. 142. *Pseudobranchus striatus striatus* (Le Conte). (1) Adult female, lateral view; actual length $5\frac{27}{32}$" (150 mm.). (2) Same, ventrolateral view. Georgia. [Photographs of a preserved specimen. U.S.N. Mus. No. 5051 a.]

pletely covered by the skin, through which the rachises and filaments may be faintly seen. Usually a single open gill slit behind the first gill, but when gills are covered the slit not open to the outside. The trunk is rather short and stout, thicker in proportion to its length than in *axanthus.* There are usually 34 costal grooves, but the number may vary from 33 to 36. The tail is nearly circular in cross section at the base, becoming compressed and sharp-edged above and below at about the distal half. Fore legs only developed. Toes 3, usually 3–1–2 in order of length from the shortest; slender, pointed, often with sharp, brown or blackish, horny tips. Tongue small, pointed, free at the tip and sides.

Palatine teeth small, comparatively slender, in 2 slightly divergent single or double lines. Horny sheath at tip of upper jaw scarcely developed in the specimens examined, that of lower jaw narrow, and brownish rather than black as in *axanthus*.

COLOR. The general ground color is brownish in preserved specimens. Of the two lateral light stripes, the upper is broad, yellow or buff in color, and extends from the gills along the trunk and basal half of the tail without much decrease in width. On the distal half of the tail it tapers to the tip. Lower light stripe narrower and extending only from the legs to the vent. Sides between the lateral light stripes brown or grayish-green, with many small, light yellow or buff flecks. Venter lighter than back, the ground color grayish-green, with many small, irregular, light yellow flecks somewhat concentrated at the sides and imparting a mottled appearance. Mid-line of tail beneath, sometimes with a narrow, light line. Head sometimes with a light stripe through the eye to the base of the gills. In some specimens there is a narrow light vertebral line.

BREEDING. Nothing has been recorded of the mating habits. Eggs apparently of this subspecies were described by Noble (1930, p. 52). A female collected at Lakeland, Georgia, Feb. 23, 1930, and confined in a crystalizing dish with two apparent males and a clump of water hyacinth, deposited a total of 11 eggs over a period of several weeks. The eggs were deposited singly or in pairs and attached to the sides of the dish or to the hyacinth roots. The individual egg had a diameter of 3 mm. and was surrounded by envelopes as follows: in addition to the vitelline membrane, which had a diameter of 3.8 mm., there were two envelopes, the inner with a diameter of 4.5 mm., the outer 5.5 mm.

**NARROW-STRIPED MUD-SIREN.** *Pseudobranchus striatus axanthus* Netting and Goin. Figs. 141, 143–144. Map 56.

TYPE LOCALITY. Payne's Prairie, 5 miles southeast of Gainesville, Alachua County, Florida.

RANGE. Peninsular Florida from the area of intergradation with typical *striatus* in the Okefinokee Swamp region, Georgia, southward to Dade County.

HABITAT. Inhabits swamps, bogs, and marshes, where it is found in submerged vegetation. Often most abundant in water-hyacinth beds, in shallow water, where it may be collected by the effective expedient of rolling up masses of the plants upon a sloping shore and catching the animals as they wriggle down toward the water.

SIZE. Attains a maximum length of approximately 8¼″ (210 mm.).

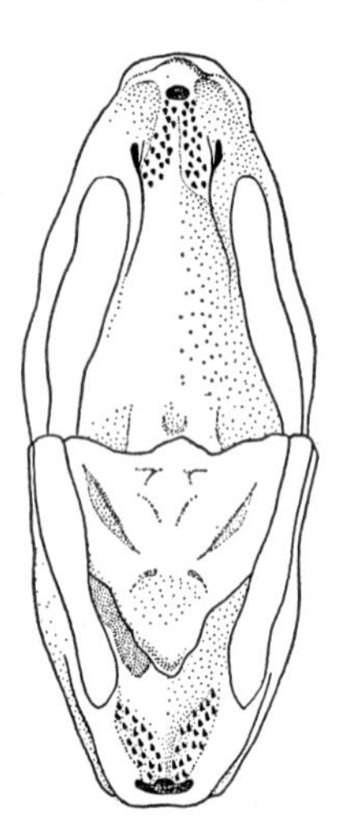

FIG. 143. Open mouth of *Pseudobranchus striatus axanthus* to show the character of the tongue, teeth, and horny sheaths at the tip of the jaws. [M.L.S. *del.*]

The average length of 8 sexually mature individuals of both sexes is 6¼″ (159 mm.) with extremes of 7 13/16″ (198 mm.) and 5″ (127 mm.). The proportions of an adult female from near Gainesville, Florida, are: total length 6 27/32″ (173 mm.), tail 2 25/32″ (71 mm.); head length 13/32″ (11 mm.), width ¼″ (6 mm.).

DESCRIPTION. The head is narrow, widest at a point about midway between the eyes and the anterior gills, the sides behind this point parallel and in front tapering rapidly to the bluntly pointed snout. Viewed from the side the head is wedge-shaped, the upper slope more pronounced than the lower. Eyes small, without lids, iris reddish-brown; the interorbital distance 3–3⅔ in the head from nostril to base of the first gills. Nostrils small and slit-like and placed at the ventrolateral angles of the

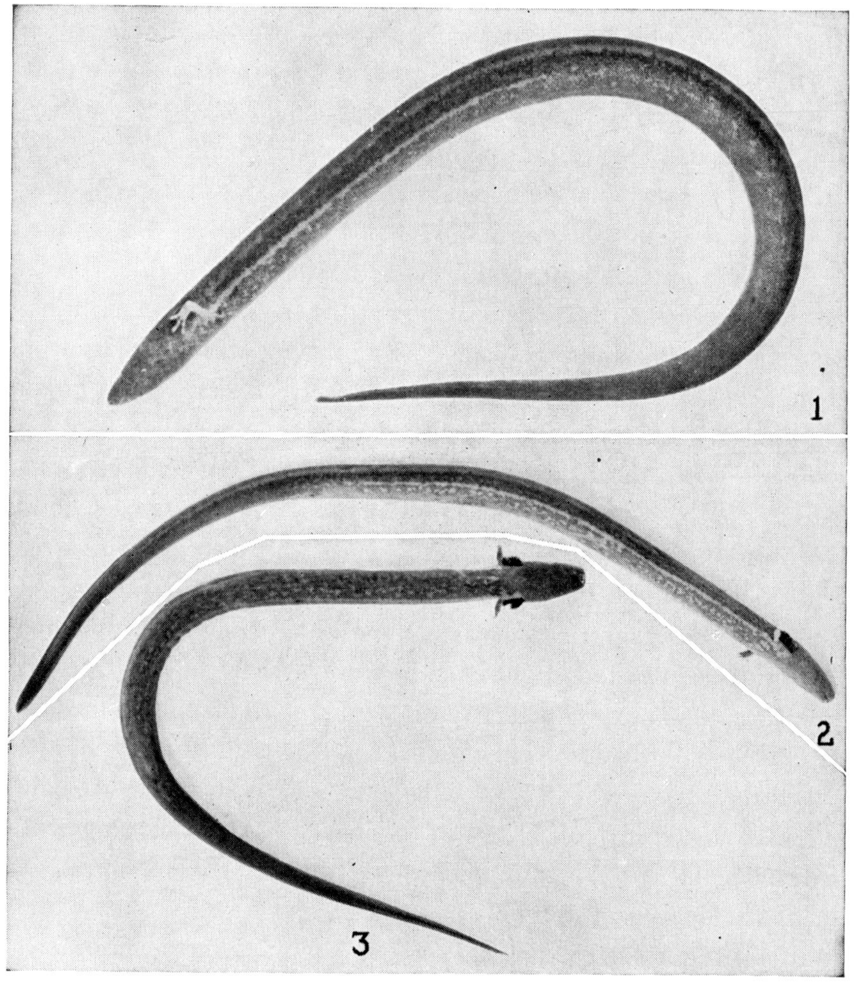


FIG. 144. *Pseudobranchus striatus axanthus* Netting and Goin. (1) Adult female, actual length 6³⁄₁₆″ (157 mm.). (2) Same, ventrolateral view. (3) Same, ventral view. Brooksville, Florida.

snout. Mouth small, crescentic, and subterminal. Gills short and stubby or moderately long and plumose, with the lateral branches basally long, apically short, and with short double fringes; gills well pigmented above. A single open gill slit. The trunk is subcylindrical in section, flattened below. There are usually 35 grooves between the axilla and the vent, but they are not strongly developed and the number may vary in either direction by 1 or 2 grooves. The tail is broadly oval in section at base, becoming more strongly compressed distally. The tail fins are narrow, the dorsal in adults arising about at the end of the proximal third and continuing to the tip, the ventral mainly confined to the distal third and sometimes continued as a low ridge toward the vent. Fore legs only developed, slender and weak. Toes 3, usually 3–1–2 in order of length from the shortest, tips black, pointed, and horny. Tongue small, pointed, and free anteriorly and at the sides. Palatine teeth in 2 scarcely divergent V-shaped patches of 2 or 3 rows. Premaxillary with a short, black, horny sheath; a longer, narrow, black, sharp-edged sheath covering the dentary.

COLOR. The general ground color above varies from light greenish-gray to dark gray or slate. In the adults there is a poorly defined, dark-margined, light vertebral line extending from between the eyes to the end of the tail and on each side of the trunk, and 2 narrow, dark-bordered, dull yellow or buff stripes, the upper often broken, extending from the base of the 3rd gills well onto the sides of the tail, and the lower, usually narrower and frequently broken, originating at the base of the legs and extending only to the vent. The back and the sides between the lateral light stripes are dark, irregularly mottled with light greenish-gray. Belly pale slate-gray, often with a few small pigment-free spots, the throat usually somewhat darker, the ventral surface of the tail orange-yellow or with a fleshy tinge, and the tail fin above conspicuously lighter than adjoining parts, yellowish. On some individuals there is an incomplete line of light dots on either side of the vertebral line.

BREEDING. Nothing is known of the mating habits. Noble and Richards (1932, p. 14) induced egg-laying by pituitary transplants. The eggs were deposited singly or in small groups and adhered to one another and to vegetation by the adhesiveness of the outer envelopes. Eggs average about 2.55 mm. in diameter and are provided with three envelopes in addition to the vitelline membrane, the inner having a diameter of 3 mm., the middle of 4.2 mm., and the outer of 5.6 mm. The upper ⅓ or ⅖ of the egg is heavily pigmented with brown. In Florida Carr (1940, p. 52) has found the eggs attached singly to the filamentous leaves of *Cabomba* and *Ceratophyllum* throughout the spring months. The recently hatched larvae have not been described.

# BIBLIOGRAPHY

# BIBLIOGRAPHY

## GENERAL WORKS


Baird, S. F., *Zool. iconogr. encycl.,* 2: 257. 1849.

——, Revision of the North American Tailed-Batrachia, with description of new genera and species. *Jour. Acad. Nat. Sci. Phila.,* 2nd ser., 1 (4): 281–292. 1850.

——, in J. G. Heck, *Iconographic encyclopædia of science, literature and arts.* Batrachia, 2: 249–261. 1851.

——, Report on reptiles collected on the survey. In E. G. Beckwith, *Report of exploration for railroad route near the 38th and 39th parallels of N. latitude,* 10 (4 no. 3): 17–20, pls. 17–18, 23–24. Washington, 1857.

——, Atlas herpetology. *U.S. expl. exped., 1838–1842, under command of Charles Wilkes,* pl. 1, figs. 1–8. 1858.

Barbour, Thomas, and Arthur Loveridge, Typical reptiles and amphibians. *Bul. Mus. Comp. Zool.,* 69 (10): 203–360. 1929.

Barnes, D. H., An arrangement of the genera of batracian animals, with a description of the more remarkable species: including a monograph of the doubtful reptils. *Amer. Jour. Sci. and Arts,* 11: 268–297. 1826.

Bishop, S. C., Notes on salamanders, with descriptions of several new forms. *Occ. Papers Mus. Zool. Univ. Mich.,* no. 451, pp. 1–27, pls. 1–2. 1941.

Bomare, J. C. V., *Dictionnaire raisonné universel d'histoire naturelle,* 4th ed., 12: 511–522. 1791.

Bosc, L. A. G., *Nouveau dictionnaire d'histoire naturelle,* 20: 41–50. Paris, 1803.

Boulenger, E. G., Class Batrachia (or Amphibia), in Tate Regan, *Natural History,* pp. 297–316, 24 figs. 1936.

Boyden, Alan, and G. K. Noble, The relationships of some common Amphibia as determined by serological study. *Amer. Mus. Novitates,* no. 606, pp. 1–24. 1933.

Brown, A. E., Generic types of Nearctic Reptilia and Amphibia. *Proc. Acad. Nat. Sci. Phila.,* 60: 112–127. 1908.

Cope, E. D., On the primary divisions of the Salamandridae, with descriptions of two new species. *Proc. Acad. Nat. Sci. Phila.,* 11: 122–128. 1859.

——, On the structures and distribution of the genera of the arciferous Anura. Suppl. on the osseus structures of the types of the Urodela. *Jour. Acad. Nat. Sci. Phila.,* 2nd ser., 6 (1): 67–112, pl. 25. Salamanders, pp. 97–108. 1866.

Cope, E. D., A review of the species of the *Amblystomidae. Proc. Acad. Nat. Sci. Phila.*, 19: 166–211. 1867.
——, A review of the species of the *Plethodontidae* and *Desmognathidae.—Proc. Acad. Nat. Sci. Phila.*, 21: 93–118. 1869.
——, Notes on the geographical distribution of Batrachia and Reptilia in western North America. *Proc. Acad. Nat. Sci. Phila.*, 35: 10–35. 1883.
——, Batrachia. In *Standard natural history*, 3: 303–344. 1885.
——, Synonymic list of the North American species of *Bufo* and *Rana*, with descriptions of some species of Batrachia, from specimens in the National Museum. *Proc. Amer. Phil. Soc.*, 23 (124): 514–526. 1886.
——, The hyoid structure in the amblystomid salamanders. *Amer. Nat.*, 21: 87–88, figs. 1–3. 1887.
——, The Batrachia of North America. *U.S. Nat. Mus. Bul.*, 34, pp. 3–525, pls. 1–79, 83, 86, figs. 1–119. 1889.
——, The geographical distribution of Batrachia and Reptilia in North America. *Amer. Nat.*, 30: 886–902, 1003–1026. 1896.
Cuvier, Georges, *The Animal Kingdom*. Order Batrachia, 9: 388–481, 5 pls. 1831.
Daudin, F. M., *Histoire naturelle des reptiles*, 8: 1–439, pls. 93–100. Paris, 1803.
Dumeril, A. M. C., and G. Bibron, *Histoire naturelle des reptiles*, 9: i–xx, 1–440. Les salamandrides, pp. 66–205. 1854.
Dunn, E. R., The salamanders of the genera *Desmognathus* and *Leurognathus.—Proc. U.S. Nat. Mus.*, 53: 393–433. 1917.
——, The collection of Amphibia Caudata of the Museum of Comparative Zoology. *Bul. Mus. Comp. Zool.*, 62: 445–471. 1918.
——, The geographical distribution of amphibians. *Amer. Nat.*, 57 (649): 129–136. 1923.
——, Mutanda herpetologica. *Proc. N. Eng. Zool. Club*, 8: 39–40. 1923 a.
——, The breeding habits of salamanders and their bearing on phylogeny. *Copeia*, no. 115, pp. 25–28. 1923 b.
——, The salamanders of the family *Plethodontidae.—Smith College Anniv. Publ.*, pp. 1–441, 3 pls., 86 maps. 1926.
——, On the relationships of certain plethodont salamanders. *Copeia*, no. 165, pp. 102–106. 1927.
——, The herpetological fauna of the Americas. *Copeia*, no. 3, pp. 106–119, 6 maps. 1931.
——, The races of *Ambystoma tigrinium.—Copeia*, no. 3, pp. 154–162. 1940.
Eigenmann, C. H., Cave vertebrates of America. *Carnegie Inst. Publ.*, 104, pp. i–ix, 1–241, pls. 1–29, 72 figs., frontispiece. The cave salamanders, pp. 29–41, pls. 1–2, figs. 3–11. 1909.
Fitzinger, L. J., *Neue Classification der Reptilien*, pp. i–iv, 1–66, 1 pl. Vienna, 1826.
——, *Systema reptilium*, 1st sec., Amblyglossae, pp. 1–106, iii–vi. 1843.
Fowler, H. W., and E. R. Dunn, Notes on salamanders. *Proc. Acad. Nat. Sci. Phila.*, 69: 7–28, pls. 3–4. 1917.
Gadow, Hans, Amphibia and reptiles. *Cambridge Nat. Hist.*, 8: v–xiii, 1–668, 181 figs., map. 1923.

Garman, Samuel, The reptiles and batrachians of North America. *Mem. Mus. Comp. Zool.,* 8 (3): i–xxxi, 1–185, pls. 1–9. Batrachia, pp. xxvii–xxxi. 1883.

Gervais, Paul, *Dictionnaire d'histoire naturelle.* Salamandre, 11: 304–311. 1849.

Gilmore, C. W., and D. M. Cochran, Amphibians and reptiles. In Cold-blooded vertebrates, *Smiths. Sci. Ser.,* 8 (2–3): 157–358, pls. 31–81, figs. 50–87. 1930.

Gmelin, C. C., *Gemeinnüssige systematische Naturgeschichte der Amphibien für gebildete Leser,* pt. 3, pp. i–x, 1–224, pls. 1–36. Salamanders, pp. 77–78. 1815.

Gravenhorst, J. L. C., *Vergleichende Uebersicht zoologischen Systeme,* pp. i–xx, 1–476. Göttingen, 1807.

——, Chelonios et batrachia (reptilia seu amphibia). *Rept. Musei Zool. Vratislav.,* 1: i–xiv, 1–106, pls. 1–17. 1829.

Gray, J. E., A synopsis of the genera of reptiles and amphibia, with a description of some new species. *Ann. Phil.,* ser. 2, 10: 193–217. 1825.

——, A synopsis of the species of the class Reptilia. In Georges Cuvier, *The animal kingdom,* appendix, 9: 1–110. London, 1831.

Griffith, E., Order Batrachia. In *The animal kingdom,* 9: 404–414, 464–481, 3 pls. London, 1831.

Guerin, F. E., *Dictionnaire d'histoire naturelle.* Salamandre, 8: 561–564. 1839.

Hallowell, Edward, Notes on the reptiles in the collection of the Academy of Natural Sciences of Philadelphia. *Proc. Acad. Nat. Sci. Phila.,* 8: 221–238. 1856.

——, Description of several new North American reptiles. *Proc. Acad. Nat. Sci. Phila.,* 9: 215–216. 1857.

——, On the caducibranchiate urodele batrachians. *Jour. Acad. Nat. Sci. Phila.,* 2nd ser., 3: 337–366. 1858.

Harlan, Richard, Observations on the genus *Salamandra,* with the anatomy of the *Salamandra gigantea* (Barton) or *S. Alleghaniensis* (Michaux) and two genera proposed. *Ann. Lyceum Nat. Hist. New York,* 1: 222–234, pls. 16–18. 1825.

——, Genera of North American Reptilia, and a synopsis of the species. *Jour. Acad. Nat. Sci. Phila.,* 5 (2): 317–372; 6 (1): 2–55. 1826–1827.

——, Observations on the genus *Salamandra,* and the establishment of the genera *Menopoma and Menobranchus.—Med. and Phys. Researches,* pp. 164–176, 2 pls. 1835.

——, Genera of North American Reptilia, and a synopsis of the species. *Med. and Phys. Researches, Phila.,* pp. 84–163, figs. 1–4. 1835 a.

Harper, Francis, Some works of Bartram, Daudin, Latreille, and Sonnini, and their bearing upon North American herpetological nomenclature. *Amer. Midland Nat.,* 23 (3): 692–723. 1940.

Heck, J. G., *Iconographic encyclopædia of science, literature, and art* (Eng. trans. by S. F. Baird). Batrachia, 2: 453–465. 1851.

Hoffman, C. K., in Bronn, *Klassen und Ordnungen Amphibien,* bd. b, abth. 2, pp. 1–726, pls. 1–72. 1873–1878.

Holbrook, J. E., *North American herpetology,* 1st ed., 1: i–viii, 9–120, 23 pls., 1836; 2: i–iv, 5–100, 30 pls., 1838; 3: i–viii, 9–122, 30 pls., 1838; 4: i–ix, 9–126, 28 pls., 1840. Philadelphia, 1836–1840.

——, *North American herpetology,* 2nd ed., 1: i–xv, 17–152, pls. 1–24; 2: i–vi,

9–142, pls. 1–20; 3: i–ii, 3–128, pls. 1–30; 4: i–vi, 7–138, pls. 1–35; 5: i–vi, 5–118, pls. 1–38. Salamanders, 5: 29–117, 32 pls. Philadelphia, 1842.

Jordan, D. S., *Manual of the vertebrate animals of the northeastern United States*, 13th ed., pp. i–xxxi, 1–446, figs. 1–15, map. Yonkers, N.Y.: World Book Co., 1929.

Latreille, P. A., in, Sonnini and Latreille. *Hist. nat. rept.*, 4: 305. 1801.

Laurenti, Joseph Nicolai, *Specimen medicum, exhibens synopsin reptilium*, pp. 1–214, pls. 1–5. Vienna, 1768.

Leuckart, Sigismund, and Hofrath Oken, Einiges über die fischartigen Amphibien. *Isis*, band 1, hefte 6, pp. 260–265, taf. 5, fig. A, a. 1821.

Leydekker, Richard, J. T. Cunningham, G. A. Boulenger, and J. A. Thompson, Reptiles, Amphibia, Fishes and lower Chordata, pp. i–xvi, 1–510, 37 pls., figs. 1–32. London, 1912.

Macauley, James, *The natural, statistical and civil history of the State of New-York*, 1: i–xxiv, 1–539. 1829.

Matthes, Benno, Die Hemibatrachier im Allgemeinen und die Hemibatrachier von Nord-Amerika im Speciellen. *Allg. Deutsche naturh. Zeitung*, n.s., 1: 249–280. 1855.

Merrem, Blasius, *Versuch eines Systems der Amphibien*, pp. i–xv, 1–191. Batrachia, Ordo III, Gradientia, pp. 184–188. Marburg, 1820.

Noble, G. K., An outline of the relation of ontogeny to phylogeny within the Amphibia. II. *Amer. Mus. Novitates*, no. 166, pp. 1–10. 1925.

——, The value of life history data in the study of the evolution of the Amphibia. *Ann. N. Y. Acad. Sci.*, 30: 31–128, pl. 9. 1927.

——, *The biology of the Amphibia*, pp. i–xiii, 1–577, figs. 1–174, frontispiece. New York: McGraw-Hill Book Co., 1931.

Oken, Hofrath, *Allgemeine Naturgeschichte für alle Stande* (6 bande). Amphibien oder Lurche, pp. 419–464. 1836.

Oppel, M. M., Le classification des reptiles. *Ann. mus. d'hist. nat.*, 16: 394–418. Paris, 1810.

Pratt, H. S., *A manual of the land and fresh water vertebrate animals of the United States (exclusive of birds)*, pp. i–xv, 1–422, figs. 1–184, map. Philadelphia: P. Blakiston's Son & Co., 1923.

Rafinesque, C. S., *Amer. Month. Mag. Crit. Rev.*, 4 (1): 41. 1818.

——, Necturus. *Jour. Phys. Chim. Hist. Nat.*, 88: 418. 1819.

Ruschenberger, W. S. W., *Elements of herpetology and of ichthyology*. Batrachians, pp. 65–76. 1846.

Shaw, George, *General Zoology*. Amphibia, 3 (1): i–ix, 1–615, pls. 1–140. Salamanders, pp. 291–305, 601–615. 1802.

Shufeldt, R. W., *Chapters on the natural history of the United States*, pp. 1–472. Class Amphibia, pp. 91–100. New York, 1897.

Slevin, J. R., The amphibians of western North America. *Occ. Papers Cal. Acad. Sci.*, 16: 1–152, pls. 1–23. 1928.

——, A handbook of reptiles and amphibians of the Pacific States. *Spec. Pub. Cal. Acad. Sci.*, pp. 1–173, pls. 1–11, figs. 1–9. 1934.

Smith, W. H., *The tailed batrachians, including the cæcilians*, pp. 1–158. 1877.

Stark, John, *Elements of natural history*. Vertebrata, 1: i–vi, 1–527, pls. 1–4. Salamanders, pp. 368–370. 1828.

Strauch, Alexander, Revision der Salamandriden-Gattungen. *Mémoires de l'acad. imp. sci. St. Petersbourg,* ser. 7, 16 (4): 1–110, pls. 1–2. 1870.

Swainson, William, On the natural history and classification of fishes, amphibians, and reptiles. *The Cabinet Cyclopædia,* 2: 1–452, figs. 1–135. Salamanders, pp. 91–100. 1839.

Wagler, Joannes, *Natürliches System der Amphibien,* pp. 129–338. München, 1830.

——, *Descriptiones et icones amphibiorum,* pts. 1–3. 1833.

Watson, D. M. S., The evolution and origin of the Amphibia. *Phil. Trans. Royal Soc. London,* ser. B, 214: 189–257, figs. 1–39. 1926.

Werner, Franz, Die Lurche und Kriechtiere, band 1. In Brehm, Tierleben, 4: i–xiv, 1–572, 37 pls., 127 figs. 1920.

——, Amphibien. *Das Tierreich,* III, Reptilien und Amphibien, 2: 1–80, figs. 1–39. 1922.

Wolterstorff, W., and W. Herre, Die Gattungen der Wassermolche der Familie *Salamandridae.—Arch. Naturgesch.,* n.s., 4: 217–229. Leipzig, 1935.

Wright, A. H., The vertebrate life of Okefinokee Swamp in relation to the Atlantic Coastal Plain. *Ecology,* 7 (1): 77–95, pls. 2–6. 1926.

## CHECKLISTS AND CATALOGS

Boulenger, G. A., *Catalogue of the Batrachia Gradientia S. Caudata and Batrachia Apoda in the collection of the British Museum,* 2nd ed., pp. i–viii, 1–127, pls. 1–9. 1882.

Cope, E. D., Check-list of North American Batrachia and Reptilia. *U. S. Nat. Mus., Bul. 1,* pp. 1–104. 1875.

Davis, N. S., Jr., and F. L. Rice, Descriptive catalogue of North American Batrachia and Reptilia found east of the Mississippi River. *Ill. State Lab. Nat. Hist., Bul.* 5, pp. 1–66. 1883.

Garman, Samuel, The North American reptiles and batrachians. *Bul. Essex Inst.,* 16: 3–46, 6 figs. 1884.

Gray, J. E., *Catalogue of the specimens of amphibians in the collection of the British Museum.* Pt. 2, Batrachia gradientia, etc., pp. 1–72, pls. 3–4. London, 1850.

Scharlinski, Hans, Nachtrag zum Katalog der Wolterstorff-Sammlung im Museum für Naturkunde und Vorgeschichte zu Magdeburg. *Abh. Ber. Mus. Nat.,* 7: 31–57. Magdeburg, 1939.

Stejneger, Leonhard, and Thomas Barbour, *A check list of North American amphibians and reptiles,* 4th ed., pp. i–xvi, 1–207. 1939.

Wolterstorff, W., Katalog der Amphibien-Sammlung im Museum für Natur.- und Heimatkunde zu Magdeburg. *Festsch. Abh. u. Ber. Mus. f. Natur.- und Heimatkunde,* 4 (2): 231–310. Magdeburg, 1925.

Yarrow, H. C., Check list of North American Reptilia and Batrachia, *Bul. U.S. Nat. Mus.,* no. 24, pp. 1–249. 1882.

Yarrow, H. C., Check list of North American Reptilia and Batrachia, *Smiths. Misc. Coll.*, 517, pp. 1–28. 1883.

## LIFE HISTORY AND HABITS

Abbott, C. C., Hibernation of the lower vertebrates. *Sci.*, 4 (75): 36–39. 1884.
Baker, Louise C., Mating habits and life history of *Amphiuma tridactylum* Cuvier and effect of pituitary injections. *Jour. Tenn. Acad. Sci.*, 12 (2): 206–218, figs. 1–3. 1937.
Cope, E. D., The retrograde metamorphosis of *Siren.—Amer. Nat.*, 19: 1226–1227. 1885.
——, A careless writer on *Amphiuma.—Amer. Nat.*, 29 (348): 1108–1110. 1895.
Dunn, E. R., The breeding habits of *Ambystoma opacum* (Gravenhorst). *Copeia*, no. 43, pp. 40–43. 1917.
Eycleschymer, A. C., The habits of *Necturus maculosus.—Amer. Nat.*, 40 (470): 123–136. 1906.
Flower, S. S., Further notes on the duration of life in animals. II. Amphibians. *Proc. Zool. Soc. London*, pt. 1, pp. 369–394. 1936.
Frear, William, Vitality of the mud puppy. *Amer. Nat.*, 16 (4): 325–326. 1882.
Gage, S. H., The changes of the salamander *Diemyctylus viridescens.—Amer. Nat.*, 25 (294): 380. 1891.
Gaines, Angus, The habits of *Amblystoma opacum.—Amer. Nat.*, 28 (335): 969. 1894.
Geyer, Hans, *Eurycea bislineata bislineata* (Green). *Blätter für Aquarien.- und Terrarienkunde*, 49 (8): 113–114, 2 figs. 1938.
Goodale, H. D., The early development of *Spelerpes bilineatus* (Green). *Amer. Jour. Anat.*, 12 (2): 173–247, figs. 1–77, pl. 1. 1911.
Hamilton, W. J., Crows feeding on larval amphibians. *Auk*, 59 (3): 446. 1942.
Hilton, W. A., A structural feature connected with the mating of *Diemyctylus viridescens.—Amer. Nat.*, 36 (428): 643–651, figs. 1–11. 1902.
Hoy, P. R., The development of *Amblystoma lurida* Sager. *Amer. Nat.*, 5: 578–579, figs. 108–109. 1871.
Humphrey, R. R., Ovulation in the four-toed salamander, *Hemidactylium scutatum*, and the external features of cleavage and gastrulation. *Biol. Bul.*, 54 (4): 307–323, figs. 1–8. 1928.
Jordan, E. O., The spermatophores of *Diemyctylus.—Jour. Morph.* 5 (2): 263–270. 1891.
Kingsbury, B. F., The spermatheca and methods of fertilization in some American newts and salamanders. *Proc. Amer. Micros. Soc.*, 17: 261–304, pls. 1–4. 1895.
Lantz, L. A., Einiges über Lebensweise und Fortpflanzung von *Ambystoma opacum* Grav. *Blätt. f. Aquar.- u. Terrarienkunde*, 41: 63–67, taf. 9–10. 1930.
——, Notes on the breeding habits and larval development of *Ambystoma opacum* Grav. *Ann. Mag. Nat. Hist.*, ser. 10, 5: 322–325, figs. a–b. 1930 a.
Matheson, Robert, *A handbook of the mosquitoes of North America*, pp. i–xvii, 1–268, 25 pls., 23 figs. 1929.

Monks, S. P., The spotted salamander. *Amer. Nat.,* 14 (5): 371–374. 1880.
Morgan, A. H., and M. C. Grierson, Winter habits and yearly food consumption of adult spotted newts, *Triturus viridescens.—Ecology,* 13 (1): 54–62, figs. 1–4. 1932.
Netting, M. G., The food of the hellbender *Cryptobranchus alleganiensis* (Daudin). *Copeia,* no. 170, pp. 23–24. 1929.
Noble, G. K., The value of life-history data in the study of the evolution of the Amphibia. *Ann. New York Acad. Sci.,* 30: 31–128, pl. 9. 1927.
——, The relation of courtship to the secondary sexual characters of the two-lined salamander, *Eurycea bislineata* (Green). *Amer. Mus. Novitates,* no. 362, pp. 1–5, 4 figs. 1929.
——, and M. K. Brady, The courtship of the plethodontid salamanders. *Copeia,* no. 2, pp. 52–54. 1930.
——, and M. K. Brady, Observations on the life history of the marbled salamander, *Ambystoma opacum* Gravenhorst. *Zoologica,* 11 (8): 89–132, figs. 83–93. 1933.
Palmer, E. L., Amphibia and Reptilia. *Cornell Rural Sch. Leaflet,* 15 (4): 303–364, many figs. 1922.
Piersol, W. H., The egg-laying habits of *Plethodon cinereus.—Univ. Toronto Studies, biol. ser.,* 10 (2): 121–126. 1914.
——, Pathological polyspermy in eggs of *Ambystoma jeffersonianum* (Green). *Tran. Royal Can. Inst.,* 17 (1): 57–74, 1 pl., 1 fig. 1929.
Pike, Nicholas, Notes on the life history of *Amblystoma opacum.—Bul. Amer. Mus. Nat. Hist.,* no. 7, art. 13, pp. 209–212. 1886.
Pope, P. H., A note on the length of life of some batrachians in captivity. *Copeia,* no. 72, pp. 66–67. 1919.
——, The life-history of the common water-newt (*Notophthalmus viridescens*), together with observations on the sense of smell. *Ann. Carnegie Mus.,* 15 (2–3): 305–368, pls. 41–51. 1924.
——, The life-history of *Triturus viridescens*—some further notes. *Copeia,* no. 168, pp. 61–73, figs. 1–2. 1928.
——, The longevity of *Ambystoma maculatum* in captivity. *Copeia,* no. 169, pp. 99–100. 1928.
——, Notes on the longevity of an *Ambystoma* in captivity. *Copeia,* no. 2, pp. 140–141. 1937.
Ryder, J. A., On a brood of larval *Amphiuma.—Amer. Nat.,* 23 (274): 927–928. 1889.
Shufeldt, R. W., The habits of *Murænopsis tridactylus* in captivity; with observations on its anatomy. *Sci.,* 2 (27): 159–163, 4 figs. 1883.
——, Notes on the diet of *Amblystomas.—Sci.,* 9 (216): 298. 1887.
Smith, B. G., The life history and habits of *Cryptobranchus allegheniensis.—Biol. Bul.,* 13 (1): 5–39, figs. 1–14. 1907.
——, The structure of the spermatophores of *Ambystoma punctatum.—Biol. Bul.,* 18 (4): 204–211, figs. 1–5. 1910.
——, The embryology of *Cryptobranchus allegheniensis,* part 1. *Jour. of Morph.,* 23 (1): 61–160, pls. 1–2, figs. 1–51. 1912.

Smith, B. G., The process of ovulation in Amphibia. *18th Ann. Rept. Mich. Acad. Sci.*, pp. 102–105. 1916.
Smith, J., Zerlegung des Proteus of the Lakes (*Menobranchus*).—*Isis*, p. 1088. 1832.
Wilder, I. W., The life history of *Desmognathus fusca.*—*Biol. Bul.*, 24 (4): 251–292, figs. 1–14; 24 (5): 293–342, figs. 15–25, pls. 1–6. 1913.
——, On the breeding habits of *Desmognathus fuscus.*—*Biol. Bul.*, 32 (1): 13–20, fig. 1. 1917.

## POPULAR

Alexander, W. P., The Allegheny hellbender and its habitat. *Hobbies, Buffalo Soc. Nat. Hist.*, 7 (10): 13–18, 2 figs. 1927.
Bateman, G. C., *The Vivarium*. Salamanders, pp. 358–397. London, 1897.
Berridge, W. S., *All about reptiles*, pp. 1–271. Salamanders, chap. xxi, pp. 239–248; Newts, chap. xxii, pp. 249–252. 1935.
Boulenger, E. G., Reptiles and batrachians. *Urodela*, pp. 245–269, 6 pls. New York: E. P. Dutton & Co., 1914.
Comstock, A. B., The newt, eft, or salamander. In *Handbook of Nature-Study*, 14th ed., pp. 197–199, 1 fig. Ithaca, N.Y.: Comstock Pub. Co., 1923.
Conant, Roger, Salamanders of the northeastern states. *Fauna*, 4 (1): 8–11, 20 figs., map. 1942.
Hodge, C. F., *Nature study and life*, pp. vii-xv, 1–514, 196 figs., frontispiece. Salamanders, pp. 301–303. New York: Ginn & Co., 1902.
Jordan, D. S., and B. H. Van Vleck, *A popular key to the birds, reptiles, batrachians and fishes of the northern United States east of the Mississippi River*, pp. 1–88. Appleton, Wis.: Reid & Miller, 1874.
Mellen, Ida, The amphibians. *Bul. N. Y. Zool. Soc.*, 30 (6): 172–205, 27 pls. 1927.
Mohr, C. E., Salamanders. *Natural History*, 36: 165–173, 10 photos. 1935.
——, Spring lizards. *Fauna*, 4 (1): 3–7, 4 figs. 1942.
Strecker, J. K., Reptiles of the South and Southwest in folk-lore. Salamanders, spring lizards, and ground puppies. *Pub. Tex. Folk-Lore Soc.*, no. 5, pp. 61–63. 1926.
——, Reptiles of the South and Southwest in folk-lore. The hellbender. *Pub. Tex. Folk-Lore Soc.*, no. 5, p. 63. 1926 a.

## STATE AND PROVINCE LISTS

### United States of America

#### ALABAMA

Brimley, C. S., Records of some reptiles and batrachians from the southeastern United States. *Proc. Biol. Soc. Wash.*, 23: 9–18. 1910.
Dunn, E. R., *see* Virginia. 1920.

Goin, C. J., *see* Virginia. 1938.
Holt, E. G., Additional records for the Alabama herpetological catalogue. *Copeia,* no. 135, pp. 93–95. 1924.
Löding, H. P., A preliminary catalogue of Alabama amphibians and reptiles. *Ala. Mus. Nat. Hist.,* paper no. 5, pp. 1–59. 1922.
Penn, G. H., Jr., Notes on the summer herpetology of DeKalb County, Alabama. *Jour. Tenn. Acad. Sci.,* 15 (3): 352–355.
Viosca, Percy, Jr., *see* Louisiana. 1937.
Wright, A. H., Some rare amphibians and reptiles of the United States. *Proc. Nat. Acad. Sci.,* 21 (6): 340–345. 1935.

### ALASKA

Cope, E. D., *see above,* General works. 1889.
Dunn, E. R., Note on *Ambystoma decorticatum.—Copeia,* no. 3, pp. 87–88. 1930.
Myers, G. S., *see* Oregon. 1942.
Swarth, H. S., Origins of the fauna of the Sitkan district. *Proc. Cal. Acad. Sci.,* 4th ser., 23 (3): 59–78, 1 fig. 1936.

### ARIZONA

Cope, E. D., On the Reptilia and Batrachia of the Sonoran province of the Nearctic region. *Proc. Acad. Nat. Sci. Phila.,* pp. 300–314. 1866.
Coues, Elliott, Synopsis of the reptiles and batrachians of Arizona. In *Rept. Geog. and Geol. Exp. and Surv. West 100th Meridian in charge of G. M. Wheeler,* 5: 585–633, pl. 18, figs. 1, 3; 19; 20, fig. 2; 21; 22; 23, figs. 1–2. 1875.
Dodge, N. N., Amphibians and reptiles of Grand Canyon National Park. *Nat. Hist. Assn. Bul.,* no. 9, pp. 1–55, 33 figs. 1938.
Eaton, T. H., Amphibians and reptiles of the Navaho country. *Copeia,* no. 3, pp. 150–151. 1935.
Franklin, D., Notes on *Amblystoma tigrinum* of Flagstaff, Arizona. *Copeia,* no. 21, pp. 30–31. 1915.
McKee, E. D., and C. M. Bogert, The amphibians and reptiles of Grand Canyon National Park. *Copeia,* no. 4, pp. 178–180. 1934.
Merriam, C. H., Results of a biological survey of the San Francisco Mountain region and desert of the Little Colorado in Arizona. *North Amer. Fauna,* no. 3, *U.S. Dept. Agr.,* pp. i–vii, 1–136, figs. 1–2, 5 maps. 1890.
Rüthling, P. D. R., *see* California. 1917.
Stejneger, Leonhard, Annotated list of reptiles and batrachians collected by Dr. C. H. Merriam and Vernon Bailey on the San Francisco Mountain plateau and desert of the Little Colorado, Arizona, with descriptions of new species. *North Amer. Fauna,* no. 3, pp. 103–118. 1890.
Van Denburgh, John, and J. A. Slevin, A list of the amphibians and reptiles of Arizona, with notes on the species in the collection of the Academy. *Proc. Cal. Acad. Sci.,* 4th ser., 3: 391–454, pls. 17–28. 1913.

## ARKANSAS

Baird, S. F., Report upon the reptiles of the route. In *U.S.P.R.R. exp. and surv. 35th parallel,* 10 (pt. 6, no. 4): 37–45, pls. 25–27. Washington, 1859.

Bishop, S. C., *see above,* General works. 1941.

Black, J. D., and S. C. Dillinger, Herpetology of Arkansas, pt. 2, The amphibians. *Occ. Papers Univ. Ark. Mus.,* no. 2, pp. 1–30. 1938.

Brimley, H. H., and C. S. Brimley, The habitat of the salamander *Linguelapsus annulatus* Cope. *Amer. Nat.,* 29 (338): 168. 1895.

Burt, C. E., Further records of the ecology and distribution of amphibians and reptiles in the Middle West. *Amer. Mid. Nat.,* 16 (3): 311–336. 1935.

Cope, E. D., *see above,* General works. 1886.

Dunn, E. R., and A. A. Heinze, A new salamander from the Ouachita Mountains. *Copeia,* no. 3, pp. 121–122. 1933.

Dury, Ralph, *see* Indiana. 1932.

Goin, C. J., Description of a new race of *Siren intermedia* Le Conte. *Ann. Carnegie Mus.,* 29: 211–217. 1942.

Hay, O. P., Observations on *Amphiuma* and its young. *Amer. Nat.,* 22 (256): 315–321, 1 fig. 1888.

——, *see* Indiana. 1892.

Hurter, Julius, and J. K. Strecker, Jr., The amphibians and reptiles of Arkansas. *Trans. Acad. Sci. St. Louis,* 18 (2): 11–27. 1909.

Noble, G. K., and B. C. Marshall, The breeding habits of two salamanders. *Amer. Mus. Nov.,* no. 347, pp. 1–12, figs. 1–4. 1929.

——, and B. C. Marshall, The validity of *Siren intermedia* Le Conte, with observations on its life history. *Amer. Mus. Novitates,* no. 532, pp. 1–17, figs. 1–4. 1932.

Ortenburger, A. I., Reptiles and amphibians from southeastern Oklahoma and southwestern Arkansas. *Copeia,* no. 170, pp. 8–12. 1929.

Smith, Hobart M., An addition to the amphibian fauna of Arkansas. *Tran. Kan. Acad. Sci.,* 36: 321–322. 1933.

Stejneger, Leonhard, Description of a new salamander from Arkansas, with notes on *Ambystoma annulatum.—Proc. U.S. Nat. Mus.,* 17: 597–599. 1895.

Stone, Witmer, A collection of reptiles and batrachians from Arkansas, Indian Territory and western Texas. *Proc. Acad. Nat. Sci. Phila.,* 55: 538–542. 1903.

Strecker, J. K., Jr., Notes on the habits of two Arkansas salamanders and a list of batrachians and reptiles collected at Hot Springs. *Proc. Biol. Soc. Wash.,* 21: 85–89. 1908.

——, Notes on the herpetology of Hot Springs, Arkansas. *Baylor Bul.* 27 (3): 29–47. 1924.

Taylor, E. H., Arkansas amphibians and reptiles in the Kansas University museum. *Univ. Kan. Sci. Bul.,* 22 (10): 207–218. 1935.

## CALIFORNIA

Adams, Lowell, A new locality record for the Mount Lyell salamander. *Copeia,* no. 1, p. 56. 1942.

——, The natural history and classification of the Mount Lyell salamander, *Hydromantes platycephalus.—Univ. Cal. Pub. Zool.,* 46 (2): 179–204, pls. 21–22, figs. 1–9. 1942 a.

Baird, S. F., Report on reptiles collected on the survey. In *Rept. exp. and surv. for a railroad route from Miss. River to Pacific Ocean.* Routes in California and Oregon, 10 (pt. 4. No. 4): 5–13, pls. 11, 28, 30, 44. 1859.

——, and Charles Girard, List of reptiles collected in California by Dr. John L. Le Conte, with description of new species. *Proc. Acad. Nat. Sci. Phila. for 1853,* 6: 300–302. 1853.

Bishop, S. C., *see above,* General works. 1941.

Blankinship, J. W., and C. A. Keeler, On the natural history of the Farallon Islands. *Zoe,* 3 (2): 144–165. San Francisco, 1892.

Bogert, C. M., An annotated list of the amphibians and reptiles of Los Angeles County, California. *Bul. Southern Calif. Acad. Sci.,* 29 (1): 3–14, map. 1930.

[Borland, J. N.] [List of reptiles collected in California by Mr. E. Samuels.] *Proc. Bost. Soc. Nat. Hist.,* 6: 192–194. 1857.

Breder, R. B., Notes on the behavior of the western newt in captivity. *Copeia.* No. 119, pp. 71–76, 1 fig. 1923.

Burke, C. V., Note on *Batrachoseps attenuatus* Esch. *Amer. Nat.,* 45: 413–414. 1911.

Burt, C. E., and M. D. Burt, *see* Utah. 1929.

Camp, C. L., *Batrachoseps major* and *Bufo cognatus californicus,* new Amphibia from southern California. *Univ. Cal. Pub. Zool.,* 12 (12): 327–334. 1915.

——, *Spelerpes platycephalus,* a new alpine salamander from the Yosemite National Park, California. *Univ. Cal. Pub. Zool.,* 17 (3): 11–14, figs. 1–5. 1916.

Campbell, Berry, Notes on *Batrachoseps.—Copeia,* no. 3, pp. 131–134. 1931.

Cooper, J. G., Natural history. In T. F. Cronise, *The natural wealth of California,* pp. i–xvi, 1–696, many figs. San Francisco, 1868.

——, The fauna of California and its geographical distribution. *Proc. Cal. Acad. Sci.,* 4: 61–81. 1870.

——, Monterey in the dry season. *Amer. Nat.,* 4 (12): 756–758. 1871.

Cope, E. D., Third contribution to the herpetology of tropical America. *Proc. Acad. Nat. Sci. Phila.,* 17: 185–198. 1865.

——, *see above,* General works. 1867.

Dunn, E. R., Two new insular *Batrachoseps.—Copeia,* no. 109, pp. 60–63. 1922.

——, The Yosemite fauna. *Copeia,* no. 133, pp. 75–76. 1924.

——, A new salamander from southern California. *Proc. U.S. Nat. Mus.,* 74: 1–3. 1929.

——, An egg cluster of *Aneides ferreus.—Copeia,* no. 1, p. 52. 1942.

Eaton, T. H., The occurrence of streptostyly in the *Ambystomidae.—Univ. Cal. Pub. Zool.,* 37 (17): 521-526, 1 fig. 1933.

Eisen, Gustav, The spermatogenesis of *Batrachoseps.—Jour. Morph.,* 17 (1): 1–117, pls. 1–14. 1901.

Eschscholtz, Friedrich, *Zoologischer Atlas,* 5: i–viii, 1–28, pls. 21–25. 1833.

Geyer, Hans, *Ensatina eschscholtzii* Gray. *Bl. Aquar. Terrar. Braunschweig,* 49 (9): 131–132, 1 fig. 1938.

Girard, Charles, *U.S. exploring expedition during the years 1838–1842 under the command of Charles Wilkes.* Herpetology, pp. i–xiv, 1–496. Salamanders, pl. 1, figs. 1–33. 1858.

Gray, J. E., Batrachians. In *Zoology of Captain Beechey's Voyage,* p. 99, pl. 31, fig. 3. 1839.

——, *see above,* Checklists and Catalogs. 1850.

——, On a new species of salamander from California. *Proc. Zool. Soc. London,* 21 (2): 11, pl. 7. 1853.

Grinnell, Joseph, and C. L. Camp, A distributional list of the amphibians and reptiles of California. *Univ. Cal. Pub. Zool.,* 17 (10): 127–208, figs. 1–14. 1917.

——, J. Dixon, and J. M. Linsdale, Vertebrate natural history of a section of northern California through the Lassen Peak region. *Univ. Calif. Pub. Zool.,* 35: i–v, 1–594, 181 figs. 1930.

——, and J. M. Linsdale, Vertebrate animals of Point Lobos Reserve, 1934–35. *Car. Inst. Wash.,* pub. no. 481, pp. i–iv, 1–159, pls. 1–39, fig. 1. 1936.

——, and T. I. Storer, Reptiles and amphibians of Yosemite National Park. Reprint from Hall's *Handbook of Yosemite Nat. Park,* pp. 175–182. 1921.

——, and T. I. Storer, *Animal life in the Yosemite,* pp. i–xviii, 1–752, 62 pls., 64 text figs. Berkeley: University of California Press, 1924.

Hall, H. M., and Joseph Grinnell, Life-zone indicators in California. *Proc. Cal. Acad. Sci.,* 9 (2): 37–67. 1919.

Hallowell, Edward, Description of a new species of salamander from Upper California. *Proc. Acad. Nat. Sci. Phila.,* 4: 126. 1849.

——, On some new reptiles from California. *Proc. Acad. Nat. Sci. Phila. for 1853,* 6: 236–238. 1853.

——, Report on the reptiles collected on the survey. In *Report of explorations in California for railroad routes, U.S.P.R.R. exp. and surv. 32nd parallel,* 10 (Pt. 4, No. 1): 1–25, pls. 1–10. Washington, 1859.

——, Report upon the Reptilia of the North Pacific exploring expedition, under command of Capt. John Rogers, U.S.N. *Proc. Acad. Nat. Sci. Phila.,* 12: 480–510. 1860.

Henry, W. V., and V. C. Twitty, Contributions to the life histories of *Dicamptodon ensatus* and *Ambystoma gracile.—Copeia,* no. 4, pp. 247–250. 1940.

Hilton, W. A., The occurrence of *Batrachoseps attenuatus* and *Autodax lugubris* in southern California. *Amer. Nat.,* 43: 53–54. 1909.

Hubbard, M. E., Correlated protective devices in some California salamanders. *Univ. Cal. Pub. Zool.,* 1 (4): 157–170, pl. 16. 1903.

Jacob, E., *Batrachoseps attenuatus.—Blätter für Aquar. und Terrarienkunde,* 21: 280–282, 298–300, 329–330, figs. 1–3. 1910.

Klauber, L. M., Notes on the salamanders of San Diego County, California. *Bul. Zool. Soc. San Diego,* no. 3, pp. 1–4, fig. 1. 1927.

——, A list of the amphibians and reptiles of San Diego County, California. *Bul. Zool. Soc. San Diego,* no. 4, pp. 1–8. 1928.

——, Range extensions in California. *Copeia,* no. 170, pp. 15–22. 1929.

Klauber, L. M., A list of the amphibians and reptiles of San Diego County, California, 2nd ed. *Bul. Zool. Soc. San Diego,* no. 5, pp. 2–8. 1930.

——, Annotated list of the amphibians and reptiles of the southern border of California. *Bul. Zool. Soc. San Diego,* no. 11, pp. 1–28, figs. 1–8, map. 1934.

Lebrun, Hector, Maturation of the eggs of *Diemyctilus torosus.*—*Biol. Bul.* 3 (1): 1–2. 1902.

Lockington, W. N., List of Californian reptiles and batrachia collected by Mr. Dunn and Mr. W. J. Fisher in 1876. *Amer. Nat.,* 14: 295–296. 1880.

Maslin, T. P., Egg-laying of the slender salamander (*Batrachoseps attenuatus*).—*Copeia,* no. 4, pp. 209–212. 1939.

Mearns, E. A., Mammals of the Mexican boundary of the United States. *Bul. U.S. Nat. Mus.,* no. 56, pt. 1, pp. i–xi, 1–530, pls. 1–13, figs. 1–126. 1907.

Miller, L. H., Capture of the salamander, *Autodax lugubris,* at Los Angeles, California. *Amer. Nat.,* 40: 741–742. 1906.

Müller, F., Dritter Nachtrag zum Katalog der herpetologischen Sammlung des Basler Museums. *Verhandlungen der naturforschende Gesellschaft in Basel,* 7: 274–299, pl. 5. 1884.

Myers, G. S., Notes on some amphibians in western North America. *Proc. Biol. Soc. Wash.,* 43: 55–64. 1930.

——, *Hydromantes platycephalus* in Sonora Pass, California. *Copeia,* no. 2, p. 91. 1938.

——, *see* Oregon. 1942.

[Orcutt, C. R.] [Note on *Aneides lugubris* Hallowell.] *West-Amer. Scientist,* 1 (2): 5. 1885.

Reid, H. A., *History of Pasadena,* pp. 1–675, illustrated. Pasadena, Cal., 1895.

Ritter, W. E., The life-history and habits of the Pacific Coast newt (*Diemyctylus torosus* Esch.) *Proc. Cal. Acad. Sci.,* 3rd ser., 1 (2): 73–114, pl. 3. 1897.

——, On the reproductive habits and development of the Californian land salamander, *Autodax.*—*Sci.,* n.s., 9 (218): 311–312. 1899.

——, Further notes on the habits of *Autodax lugubris.*—*Amer. Nat.,* 37: 883–886. 1903.

——, and Loye Miller, A contribution to the life-history of *Autodax lugubris.*—*Amer. Nat.,* 33 (393): 691–704, figs. 1–7. 1899.

Rüthling, P. D. R., The water dog of California. *Aquatic Life,* 1 (3): 47–48, 1 fig. 1915.

——, Los Angeles salamanders. *Copeia,* no. 25, pp. 61–62. 1915 a.

——, Collecting trip to Verdugo Peak. *Lorquinia,* 1 (1): 6. 1916.

——, Some feeding habits of the desert rough-scaled swift (*Sceloporus magister* Hallowell). *Lorquinia,* 2 (2): 9–11. 1917.

Slevin, J. R., Further notes on the genus *Ensatina* in California. *Copeia,* no. 3, pp. 77–78. 1930.

Smith, R. E., The spermatophores of *Triturus torosus* and *Triturus rivularis.*—*Proc. Nat. Acad. Sci.,* 27 (6): 261–264, 1 fig. 1941.

Snook, H. J., and J. A. Long, Parasynaptic stages in the testis of *Aneides lugubris*

(Hallowell). *Univ. Cal. Pub. in Zool.,* 11 (15): 511–528, pls. 25–26, 1 fig. 1914.
Snyder, J. O., Eggs of *Batrachoseps attenuatus.—Copeia,* no. 121, pp. 86–88. 1923.
Stephens, Frank, An annotated list of the amphibians and reptiles of San Diego County, California. *Trans. San Diego Soc. Nat. Hist.,* 3 (4): 59–69. 1921.
Storer, T. I., Additional records of the tiger salamander in California. *Copeia,* no. 24, p. 56. 1915.
——, A synopsis of the Amphibia of California. *Univ. Cal. Pub. Zool.,* 27: 1–342, 18 pls., 42 figs. 1925.
——, Notes on the genus *Ensatina* in California, with description of a new species from the Sierra Nevada. *Univ. Cal. Pub. Zool.,* 30 (16): 443–452. 1929.
Strauch, Alexander, *see above,* General works. 1870.
Townsend, C. H., Field-notes on the mammals, birds and reptiles of northern California. *Proc. U.S. Nat. Mus.,* 10: 159–241, pl. 5. 1887.
Twitty, V. C., Two new species of *Triturus* from California. *Copeia,* no. 2, pp. 73–80, figs. 1–5. 1935.
——, Data on the life history of *Ambystoma tigrinum californiense* Gray. *Copeia,* no. 1, pp. 1–4. 1941.
——, The species of California *Triturus.—Copeia,* no. 2, pp. 65–76, pls. 1–5. 1942.
Van Denburgh, John, Notes on the habits and distribution of *Autodax iëcanus.—Proc. Cal. Acad. Sci.,* 2nd ser., 5: 776–778. 1895.
——, Herpetological notes. *Proc. Amer. Philos. Soc.,* 37 (157): 139–141. 1898.
——, The reptiles and amphibians of the islands of the Pacific Coast of North America from the Farallons to Cape San Lucas and the Revilla Gigedos. *Proc. Cal. Acad. Sci.,* 3rd ser., 4 (1): 1–34, pls. 1–8. 1905.
——, Four species of Salamanders new to the State of California, with a description of *Plethodon elongatus,* a new species, and notes on other salamanders. *Proc. Cal. Acad. Sci.,* 4th ser., 6 (7): 215–221. 1916.
——, and J. R. Slevin, Reptiles and amphibians of the islands of the west coast of North America. *Proc. Cal. Acad. Sci.,* 4th ser., 4: 129–152. 1914.
Wolterstorff, Willy, Ueber eine eigentümliche Form des kalifornischen Wassermolches, *Taricha torosa* (Rathke). *Blätter für Aquarien.- und Terrarienkunde,* 46 (8): 178–184, 1 fig. 1935.
Wood, W. F., Notes on the salamander, *Plethodon elongatus.—Copeia,* no. 4, p. 191. 1934.
——, *Aneides flavipunctatus* in burnt-over areas. *Copeia,* no. 3, p. 171. 1936.
——, Amphibian records from northwestern California. *Copeia,* no. 2, p. 110. 1939.
——, A new race of salamander, *Ensatina eschscholtzii picta,* from northern California and southern Oregon. *Univ. Cal. Pub. Zool.,* 42 (10): 425–428. 1940.
Anonymous, *Autodax iëcanus.—Amer. Nat.,* 30 (352): 325. 1896.

## COLORADO

Burt, C. E., Amphibians from the Great Basin of the West and adjacent areas (1932). *Amer. Midland Nat.,* 14 (4): 350–354. 1933.

Cary, Merritt, A biological survey of Colorado. *North American Fauna,* U. S. Dept. Agr., no. 33, pp. 1–256, 17 pls., 39 figs., map. 1911.

Cockerell, T. D. A., Reptiles and amphibians of the University of Colorado expedition of 1909. *Univ. Colo. Studies,* 7: 130–131. 1910.

——, Zoology of Colorado. *Univ. Colo. Semicentennial Series,* 3: i–vii, 1–262, 50 figs., many pls. 1927.

Ellis, M. M., and Junius Henderson, The Amphibia and Reptilia of Colorado, part i. *Univ. Colo. Studies,* 10: 39–129, pls. 1–8. 1913.

——, The Amphibia and Reptilia of Colorado, part ii. *Univ. Colo. Studies,* 11 (4): 253–263, pls. 1–2. 1915.

Elrod, M. J., Among the Rockies. *The Museum,* 1 (9): 261–266. Albion, N. Y., 1895.

Gadow, Hans, *see* Baja California. 1905.

Johnson, R. D. O., Reversion of *Amblystoma.—Sci.,* n.s., 36 (931): 594–595. 1912.

Osborn, H. L., On some points in the anatomy of a collection of axolotls from Colorado, and a specimen from North Dakota. *Amer. Nat.,* 35 (419): 887–904, figs. 1–6. 1901.

Prosser, D. T., Habits of "Amblystoma tigrinum" at Tolland, Colorado. *Univ. Colo. Studies,* 8 (4): 257–263, figs. 1–3. 1911.

Yarrow, H. C., Report upon the collections of batrachians and reptiles made in portions of Nevada, Utah, California, Colorado, New Mexico, and Arizona during the years 1871, 1872, 1873, and 1874. *Rept. Geog. and Geol. Exp. and Surv. West 100th Meridian in charge of G. M. Wheeler,* 5: 509–584, pls. 19, fig. 1; 20, fig. 1; 23, fig. 1; 24, figs. 1–2; 25, figs. 1–8; 26, figs. 1–3. 1875.

## COLUMBIA, DISTRICT OF

Hay, W. P., A list of the batrachians and reptiles of the District of Columbia and vicinity. *Proc. Biol. Soc. Wash.,* 15: 121–145, figs. 1–3. 1902.

McAtee, W. L., A sketch of the natural history of the District of Columbia. *Bul. Biol. Soc. Wash.,* no. 1, pp. 1–142, maps. 1918.

Shufeldt, R. W., The spotted salamander. *Aquatic Life,* 1 (1): 13–14, 2 figs. 1915.

——, The gray and the dusky salamanders. *Aquatic Life,* 1 (5): 69–71, 2 figs. 1916.

Stejneger, Leonhard, A salamander new to the District of Columbia. *Proc. Biol. Soc. Wash.,* 15: 239–240. 1902.

True, F. W., [Note on *Siren lacertina.*]—*Proc. Biol. Soc. Wash.,* 1: 36. 1882.

## CONNECTICUT

Allen, J. A., *see* Massachusetts. 1869.

Babbitt, L. H., Some remarks on Connecticut herpetology. *Bul. Bost. Soc. Nat. Hist.,* no. 63, pp. 23–28. 1932.

——, The Amphibia of Connecticut. *Conn. State Geol. and Nat. Hist. Survey,* bul. 57, pp. 1–50, pls. 1–20, figs. 1–4. 1937.

Bumpus, H. C., *see* Rhode Island. 1886.

Doolittle, A. A., The plankton environment in the Connecticut lakes. *U. S. Bu. Fish.*, doc. 633, pp. 65–77, map. 1908.
Hausman, S. A., The spotted newt. *Turtox News,* 14 (9): 94. 1936.
Hay, O. P., *see* Indiana. 1892.
Kendall, W. C., and E. L. Goldsborough, The fishes of the Connecticut lakes and neighboring waters. *U. S. Bu. Fish.,* doc. 633, p. 54. 1908.
Linsley, J. H., A catalogue of the reptiles of Connecticut, arranged according to their natural families. *Amer. Jour. Sci. and Arts,* 46: 37–51. 1844.
Noble, G. K., *see* New York. 1926.
Anonymous, Salamander found two feet underground at Shippan. *Guide to Nature,* 19 (12): 190. 1927.

**DELAWARE**

Fowler, H. W., Records of amphibians and reptiles for Delaware, Maryland and Virginia. I. Delaware. *Copeia,* no. 145, pp. 57–61. 1925.
Higbee, Edna, Experimental modification of the secondary sexual characters of the red-spotted newt, *Triturus viridescens.—Proc. Penn. Acad. Sci.,* 8: 133–139. 1934.
Stone, Witmer, *see* Pennsylvania. 1906.

**FLORIDA**

Barbour, Thomas, A red eft from Florida. *Copeia,* no. 3, p. 175. 1939.
Brimley, C. S., *see* Alabama. 1910.
Carr, A. F., Jr., A contribution to the herpetology of Florida. *Univ. Florida, Biol. Science Ser.,* 3 (1): 1–118. 1940.
Cope, E. D., Catalogue of the Reptilia and Batrachia obtained by C. J. Maynard in Florida. *Third Ann. Rept. Peabody Acad. Sci.,* pp. 82–85. 1871.
——, *see* North Carolina. 1877.
Dunn, E. R., *Siren,* a herbivorous salamander? *Science,* 59 (1519): 145. 1924.
Dury, Ralph, *see* Indiana. 1932.
Fowler, H. W., Cold-blooded vertebrates from Florida, the West Indies, Costa Rica, and eastern Brazil. *Proc. Acad. Nat. Sci. Phila.,* 67: 244–269, figs. 1–4. 1915.
Goin, C. J., *see* Virginia. 1938.
——, Notes on *Pseudotriton ruber vioscai* Bishop. *Copeia,* no. 4, p. 231. 1939.
Harlan, Richard, Further observations on the *Amphiuma means.—Ann. Lyceum Nat. Hist. N. Y.,* 1: 269–270, pl. 22. 1825.
——, Note on the *Amphiuma means,* described in vol. iii of this journal. *Jour. Acad. Nat. Sci. Phila.,* 6 (1): 147–148. 1829.
Harper, Francis, *see* Georgia. 1935.
Holbrook, J. E., *see above,* General works. 1842.
Lönnberg, Einar, Notes on reptiles and batrachians collected in Florida in 1892 and 1893. *Proc. U. S. Nat. Mus. 1894,* 17: 317–339, figs. 1–3. 1894.
——, Notes on tailed batrachians without lungs. *Zool. Anz.,* 19 (494): 33–37. 1896.

Netting, M. G., and C. J. Goin, Descriptions of two new salamanders from peninsular Florida. *Ann. Carnegie Mus.*, 29: 175–196, pl. 1. 1942.
Noble, G. K., and L. B. Richards, *see* North Carolina. 1932.
Van Hyning, O. C., Food of some Florida snakes. *Copeia,* no. 1, p. 37. 1932.
——, Batrachia and Reptilia of Alachua County, Florida. *Copeia,* no. 1, pp. 3–7. 1933.

## GEORGIA

Allard, H. A., Notes on some salamanders and lizards of north Georgia. *Science,* n.s., 30 (760): 122–124. 1909.
Bailey, J. R., Notes on plethodont salamanders of the southeastern United States. *Occ. Papers Mus. Zool. Univ. Mich.*, no. 364, pp. 1–10. 1937.
Bishop, S. C., *see* North Carolina. 1928.
——, *see above,* General works. 1941.
Brimley, C. S., *see* Alabama. 1910.
Burnett, W. I., [List of Reptilia of Aiken, S. C., region]. *Proc. Bost. Soc. Nat. Hist.,* 4: 146–147. 1852.
Burt, C. E., *see* New York. 1932.
Carr, A. F., *Haideotriton wallacei,* a new subterranean salamander from Georgia. *Occ. Papers Bost. Soc. Nat. Hist.,* 8: 333–336, pl. 11–12. 1939.
Dunn, E. R., *see* North Carolina. 1916.
Fountain, Paul, *The great deserts and forests of North America,* pp. i–ix, 1–295. London: Longmans, Green & Co., 1901.
Goin, C. J., *see* Virginia. 1938.
Hallowell, Edward, Description of two new species of urodeles, from Georgia. *Proc. Acad. Nat. Sci. Phila. for 1856,* 8: 130–131. 1856.
Harlan, Richard, Dissection of a batracian animal in the living state. *Jour. Acad. Nat. Sci. Phila.,* 3 (1): 54–59, figs. 1–5. 1823.
Harper, Francis, Notes on fishes, amphibians and reptiles of Randolph County, Georgia. *Copeia,* no. 4, pp. 152–154. 1930.
——, Records of amphibians in the southeastern states. *Amer. Mid. Nat.,* 16 (3): 275–310, figs. 1–17. 1935.
Holbrook, J. E., *see above,* General works. 1836–1840.
——, *see above,* General works. 1842.
——, Reptiles. In George White, *Statistics of the State of Georgia,* appendix, pp. 13–15. 1849.
Howell, A. H., Notes on the summer birds of north Georgia. *Auk,* n.s., 26 (2): 129–137. 1909.
Le Conte, John, Description of a new species of *Siren,* with some observations on animals of a similar nature. *Ann. Lyceum Nat. Hist. N. Y.,* 1 (1): 52–58, pl. 4. 1824. [*See* F. Harper, *Amer. Mid. Nat.,* 16 (3): 279. 1935.]
——, Description of a new species of *Siren.—Ann. Lyceum Nat. Hist. N. Y.,* 2: 133–134, pl. 1. 1828.

Lönnberg, Einar, *see* Florida. 1896.
Mitchill, S. L., Description of a batracian animal from Georgia, different from the reptiles of that order known. *Amer. Med. Recorder,* 5: 499–503. 1822.
Mohr, C. E., *see* Pennsylvania. 1937.
Neill, W. T., A collection of salamanders from Georgia. *Copeia,* no. 3, p. 177. 1941.
Noble, G. K., The eggs of *Pseudobranchus.—Copeia,* no. 2, p. 52. 1930.
——, and B. C. Marshall, *see* Arkansas. 1932.
Viosca, Percy, Jr., *see* Louisiana. 1937.

## IDAHO

Cope, E. D., *see* Montana. 1879.
Erwin, R. P., List of Idaho reptiles and amphibians in the Idaho state historical museum. *11th Bien. Rept. State Historical Soc. Idaho,* pp. 31–33. 1928.
Slater, J. R., *see* Washington. 1937.
——, The distribution of amphibians and reptiles in Idaho. *Occ. Papers Dept. Biol. College Puget Sound,* no. 14, pp. 78–109. 1941.
——, and J. W. Slipp, A new species of *Plethodon* from northern Idaho. *Occ. Papers Dept. Zool. College Puget Sound,* no. 8, pp. 38–43, figs. 1–3. 1940.
——, The Pacific giant salamander in Idaho. *Occ. Papers Dept. Zool. College Puget Sound,* no. 11, p. 69. 1940 a.
Stejneger, Leonhard, Annotated list of reptiles and batrachians collected by Dr. C. Hart Merriam and party in Idaho, 1890. *North Amer. Fauna,* no. 5, pp. 109–117. 1891.
Tanner, W. W., The reptiles and amphibians of Idaho, no. 1. *Great Basin Nat.,* 2 (2): 87–97. 1941.
Van Denburgh, John, and J. R. Slevin, A list of the amphibians and reptiles of Idaho, with notes on the species in the collection of the Academy. *Proc. Calif. Acad. Sci.,* 4th ser., 11 (3): 39–47. 1921.

## ILLINOIS

Blanchard, F. N., A collection of amphibians and reptiles from southeastern Missouri and southern Illinois. *Papers of the Michigan Acad. Sci., Arts and Letters,* 4: 533–541. 1924.
Burt, C. E., *see* Arkansas. 1935.
Cagle, F. R., Key to the reptiles and amphibians of Illinois. *Contr. No. 5, Mus. Nat. and Social Sci., Southern Ill. Normal Univ.,* pp. i–iv, 1–32, pls. 1–3, figs. 1–66. 1941.
——, Herpetological fauna of Jackson and Union Counties, Illinois. *Amer. Mid. Nat.,* 28 (1): 164–200. 1942.
Cagle, F. R., and P. E. Smith, A winter aggregation of *Siren intermedia* and *Triturus viridescens.—Copeia,* no. 4, pp. 232–233, fig. 1. 1939.
Cahn, A. R., A set of albino eggs of *Ambystoma microstomum.—Copeia,* no. 1, pp. 18–19. 1930.

Cockrum, Lendell, Notes on *Siren intermedia.—Copeia,* No. 4, p. 265. 1941.
Davis, N. S., and F. L. Rice, List of Batrachia and Reptilia of Illinois. *Bul. Chicago Acad. Sci.,* 1 (3): 23–32. 1883.
Gaige, H. T., A list of the amphibians and reptiles observed in Richland County, Illinois, in May, 1913. *Copeia,* no. 11, p. 4. 1914.
Garman, Harrison, Notes on Illinois reptiles and amphibians, including several species not before recorded from the Northern States. *Bul. Ill. State Lab. Nat. Hist.,* 3 (art. 10): 185–190. 1890.
——, A synopsis of the reptiles and amphibians of Illinois. *Bul. Ill. State Lab. Nat. Hist.,* 3 (art. 13): 215–385, pls. 9–15. 1892.
Goodnight, C. J., A key to the adult salamanders of Illinois. *Trans. Ill. State Acad. Sci.,* 30 (2); 300–302. 1937.
Hankinson, T. L., The vertebrate life of certain prairie and forest regions near Charleston, Illinois. *Bul. Ill. Lab. Nat. Hist.,* 11: 281–301, 16 pls. 1915.
——, Amphibians and reptiles of the Charleston region. *Trans. Ill. Acad. Sci.,* 10: 322–330. 1917.
Hay, O. P., *see* Indiana. 1892.
Kennicott, Robert, Catalogue of animals observed in Cook County, Illinois. *Trans. Ill. State Agr. Soc.,* 1: 577–595. 1855.
Milner, J. W., *see* Michigan. 1874.
Myers, G. S., *see* Indiana. 1926.
Necker, W. L., Check list of reptiles and amphibians of the Chicago region. *Chi. Acad. Sci., Leaflet No. 1,* pp. 1–4, map. 1938.
——, Records of amphibians and reptiles of the Chicago region, 1935–1938. *Bul. Chi. Acad. Sci.,* 6 (1): 1–10. 1939.
Noble, G. K., and B. C. Marshall, *see* Arkansas. 1932.
O'Donnell, D. J., Natural history of the ambystomid salamanders of Illinois. *Amer. Mid. Nat.,* 18 (6): 1063–1071. 1937.
Owens, D. W. Some amphibians and reptiles from southern Illinois. *Copeia,* no. 3, p. 183. 1941.
Pearse, A. S., *see* Wisconsin. 1921.
Schmidt, K. P., The salamanders of the Chicago area. *Field Mus. Nat. Hist. Zool., Leaflet 12,* pp. 1–16, pls. 1–4. 1930.
——, and W. L. Necker, Amphibians and reptiles of the Chicago region. *Bul. Chi. Acad. Sci.,* 5 (4): 57–77, map. 1935.
Smith, B. G., *see* Wisconsin. 1911.
Stein, H. A., *Ambystoma talpoideum* (Gray) in Illinois. *Trans. Ill. Acad. Sci.,* 26 (3): 135. 1934.
Weed, A. C., Notes on reptiles and batrachians of central Illinois. *Copeia,* no. 116, pp. 45–50. 1923.

**INDIANA**

Allyn, W. P., Representative animal life of Indiana. *Teachers College Jour., Terre Haute,* 6 (2): 95–125, figs. 1–40. 1934.

Allyn, W. P., and Clarence Shockley, A preliminary survey of the surviving species of Caudata of Vigo County and vicinity. *Proc. Ind. Acad. Sci.,* 48: 238–243, figs. 1–8. 1939.
Atkinson, Curtis, Batrachia. In First rept. Ind. Univ. Biol. Sta., Turkey Lake, Indiana. *Proc. Ind. Acad. Sci.,* 5: 258–261. 1896.
Banta, A. M., The fauna of Mayfield's Cave. *Carnegie Inst. Wash.,* pub. 67, pp. 1–114, pls. 1–2, figs. 1–13. 1907.
——, and W. L. McAtee, The life history of the cave salamander, *Spelerpes maculicaudus* (Cope). *Proc. U. S. Nat. Mus.,* 30: 67–73, fig. 1, 1906.
Blanchard, F. N., A collection of amphibians and reptiles from southern Indiana and adjacent Kentucky. *Papers Mich. Acad. Sci., Arts and Letters,* 5: 367–388, pls. 22–23. 1926.
Blatchley, W. S., Notes on the batrachians and reptiles of Vigo County, Indiana. *Jour. Cin. Soc. Nat. Hist.,* 14 (1): 22–35. 1891.
——, Indiana caves and their fauna. *Ind. Dept. Geol. and Nat. Resources, 21st Ann. Rept.,* pp. 121–212, pls. 4–13, figs. 1897.
——, Notes on the batrachians and reptiles of Vigo County, Indiana (II). *Ind. Dept. Geol. and Nat. Resources, 24th Ann. Rept.,* pp. 537–552. 1900.
——, On some reptilian freaks from Indiana. *Proc. Acad. Nat. Sci. Phila.,* 58: 419–422. 1906.
Brimley, C. S., Batrachia of Vincennes, Indiana. *Amer. Nat.,* 29 (337): 53–56. 1895.
Butler, A. W., Herpetology. *Jour. Cin. Soc. Nat. Hist.,* 9 (4): 263–265. 1887.
——, Some notes on Indiana amphibians and reptiles. No. 2. *Jour. Cin. Soc. Nat. Hist.,* 10 (3): 147–148. 1887.
——, Contributions to Indiana herpetology. No. 3. *Jour. Cin. Soc. Nat. Hist.,* 14 (4): 169–179. 1892.
Cope, E. D., On a new species of salamander from Indiana. *Amer. Nat.,* 24 (286): 966–967, figs. 1–7. 1890.
Dury, Ralph, Recent acquisitions to the department of herpetology. *Proc. Junior Soc. Nat. Sci.,* 3 (2): 26–28. 1932.
——, Notes on reptiles and amphibians from Clifty Falls State Park, Jefferson County, Indiana. *Proc. Junior Soc. Nat. Sci.,* 3 (2): 23–26. 1932 a.
Eigenmann, C. H., Description of a new cave salamander, *Spelerpes stejnegeri,* from the caves of southwestern Missouri. *Trans. Am. Micros. Soc.,* 22: 189–192, pls. 27–28. 1901.
Evermann, B. W., and H. W. Clark, The turtles and batrachians of the Lake Maxinkuckee region. *Proc. Ind. Acad. Sci.,* 26: 472–518. 1917.
——, and H. W. Clark, Lake Maxinkuckee, a physical and biological survey. *Dept. Cons. Ind., Pub. No. 7,* 1: 1–660, pls. 1–38, 23 figs., map. 1920.
Gaines, Angus, Batrachia of Vincennes, Indiana. *Amer. Nat.,* 29: 53–56. 1895.
Grant, Chapman, Herpetological notes from northern Indiana. *Proc. Ind. Acad. Sci.,* 45: 323–333. 1936.
Grave, B. H., The Amphibia of Montgomery County, Indiana. *Proc. Ind. Acad. Sci.,* 40: 339. 1931.

Hahn, W. L., Notes on the mammals and cold-blooded vertebrates of the Indiana University farm, Mitchell, Indiana. *Proc. U. S. Nat. Mus.*, 35: 545–581. 1908.

Hay, O. P., Description of a new species of *Amblystoma* (*Amblystoma copeianum*) from Indiana. *Proc. U. S. Nat. Mus.*, 8 (14): 209–213, pl. 14. 1885.

——, A preliminary catalogue of the Amphibia and Reptilia of the State of Indiana. *Jour. Cin. Soc. Nat. Hist.*, 10: 59–69. 1887.

——, Notes on the habits of some *Amblystomas*. *Amer. Nat.*, 23 (271): 602–612. 1889.

——, Note on *Gyrinophilus maculicaudus* Cope. *Amer. Nat.*, 25 (300): 1133–1135. 1891.

——, The batrachians and the reptiles of Indiana. *Ind. Dept. Geol. and Nat. Res., 17th Ann. Rept.*, pp. 409–602, pls. 1–3. 1892.

Hughes, Edward, Preliminary list of reptiles and batrachians of Franklin County. *Brookville Soc. Nat. Hist.*, bul. 2, pp. 40–45. 1886.

Kirsch, P. H., A report upon explorations made in Eel River basin in the northeastern part of Indiana in the summer of 1892. *Bul. U. S. Fish. Comm.*, 14 (6): 31–41. 1895.

McAtee, W. L., Development of the color pattern in the larvae of *Spelerpes maculicaudus*.—*Proc. U. S. Nat. Mus.*, 30 (1443): 74–83, pls. 8–10, figs. 2–3. 1906.

——, A list of the mammals, reptiles and batrachians of Monroe County, Indiana. *Proc. Biol. Soc. Wash.*, 20: 1–16. 1907.

——, Notes on the banded salamander (*Ambystoma opacum*).—*Copeia*, no. 4, pp. 218–219. 1933.

Myers, G. S., A synopsis for the identification of the amphibians and reptiles of Indiana. *Proc. Ind. Acad. Sci.*, 35: 277–294, fig. 1. 1926.

——, Notes on Indiana amphibians and reptiles. *Proc. Ind. Acad. Sci.*, 36: 337–340. 1927.

Necker, W. L., *see* Illinois. 1939.

Ortenburger, A. I., A list of Amphibia and Reptilia collected in Indiana. *Copeia*, no. 99, pp. 73–76. 1921.

Piatt, Jean, Herpetological report of Morgan County, Indiana. *Proc. Ind. Acad. Sci.*, 40: 361–368. 1931.

——, An albino salamander. *Copeia*, no. 1, p. 29. 1931 a.

Quick, E. R., Zoological miscellany, Herpetology. *Jour. Cin. Soc. Nat. Hist.*, 5 (2): 89–96. 1882.

Ramsey, E. R., The cold-blooded vertebrates of Winona Lake and vicinity. *Proc. Ind. Acad. Sci. for 1900*, pp. 218–224. 1901.

Reynolds, A. E., Additional observations on the salamanders of Putnam County and vicinity. *Proc. Ind. Acad. Sci.*, 46: 225–229, fig. 1. 1937.

——, and E. G. Black, The salamanders of Putnam County (Indiana). *Proc. Ind. Acad. Sci.*, 45: 287–294. 1936.

Schmidt, K. P., *see* Illinois. 1930.

——, and W. L. Necker, *see* Illinois. 1935.

Springer, Stewart, A list of reptiles and amphibians taken in Marion County, Indiana, in 1924–1927. *Proc. Ind. Acad. Sci.*, 37: 491–492. 1928.

Swanson, P. L., Herpetological notes from Indiana. *Amer. Mid. Nat.*, 22 (3): 684–695, pls. 1–3, map. 1939.

### IOWA

Blanchard, F. N., The amphibians and reptiles of Dickinson County, Iowa. *Univ. Iowa Studies in Nat. Hist.*, 10 (2): 19–26. 1923.

Künze, Albert, and José Zozaya, The feeding reactions of *Ambystoma tigrinum* (Green). *Univ. Iowa Studies in Nat. Hist.*, 10 (2): 51–60. 1923.

McMullen, D. B., and R. L. Roudabush, A new species of trematode, *Cercorchis cryptobranchi,* from *Cryptobranchus alleganiensis.—Jour. Parasit.*, 22 (5): 516–517, fig. 1. 1936.

Osborn, Herbert, A partial catalogue of the animals of Iowa. In *Coll. Iowa Agr. College,* pp. 1–39. Ames, Iowa, 1892.

Ruthven, A. G., Contributions to the herpetology of Iowa. *Proc. Iowa Acad. Sci.*, 17: 198–209, figs. 1–11. 1910.

——, Contributions to the herpetology of Iowa (II). *Proc. Iowa Acad. Sci.*, 19: 207. 1912.

——, Description of a new salamander from Iowa. *Proc. U. S. Nat. Mus.*, 41: 517–519. 1912 a.

——, Contributions to the herpetology of Iowa (III). *Occ. Papers Mus. Zool. Univ. Mich.*, no. 66, pp. 1–3. 1919.

### KANSAS

Brennan, L. A., A check list of the amphibians and reptiles of Ellis County, Kansas. *Trans. Kan. Acad. Sci.*, 37: 189–191. 1934.

——, A study of the habitat of the reptiles and amphibians of Ellis County, Kansas. *Trans. Kan. Acad. Sci.*, 40: 341–347. 1938.

Breukelman, John, and Allen Downs, A list of Amphibia and reptiles of Chase and Lyon Counties, Kansas. *Trans. Kan. Acad. Sci.*, 39: 267–268. 1936.

Burt, C. E., An annotated list of the amphibians and reptiles of Riley County, Kansas. *Occ. Papers Mus. Zool. Univ. Mich.*, no. 189, pp. 1–9. 1927.

——, *see* New York. 1932.

——, *see* Arkansas. 1935.

Cragin, F. W., A preliminary catalogue of Kansas reptiles and batrachians. *Trans. Kans. Acad. Sci. for 1879–1880*, 7: 114–123. 1881.

——, Second contribution to the herpetology of Kansas, with observations on the Kansas fauna. *Trans. Kans. Acad. Sci. for 1883–1884*, 9: 136–140. 1885.

——, Recent additions to the list of Kansas reptiles and batrachians, with further notes on species previously reported. *Bul. Washburn College Lab.*, 1 (3): 100–103. 1885 a.

Gloyd, H. K., The amphibians and reptiles of Franklin County, Kansas. *Trans. Kans. Acad. Sci.*, 31: 115–141. 1928.

——, The herpetological fauna of the Pigeon Lake region, Miami County, Kansas. *Papers Mich. Acad. Sci., Arts and Letters,* 15: 389–409, pls. 30–32, map. 1932.
Hallowell, Edward, Notice of a collection of reptiles from Kansas and Nebraska, presented to the Academy of Natural Sciences, by Dr. Hammond, U.S.A. *Proc. Acad. Nat. Sci. Phila. for 1856,* 8: 238–253. 1856.
——, Note on the collection of reptiles from the neighborhood of San Antonio, Texas, recently presented to the Academy of Natural Sciences by Dr. A. Heermann. *Proc. Acad. Nat. Sci. Phila. for 1856,* 8: 306–310. 1856 a.
Hartman, F. A., Food habits of Kansas lizards and batrachians. *Trans. Kan. Acad. Sci.,* 20 (2): 225–229. 1907.
Smith, Hobart M., A report upon amphibians hitherto unknown from Kansas. *Trans. Kans. Acad. Sci.,* 35: 93–96. 1932.
——, The amphibians of Kansas. *Amer. Mid. Nat.,* 15 (4): 377–528, pls. 12–20, maps 1–24. 1934.
Taylor, E. H., List of reptiles and batrachians of Morton County, Kansas, reporting species new to the fauna. *Univ. Kan. Sci. Bul.,* 19 (6): 63–65. 1929.
Tihen, J. A., Additional distributional records of amphibians and reptiles in Kansas counties. *Trans. Kan. Acad. Sci.,* 40: 401–409. 1937.

### KENTUCKY

Bailey, Vernon, Cave life in Kentucky. *Amer. Mid. Nat.,* 14 (5): 385–635, figs. 1–90. 1933.
Banta, A. M., and W. L. Mc Atee, *see* Indiana. 1906.
Bishop, S. C., Records of some amphibians and reptiles from Kentucky. *Copeia,* no. 152, pp. 118–120. 1926.
——, *see above,* General works. 1941.
Blanchard, F. N., *see* Indiana. 1926.
Burt, C. E., A contribution to the herpetology of Kentucky. *Amer. Mid. Nat.,* 14 (6): 669–679. 1933.
——, and M. D. Burt, A collection of amphibians and reptiles from the Mississippi Valley, with field observations. *Amer. Mus. Novitates,* no. 381, pp. 1–14. 1929.
Cope, E. D., *see above,* General works. 1889.
Dury, Ralph, *see* Indiana. 1932.
Dury, Ralph, and William Gessing, Additions to the herpetofauna of Kentucky. *Herpetologica,* 2 (2): 31–32. 1940.
Dury, Ralph, and R. S. Williams, Notes on some Kentucky amphibians and reptiles. *Baker-Hunt Foundation Mus.,* bul. 1, pp. 1–22. 1933.
Eigenmann, C. H., Explorations in the caves of Missouri and Kentucky. *Proc. Ind. Acad. Sci.,* 8: 58–61. 1899.
Garman, Harrison, A preliminary list of the vertebrate animals of Kentucky. *Bul. Essex Inst.,* 26 (1): 1–63. 1894.
Grant, Chapman, Herpetological notes from Fort Knox, Kentucky. *Proc. Ind. Acad. Sci.,* 45: 334. 1936.
Green, N. B., The four-toed salamander in Kentucky. *Copeia,* no. 1, p. 53. 1941.

Hibbard, C. W., The amphibians and reptiles of Mammoth Cave National Park proposed. *Trans. Kan. Acad. Sci.,* 39: 277–281. 1936.
Matthes, Benno, *see above,* General works. 1855.
Mohr, C. E., *see* Pennsylvania. 1937.
Netting, M. G., *see* Pennsylvania. 1939.
Packard, A. S., The cave fauna of North America, with remarks on the anatomy of the brain and origin of the blind species. *Mem. Nat. Acad. Sci.,* 4 (1): 1–156, pls. 1–27, figs. 1–21, map. 1888.
Pope, C. H., *see* North Carolina. 1928.
Rafinesque, C. S., *see* New York. 1820.
——, *Kentucky Gazette,* n.s., 1 (9): 3. Lexington, 1822.
——, Description of the *Spelerpes* or salamander of the caves of Kentucky. *Atlantic Jour.,* 1 (1): 22. 1832.
——, The caves of Kentucky. *Atlantic Jour.,* 1 (1): 27–30. 1832 a.
——, On the salamander of the hills of east Kentucky. *Atlantic Jour.,* 1 (2): 63–64. 1832 b.
——, On three new salamanders of Kentucky. *Atlantic Jour.,* 1 (3): 121. 1832 c.
Walker, C. F., and W. H. Weller, The identity and status of *Pseudotriton duryi.* —*Copeia,* no. 2, pp. 81–83. 1932.
Weller, W. H., Notes on amphibians collected in Carter County, Kentucky. *Proc. Junior Soc. Nat. Sci.* [Cincinnati], 1 (5–6): 6–9. 1930.
——, Corrections to herpetological notices. *Proc. Junior Soc. Nat. Sci.,* 2 (1): 7–9. 1931.
Welter, W. A., and R. W. Barbour, Additions to the herpetofauna of northeastern Kentucky. *Copeia,* no. 2, pp. 132–133. 1940.
——, and Katherine Carr, Amphibians and reptiles of northeastern Kentucky. *Copeia,* no. 3, pp. 128–130. 1939.

## LOUISIANA

Beyer, G. E., Louisiana herpetology. *Proc. La. Soc. Nat.,* Appendix 1, pp. 25–46. 1900.
Bishop, S. C., A new subspecies of the red salamander from Louisiana. *Occ. Papers Bost. Soc. Nat. Hist.,* 5: 247–249, pl. 15. 1928.
Burt, C. E., *see* Arkansas. 1925.
Chandler, A. C., Three new trematodes from *Amphiuma means.*—*Proc. U. S. Nat. Mus.,* 63 (3): 1–7, pls. 1–2. 1923.
Goin, C. J., *see* Virginia. 1938.
Green, Jacob, Description of two new species of salamander. *Jour. Acad. Nat. Sci. Phila.,* 6 (2): 253–254. 1831.
Hargitt, C. W., On some habits of *Amphiuma means.*—*Sci.,* 20 (502): 159–160. 1892.
Le Conte, John, Notice of American animals, formerly known but now forgotten or lost. *Proc. Acad. Nat. Sci. Phila. for 1854,* 7: 8–14. 1856.
Meade, G. P., Feeding *Farancia abacura* in captivity. *Copeia,* no. 2, pp. 91–92. 1934.

Osborn, H. F., Preliminary observations upon the brain of *Amphiuma.—Proc. Acad. Nat. Sci. Phila.,* 35: 177. 1883.
Schmidt, K. P., On the common name of *Amphiuma.—Copeia,* no. 82, pp. 41–42. 1920.
Strecker, J. K., Reptile myths in northwestern Louisiana. Snakes with legs. *Pub. Texas Folk-Lore Soc.,* no. 4, p. 50. 1925.
——, and L. S. Frierson, Jr., The herpetology of Caddo and De Soto Parishes, Louisiana. *Contr. Baylor Univ. Mus.,* no. 5, pp. 1–10. 1926.
——, and L. S. Frierson, Jr., The herpetology of Caddo and De Soto Parishes, Louisiana. *Baylor Bul.,* 38 (3): 33–34. 1935.
Viosca, Percy, Jr., An ecological study of the cold blooded vertebrates of southeastern Louisiana. *Copeia,* no. 115, pp. 35–44. 1923.
——, Observations on the life history of *Ambystoma opacum.—Copeia,* no. 134, pp. 86–88. 1924.
——, A terrestrial form of *Siren lacertina.—Copeia,* no. 136, pp. 102–104. 1925.
——, A tentative revision of the genus *Necturus,* with descriptions of three new species from the southern Gulf drainage area. *Copeia,* no. 2, pp. 120–138, figs. 1–7. 1937.
——, A new waterdog from central Louisiana. *Proc. Biol. Soc. Wash.,* 51: 143–145, pls. 1–2, 1938.
Wolterstorff, Willy, Ueber *Diemyctylus viridescens* Raf. subsp. *louisianensis* n. subsp. *Abhandl. u. Ber. Mus. f. Natur- und Heimatkunde u. Naturwiss. Ver. Magdeburg,* 2 (4): 383–392, pl. 8. 1914.
Zeliff, C. C., A new species of Cestode, *Crepidobothrium amphiumae* from *Amphiuma tridactylum.—Proc. U. S. Nat. Mus.,* 81 (3): 1–3, pl. 1. 1932.

## MAINE

Bishop, S. C., and N. T. Clarke, A scientific survey of Turners Lake, Isle-au-Haut, Maine. *N. Y. State Mus.,* pp. 1–29, 20 pls., 2 maps. Privately printed, 1923.
Boardman, G. A., List of Urodella [*sic*] [of Maine]. In S. L. Boardman, *The Naturalist of the Saint Croix,* pp. i–xi, 3–351. Bangor, Maine. (This list is said to have been reprinted from the *Calais Weekly Times.*)
Burt, C. E., *see* New York. 1932.
Dury, Ralph, *see* Indiana. 1932.
Fogg, B. F., List of reptiles and amphibians found in the State of Maine. *Proc. Portland Soc. of Nat. Hist.,* 1 (1): 86. 1862.
——, List of reptiles and amphibians found in the State of Maine. *Maine Board Agr., 7th Ann. Rept. Sec'y,* pp. 141–142. 1862 a.
Fowler, J. A., In the field. *New Eng. Nat.,* 2 (13): 24. 1941.
Manville, R. H., Notes on the herpetology of Mount Desert Island, Maine. *Copeia,* no. 3, p. 174. 1939.
Noble, G. K., *see* Massachusetts. 1929.
Norton, A. H., Notes on the history of herpetology in Maine. *Maine Nat.,* 9 (2): 53–63. 1929.

Pope, P. H., Some new records for *Gyrinophilus porphyriticus* (Green). *Copeia,* no. 19, pp. 14–15. 1915.
——, Winter activity of *Amblystoma punctatum* Baird. *Copeia,* no. 30, p. 35. 1916.
——, Some doubtful points in the life-history of *Notophthalmus viridescens.—Copeia,* no. 91, pp. 14–15. 1921.
Storer, H. R., [Memoranda on certain North American reptiles]. *Proc. Bost. Soc. Nat. Hist.,* 4: 137–138. 1852.
Verrill, A. E., Catalogue of the reptiles and batrachians found in the vicinity of Norway, Oxford County, Maine. *Proc. Bost. Soc. Nat. Hist.,* 9: 195–199. 1863.
——, Notice of the eggs and young of a salamander *Desmognathus fusca* Baird, from Maine. *Proc. Bost. Soc. Nat. Hist.,* 9: 253–255. [Described eggs of *Eurycea bislineata,* NOT *Desmognathus fuscus.*] 1863 a.
——, Breeding habits of salamanders and frogs. *Amer. Nat.,* 3 (3): 157–158. 1869.

## MARYLAND

Andrews, E. A., Breeding habits of the spotted salamander. *Amer. Nat.,* 31: 635–637. 1897.
Brady, M. K., Natural history of Plummers Island, Maryland. VI. Reptiles and amphibians. *Proc. Biol. Soc. Wash.,* 50: 137–139. 1937.
Davis, W. T., Report of the section of biology. *Proc. Staten Island Assn. Arts and Sci.,* 4 (1–2): 82–85. 1912.
Fowler, J. A., The occurrence of *Pseudotriton montanus montanus* in Maryland. *Copeia,* no. 3, p. 181. 1941.
Fowler, H. W., Some amphibians and reptiles of Cecil County, Maryland. *Copeia,* no. 22, pp. 37–40. 1915.
——, Records of amphibians and reptiles for Delaware, Maryland and Virginia. II. Maryland. *Copeia,* no. 145, pp. 61–64. 1925.
Keim, T. D., Amphibians and reptiles at Jennings, Maryland. *Copeia,* no. 2, p. 2. 1914.
Lynn, W. G., Some salamanders of the Baltimore region. The Maryland Naturalist. *Maryland Acad. Sci.,* 2 (1): 1. 1936.
McCauley, R. H., and C. S. East, Amphibians and reptiles from Garrett County, Maryland. *Copeia,* no. 2, pp. 120–123. 1940.
Mansueti, Romeo, Sounds produced by the slimy salamander. *Copeia,* no. 4, pp. 266–267. 1941.
Moment, Gairdner, Springtime and salamanders. *Scientific Monthly,* 46: 561–564. 1938.
Netting, M. G., *see* Pennsylvania. 1938.
Robertson, H. C., A preliminary report on the reptiles and amphibians of Catoctin National Park, Maryland. *Bul. Nat. Hist. Soc. Maryland,* 9 (10): 88–93, figs. 1–5. 1939.

## MASSACHUSETTS

Allen, J. A., Catalogue of reptiles and batrachians found in the vicinity of Springfield, Massachusetts, with notices of all the other species known to inhabit the state. *Proc. Bost. Soc. Nat. Hist.,* 12: 171–204. 1869.

——, The reptiles and batrachians of Massachusetts. *Proc. Bost. Soc. Nat. Hist.,* 12: 248–250. 1869 a.

——, Notes on Massachusetts reptiles and batrachians. *Proc. Bost. Soc. Nat. Hist.,* 13: 260–263. 1871.

Barbour, Thomas, An unusual red salamander. *Copeia,* no. 10, pp. 3–4. 1914.

——, A salamander new to Cape Cod. *Bul. Bost. Soc. Nat. Hist.,* no. 78, p. 9. 1936.

Bumpus, H. C., *see* Rhode Island. 1886.

Cochran, M. E., The biology of the red-backed salamander (*Plethodon cinereus erythronotus* Green). *Biol. Bul.,* 20 (6): 332–349, 11 figs. 1911.

Davis, W. T., *see* New York. 1933.

Dunn, E. R., *Ambystoma opacum* at Florence, Massachusetts. *Copeia,* no. 70, pp. 51–52. 1919.

——, Reptiles and amphibians of Northampton and vicinity. *Bul. Bost. Soc. Nat. Hist.,* 57: 3–8, 1 fig. 1930.

Dury, Ralph, *see* Indiana. 1932.

Garman, Samuel, Prehensile-tailed salamanders. *Sci.,* 8 (178): 13–14. 1886.

Hitchcock, H. B., Notes on the newt, *Triturus viridescens.—Herpetologica,* 1 (6): 149–150. 1939.

Jordan, E. O., The habits and development of the newt (*Diemyctylus viridescens*). *—Jour. Morph.,* 8 (2): 269–366, pls. 14–18. 1893.

Lynn, W. G., and J. N. Dent, Notes on *Plethodon cinereus* and *Hemidactylium scutatum* on Cape Cod. *Copeia,* no. 2, pp. 113–114. 1941.

Noble, G. K., *see* New York, 1926.

——, Further observations on the life-history of the newt, *Triturus viridescens.—Amer. Mus. Nov.,* no. 348, pp. 1–22, figs. 1–7. 1929.

Putnam, F. W., Red-backed salamander. *Proc. Bost. Soc. Nat. Hist.,* 9: 173–174. 1863.

Smith, D. S. C. H., in Hitchcock, *A Catalogue of the animals and plants in Massachusetts; pt. 4, Rept. on the geology, mineralogy, botany, and zoology of Massachusetts,* pp. 543–652. 1833.

Smith, Louise, Some notes on *Notophthalmus viridescens.—Copeia,* no. 80, pp. 22–24. 1920.

——, A note on the eggs of *Ambystoma maculatum.—Copeia,* no. 97, p. 41. 1921.

Stein, K. F., Migration of *Triturus viridescens.—Copeia,* no. 2, pp. 86–88. 1938.

Storer, D. H., Reptiles of Massachusetts. In *Rept. on the fishes, reptiles and birds of Massachusetts to the legislature of Massachusetts,* pp. 202–253. Boston, 1839.

——, Report on the reptiles of Massachusetts. *Bost. Jour. Nat. Hist.,* 3 (1): 1–64. 1840.

Storer, H. R., *see* Maine. 1852.

Storer, H. R., [Amphibians and reptiles presented to the Society]. *Proc. Bost. Soc. Nat. Hist.,* 4: 149. 1852 a.

Uhlenhuth, E., and Hilda Karns, The morphology and physiology of the salamander thyroid gland. III. The relation of the number of follicles to development and growth of the thyroid in *Ambystoma maculatum. Biol. Bul.,* 54 (2): 128–164. 1928.

Warfel, H. E., Notes on the occurrence of *Necturus maculosus* (Rafinesque) in Massachusetts. *Copeia,* no. 4, p. 237. 1936.

——, A new locality record for *Gyrinophilus porphyriticus* (Green) in Massachusetts. *Copeia,* no. 1, p. 74. 1937.

Wilder, I. W., Spermatophores of *Desmognathus fusca.—Copeia,* no. 121, pp. 88–92. 1923.

——, The relation of growth to metamorphosis in *Eurycea bislineata* (Green). *Jour. Exp. Zool.,* 40: 1–112, figs. 1–18. 1924.

——, Variations in the premaxillary of *Eurycea bislineata.—Amer. Nat.,* 58: 538–543, figs. 1–3. 1924 a.

——, The developmental history of *Eurycea bislineata* in western Massachusetts. *Copeia,* no. 133, pp. 77–80. 1924 b.

## MICHIGAN

Allen, D. L., Some notes on the Amphibia of a waterfowl sanctuary, Kalamazoo County, Michigan. *Copeia,* no. 3, pp. 190–191. 1937.

——, Ecological studies on the vertebrate fauna of a 500-acre farm in Kalamazoo County, Michigan. *Ecol. Monog.,* 8 (3): 347–436, figs. 1–28. 1938.

Bishop, S. C., *see above,* General works. 1941.

Blanchard, F. C., Length of life in the tiger salamander *Ambystoma tigrinum* (Green). *Copeia,* no. 2, pp. 98–99. 1932.

Blanchard, F. N. Discovery of the eggs of the four-toed salamander in Michigan. *Occ. Papers Mus. Zool. Univ. Mich.,* no. 126, pp. 1–3. 1922.

——, The life history of the four-toed salamander. *Amer. Nat,* 57: 262–268. 1923.

——, Topics from the life-history and habits of the red-backed salamander in southern Michigan. *Amer. Nat.,* 62: 156–164, figs. 1–4. 1928.

——, Amphibians and reptiles of the Douglas Lake region in northern Michigan. *Copeia,* no. 167, pp. 42–51. 1928.

——, The stimulus to the breeding migration of the spotted salamander, *Ambystoma maculatum* (Shaw). *Amer. Nat.,* 64: 154–167. 1930.

——, Late autumn collections and hibernating situations of the salamander *Hemidactylium scutatum* (Schlegel) in southern Michigan. *Copeia,* no. 4, p. 216. 1933.

——, Spermatophores and the mating season of the salamander *Hemidactylium scutatum* (Schlegel). *Copeia,* no. 1, p. 40. 1933 a.

——, Natural history of the four-toed salamander, *Hemidactylium scutatum* (Schlegel). *Anat. Rec.,* 57 (suppl.): 100–101. 1933 b.

Blanchard, F. N., The spring migration of the four-toed salamander, *Hemidactylium* (Schlegel). *Copeia,* no. 1, p. 50. 1934.

——, The relation of the female four-toed salamander to her nest. *Copeia,* no. 3, pp. 137–138. 1934 a.

——, The date of egg-laying of the four-toed salamander *Hemidactylium scutatum* (Schlegel) in southern Michigan. *Papers of Mich. Acad. Sci., Arts and Letters,* 19: 571–575. 1934 b.

——, The sex ratio in the salamander *Hemidactylium scutatum* (Schlegel). *Copeia,* no. 2, p. 103. 1935.

——, The number of eggs produced and laid by the four-toed salamander, *Hemidactylium scutatum* (Schlegel), in southern Michigan. *Papers Mich. Acad. Sci., Arts and Letters,* 21: 567–573, pl. 57, figs. 24–26. 1936.

——, and F. C. Blanchard, Size groups and their characteristics in the salamander *Hemidactylium scutatum* (Schlegel). *Amer. Nat.,* 65: 149–164. 1931.

Branin, M. L., Courtship activities and extra-seasonal ovulation in the four-toed salamander, *Hemidactylium scutatum* (Schlegel). *Copeia,* no. 4, pp. 172–175. 1935.

Clanton, Wesley, An unusual situation in the salamander *Ambystoma jeffersonianum* (Green). *Occ. Papers Mus. Zool. Univ. Mich.,* no. 290, pp. 1–14, pl. 1, fig. 1. 1934.

Clark, H. L., Notes on the reptiles and batrachians of Eaton County. *4th Rept. Mich. Acad. Sci.,* pp. 192–194. 1904.

Cope, E. D., Partial catalogue of the cold-blooded Vertebrata of Michigan. *Proc. Acad. Nat. Sci. Phila.,* 17 (2): 78–88. 1865.

Dempster, W. T., The growth of larvae of *Ambystoma maculatum* under natural conditions. *Biol. Bul.,* 58 (2): 182–192, figs. 1–3, 1930.

Dury, Ralph, *see* Indiana. 1932.

Ellis, M. M., Amphibians and reptiles of the Douglas Lake region. *19th Ann. Rept. Mich. Acad. Sci.,* pp. 45–63. 1917.

Evans, A. T., A collection of amphibians and reptiles from Gogebic County, Michigan. *Proc. U. S. Nat. Mus.,* 49: 351–354. 1916.

Gaige, H. T., Some amphibians new to Whitefish Point, Michigan. *Copeia,* no. 14, p. 2. 1915.

——, The amphibians and reptiles collected by the Bryant Walker expedition to Schoolcraft County, Michigan. *Occ. Papers Mus. Zool. Univ. Mich.,* no. 17, pp. 1–5. 1915.

Gibbs, Morris, The reptiles and batrachians of Michigan. *Amer. Field,* 44 (1–7): 6–7, 30–31, 55, 79–80, 106, 132, 155–156. (July 6—Aug. 17) 1895.

——, F. N. Notestein, and H. L. Clark, A preliminary list of the Amphibia and Reptilia of Michigan. *7th Rept. Mich. Acad. Sci.,* pp. 109–110. 1905.

Hallowell, Edward, Description of several species of Urodela, with remarks on the geographical distribution of the caducibranchiate division of these animals and their classification. *Proc. Acad. Nat. Sci. Phila. for 1856,* 8: 6–11. 1856.

Hatt, R. T., The land vertebrate communities of western Leelanau County,

Michigan, with an annotated list of the mammals of the county. *Papers Mich. Acad. Sci., Arts and Letters,* 3: 369–402, pls. 24–26. 1923.

Higley, D. J., Salamanders. *Sci. News,* 1 (11): 175. 1879.

Kneeland, Samuel, On a supposed new species of *Siredon* from Lake Superior. *Proc. Bost. Soc. Nat. Hist.,* 6: 152–154, 218. 1857.

Lagler, K. F., and K. E. Goellner, The mud puppy, an enemy of fish? *Mich. Conservation,* 8 (7): 3, figs. 1–2. 1939.

——, and K. E. Goellner, Notes on *Necturus maculosus* (Rafinesque). *Copeia,* no. 2, pp. 96–98. 1941.

Langlois, T. H., [*Necturus maculosus* in Michigan.] *Copeia,* no. 158, p. 167. 1927.

McCurdy, H. M., On certain relations of the flora and vertebrate fauna of Gratiot County, Michigan, with an appended list of mammals and amphibians. *14th Rept. Mich. Acad. Sci.,* pp. 217–225. 1912.

Miles, M., A catalogue of the mammals, birds, reptiles and mollusks of Michigan. *1st Biennial Rept. Geol. Survey of Mich.,* pp. 213–241. 1861.

Milner, J. W., Report on the fisheries of the Great Lakes; the results of inquiries prosecuted in 1871 and 1872. *Rept. Comm. U. S. Comm. Fish and Fisheries, Pt. 2, for 1872 and 1873,* pp. 1–78. 1874.

Necker, W. L., *see* Illinois. 1939.

Pearse, A. S., *see* Wisconsin. 1921.

Potter, Doreen, Reptiles and amphibians collected in central Michigan in 1919. *Copeia,* no. 82, pp. 39–41. 1920.

Ruthven, A. G., Notes on the molluscs, reptiles and amphibians of Ontonagon County, Michigan. *6th Rept. Mich. Acad. Sci.,* pp. 188–192. 1904.

——, The cold-blooded vertebrates of the Porcupine Mountains in Isle Royale, Michigan. *Rept. State Board Geol. Surv. Mich.,* pp. 107–112. 1906.

——, The cold-blooded vertebrates of Isle Royale. In An ecological survey of Isle Royale, Lake Superior. *Rept. Univ. Mich. Mus.,* pp. 329–333. 1909.

——, Amphibians and reptiles. In A biological survey of the sand dune region on the south shore of Saginaw Bay, Michigan. *Mich. Geol. and Biol. Surv.,* pub. 4, biol. ser. 2, pp. 257–272. 1911.

——, Notes on Michigan reptiles and amphibians. III. *13th Rept. Mich. Acad. Sci.,* pp. 114–115, fig. 1. 1911 a.

——, Crystal Thompson, and Helen Thompson, The herpetology of Michigan. *Mich. Geol. and Biol. Surv.,* pub. 10, biol. ser. 3, pp. 1–166, figs. 1–55, pls. 1–20. 1912.

——, Crystal Thompson, and H. T. Gaige, The herpetology of Michigan. *Univ. Mich. Handbook,* Ser. 3, pp. i–ix, 1–228, figs. 1–52, pls. 1–19, frontispiece. 1928.

Sagar, Abram, Report of the State Zoologist to the State Geologist. Catalog of mammals, birds, reptiles, amphibians and mollusks. *2nd Ann. Rept. State Geol. Mich.,* sen. doc. 13, pp. 1–15. 1839.

Schmidt, K. P., and W. L. Necker, *see* Illinois. 1935.

Smith, B. G., The breeding habits of *Amblystoma punctatum* Linn. *Amer. Nat.,* 41: 381–390. 1907.

——, *see* Wisconsin. 1911.

Smith, B. G., An adult *Diemyctylus viridescens* with bifurcated tail. *15th Rept. Mich. Acad. Sci.,* p. 105, fig. 1. 1913.
Smith, W. H., Catalogue of the Reptilia and Amphibia of Michigan. *Science News,* 1 (23): i–viii (supplement). 1879.
Thompson, Crystal, and Helen Thompson, The amphibians of Michigan. *Mich. Geol. and Biol. Surv.,* pub. 10, biol. ser. 3, pp. 13–62, figs. 1–20. 1912.
——, Results of the Merson expedition to the Charity Islands, Lake Huron. *14th Rept. Mich. Acad. Sci.,* pp. 156–158. 1912.
——, Results of the Shiras expeditions to Whitefish Point, Michigan. *15th Rept. Mich. Acad. Sci.,* pp. 215–217. 1913.
Thompson, Crystal, The reptiles and amphibians of Manistee County, Michigan. *Occ. Papers Mus. Zool. Univ. Mich.,* no. 18, pp. 1–6. 1915.
——, The reptiles and amphibians of Monroe County, Michigan. *Mich. Geol. and Biol. Surv.,* pub. 20, biol. ser. 4, pp. 61–63. 1916.
Woodhead, A. E., *Haptophyra michiganensis* sp. nov. A protozoan parasite of the four-toed salamander. *Jour. Parasitology,* 14: 177–182. 1928.

#### MINNESOTA

Osborn, H. L., Some biological notes on *Amblystoma tigrinum* I. *Sci.,* 20 (517): 366–368. 1892.
——, *see* Colorado. 1901.
Swanson, Gustav, A preliminary list of Minnesota amphibians. *Copeia,* no. 3, pp. 152–154. 1935.

#### MISSISSIPPI

Allen, M. J., A survey of the amphibians and reptiles of Harrison County, Mississippi. *Amer. Mus. Novitates,* no. 542, pp. 1–20. 1932.
Baird, S. F., *see* Oregon. 1850.
Brimley, C. S., *see* Alabama. 1910.
——, *see* North Carolina. 1939–1940.
Corrington, J. D., Field notes on some amphibians and reptiles at Biloxi, Mississippi. *Copeia,* no. 165, pp. 98–102. 1927.
Goin, C. J., *see* Virginia. 1938.
Potter, Doreen, Reptiles and amphibians collected in northern Mississippi in 1919. *Copeia,* no. 86, pp. 82–83. 1920.
Ryder, J. A., Morphological notes on the limbs of the *Amphiumidae,* as indicating a possible synonymy of the supposed genera. *Proc. Acad. Nat. Sci. Phila. for 1879,* pp. 14–15. 1880.
Viosca, Percy, Jr., *see* Louisiana. 1937.

#### MISSOURI

Alt, Adolf, On the histology of the eye of *Typhlotriton spelæus* from Marble Cave, Missouri. *Trans. Acad. Sci. St. Louis,* 19 (6): 83–98, pls. 26–34. 1910.

Anderson, Paul, Amphibians and reptiles of Jackson County, Missouri. *Bul. Chi. Acad. Sci.*, 6 (11): 203–220. 1942.

Banta, A. M., and W. L. McAtee, *see* Indiana. 1906.

Bishop, S. C., *see above*, General works. 1941.

Blanchard, F. N., *see* Illinois. 1924.

Boyer, D. A., and A. A. Heinze, An annotated list of the amphibians and reptiles of Jefferson County, Missouri. *Trans. Acad. Sci. St. Louis*, 28 (4): 185–200, figs. 1–2. 1934.

Burt, C. E., A collection of amphibians and reptiles from southern Missouri. *Amer. Mid. Nat.*, 14 (2): 170–173. 1933.

Cope, E. D., *see above*, General works. 1886.

——, On a collection of Batrachia and Reptilia from southwest Missouri. *Proc. Acad. Nat. Sci. Phila. for 1893*, pp. 383–385. 1894.

——, On the Batrachia and Reptilia of the plains at latitude 36° 30′. *Proc. Acad. Nat. Sci. Phila. for 1893*, pp. 386–387. 1894 a.

Eigenmann, C. H., *see* Kentucky. 1899.

——, The blind fishes of North America. *Popular Sci. Monthly*, 56: 437–486, figs. 1–20. 1900.

——, The eyes of the blind vertebrates of North America. II. The eyes of *Typhlomolge rathbuni* Stejneger. *Trans. Amer. Micros. Soc.*, 21: 49–60, pls. 3–4. 1900 a.

——, Degeneration in the eyes of the cold-blooded vertebrates of the North American caves. *Sci.*, n.s., 11 (274): 492–503, 2 pls. 1900 b.

——, Degeneration in the eyes of the cold-blooded vertebrates of the North American caves. *Proc. Ind. Acad. Sci.*, 9: 31–46, figs. a–j. 1900 c.

——, *see* Indiana. 1901.

——, and W. A. Denny, The eyes of *Typhlotriton spelæus*.—*Proc. Ind. Acad. Sci.* (1898), 8: 252–253. 1899.

——, and W. A. Denny, The eyes of the blind vertebrates of North America. III. *Biol. Bul.*, 2 (1): 33–41, figs. 1–8. 1901.

——, and Clarence Kennedy, Variation notes. *Biol. Bul.*, 4 (5): 227–230, figs. 1–5. 1903.

Hay, O. P., *see* Indiana. 1892.

Henning, W. L., Amphibians and reptiles of a 2220-acre tract in central Missouri. *Copeia*, no. 2, pp. 91–92. 1938.

Hurter, Julius, Catalogue of reptiles and batrachians found in the vicinity of St. Louis, Missouri. *Trans. Acad. Sci. St. Louis*, 6 (11): 251–261. 1893.

——, A contribution to the herpetology of Missouri. *Trans. Acad. Sci. St. Louis*, 7 (19): 499–503. 1897.

——, Second contribution to the herpetology of Missouri. *Trans. Acad. Sci. St. Louis*, 13 (3): 77–86. 1903.

——, Herpetology of Missouri. *Trans. Acad. Sci. St. Louis*, 20 (5): 59–274, pls. 18–24. 1911.

Lönnberg, Einar, Salamanders with and without lungs. *Zool. Anz.*, 22 (604): 545–548. 1899.

Noble, G. K., Creatures of perpetual night. *Nat. Hist.,* 27 (5): 405–419, 29 figs. 1927.

——, and B. C. Marshall, *see* Arkansas. 1929.

Schenkel, E., Achter Nachtrag zum Katalog der herpetologischen Sammlung des Basler Museums. *Verh. der Naturfor. Ges.,* 13 (1): 142–199. 1901.

Stejneger, Leonhard, Preliminary description of a new genus and species of blind cave salamander from North Amercia. *Proc. U. S. Nat. Mus.,* 15: 115–117, pl. 9. 1892.

Strecker, J. K., *see* Arkansas. 1924.

## MONTANA

Allen, J. A., Notes on the natural history of portions of Dakota and Montana. . . . *Proc. Bost. Soc. Nat. Hist.,* 17: 33–85. 1874.

Bishop, S. C., An older name for a recently described salamander. *Copeia,* no. 4, p. 256. 1942.

Cope, E. D., A contribution to the zoology of Montana. *Amer. Nat.,* 13 (7): 432–441. 1879.

Coues, Elliott, and H. C. Yarrow, Notes on the herpetology of Dakota and Montana. *Bul. U.S. Geol. Survey,* 4 (1): 259–291. 1878.

Peters, W., Vorlegung dreier neuen Batrachier (*Amblystoma Krausei, Nyctibatrachus sinensis, Bufo Buchneri*).—*Sitzungsbericht der Gesellschaft naturforschender Freunde, Berlin,* no. 10, pp. 145–148. 1882.

Rodgers, T. L., and W. L. Jellison, A collection of amphibians and reptiles from western Montana. *Copeia,* no. 1, pp. 10–13. 1942.

Slater, J. R., *see* Idaho. 1941.

Test, F. C., Annotated list of reptiles and batrachians collected. In Fish-cultural investigations in Montana and Wyoming. *Bul. U.S. Fish. Comm. for 1891,* 11: 57–58. 1893.

## NEBRASKA

Burt, C. E., and M. D. Burt, *see* Utah. 1929.

Hayden, F. V., Catalogue of the collections in geology and natural history. In *Preliminary rept. of explorations in Nebraska and Dakota by Lieut. G. K. Warren,* reprint, pp. 59–125. 1875.

Powers, J. H., The causes of acceleration and retardation in the metamorphosis of *Amblystoma tigrinum:* A preliminary report. *Amer. Nat.,* 37: 385–410. 1903.

——, (1) Morphological variation and its causes in *Amblystoma tigrinum.—Univ. Studies. Univ. Neb.,* 7 (3): 197–273, pls. 1–9. 1907.

## NEW ENGLAND

Dunn, E. R., The New England salamanders. *Bul. Bost. Soc. Nat. Hist.,* 57: 23–24. 1930.

Babcock, H. L., Some New England salamanders. *Bul. Bost. Soc. Nat. Hist.*, 48: 3–7, 3 figs. 1928.

Henshaw, Samuel, Fauna of New England. 2. List of the Batrachia. *Occ. Papers Bost. Soc. Nat. Hist.*, 7: 1–10. 1904.

#### NEW HAMPSHIRE

Allen, G. M., Notes on the reptiles and amphibians of Intervale, New Hampshire. *Proc. Bost. Soc. Nat. Hist.*, 29 (3): 63–75. 1899.

——, Notes on the reptiles and batrachians of Intervale, New Hampshire. *Proc. Bost. Soc. Nat. Hist.*, 29: 73–74. 1901.

Burt, C. E., *see* New York. 1932.

Evermann, B. W., Notes on some reptiles and amphibians of Waterville, New Hampshire. *Copeia,* no. 61, pp. 81–83. 1918.

Hoopes, Isabel, Marbled salamander from New Hampshire. *Bul. New England Mus. Nat. Hist.*, no. 87, pp. 16–17, 1 fig. 1938.

Howe, R. H., *Spelerpes porphyriticus* in New Hampshire. *Proc. Biol. Soc. Wash.*, 17: 102. 1904.

Kendall, W. C., Fishes and fishing in Sunapee Lake. *U. S. Bur. Fish.*, doc. 783, p. 88. 1913.

MacCoy, C. F., Key for identification of New England amphibians and reptiles. *Bul. Bost. Soc. Nat. Hist.*, no. 59, pp. 25–33, figs. 1–2. 1931.

Murphy, R. C., The jumping ability of *Plethodon* and its possible bearing upon the origin of saltation in the ancestors of the Anura. Copeia, no. 51, pp. 105–106. 1917.

Oliver, J. A., and J. R. Bailey, Amphibians and reptiles of New Hampshire. *Biol. Surv. Conn. Watershed,* rept. no. 4, pp. 195–217, figs. 1–19. 1939.

Orton, Grace, Key to New Hampshire amphibian larvae. *Biol. Surv. Conn. Watershed,* rept. no. 4, pp. 218–221, pl. 2. 1939.

Putnam, F. W., *see* Massachusetts. 1863.

Speck, F. G., Reptile and amphibian notes from Intervale, New Hampshire. *Copeia,* no. 70, pp. 46–48. 1919.

Storer, H. R., *see* Maine. 1852.

#### NEW JERSEY

Abbott, C. C., Catalogue of vertebrate animals of New Jersey. *Geol. of New Jersey,* pp. 751–830. 1868.

de Beauvois, M., On a new species of *Siren.—Trans. Amer. Phil. Soc.*, 4: 277–281, figs. 1–4. 1799.

Breder, R. B., On the manner of food taking in metamorphosing newts. *Copeia,* no. 129, pp. 47–48. 1924.

——, The courtship of the spotted salamander. *N. Y. Zool. Soc.*, bul. 30: 51–56, 1 pl. 12 figs. 1927.

Burt, C. E., *see* New York. 1931.

Conant, Roger, and R. M. Bailey, Some herpetological records from Monmouth and Ocean Counties, New Jersey. *Occ. Papers Mus. Zool. Univ. Mich.*, no. 328, pp. 1–10. 1936.

Crozier, W. J., Opisthotonic death in a salamander. *Amer. Nat.*, 58 (658): 479–480, 1 fig. 1924.

Davis, W. T., Notes on New Jersey amphibians and reptiles. *Proc. S. I. Assn. Arts and Sci.*, 2 (2): 47–52. 1908.

——, *see* New York. 1931.

——, *see* New York. 1935.

Fowler, H. W., The amphibians and reptiles of New Jersey. *Ann. Rept. N. J. State Mus.*, pt. 2, pp. 29–250, pls. 1–69, 70 figs. 1907.

——, A supplementary account of New Jersey amphibians and reptiles. *Ann. Rept. N. J. State Mus. for 1907*, pp. 190–202, pl. 69. 1908.

——, Notes on New Jersey amphibians and reptiles. *Ann. Rept. N. J. State Museum for 1908*, pp. 393–408. 1909.

Green, Jacob, Description of several species of North American Amphibia, accompanied with observations. *Jour. Acad. Nat. Sci. Phila.*, 1 (2): 348–358. 1818.

——, Description of a new species of salamander. *Jour. Acad. Nat. Sci. Phila.*, 5: 116–118. 1825.

Green, H. T., *see* New York. 1923.

Hay, O. P., Our present knowledge concerning the green *Triton, Diemyctylus viridescens.—Proc. Ind. Acad. Sci.*, 1: 145–147. 1891.

Miller, W. De W., Notes on New Jersey batrachians and reptiles. *Copeia*, no. 34, pp. 67–68. 1916.

Myers, G. S., *see* California. 1930.

——, *see* New York. 1930 a.

Nelson, Julius, Descriptive catalogue of the vertebrates of New Jersey. *Geol. Surv. N. J.: Final Rept. State Geol.*, 2 (2): 489–824. 1890.

Netting, M. G., *see* Pennsylvania. 1938.

Noble, G. K., *see* New York. 1926.

——, *see* New York. 1927.

——, and J. A. Weber, *see* Pennsylvania. 1929.

Pike, Nicolas, *see* New York. 1886.

Sherwood, W. L., *see* New York. 1895.

Stone, Witmer, *see* Pennsylvania. 1906.

Street, J. F., Amphibians and reptiles observed at Beverly, New Jersey. *Copeia*, no. 4, p. 2. 1914.

## NEW MEXICO

Bailey, Vernon, Life zones and crop zones of New Mexico. *Bureau Biol. Surv. North Amer. Fauna*, no. 35, pp. 1–100, pls. 1–16, figs. 1–6. 1913.

——, Mammals of New Mexico. *North Amer. Fauna*, no. 53, p. 320. 1931.

Baird, S. F., *see* Oregon. 1850.

——, and Charles Girard, In Howard Stansbury, Exploration and survey of the valley of the Great Salt Lake of Utah, Appendix C. *Senate Doc. Spec. Session Mar. 1851,* exec. no. 3, pp. 1–495. 1853.

Baird, S. F., and Charles Girard, Characteristics of some new reptiles in the museum of the Smithsonian Institution. *Proc. Acad. Nat. Sci. Phila. for 1852,* 6: 68–70. 1853 a.

Burt, C. E., *see* Colorado. 1933.

Cockerell, T. D. A., Reptiles and batrachians of Mesilla Valley, New Mexico. *Amer. Nat.,* 30 (352): 325–327. 1896.

Hallowell, Edward, Reptiles. In Report of an expedition down the Zuni and Colorado Rivers by Capt. L. Sitgreaves. *Senate Document, 33rd Congress,* pp. 106–147, pls. 1–20. 1854.

Little, E. L., Jr., and J. G. Keller, Amphibians and reptiles of the Jornada experimental range, New Mexico. *Copeia,* no. 4, pp. 216–222. 1937.

Mosauer, Walter, The amphibians and reptiles of the Guadalupe Mountains of New Mexico and Texas. *Occ. Papers Mus. Zool. Univ. Mich.,* no. 246, pp. 1–18, pl. 1. 1932.

Sager, Abram, Description of a new genus of perenni-branchiate amphibians. *Peninsular Jour. Medicine,* 5 (8): 428–429. 1858.

Shufeldt, R. W., The Mexican axolotl and its susceptibility to transformations. *Sci.,* 6 (138): 263–264. 1885.

Taylor, E. H., A new plethodont salamander from New Mexico. *Proc. Biol. Soc. Wash.,* 54: 77–79. 1941.

Van Denburgh, John, Notes on the herpetology of New Mexico, with a list of species known from that state. *Proc. Cal. Acad. Sci.,* 13 (12): 189–230. 1924.

#### NEW YORK

Baird, S. F., List of living salamanders presented to the State cabinet of natural history. *5th Ann. Rept. State Cab. Nat. Hist.,* pp. 21–22. 1852.

Banta, A. M., Some notes on albinism. *Sci.,* n.s., 41 (1059): 577–578. 1915.

——, and R. A. Gortner, An albino salamander, *Spelerpes bilineatus.—Proc. U.S. Nat. Mus.,* 49: 377–379, pls. 54–55. 1915.

Bicknell, E. P., A review of the summer birds of a part of the Catskill Mountains. *Trans. Linn. Soc. N. Y.,* 1: 115–168. 1882.

Bishop, S. C., Notes on the habits and development of the four-toed salamander, *Hemidactylium scutatum* (Schlegel). *N. Y. State Mus. Bul.,* 219–220, pp. 251–282, pls. 1–7. 1919.

——, Notes on the herpetology of Albany County, New York. *Copeia,* no. 118, pp. 64–68. 1923.

——, The life of the red salamander. *Nat. Hist.,* 25 (4): 385–389, 3 figs. 1925.

——, Notes on the habits and development of the mudpuppy, *Necturus maculosus* (Rafinesque). *N. Y. State Mus. Bul.,* 268, pp. 5–60, pls. 1–11. 1926.

——, The spermatophores of *Necturus maculosus* Rafinesque. *Copeia,* no. 1, pp. 1–3, figs. 1–3. 1932.

Bishop, S. C., The salamanders of New York. *N. Y. State Mus. Bul.*, 324, pp. 1–365, figs. 1–66. 1941.

——, and W. P. Alexander, The amphibians and reptiles of Alleghany State Park. *N. Y. State Mus. Handbook*, 3, pp. 3–141, figs. 1–58, map. 1927.

——, and H. P. Chrisp, The nests and young of the Allegheny salamander. *Desmognathus fuscus ochrophæus* (Cope). *Copeia*, no. 4, pp. 194–198, figs. 1–5. 1933.

Boyle, H. S., Four-toed salamander on Long Island. *Copeia*, no. 9, p. 4. 1914.

Branin, M. L., *Hemidactylium scutatum* from northern Dutchess County. *Copeia*, no. 2, p. 130. 1940.

Breder, R. B., *see* New Jersey. 1927.

Britcher, H. W., An occurrence of albino eggs of the spotted salamander, *Amblystoma punctatum L.—Trans. Amer. Micros. Soc.*, 20: 69–74, pl. 6. 1899.

——, Batrachia and Reptilia of Onondaga County. *Proc. Onondaga Academy of Sci.*, 1: 120–122. 1903.

Burt, C. E., A report on some amphibians and reptiles from New York and New Jersey. *Jour. Wash. Acad. Sci.*, 21 (9): 198–203. 1931.

——, Records of amphibians from the eastern and central United States (1931). *Amer. Mid. Nat.*, 13 (2): 75–85. 1932.

Coker, C. M., Hermit thrush feeding on salamanders. *Auk*, 48 (2): 277. 1931.

Corrington, J. D., Abnormal herpetological specimens from Syracuse, New York. *Amer. Nat.*, 65: 77–85. 1931.

Davis, W. T., The reptiles and batrachians of Staten Island. *Proc. Nat. Sci. Assn. Staten Island*, 1: 13. 1884.

——, The four-toed salamander on Staten Island. *Proc. Nat. Sci. Assn. Staten Island*, 9 (2): 4. 1903.

——, *see* New Jersey. 1908.

——, Natural history records from the meetings of the Staten Island nature club. *Proc. Staten Island Inst. Arts and Sci.*, 2 (2–4): 139–158. 1924.

——, Natural history records from the meetings of the Staten Island nature club. *Proc. Staten Island Inst. Arts and Sci.*, 4 (1–3): 65–76. 1927.

——, Natural history records from the meetings of the Staten Island nature club. *Proc. Staten Island Inst. Arts and Sci.*, 4 (4): 116–123. 1928.

——, Natural history notes from the meetings of the Staten Island nature club. *Proc. Staten Island Inst. Arts and Sci.*, 6 (1): 28–51. 1931.

——, Natural history records from the meetings of the Staten Island nature club. *Proc. Staten Island Inst. Arts and Sci.*, 6 (2–4): 165–187. 1933.

——, Natural history notes from the meetings of the Staten Island nature club. *Proc. Staten Island Inst. Arts and Sci.*, 7 (1–2): 25–50. 1933 a.

——, Natural history records from the meetings of the Staten Island nature club *Proc. Staten Island Inst. Arts and Sci.*, 7 (3–4): 114–139. 1935.

——, A very large red-backed salamander. *Copeia*, no. 4, p. 258. 1942.

Deckert, R. F., Salamanders collected in Westchester County, New York. *Copeia*, no. 13, pp. 3–4. 1914.

Deckert, R. F., Note on *Amblystoma opacum* Grav. *Copeia,* no. 28, pp. 23–24. 1916.
——, Another record of *Amblystoma opacum* from Long Island. *Copeia,* no. 41, p. 24. 1917.
DeKay, J. E., *Zoology of New York (part 3, reptiles and amphibia)*, pp. i–vii, 1–415, pls. 1–79. Text and plates separate. 1842.
Ditmars, R. L., The batrachians of the vicinity of New York City. *Amer. Mus. Jour.,* 5 (4): 161–206, figs. 1–28. 1905.
Eckel, E. C., and F. C. Paulmier, Catalogue of New York reptiles and batrachians. *N. Y. State Mus. Bul.,* 51, pp. 353–414, 1 pl., figs. 1–24. 1902.
Eights, James, Naturalist's every day book. *The Zodiac,* 1 (1): 4–8. 1835.
Engelhardt, G. P., *Amblystoma* of Long Island. *Copeia,* no. 8, pp. 2–4. 1914.
——, *Amblystoma tigrinum* on Long Island (I). *Copeia,* no. 28, pp. 20–22. 1916.
——, *Amblystoma tigrinum* on Long Island (II). *Copeia,* no. 30, pp. 32–35. 1916 a.
——, *Amblystoma tigrinum* on Long Island. *Copeia,* no. 32, pp. 48–51. 1916 b.
——, *Amblystoma opacum* on Long Island. *Copeia,* no. 37, pp. 88–89. 1916 c.
——, Another Long Island record for *Amblystoma jeffersonianum* (Green). *Copeia,* no. 50, pp. 98–99. 1917.
——, The habitat of *Gyrinophilus porphyriticus.—Copeia,* no. 69, pp. 20–21. 1919.
Evermann, B. W., Notes on some Adirondack reptiles and amphibians. *Copeia,* no. 56, pp. 48–51. 1918.
Fisher, G. C., Note on the dusky salamander. *Copeia,* no. 58, pp. 65–66. 1918.
Gadow, Hans, *see* Baja California. 1905.
Gage, S. H., Life-history of the vermilion-spotted newt (*Diemyctylus viridescens* Raf.). *Amer. Nat.,* 25 (300): 1084–1110, pl. 23. 1891.
Gage, S. H., and H. W. Norris, Notes on the Amphibia of Ithaca. *Proc. Amer. Assn. Adv. Sci.,* 39: 338–339. 1891.
Gebhard, John, Jr., Catalogue of the quadrupeds, birds, reptiles, amphibians, etc., added to the State cabinet of natural history, from January 1, 1852, to January 1, 1853. *6th Ann. Rept. State Cab. Nat. Hist.,* pp. 17–24; appendix, p. 23. 1853.
——, Appendix A: Catalogue of the quadrupeds, birds, reptiles, amphibians, etc., added to the State cabinet of natural history, from January 1, 1853, to January 1, 1854. *7th Ann. Rept. State Cab. Nat. Hist.,* pp. 25–26. 1854.
[——], *Menobranchus maculatus* Barnes [in Hudson River]. *9th Ann. Rept. State Cab. Nat. Hist.,* pp. 16–18. 1856.
Gilbert, P. W., Eggs and Nests of *Hemidactylium scutatum* in the Ithaca region. *Copeia,* no. 1, p. 47. 1941.
——, Observations on the eggs of *Ambystoma maculatum,* with special reference to the green algae found within the egg envelopes. *Ecology,* 23 (2): 215–227, pls. 1–2, figs. 1–5. 1942.
Gravenhorst, J. L. C., *see above,* General works. 1807.
Green, H. T., Notes on Middle States amphibians and reptiles. *Copeia,* no. 122, pp. 99–100. 1923.
Green, Jacob, *see* New Jersey. 1818.
Grote, A. R., A preliminary note on *Menopoma alleghaniense* of Harlan. *Proc. Amer. Assn. Adv. Sci., 25th Meeting,* pp. 255–257. 1877.

Hall, James, *23rd Ann. Rept. State Cab. Nat. Hist.,* p. 21. 1873.
——, *27th Ann. Rept. N.Y. State Mus.,* p. 23. 1875.
——, *29th Ann. Rept. N.Y. State Mus.,* p. 20. 1878.
——, *41st Ann. Rept. N.Y. State Mus.,* p. 29. 1888.
Hamilton, W. J., The food and feeding habits of some eastern salamanders. *Copeia,* no. 2, pp. 83–86. 1932.
——, Seasonal foods of skunks in New York. *Jour. Mammalogy,* 17 (3): 240–246, tab. 1–3. 1936.
——, The feeding habits of larval newts, with reference to availability and predilection of food items. *Ecology,* 21 (3): 351–356. 1940.
Hassler, W. G., Salamanders and water hygiene. *Nat. Hist.,* 32 (3): 303–310, 10 figs. 1932.
——, New locality records for two salamanders and a snake in Cattaraugus County, New York. *Copeia,* no. 2, pp. 94–96. 1932 a.
Hood, H. H., A note on the red-backed salamander at Rochester, New York. *Copeia,* no. 3, pp. 141–142. 1934.
Hough, F. B., Catalogue of reptiles and fishes from St. Lawrence County, procured for the State cabinet of natural history. *5th Ann. Rept. State Cab. Nat. Hist.,* pp. 23–28. 1852.
Humphrey, R. R., The interstitial cells of the urodele testis. *Amer. Jour. Anat.,* 29 (2): 213–278, pls. 1–4, figs. 1–7. 1921.
Johnson, C. E., Part I. Preliminary reconnaissance of the land vertebrates of the Archer and Anna Huntington wild life forest station. *Roosevelt Wild Life Bul.,* 6 (4): 557–609, figs. 285–300. 1937.
Kingsbury, B. F., The spermatogenesis of *Desmognathus fusca.—Amer. Jour. Anat.,* 1 (2): 99–135, pls. 1–4, fig. A. 1902.
Kline, E. K., and N. M. Fuller, Interpretation of laboratory findings in rural spring water supplies. *Amer. Jour. Pub. Health,* 22 (7): 691–699, figs. 1–5. 1932.
Kumpf, K. F., The courtship of *Ambystoma tigrinum.—Copeia,* no. 1, pp. 7–10. 1934.
——, and S. C. Yeaton, Jr., Observations on the courtship behavior of *Ambystoma jeffersonianum.—Amer. Mus. Nov.,* no. 546, pp. 1–7, figs. 1–3. 1932.
Leffingwell, D. J., Vertebrates. In A preliminary biological survey of the Lloyd-Cornell reservation. *Lloyd Library of Botany, Pharmacy and Materia Medica,* bul. 27 (ent. ser. 5), pp. 71–82. 1926.
Marshall, W. B., *Necturus maculatus* in the Hudson River. *Amer. Nat.,* 26 (309): 779–780. 1892.
——, *46th Ann. Rept. N.Y. State Mus.,* p. 21. 1893.
——, *Necturus maculatus* in the Hudson River. *47th Ann. Rept. N. Y. State Mus. for 1893,* p. 33. 1894.
Matheson, Robert, and E. H. Hinman, The vermilion spotted newt (*Diemictylus viridescens* Rafinesque) as an agent in mosquito control. *Amer. Jour. Hygiene,* 9 (1): 188–191. 1929.
Mearns, E. A., Notes on the mammals of the Catskill Mountains, New York, with

general remarks on the fauna and flora of the region. *Proc. U.S. Nat. Mus.,* 21: 341–360, figs. 1–6. 1898.

Mearns, E. A., A study of the vertebrate fauna of the Hudson Highlands. *Bul. Amer. Mus. Nat. Hist.,* 10: 303–352. 1898 a.

Merrill, F. J. H., *53rd Ann. Rept. N. Y. State Mus.,* p.r. 169. 1901.

Montgomery, Henry, Some observations of the *Menobranchus maculatus.—Can. Nat.,* n.s., 9 (3): 160–164. 1879.

Moore, J. A., Temperature tolerance and rates of development in the eggs of Amphibia. *Ecology,* 20 (4): 459–478, figs. 1–8. 1939.

Myers, G. S., Amphibians and reptiles observed in the Palisades Interstate Park, New York and New Jersey. *Copeia,* no. 173, pp. 99–103. 1930.

Needham, J. G., Report of the entomologic field station conducted at Old Forge, New York, in the summer of 1905. *N. Y. State Mus.,* bul. 124, pp. 156–263, figs. 1–28. 1908.

Netting, M. G., *see* Pennsylvania. 1938.

Noble, G. K., The Long Island newt: A contribution to the life history of *Triturus viridescens.—Amer. Mus. Nov.,* no. 228, pp. 1–11, figs. 1–3. 1926.

——, Distributional list of the reptiles and amphibians of the New York City region. *Amer. Mus. Nat. Hist., Guide Leaflet Series,* no. 69, pp. 1–9. 1927.

——, *see* Massachusetts. 1929.

——, and C. H. Pope, The modification of the cloaca and teeth of the adult salamander, *Desmognathus,* by testicular transplants and by castration. *Brit. Jour. Exp. Biol.,* 6 (4): 399–411, pl. 809, figs. 1–4. 1929.

——, and L. B. Richards, The induction of egg-laying in the salamander, *Eurycea bislineata,* by pituitary transplants. *Amer. Mus. Nov.,* no. 396, pp. 1–3. 1930.

——, and J. A. Weber, *see* Pennsylvania. 1929.

Pike, Nicolas, Some notes on the life-history of the common newt. *Sci.,* 20 (1): 17–25. 1886.

Rafinesque, C. S., Erpetia: The reptiles. *Annals of Nature,* no. 1, p. 4. 1820.

Reed, H. D., A note on the coloration of *Plethodon cinereus.—Amer. Nat.,* 42: 460–464, 1 pl. 1908.

——, and A. H. Wright, The vertebrates of the Cayuga Lake Basin, New York. *Proc. Amer. Phil. Soc.,* 48 (193): 370–459, pls. 17–20. 1909.

Sherwood, W. L., The salamanders found in the vicinity of New York City, with notes upon extra-limital or allied species. *Proc. Linn. Soc. N. Y.,* no. 7, pp. 21–41. 1895.

Skinner, Alanson, Capture of an adult *Amblystoma punctata* on Staten Island. *Proc. Staten Island Assn. Arts and Sci.,* 5 (3–4): 98. 1916.

Smallwood, W. M., Notes on the food of some Onondaga Urodela. *Copeia,* no. 169, pp. 89–98. 1928.

Smith, B. G., *see* Wisconsin. 1911.

Uhlenhuth, Eduard, Die Morphologie und Physiologie der Salamander-Schildrüse. *Arch. f. Entwick. der Organismen,* 109 (5): 611–749, pls. 3–6, figs. 1–34. 1927.

——, and S. Schwartzbach, Anterior lobe substance, the thyroid stimulator. I.

Induces precocious metamorphosis. *Proc. Soc. Exper. Biol. and Med.,* 26: 149–151. 1928.
Waite, F. C., Specific name of *Necturus maculosus.—Amer. Nat.,* 41: 23–27. 1907.
Weber, J. A., Herpetological observations in the Adirondack Mountains, New York. *Copeia,* no. 169, pp. 106–112. 1928.
Wellborn, Vera, Beschreibung eines neuen molches der Gattung *Cryptobranchus.—Zool. Anz.,* 114 (1–2): 63–64. 1936.
Wilder, I. W., *see* Massachusetts. 1924.
——, *see* Massachusetts. 1924 a.
Wilmott, G. B., The salamanders of Staten Island, New York, in 1931. *Proc. Staten Island Inst. Arts and Sci.,* 6: 161–164. 1933.
Wright, A. H., Notes on the breeding habits of *Amblystoma punctatum.—Biol. Bul.,* 14 (4): 284–289. 1908.
——, Notes on Clemmys. *Proc. Biol. Soc. Wash.,* 31: 51–58, pl. 1. 1918.
——, Notes on the Muhlenbergs' turtle. *Copeia,* no. 52, pp. 5–7. 1918 a.
——, and A. A. Allen, The early breeding habits of *Amblystoma punctatum.—Amer. Nat.,* 43: 687–692. 1909.
——, and A. A. Allen, The fauna of Ithaca, New York: Salamanders. In *Field note-book of fishes, amphibians, reptiles and mammals.* Ithaca, 1913.
——, and J. M. Haber, The carnivorous habits of the purple salamander. *Copeia,* no. 105, pp. 31–32. 1922.
——, and Julia Moesel, The salamanders of Monroe and Wayne Counties, New York. *Copeia,* no. 72, pp. 63–64. 1919.
Anonymous, *Amer. Nat.,* 24 (282): 598. 1890.
——, *Proc. Nat. Sci. Assn. Staten Island,* 2: 47. 1890.
——, *Proc. Staten Island Assn. Arts and Sci.,* 2 (4): 227. 1910.
——, *Proc. Staten Island Inst. Arts and Sci.,* 2 (2–4): 156. 1924.
——, *62nd Ann. Rept. N. Y. State Mus., 1909,* 1: 99–100. 1909.
——, *State Mus. Bul.,* 158, p. 101. 1912.

## NORTH CAROLINA

Bailey, J. R., *see* Georgia. 1937.
Bishop, S. C., Notes on salamanders. *N. Y. State Mus. Bul.,* 253, pp. 87–102, pls. 1–3. 1924.
——, Records of some salamanders from North Carolina and Pennsylvania. *Copeia,* no. 139, pp. 9–12. 1925.
——, Notes on some amphibians and reptiles from the southeastern states, with a description of a new salamander from North Carolina. *Jour. Elisha Mitchell Sci. Soc.,* 43 (3–4): 153–170, pls. 23–26. 1928.
Brady, M. K., The habitat of *Stereochilus.—Copeia,* no. 2, p. 58. 1930.
Breder, C. M., Jr., and R. B. Breder, A list of fishes, amphibians and reptiles collected in Ashe County, North Carolina. *Zoologica,* 4 (1): 1–23, figs. 1–8. 1923.

Brimley, C. S., Batrachia found at Raleigh, North Carolina. *Amer. Nat.*, 30: 500–501. 1896.
——, The salamanders of North Carolina. *Jour. Elisha Mitchell Sci. Soc.*, 23 (4): 150–156. 1907.
——, Some notes on the zoology of Lake Ellis, Craven County, North Carolina, with special reference to herpetology. *Proc. Biol. Soc. Wash.*, 22: 129–137. 1909.
——, Notes on the salamanders of the North Carolina mountains, with descriptions of two new forms. *Proc. Biol. Soc. Wash.*, 25: 135–140, pls. 6–7. 1912.
——, List of reptiles and amphibians of North Carolina. *Jour. Elisha Mitchell Sci. Soc.*, 30: 195–206. 1915.
——, The two Raleigh *Amblystomae* compared. *Jour. Elisha Mitchell Sci. Soc.*, 32 (2): 45–46. 1916.
——, The two forms of red *Spelerpes* occurring at Raleigh, North Carolina. *Proc. Biol. Soc. Wash.*, 30: 87–88. 1917.
——, Brief comparison of the herpetological faunas of North Carolina and Virginia. *Jour. Elisha Mitchell Sci. Soc.*, 34 (3): 146–147. 1918.
——, Eliminations from and additions to the North Carolina list of reptiles and amphibians. *Jour. Elisha Mitchell Sci. Soc.*, 34 (3): 148–149. 1918 a.
——, Notes on *Amphiuma* and *Necturus.*—*Copeia,* no. 77, pp. 5–7. 1920.
——, Reproduction of the marbled salamander. *Copeia,* no. 80, p. 25. 1920 a.
——, The life history of the American newt. *Copeia,* no. 94, pp. 31–32. 1921.
——, Breeding dates of *Ambystoma maculatum* at Raleigh, North Carolina. *Copeia,* no. 93, pp. 26–27. 1921 a.
——, Herpetological notes from North Carolina. *Copeia,* no. 107, pp. 47–48. 1922.
——, Herpetological notes from North Carolina (II). *Copeia,* no. 109, pp. 63–64. 1922 a.
——, North Carolina herpetology. *Copeia,* no. 114, pp. 3–4. 1923.
——, The dwarf salamander at Raleigh, North Carolina. *Copeia,* no. 120, pp. 81–83. 1923 a.
——, The water dogs (*Necturus*) of North Carolina. *Jour. Elisha Mitchell Sci. Soc.*, 40 (3–4): 166–168. 1924.
——, Revised key and list of the amphibians and reptiles of North Carolina. *Jour. Elisha Mitchell Sci. Soc.*, 42: 75–93. 1926.
——, An apparently new salamander (*Plethodon clemsonae*) from South Carolina. *Copeia,* no. 164, pp. 73–75. 1927.
——, Some records of amphibians and reptiles from North Carolina. *Copeia,* no. 162, pp. 10–12. 1927 a.
——, Yellow-cheeked *Desmognathus* from Macon County, North Carolina. *Copeia,* no. 166, pp. 21–23. 1928.
——, The amphibians and reptiles of North Carolina. *Carolina Tips, Elon College, N. C.*, 2 (1–7): 1–4, 6–7, 10–11, 14–15, 18–19, 22–23, figs. 12–22; 3 (1–2): 2–3, 6–7, figs. 1–12; 4 (1): 2–3. 1939–1940.
——, Reptiles and amphibians of North Carolina. *Carolina Tips,* 4 (1): 2–3. 1941.
——, and W. B. Mabee, Reptiles, amphibians and fishes collected in eastern North Carolina in the autumn of 1923. *Copeia,* no. 139, pp. 14–16. 1925.

Brimley, C. S., and Franklin Sherman, Notes on the life-zones in North Carolina. *Jour. Elisha Mitchell Sci. Soc.*, 24 (1): 14–22. 1908.
Chadwick, C. S., Some notes on the burrows of *Plethodon metcalfi.—Copeia,* no. 1, p. 50. 1940.
Coker, W. C., Opportunities for biological work at Highlands and report of progress. *Highlands Mus. and Biol. Lab.*, pub. 4, pp. 1–16, 1 fig. 1939.
Cope, E. D., On some new and little known reptiles and fishes from the Austroriparian region. *Proc. Amer. Phil. Soc.*, 17 (100): 63–68. 1877.
Coues, Elliott, Notes on the natural history of Fort Macon, North Carolina, and vicinity (No. 1). *Proc. Acad. Nat. Sci. Phila.*, 23: 12–49. 1871.
——, and H. C. Yarrow, Notes on the natural history of Fort Macon, North Carolina, and vicinity (No. 4). *Proc. Acad. Nat. Sci. Phila. for 1878,* pp. 21–28. 1879.
Dunn, E. R., Two new salamanders of the genus *Desmognathus.—Proc. Biol. Soc. Wash.*, 29: 73–76. 1916.
——, Reptile and amphibian collections from the North Carolina mountains, with especial reference to salamanders. *Bul. Amer. Mus. Nat. Hist.*, 37: 593–634, pls. 57–61, figs. 1–7. 1917.
——, *see* Virginia. 1920.
——, A new specimen of *Leurognathus marmorata.—Copeia,* no. 127, pp. 31–32. 1924.
——, A new mountain race of *Desmognathus.—Copeia,* no. 164, pp. 84–86. 1927.
——, *see* Virginia. 1928.
Dury, Ralph, *see* Michigan. 1932.
Garman, Harrison, *Diemyctylus viridescens* var. *vittatus,* a new variety of the red-spotted *Triton.—Jour. Cin. Soc. Nat. Hist.*, 19 (2): 49–51, 3 figs. 1897.
Goin, C. J., *see* Virginia. 1938.
Gray, I. E., Amphibians and reptiles of the Duke Forest and vicinity. *Amer. Mid. Nat.*, 25 (3): 652–658. 1941.
Grobman, A. B., Variation of the salamander *Pseudotriton ruber.—Copeia,* no. 3, p. 179. 1941.
Harper, Francis, *see* Georgia. 1935.
——, *see above,* General works. 1940.
Hassler, W. G., *see* Tennessee. 1929.
Holl, F. J., New trematodes from the newt *Triturus viridescens.—Jour. Helm.*, 6 (3): 175–182, figs. 1–9. 1928.
——, A new trematode from the newt *Triturus viridescens.—Jour. Elisha Mitchell Sci. Soc.*, 43 (3–4): 181–183, pl. 27. 1928 a.
——, Two new nematode parasites. *Jour. Elisha Mitchell Sci. Soc.*, 43 (3–4): 184–186, pl. 28. 1928 b.
——, The ecology of certain fishes and amphibians, with special reference to their helminth and lingualid parasites. *Ecol. Monographs,* 2: 83–107, figs. 1–9. 1932.
King, Willis, *see* Tennessee. 1939.
Mittleman, M. B., and H. G. M. Jopson, *see* South Carolina. 1941.
Moore, J. P., *Leurognathus marmorata,* a new genus and species of salamander of

the family Desmognathidae. *Proc. Acad. Nat. Sci. Phila.,* 51: 316–323, pl. 14. 1899.

Myers, G. S., Amphibians and reptiles from Wilmington, North Carolina. *Copeia,* no. 131, pp. 59–62. 1924.

——, *see* Indiana. 1927.

Noble, G. K., The plethodontid salamanders: Some aspects of their evolution. *Amer. Mus. Nov.,* no. 249, pp. 1–26, figs. 1–10. 1927.

——, and C. H. Pope, *see* New York. 1929.

——, and L. B. Richards, Experiments on the egg-laying of salamanders. *Amer. Mus. Nov.,* no. 513, pp. 1–25, figs. 1–7. 1932.

Pope, C. H., Notes on North Carolina salamanders, with especial reference to the egg-laying habits of *Leurognathus* and *Desmognathus.—Amer. Mus. Nov.,* no. 153, pp. 1–15, figs. 1–2. 1924.

——, Some plethodontid salamanders from North Carolina and Kentucky, with the description of a new race of *Leurognathus.—Amer. Mus. Nov.,* no. 306, pp. 1–19, fig. 1. 1928.

Rankin, J. S., An ecological study of parasites of some North Carolina salamanders. *Ecol. Monog.,* 7 (2): 169–269, figs. 1–15. 1937.

Reinke, E. E., and C. S. Chadwick, The origin of the water drive in *Triturus viridescens.—Jour. Exp. Zool.,* 83 (2): 223–233. 1940.

Schmidt, K. P., Notes on the herpetology of North Carolina. *Jour. Elisha Mitchell Sci. Soc.,* 32 (1): 33–37. 1916.

Stejneger, Leonhard, Rediscovery of one of Holbrook's salamanders. *Proc. U.S. Nat. Mus.,* 26: 557–558. 1903.

——, A new salamander from North Carolina. *Proc. U.S. Nat. Mus.,* 30: 559–562, figs. 1–6. 1906.

Uhlenhuth, Eduard, *see* New York. 1927.

Unterstein, W., Bemerkungen zur Fortpflanzung von *Desmognathus fuscus* Raf. *Sitzungsbericht d. Ges. naturfor. Freunde zu Berlin,* nos. 8–10, pp. 245–249, figs. 1–3. 1929.

Viosca, Percy, Jr., *see* Louisiana. 1927.

Walker, C. F., Description of a new salamander from North Carolina. *Proc. Junior Soc. Nat. Sci.,* 2: 48–51. 1931.

Weller, W. H., A new salamander from the Great Smoky Mountain national park. *Proc. Junior Soc. Nat. Hist.,* 1 (7): 3–4. 1930.

——, Records of some reptiles and amphibians from Chimney Rock Camp, Chimney Rock, North Carolina, and vicinity. *Proc. Junior Soc. Nat. Sci.,* 1 (8–9): 9–12. 1930 a.

——, A preliminary list of the salamanders of the Great Smoky Mountains of North Carolina and Tennessee. *Proc. Junior Soc. Nat. Sci.,* 2 (1): 21–32. 1931.

#### NORTH DAKOTA

Osborn, H. L., A remarkable axolotl from North Dakota. *Amer. Nat.,* 34: 551–562, figs. 1–4. 1900.

Osborn, H. L., *see* Colorado. 1901.
Pope, T. E. B., Devils Lake, North Dakota. *U. S. Bureau Fish.*, doc. 634, pp. 1–22, pls. 1–3, map. 1908.
Young, R. T., *Cryptobranchus alleghaniensis, Larus atricilla* and *Larus marinus* in North Dakota. *Sci.*, n.s., 35 (895): 308–309. 1912.
——, The life of Devils Lake, North Dakota. *N. D. Biol. Station,* pp. 1–114, pls. 1–23, figs. 1–25. 1924.

## OHIO

Bishop, S. C., *see above,* General works. 1941.
Burt, C. E., and M. D. Burt, *see* Kentucky. 1929.
Cope, E. D., [Remarks on salamanders.] *Proc. Acad. Nat. Sci. Phila.*, 13: 123–124. 1861.
Dury, Ralph, Director's notes. *Proc. Junior Soc. Nat. Sci.,* 1 (8–9): 3. 1930.
——, *see* Indiana. 1932.
James, J. F., Catalogue of the mammals, birds, reptiles, batrachians and fishes in the collection of the Cincinnati Society of Natural History. *Jour. Cin. Soc. Nat. Hist.,* 10 (1): 34–48. 1887.
King, Willis, Ecological observations on *Ambystoma opacum.—Ohio Jour. Sci.,* 35 (1): 2–14, pl. 1. 1935.
Kirsch, P. H., A report upon investigations in the Maumee River basin during the summer of 1893. *Bul. U. S. Fish. Comm. for 1894,* 14 (20): 315–337. 1895.
Kirtland, J. P., A catalogue of the Mammalia, Birds, Reptiles, Fishes, Testacea, and Crustacea of Ohio. *2nd Ann. Rept. Geol. Surv. Ohio,* pp. 160–200. Batrachia, pp. 167–168. 1838.
Maximilian, A. P. (Prince of Wied), Part I. Travels in the interior of North America, 1832–1834. In R. G. Thwaites, *Early Western Travels, 1748–1846,* vol. 22. 1906.
Mittleman, M. B., Notes on salamanders of the genus *Gyrinophilus.—Proc. N. Eng. Zool. Club,* 20: 25–42, figs. 1–5, pls. 3–6. 1942.
Morse, Max, Ohio Batrachia in the zoological museum of the Ohio State Univ. *Ohio Nat.,* 1 (7): 114–115. 1901.
——, Salamanders taken at Sugar Grove, Ohio. *Ohio Nat.,* 2 (2): 164. 1901 a.
——, Batrachians and reptiles of Ohio. *Proc. Ohio State Acad. Sci.,* 4 (3): 95–144, 1 pl., figs. 1–3. 1904.
Myers, G. S., *see* Indiana. 1926.
Netting, M. G., *see* Pennsylvania. 1938.
——, *see* Pennsylvania. 1939.
——, and M. B. Mittleman, *see* West Virginia. 1938.
Okey, K. W., The relation of the profundus and gasserian ganglia in the embryo of the urodele, *Plethodon glutinosus.—Ohio Jour. Sci.,* 17 (2): 25–51, pls. 2–5. 1916.
Rafinesque, C. S., *see* New York. 1820.

Smith, Frank, Some additional data on the position of the sacrum in *Necturus.—Amer. Nat.*, 34: 635–638. 1900.
Smith, W H., Report on the amphibians and reptiles of Ohio. *Rept. Geol. Survey of Ohio*, 4: 629–734, figs. 1–8. 1882.
Walker, C. F., Some new amphibian records for Ohio. *Copeia,* no. 4, p. 224. 1933.
——, and Woodrow Goodpaster, The green salamander, *Aneides æneus,* in Ohio. *Copeia,* no. 3, p. 178. 1941.
Williamson, E. B., An abnormal salamander. *Ohio Nat.*, 2 (2): 141. 1901.

## OKLAHOMA

Blair, A. P., Records of the salamander *Typhlotriton.—Copeia,* no. 2, pp. 108–109. 1939.
——, *Ambystoma talpoideum* in Oklahoma. *Copeia,* no. 3, p. 184. 1941.
Burt, C. E., *see* Arkansas. 1935.
Chase, H. D., and A. P. Blair, Two new blind isopods from northeastern Oklahoma. *Amer. Mid. Nat.*, 18 (2): 220–222. 1937.
Cope, E. D., *see above,* General works. 1869.
——, *see* Missouri. 1894.
Dunn, E. R., and A. A. Heinze, *see* Arkansas. 1933.
Force, E. R., A preliminary check list of amphibians and reptiles of Tulsa County, Oklahoma. *Proc. Okla. Acad. Sci.*, 8: 78–79. 1928.
——, The amphibians and reptiles of Tulsa County, Oklahoma, and vicinity. *Copeia,* no. 2, pp. 25–39. 1930.
Moore, G. A., and W. J. Carter, A new record for *Eurycea lucifuga* in Oklahoma. *Copeia,* no. 1, p. 52. 1942.
——, and R. C. Hughes, A new plethodontid from eastern Oklahoma. *Amer. Mid. Nat.*, 22 (3): 696–699, figs. 1–2. 1939.
——, and R. C. Hughes, A new plethodont salamander from Oklahoma. *Copeia,* no. 3, pp. 139–142, figs. 1–2. 1941.
Ortenburger, A. I., A list of reptiles and amphibians from the Oklahoma panhandle. *Copeia,* no. 163, pp. 46–48. 1927.
——, A report on the amphibians and reptiles of Oklahoma. *Proc. Okla. Acad. Sci.*, 6 (1): 89–100. 1927 a.
——, *see* Arkansas. 1929.
——, Reptiles and amphibians from northeastern Oklahoma. *Copeia,* no. 170, pp. 26–28. 1929 a.
——, and Beryl Freeman, Notes on some reptiles and amphibians from western Oklahoma. *Pub. Univ. Okla. Biol. Surv.*, 2 (4): 175–188. 1930.
Trowbridge, A. H., New records of Amphibia for Oklahoma. *Copeia,* no. 1, pp. 71–72. 1937.
——, Ecological observations on amphibians and reptiles collected in southeastern Oklahoma during the summer of 1934. *Amer. Mid. Nat.*, 18 (2): 285–303. 1937 a.

## OREGON

Baird, S. F., Descriptions of four new species of North American salamanders, and one new species of scink. *Jour. Acad. Nat. Sci. Phila.,* 2nd ser., 1 (4): 292–294. 1850.

——, *see* California. 1859.

——, and Charles Girard, Descriptions of new species of reptiles, collected by the U.S. exploring expedition under the command of Capt. Charles Wilkes, U.S.N. *Proc. Acad. Nat. Sci. Phila. for 1852–1853,* 6: 174–177. 1853.

Bishop, S. C., Description of a new salamander from Oregon, with notes on related species. *Proc. Biol. Soc. Wash.,* 47: 169–171, 1 pl. 1934.

——, A remarkable new salamander from Oregon. *Herpetologica,* 1 (3): 93–95, pl. 9. 1937.

Chandler, A. C., The western newt or water-dog (*Notophthalmus torosus*) a natural enemy of mosquitoes. *Ore. Agr. Col. Exp. Sta.,* bul. 152, pp. 1–24, figs. 1–6. 1918.

Cooper, J. G., *Rept. expl. and surv. from Miss. River to Pacific Ocean,* 12 (2), pt. 4: 292–306, pl. 31, fig. 4. 1860.

Cope, E. D., *see above,* General works. 1869.

Diller, J. S., A salamander-snake fight. *Sci.,* n.s., 26 (678): 907–908. 1907.

Dunn, E. R., *see* Washington. 1926.

Evermann, B. W., U.S. Fish Commission investigations at Crater Lake. *Mazama,* 1 (2): 230–238, pl. 27. 1897.

Fitch, H. S., Amphibians and reptiles of the Rogue River basin, Oregon. *Amer. Mid. Nat.,* 17 (3): 634–652. 1936.

——, An older name for *Triturus similans* Twitty. *Copeia,* no. 3, pp. 148–149. 1938.

Girard, Charles, On a new genus and species of Urodela from the collections of the U.S. expl. exped., under the command of Capt. Charles Wilkes, U.S.N. *Proc. Acad. Nat. Sci. Phila.,* 8: 140–141. 1856.

——, *see* California. 1858.

Gordon, Kenneth, The Amphibia and Reptilia of Oregon. *Ore. State Mon. Studies in Zool.,* no. 1, pp. 1–82, figs. 1–54. 1931.

Graf, William, Records of amphibians and reptiles from Oregon. *Copeia,* no. 2, pp. 101–104. 1939.

Grinnell, Joseph, and C. L. Camp, *see* California. 1917.

Hardy, G. A., *see* British Columbia. 1926.

Jewett, S. G., Jr., Notes on the amphibians of the Portland, Oregon, area. *Copeia,* no. 1, pp. 71–72. 1936.

——, H. L. Gordon, L. E. Griffin, and Henry Wu, The Mullerian duct of the male *Triturus torosus.—Copeia,* no. 4, pp. 183–184. 1934.

Müller, F., *see* Texas. 1885.

Myers, G. S., *see* California. 1930.

——, Notes on Pacific Coast *Triturus.—Copeia,* no. 2, pp. 77–82. 1942.

Skilton, A. J., Description of two reptiles from Oregon. *Amer. Jour. Sci. Arts,* 2nd ser., 7: 202, 1 pl. 1849.

Slater, J. R., *Rhyacotriton olympicus* (Gaige) in northern Oregon and southern Washington. *Herpetologica,* 1 (5): 136. 1938.

——, *Plethodon dunni* in Oregon and Washington. *Herpetologica,* 1 (6): 154. 1939.

Stone, Witmer, On some collections of reptiles and batrachians from the western United States. *Proc. Acad. Nat. Sci. Phila.,* 63: 222–232. 1911.

Storer, T. I., *Ambystoma paroticum* in Oregon. *Copeia,* no. 151, p. 111. 1926.

Taylor, Alfred, The susceptibility of the newt *Triturus torosus* to its own poison. *Copeia,* no. 4, p. 183. 1934.

Van Denburgh, John, *see* California. 1916.

Wood, W. F., *see* California. 1940.

## PENNSYLVANIA

Baird, S. F., *see* Oregon. 1850.

Barton, B. S., Some account of a new species of North American lizard. *Trans. Amer. Phil. Soc.,* 6: 108–112, pl. 4, fig. 6. 1809.

Bishop, S. C., *see* North Carolina. 1925.

——, *see* New York. 1926.

Bumpus, H. C., *see* Rhode Island. 1886.

Burger, J. W., A preliminary list of the amphibians of Lebanon County, Pennsylvania, with notes on habits and life history. *Copeia,* no. 2, pp. 92–94. 1933.

——, *Plethodon cinereus* (Green) in eastern Pennsylvania and New Jersey. *Amer. Nat.,* 69 (725): 578–586. 1935.

Cochran, M. E., *see* Massachusetts. 1911.

Collins, H. H., Observations on the life history of the amphibian, *Triturus viridescens,* in western Pennsylvania. *Proc. Penn. Acad. Sci.,* 6: 164–165. 1932.

Conant, Roger, Amphibians and reptiles from Dutch Mountain (Pennsylvania) and vicinity. *Amer. Mid. Nat.,* 27 (1): 154–170. 1942.

Cope, E. D., *see above,* General works. 1859.

——, *see* Ohio. 1861.

——, A new species of *Eutania* from western Pennsylvania. *Amer. Nat.,* 26 (311): 964–965. 1892.

Dunn, E. R., Some amphibians and reptiles of Delaware County, Pennsylvania. *Copeia,* no. 16, pp. 2–4. 1915.

——, The transformation of *Spelerpes ruber* (Daudin). *Copeia,* no. 21, pp. 28–30. 1915 a.

——, The red salamander. *Aquatic Life,* 5 (11): 123–124, 1 fig. 1920.

Engelhardt, G. P., *see* New York. 1919.

Evermann, B. W., Notes on some reptiles and amphibians of Pike County, Pennsylvania. *Copeia,* no. 58, pp. 66–67. 1918.

Fankhauser, Gerhard, Triploidy in the newt, *Triturus viridescens.—Proc. Amer. Phil. Soc.,* 79 (4): 715–739, pls. 1–4, figs. 1–7. 1938.

Fowler, H. W., *Necturus maculosus* Rafinesque in the lower Delaware River. *Amer. Nat.,* n.s., 11 (275): 555. 1900.

——, Notes on Pennsylvania fishes. *Amer. Nat.,* 40 (476): 595–596. 1906.

Fowler, H. W., *Pimephales notatus* in the lower Susquehanna. *Amer. Nat.,* 40 (478): 743. 1906 a.

——, Note on the dusky salamander. *Proc. Acad. Nat. Sci. Phila.,* 58: 356–357, pl. 13, figs. 1–5. 1906 b.

——, An annotated list of the cold-blooded vertebrates of Delaware County, Pennsylvania. *Proc. Delaware Co. Inst. Sci.,* 7 (2): 33–45. 1915.

——, Some amphibians and reptiles from Bucks County, Pennsylvania. *Copeia,* no. 40, pp. 14–15. 1917.

——, and E. R. Dunn, *see above,* General works. 1917.

Green, H. T., *see* New York. 1923.

——, The egg laying of the purple salamander. *Copeia,* no. 141, p. 32. 1925.

Green, Jacob, An account of some new species of salamanders. *Contr. Maclur. Lyc.,* 1: 3–7, pl. 1, fig. 1. 1827.

Hallowell, Edward, [Urodeles in the neighborhood of Philadelphia.] *Proc. Acad. Nat. Sci. Phila. for 1856,* 8: 101. 1856.

Harlan, R., Description of a new species of *Salamandra.—Jour. Acad. Nat. Sci. Phila.,* 5 (1): 136. 1825.

——, Notice of a new species of salamander (inhabiting Pennsylvania). *Amer. Jour. Sci. and Arts.,* 10: 286–287. 1826.

Hay, O. P., *see* Indiana. 1892.

Hilsman, H. M., Notes on the life-history of the newt, *Triturus viridescens,* with a summary of experimental work on Amphibia. *Nawakwa Fireside,* n.s., no. 3–4, pp. 1–13. 1935.

Hudson, H. E., The distribution and habitat preference of the urodele amphibian *Triturus viridescens,* at Presque Isle, Erie, Pennsylvania. *Proc. Penn. Acad. Sci.,* 4: 55–58. 1930.

Keim, T. D., Notes on the fauna about the headwaters of the Allegheny, Genesee and Susquehanna Rivers in Pennsylvania. *Copeia,* no. 24, pp. 51–52. 1915.

Kelly, H. A., Identity of *Diemyctylus miniatus* with *Diemyctylus viridescens.— Amer. Nat.,* 12: 399. 1878.

Mattern, E. S., and W. I. Mattern, Amphibians and reptiles of Lehigh County, Pennsylvania. *Copeia,* no. 46, pp. 64–66. 1917.

Maximilian, A. P. (Prince of Wied), *see* Ohio. 1839.

Mohr, C. E., Habits of Amphibia in winter. *Proc. Penn. Acad. Sci.,* 3: 94–97. 1929.

——, The ambystomid salamanders of Pennsylvania. *Proc. Penn. Acad. Sci.,* 4: 50–55. 1930.

——, Observations on the early breeding habits of *Ambystoma jeffersonianum* in central Pennsylvania. *Copeia,* no. 3, pp. 102–104, figs. 1–2. 1931.

——, Notes on cave vertebrates. *Proc. Penn. Acad. Sci.,* 11: 38–42. 1937.

——, The amphibians of Berks County. *Proc. Penn. Acad. Sci.,* 13: 76–78. 1939.

Montgomery, T. H., Peculiarities of the terrestrial larva of the urodelous batrachian, *Plethodon cinereus* Green. *Proc. Acad. Nat. Sci. Phila.,* 53: 503–508. 1901.

Netting, M. G., A preliminary list of the amphibians and reptiles of Cook Forest, Clarion and Forest Counties, Pennsylvania. *Cardinal,* 2 (3): 65–67. 1928.

Netting, M. G., The food of the hellbender *Cryptobranchus alleganiensis* (Daudin). *Copeia,* no. 170, pp. 23–24. 1929.

——, The amphibians of Pennsylvania. *Proc. Penn. Acad. Sci.,* 7: 100–110. 1933.

——, A non-technical key to the amphibians and reptiles of western Pennsylvania. *Nawakwa Fireside,* n.s., no. 3–4, pp. 33–49. 1935.

——, Hand list of the amphibians and reptiles of Pennsylvania. *Carnegie Mus. Herpet., leaflet no. 1,* pp. 1–4. 1936.

——, The amphibians and reptiles of Indiana County, Pennsylvania. *Proc. Penn. Acad. Sci.,* 10: 25–28. 1936.

——, Wehrle's salamander, *Plethodon wehrlei* Fowler and Dunn, in Pennsylvania. *Proc. Penn. Acad. Sci.,* 10: 28–30. 1936.

——, The occurrence of the eastern tiger salamander, *Ambystoma tigrinum tigrinum* (Green), in Pennsylvania and near-by states. *Ann. Carnegie Mus.,* 27: 159–166. 1938.

——, Hand list of the amphibians and reptiles of Pennsylvania. *Bien. Rept. Penna. Fish. Comm.,* pp. 109–112. 1939.

——, The amphibians of Pennsylvania (2nd ed.). *Bien. Rept. Penna. Fish. Comm.,* pp. 113–122. 1939 a.

——, The ravine salamander, *Plethodon richmondi* Netting and Mittleman, in Pennsylvania. *Proc. Penn. Acad. Sci.,* 13: 50–51. 1939 b.

——, and M. B. Mittleman, *see* West Virginia.

Nigrelli, R. F., On the cytology and life-history of *Trypanosoma diemyctyli* and the polynuclear count of infested newts (*Triturus viridescens*).—*Amer. Micros. Soc.,* 48 (4): 366–387, pls. 45–46, fig. A. 1929.

Noble, G. K., and J. A. Weber, The spermatophores of *Desmognathus* and other plethodontid salamanders. *Amer. Mus. Nov.,* no. 351, pp. 1–15, figs. 1–4. 1929.

Pawling, R. O., The amphibians and reptiles of Union County, Pennsylvania. *Herpetologica,* 1 (6): 165–169. 1939.

Pope, P. H., and G. K. Noble, Purple salamander. *Copeia,* no. 100, pp. 79–80. 1921.

Reese, A. M., The sexual elements of the giant salamander, *Cryptobranchus allegheniensis.*—*Biol. Bul.,* 6 (5): 220–223. 1904.

Say, Thomas, In H. Long, *Account of an expedition from Pittsburgh to the Rocky Mountains,* 1: i–v, 1–503; 2: i–v, 1–442. Philadelphia: H. Long, 1823.

Smith, B. G., *see* Wisconsin. 1911.

Stewart, N. H., Some rare vertebrates of the Susquehanna Valley, Pennsylvania. *Proc. Penn. Acad. Sci.,* 2: 21–24. 1928.

Stone, Witmer, Notes on reptiles and batrachians of Pennsylvania, New Jersey and Delaware. *Amer. Nat.,* 40: 159–170. 1906.

Surface, H. A., The amphibians of Pennsylvania. *Bi-monthly Zool. Bul. Div. Zool. Pa. Dept. Agr.,* 3 (3–4): 65–152, pls. 1–11, figs. 1–25. 1913.

Townsend, C. H., Habits of the *Menopoma. Amer. Nat.,* 16 (2): 139–140. 1882.

Williams, S. H., Preliminary report on the animal ecology of Presque Isle, Lake Erie, Pennsylvania. *Proc. Penn. Acad. Sci.,* 5: 88–97, fig. 9. 1931.

Anonymous, Museum catalog: Sec. reptiles, batrachians. *Proc. Delaware Co. Inst. Sci.,* 7 (1): 13–18. 1914.

## RHODE ISLAND

Bumpus, H. C., Reptiles and batrachians of Rhode Island. *Random notes on natural history,* 3 (3): 21; 3 (5): 35; 3 (6): 43; 3 (7): 52; 3 (9): 69; 3 (10): 76; 3 (11): 83–84. Providence, R. I., 1886.

Drowne, F. P., Reptiles and batrachians of Rhode Island. *Monograph No. 15, Roger Williams Park Museum,* pp. 1–24. 1905.

Murphy, R. C., The food of *Plethodon cinereus.—Copeia,* no. 52, p. 8. 1918.

Anonymous, The batrachians of Rhode Island. *Park Mus. Bul., Roger Williams Park, Providence, R. I.,* 10 (3): 89–92. 1918.

## SOUTH CAROLINA

Barton, B. S., Von der Rathselhaften *Siren lacertina* Linne. *Voigt's Mag. für den neuesten Lustand der Naturkünde,* 12: 486–487. 1808.

Bishop, S. C., Records of some salamanders from South Carolina. *Copeia,* no. 161, pp. 187–188. 1927.

——, *see* North Carolina. 1928.

——, *see above,* General works. 1941.

Brimley, C. S., *see* North Carolina. 1927.

——, *see* North Carolina. 1939–1940.

Chamberlain, E. B., Some salamanders from Caesar's Head, South Carolina. *Copeia,* no. 167, pp. 51–52. 1928.

——, *Stereochilus marginatum* (Hallowell) from South Carolina. *Copeia,* no. 3, p. 88. 1930.

Cope, E. D., *see above,* General works. 1867.

Corrington, J. D., Herpetology of the Columbia, South Carolina, region. *Copeia,* no. 172, pp. 58–83. 1929.

Ellis, John, An account of an amphibious bipes. *Phil. Trans. Royal Soc. London, for 1766,* 56: 189–192, pl. 9. 1767.

Gee, N. G., South Carolina vertebrate fauna. (Salamanders.) An unpaged list printed in two columns, from the department of biology, Lander College, Greenwood, S. C., dated October 3, 1936.

Gibbes, L. R., Description of a new species of salamander. *Bost. Jour. Nat. Hist.,* 5 (1): 89–90, pl. 10. 1844.

——, On a new species of *Menobranchus,* from South Carolina. *Proc. Amer. Assn. Adv. Sci.,* 3: 159. 1850.

——, Description (with figure) of *Menobranchus punctatus.—Jour. Bost. Soc. Nat. Hist.,* 6 (3): 369–373, pl. 13. 1853.

Goin, C. J., *see* Virginia. 1938.

Harlan, Richard, *see above,* General works. 1825.

——, Description of a variety of the *Coluber fulvius,* Linn., a new species of *Scincus,* and two new species of *Salamandra.—Jour. Acad. Nat. Sci. Phila.,* 5 (1): 154–158. 1825 a.

Harlan, Richard, Description of a new species of *Salamandra. Jour. Acad. Nat. Sci. Phila.,* 6 (1): 101. 1828.

——, Description of a variety of the *Coluber fulvius;* of the *Scincus unicolor;* and two new species of *Salamandra,—S. cylindracea,* and *S. symmetrica.—Med. and Phys. Researches, Phila.,* pp. ix–xxxix, 9–653 (reprinted from *J.A.N.S. Phila., 1825*). 1835.

Harper, Francis, *see* Georgia. 1935.

Holbrook, J. E., *see above,* General works. 1838.

——, *see above,* General works. 1842.

Jopson, H. G. M., Reptiles and amphibians from Georgetown County, South Carolina. *Herpetologica,* 2 (2): 39–42. 1940.

Mittleman, M. B., and H. G. M. Jopson, A new salamander of the genus *Gyrinophilus* from the southern Appalachians. *Smiths. Miscell. Coll.,* 101 (2): 1–5, pl. 1. 1941.

Müller, F., Erster Nachtrag zum Katalog der herpetologischen Sammlung des Basler Museums. *Verhandlungen der Naturforschende Gesellschaft in Basel,* 7: 120–165, 1 pl. 1882.

Pickens, A. L., Amphibians of upper South Carolina. *Copeia,* no. 165, pp. 106–110. 1927.

Rea, P. M., *Notes from the Museum Bul.* [Charleston (S. C.) Museum], 6 (1): 35. 1909.

Schmidt, K. P., A list of amphibians and reptiles collected near Charleston, South Carolina. *Copeia,* no. 132, pp. 67–69. 1924.

Storer, H. R., *see* Maine. 1852.

True, F. W., A list of the vertebrate animals of South Carolina. In Harry Hammond, *South Carolina,* Chap. 10, pp. 209–264. Salamanders, pp. 241–243. Charleston, S. C.: State Board of Agriculture, 1883.

Viosca, Percy, Jr., *see* Louisiana. 1937.

Wright, A. H., *see* Alabama. 1935.

## SOUTH DAKOTA

Burt, C. E., *see* New York. 1932.

Over, W. H., Amphibians and reptiles of South Dakota. *S. Dakota Geol. and Nat. Hist. Surv.,* bul. 12, pp. 1–34, pls. 1–18. 1923.

Skinner, Alanson, *Ambystoma tigrinum* in South Dakota. *Copeia,* no. 12, pp. 3–4. 1914.

Visher, S. S., A preliminary list of the reptiles and amphibians of Harding County. *S. D. Geol. and Nat. Hist. Surv.,* bul. 6, pp. 92–93. 1914.

## TENNESSEE

Banta, A. M., and W. L. McAtee, *see* Indiana. 1906.

Bishop, S. C., *see* North Carolina. 1928.

——, *see above,* General works. 1941.

Blanchard, F. N., The amphibians and reptiles of western Tennessee. *Occ. Papers Mus. Zool. Univ. Mich.*, no. 117, pp. 1–18. 1922.

Blatchley, W. S., On a small collection of batrachians, with descriptions of two new species. *25th Ann. Rept. Ind. Dept. Geol. and Nat. Resources,* pp. 759–763. 1901.

Brimley, C. S., *see* North Carolina. 1939–1940.

Coker, W. C., *see* North Carolina. 1939.

Cope, E. D., The fauna of the Nickajack cave. In E. D. Cope and A. S. Packard, *Amer. Nat.,* 15 (11): 877–882, pl. 7. 1881.

Davison, Alvin, A contribution to the anatomy and phylogeny of *Amphiuma means* (Gardner). *Jour. Morph.,* 11 (2): 375–410, pls. 23–24. 1895.

Dunn, E. R., *see* North Carolina. 1916.

——, *see* Virginia. 1920.

Gentry, Glenn, Herpetological collections from counties in the vicinity of the Obey River drainage of Tennessee. *Jour. Tenn. Acad. Sci.,* 16 (3): 329–332. Reprinted as *Tenn. Dept. Cons.,* miscell. pub. no. 4, pp. 329–332. 1941.

——, Herpetological collections, 1939. In A limited biological survey of the Obey River and adjacent streams in Tennessee. *Rept. Reelfoot Lake Biol. Sta.,* 5: 74–75. See also *Jour. Tenn. Acad. Sci.,* 16 (1), 1941. See also *Tenn. Dept. Cons.,* miscell. pub. no. 3, 1941 (same account). 1941 a.

Gray, I. E., An extension of the range of *Plethodon yonahlossee.—Copeia,* no. 2, p. 106. 1939.

Hassler, W. G., Salamanders of the Great Smokies. *Nat. Hist.,* 29 (1): 95–99, 7 figs. 1929.

King, Willis, A new salamander (*Desmognathus*) from the southern Appalachians. *Herpetologica,* 1 (2): 57–60, pl. 5. 1936.

——, A survey of the herpetology of Great Smoky Mountains National Park. *Amer. Mid. Nat.,* 21 (3): 531–582, figs. 1–9, map. 1939.

McClure, G. W., The Great Smoky Mountains, with preliminary notes on the salamanders of Mt. Le Conte and Le Conte Creek. *Zoologica,* 11 (6): 53–72, figs. 74–80. 1931.

Necker, W. L., Contribution to the herpetology of the Smoky Mountains of Tennessee. *Bul. Chicago Acad. Sci.,* 5 (1): 1–4. 1934.

Packard, A. S., *see* Kentucky. 1888.

Parker, M. V., Some amphibians and reptiles from Reelfoot Lake. *Jour. Tenn. Acad. Sci.,* 12 (1): 60–86, figs. 1–18. 1937.

——, The amphibians and reptiles of Reelfoot Lake and vicinity, with a key for the separation of species and subspecies. *Rept. Reelfoot Lake Biol. Sta.,* 3: 72–101, figs. 1–14. See also *Jour. Tenn. Acad. Sci.,* 14 (1): 72–101. 1939.

Pope, C. H., *see* North Carolina. 1928.

Rhoads, S. N., Contributions to the zoology of Tennessee. No. 1, Reptiles and amphibians. *Proc. Acad. Nat. Sci. Phila.,* 47: 376–407. 1895.

Shoup, C. S., and J. H. Peyton, Collections from the drainage of the Big South Fork of the Cumberland River in Tennessee. *Tenn. Dept. Cons.,* miscell. pub. no. 2, pp. 106–116. 1940.

Troost, Gerard, List of reptiles inhabiting the State of Tennessee. *7th Geol. Rept. State Tenn.*, pp. 39–42. 1844.
Weller, W. H., Notes on *Aneides æneus,* Cope and Packard. *Proc. Junior Soc. Nat. Sci.,* 1 (1): 2–3. 1930.
——, *see* North Carolina. 1931.
Windsor, A. S., Salamanderin' in the Smokies. *Turtox News,* 9 (12): 97–99, 2 figs. 1931; 10 (1): 107–108, 2 figs. 1932. 1931–1932.

## TEXAS

Baird, S. F., Reptiles of the boundary. In *Rept. U.S. and Mexican Boundary Survey,* 2 (2): 1–35, pls. 1–41. 1859.
——, and Charles Girard, Characteristics of some new reptiles in the museum of the Smithsonian Institution. *Proc. Acad. Nat. Sci. Phila. for 1852,* 6: 173. 1853.
Bishop, S. C., *see above,* General works. 1941.
——, and M. R. Wright, A new neotenic salamander from Texas. *Proc. Biol. Soc. Wash.,* 50: 141–143, 1 fig. 1937.
Blackford, C. M., A curious salamander. *Nature,* 60: 389–390, figs. 1–2. 1899.
Boulenger, G. A., On a rare American newt, Molge meridionalis, Cope. *Ann. Mag. Nat. Hist.,* 6 Ser., 1: 24. 1888.
Burt, C. E., Contributions to Texan herpetology. VII. The salamanders. *Amer. Mid. Nat.,* 20 (2): 374–380. 1938.
Cope, E. D., On the zoological position of Texas. *Bul. U.S. Nat. Mus.,* no. 17, pp. 1–51. 1880.
——, Catalogue of Batrachia and Reptilia brought by William Taylor from San Diego, Texas. *Proc. U.S. Nat. Mus.,* 11: 395–398, pl. 36, fig. 2.
——, The Batrachia and Reptilia of northwestern Texas. *Proc. Acad. Nat. Sci. Phila.,* 44: 331–337. 1893.
Eigenmann, C. H., The eye of *Typhlomolge* from the artesian wells of San Marcos, Texas. *Proc. Ind. Acad. Sci.,* 8: 251. 1899.
——, *see* Missouri. 1900.
——, *see* Missouri. 1900 a.
——, *see* Missouri. 1900 b.
——, *see* Missouri, 1900 c.
Emerson, E. T., General anatomy of *Typhlomolge rathbuni.—Proc. Bost. Soc. Nat. Hist.,* 32 (3): 43–76, pls. 2–6. 1905.
Garman, Samuel, Reptiles and batrachians from Texas and Mexico. *Bul. Essex Inst.,* 19: 119–138. 1887.
Hallowell, Edward, *see* New Mexico. 1854.
Harwood, P. D., The helminths parasitic in the Amphibia and Reptilia of Houston, Texas, and vicinity. *Proc. U.S. Nat. Mus.,* 81: 1–71, pls. 1–5. 1932.
Matthes, Benno, *see above,* General works. 1855.
Mearns, E. A., *see* California. 1907.
Müller, F., *see* South Carolina. 1882.

Müller, F., Vierter Nachtrag zum Katalog der herpetologischen Sammlung des Basler Museums. *Verhandlungen der naturforschende Gesellschaft in Basel,* 7: 668–717, pls. 9–11. 1885.

Murray. L. T., Annotated list of amphibians and reptiles from the Chisos Mountains. *Contr. Baylor Univ. Mus.,* no. 24, pp. 1–16, 1 pl. 1939.

Noble, G. K., and B. C. Marshall, *see* Arkansas. 1932.

Norman, W. W., Remarks on the San Marcos salamander, *Typhlomolge rathbuni* Stejneger. *Amer. Nat.,* 34: 179–183, figs. 1–4. 1900.

Pope, P. H., Some notes on the amphibians of Houston, Texas. *Copeia,* no. 76, pp. 93–98. 1919.

Stejneger, Leonard, Description of a new genus and species of blind tailed batrachians from the subterranean waters of Texas. *Proc. U. S. Nat. Mus.,* 18: 619–621. See also: Review, Blind Batrachia and Crustacea from the subterranean waters of Texas. *Amer. Nat.,* 30 (354): 498–499. 1896.

Strecker, J. K., A preliminary report on the reptiles and batrachians of McLennan County, Texas. *Trans. Tex. Acad. Sci. for 1901,* 4 (2): 95–101. 1902.

——, The reptiles and batrachians of Victoria and Refugio Counties, Texas. *Proc. Biol. Soc. Wash.,* 21: 47–52. 1908.

——, A preliminary annotated list of the Batrachia of Texas. *Proc. Biol. Soc. Wash.,* 21: 53–62. 1908 a.

——, The reptiles and batrachians of McLennan County, Texas. *Proc. Biol. Soc. Wash.,* 21: 69–83. 1908 b.

——, Notes on the herpetology of Burnet County, Texas. *Baylor Univ. Bul.,* 12 (1): 1–9. 1909.

——, Notes on the Texan salamander (*Ambystoma texanum* Matthes). *Baylor Univ. Bul.,* 12 (1): 17–20. 1909 a.

——, Notes on the fauna of a portion of the canyon region of northwestern Texas. *Baylor Univ. Bul.,* 13 (4–5): 1–31, pl. 1, figs. 1–2. 1910.

——, Reptiles and amphibians of Texas. *Baylor Univ. Bul.,* 18 (4): 1–82. 1915.

——, An annotated catalogue of the amphibians and reptiles of Bexar County, Texas. *Sci. Soc. San Antonio,* bul. 4, pp. 1–31. 1922.

——, Notes on the herpetology of the east Texas timber belt. 1. Liberty County Amphibians and reptiles. *Contr. Baylor Univ. Mus.,* no. 3, pp. 1–3. 1926.

——, On the habits of some southern snakes. *Contr. Baylor Univ. Mus.,* no. 4, pp. 3–11. 1926 a.

——, Notes on the herpetology of the east Texas timber belt. 2. Henderson County amphibians and reptiles. *Contr. Baylor Univ. Mus.,* no. 3, pp. 1–3. 1926.

——, Observations on the food habits of Texas amphibians and reptiles. *Copeia,* no. 162, pp. 6–9. 1927.

——, Common English and folk names for Texas amphibians and reptiles. *Contr. Baylor Univ. Mus.,* no. 16, pp. 1–21. 1928.

——, Field notes on the herpetology of Wilbarger County, Texas. *Contr. Baylor Univ. Mus.,* no. 19, pp. 3–9. 1929.

——, A preliminary list of the amphibians and reptiles of Tarrant County, Texas. *Contr. Baylor Univ. Mus.,* no. 19, pp. 10–15. 1929 a.

Strecker, J. K., A catalogue of the amphibians and reptiles of Travis County, Texas. *Contr. Baylor Univ. Mus.*, no. 23, pp. 1–16. 1930.
——, *Siren lacertina* Linne in central Texas. *Baylor Bul.*, 38 (3): 30. 1935.
——, The reptiles of West Frio Canyon, Real County, Texas. *Baylor Bul.*, 38 (3): 32. 1935 a.
——, and J. E. Johnson, Notes on the herpetology of Wilson County, Texas. *Baylor Bul.*, 38 (3): 17–23. 1935.
——, and W. J. Williams, Herpetological records from the vicinity of San Marcos, Texas, with distributional data on the amphibians and reptiles of the Edwards Plateau region and central Texas. *Contr. Baylor Univ. Mus.*, no. 12, pp. 1–16. 1927.
——, and W. J. Williams, Field notes on the herpetology of Bowie County, Texas. *Contr. Baylor Univ. Mus.*, no. 17, pp. 1–19. 1928.
Test, F. C., Annotated list of the reptiles and batrachians collected in Missouri and Texas in the fall of 1891. *Bul. U. S. Fish. Comm. 1892*, 12: 121–122. 1894.
Uhlenhuth, Eduard, Observations on the distribution of the blind Texan cave salamander, *Typhlomolge rathbuni.—Copeia*, no. 69, pp. 26–27. 1919.
——, Observations on the distribution and habits of the blind Texan cave salamander, *Typhlomolge rathbuni.—Biol. Bul.*, 40 (2): 73–104, figs. 1–14. 1921.
Wright, A. H., and A. A. Wright, Amphibians of Texas. *Trans. Tex. Acad. Sci.*, 21: 1–38, pls. 1–3, figs. 1–6. 1938.

## UTAH

Burt, C. E. and M. D. Burt, Field notes and locality records on a collection of amphibians and reptiles, chiefly from the western half of the United States. *Jour. Wash. Acad. Sci.*, 19 (19–20): 428–460. 1929.
Engelhardt, G. P., Batrachians from southwestern Utah. *Copeia*, no. 60, pp. 77–80. 1918.
Hardy, Ross, An annotated list of reptiles and amphibians of Carbon County, Utah. *Utah Acad. Sci., Arts and Letters*, 15: 99–102. 1938.
Ruthven, A. G., Notes on the amphibians and reptiles of Utah. *Occ. Papers Mus. Zool. Univ. Mich.*, no. 243, pp. 1–4. 1932.
Tanner, V. M., An ecological study of Utah Amphibia. *Proc. Utah Acad. Sci.*, 5: 6–7, maps. 1927.
——, Distribution list of the amphibians and reptiles of Utah. *Copeia*, no. 163, pp. 54–58. 1927 a.
——, The amphibians and reptiles of Bryce Canyon National Park, Utah. *Copeia*, no. 2, pp. 41–43. 1930.
——, A synoptical study of Utah Amphibia. *Proc. Utah Acad. Sci.*, 8: 159–198, pls. 8–20. 1931.
Tanner, W. W., Notes on the herpetological specimens added to the Brigham Young University vertebrate collection during 1939. *Great Basin Nat.*, 1 (3–4): 138–146. 1940.

Telford, I. R., Histological aspects of the metamorphosis of *Ambystoma tigrinum* (Green). *Bul. Univ. Utah,* 25 (7): 3–13, pls. 1–3. 1935.
Uhlenhuth, Eduard, and S. Schwartzbach, *see* New York. 1928.
Van Denburgh, John, and J. R. Slevin, A list of the amphibians and reptiles of Utah, with notes on the species in the collection of the Academy. *Proc. Cal. Acad. Sci.,* 5 (4): 99–110, pls. 12–14. 1915.
Wolterstorff, Willy, Zur Systematik und Biologie der Urodelen Mexikos. *Abhandl. u. Ber. Mus. f. Nat.- und Heimatkunde u. Naturwiss. Ver. Magdeburg,* 6 (2): 129–149, figs. 1–13. 1930.

### VERMONT

Barnes, D. H., Note on the doubtful reptils. *Amer. Jour. Sci. and Arts,* 13: 66–70. 1828.
Burt, C. E., *see* New York. 1932.
Fowler, J. A., and H. J. Cole, Notes on some reptiles and amphibians from central Vermont. *Copeia,* no. 2, p. 93. 1938.
Loveridge, Arthur, Some herpetological records from Vermont. *Bul. Bost. Soc. Nat. Hist.,* no. 61, pp. 15–16. 1931.
Marshall, W. B., *see* New York. 1892.
Schneider, J. G., Historiae amphibiorum naturalis et literariae, pp. i–xiii, 1–264, pls. 1–2. Jena, 1799.
Thompson, Zadock, *History of Vermont, natural, civil and statistical* (1st ed.). Salamanders, pp. 123–127. 1842.
——, *History of Vermont, natural, civil, and statistical* (2nd ed.), pt. 1, pp. 1–224, figs. 1853.

### VIRGINIA

Bailey, J. R., *see* Georgia. 1937.
Bishop, S. C., *see above,* General works. 1941.
Brady, M. K., Eggs of *Desmognathus phoca* (Matthes). *Copeia,* no. 127, p. 29. 1924.
——, *Pseudotriton montanus* near Washington. *Copeia,* no. 130, pp. 54–55. 1924 a.
——, Notes on the reptiles and amphibians of the Dismal Swamp. *Copeia,* no. 162, pp. 26–29. 1927.
Brimley, C. S., *see* North Carolina. 1918.
——, *see* North Carolina. 1918 a.
——, *see* North Carolina. 1939–1940.
Drowne, F. P., A trip to Fauquier County, Virginia; with notes on the specimens obtained. *The Museum,* 6 (3): 38–45. Albion, N. Y., 1900.
Dunn, E. R., List of amphibians and reptiles observed in the summers of 1912, 1913, and 1914, in Nelson County, Virginia. *Copeia,* no. 18, pp. 5–7. 1915.
——, List of reptiles and amphibians from Clark County, Virginia. *Copeia,* no. 25, pp. 62–63. 1915 a.
——, Notes on Virginia herpetology. *Copeia,* no. 28, pp. 22–23. 1916.

Dunn, E. R., A preliminary list of the reptiles and amphibians of Virginia. *Copeia*, no. 53, pp. 16–27. 1918.
——, Two new Virginia records. *Copeia*, no. 77, p. 8. 1920.
——, Some reptiles and amphibians from Virginia, North Carolina, Tennessee and Alabama. *Proc. Biol. Soc. Wash.*, 33: 129–137. 1920 a.
——, The habitats of *Plethodontidae*.—*Amer. Nat.*, 62 (680): 236–248. 1928.
Fisher, A. K., *Spelerpes guttolineatus* Holbrook in the vicinity of Washington, D. C. *Amer. Nat.*, 21 (7): 672. 1887.
Fowler, H. W., *see* Maryland. 1925.
——, Records of amphibians and reptiles for Delaware, Maryland and Virginia. III. Virginia. *Copeia*, no. 146, pp. 65–67. 1925 a.
Goin, C. J., The status of *Amphiuma tridactylum* Cuvier. *Herpetologica*, 1 (5): 127–130. 1938.
Green, N. B., The pygmy salamander *Desmognathus wrighti* King, on White Top Mountain, Virginia. *Copeia*, no. 1, p. 49. 1939.
Mann, Charles, On the habits of a species of salamander, *Amblystoma opacum* Bd. *9th Ann. Rept. Smiths. Inst.*, pp. 294–295. 1855.
Mohr, C. E., *see* Pennsylvania. 1937.
Netting, M. G., *Desmognathus fuscus ochrophæus* in Virginia. *Copeia*, no. 2, p. 101. 1932.
——, *see* Pennsylvania. 1938.
——, and L. W. Wilson, Notes on amphibians from Rockingham County, Virginia. *Ann. Carnegie Mus.*, 28: 1–8. 1940.
Noble, G. K., and B. C. Marshall, *see* Arkansas. 1932.
Richmond, N. D., and C. J. Goin, Notes on a collection of amphibians and reptiles from New Kent County, Virginia. *Ann. Carnegie Mus.*, 27: 301–310. 1938.
Shufeldt, R. W., The slimy salamander. *Aquatic Life*, 3 (2): 25–26, 1 fig. 1917.
Smith, Hugh M., On the occurrence of *Amphiuma*, the so-called Congo snake, in Virginia. *Proc. U. S. Nat. Mus.*, 21: 379–380. 1899.
Viosca, Percy, Jr., Distributional problems of the cold-blooded vertebrates of the Gulf Coastal Plain. *Ecology*, 7 (3): 307–314. 1926.
Walker, C. F., *Plethodon welleri* at White Top Mountain, Virginia. *Copeia*, no. 4, p. 190. 1934.

### WASHINGTON

Blanchard, F. N., A collection of amphibians and reptiles from northeastern Washington. *Copeia*, no. 90, pp. 5–6. 1921.
Brown, W. C., and J. R. Slater, The amphibians and reptiles of the islands of the State of Washington. *Occ. Papers Dept. Biol. College of Puget Sound*, no. 4, pp. 6–31. 1939.
Dice, L. R., Distribution of the land vertebrates of southeastern Washington. *Univ. Cal. Pub. Zool.*, 16 (17): 293–348, pls. 24–26. 1916.
Dunn, E. R., Notes on two Pacific Coast *Ambystomidae*.—*Proc. New Eng. Zool. Club*, 7: 55–59, figs. 1–3. 1920.

Dunn, E. R., The status of *Siredon gracilis* Baird. *Copeia*, no. 154, pp. 135–136. 1926.
Eaton, T. H., *see* California. 1933.
Gaige, H. T., Description of a new salamander from Washington. *Occ. Papers Mus. Zool. Univ. Mich.*, no. 40, pp. 1–3, pl. 1. 1917.
Henry, W. V., and V. C. Twitty, *see* California. 1940.
Meek, S. E., Notes on a collection of cold-blooded vertebrates from the Olympic Mountains. *Field Columbian Mus., pub. 31, zool. ser.*, 1 (12): 225–236. 1899.
Myers, G. S., *see* California. 1930.
——, *see* Oregon. 1942.
Noble, G. K., and L. B. Richards, *see* North Carolina. 1932.
Ruthven, A. G., Two amphibians with Asiatic affinities from the State of Washington. *Copeia*, no. 53, p. 10. 1918.
Slater, J. R., *Ambystoma decorticatum* Cope rediscovered in Washington. *Copeia*, no. 3, p. 87. 1930.
——, Notes on Washington salamanders. *Copeia*, no. 1, p. 44. 1933.
——, Notes on northwestern amphibians. *Copeia*, no. 3, pp. 140–141. 1934.
——, *Ambystoma tigrinum* in the State of Washington. *Copeia*, no. 4, pp. 189–190. 1934 a.
——, Notes on *Ambystoma gracile* Baird and *Ambystoma macrodactylum* Baird. *Copeia*, no. 4, pp. 234–236. 1936.
——, Notes on the tiger salamander, *Ambystoma tigrinum*, in Washington and Idaho. *Herpetologica*, 1 (3): 81–83. 1937.
——, *see* Oregon. 1938; *also* 1939.
——, Some species of amphibians new to the State of Washington. *Occ. Papers Dept. Biol. College of Puget Sound*, no. 2, pp. 4–5. 1939 a.
——, and C. F. Brockman, Amphibians of Mt. Ranier National Park. *Mt. Rainier Nat. Park Nature Notes*, 14 (4): 111–138, pls. 1–4. 1936.
——, and W. C. Brown, Island records of amphibians and reptiles for Washington. *Occ. Papers Dept. Biol. College of Puget Sound*, no. 13, pp. 74–77. 1941.
Stejneger, Leonhard, The salamander genus *Ranodon* in North America. *Proc. Biol. Soc. Wash.*, 30: 123–124. 1917.
Svihla, Arthur, An extension of the range of *Dicamptodon ensatus*.—*Copeia*, no. 3, p. 143. 1931.
——, Extension of the ranges of some Washington Amphibia. *Copeia*, no. 1, p. 39. 1933.
——, and R. D. Svihla, Another record for the marbled salamander in Washington. *Copeia*, no. 1, p. 38. 1932.
——, and R. D. Svihla, Amphibians and reptiles of Whitman County, Washington. *Copeia*, no. 3, pp. 125–128. 1933.
Van Denburgh, John, Annotated list of reptiles and batrachians. In A report upon investigations in the Columbia River basin, with descriptions of four new species of fishes. *Bul. U. S. Fish Comm.*, 14: 169–207, pls. 16–21. 1895.
——, Description of a new species of the genus *Plethodon* (*Plethodon vandykei*)

from Mount Rainier, Washington. *Proc. Cal. Acad. Sci.*, 4 (4): 61–63 (exact reprint issued March 26, 1915). 1906.

——, Notes on *Ascaphus,* the discoglossoid toad of North America. *Proc. Cal. Acad. Sci.*, ser. 4, vol. 3, pp. 259–264. 1912.

#### WEST VIRGINIA

Banta, A M., and W. L. McAtee, *see* Indiana. 1906.

Bishop, S. C., *see above,* General works. 1941.

Bond, H. D., Some amphibians and reptiles of Monongalia County, West Virginia. *Copeia,* no. 2, pp. 53–54. 1931.

Dunn, E. R., *see* North Carolina. 1916.

Dury, Ralph, *see* Michigan. 1932.

Fowler, J. A., A note on the eggs of *Plethodon glutinosus.—Copeia,* no. 2, p. 133. 1940.

Green, N. B., *Cryptobranchus alleganiensis* in West Virginia. *Proc. W. Va. Acad. Sci.*, 7: 28–30. 1933.

——, Further notes on the food habits of the water dog, *Cryptobranchus alleganiensis* Daudin. *Proc. W. Va. Acad. Sci.*, 9: 36. 1935.

——, The amphibians of Tucker County. *Proc. W. Va. Acad. Sci.*, 10: 80–83. 1936.

——, The amphibians and reptiles of Randolph County, West Virginia. *Herpetologica,* 1 (4): 113–116. 1937.

——, The herpetological status of Randolph County, West Virginia. *Mag. of Hist. and Biography,* 9: 61–69. 1937 a.

——, A new salamander, *Plethodon nettingi,* from West Virginia. *Ann. Carnegie Mus.*, 27: 295–299. 1938.

——, Amphibians and reptiles of the Huntington region. *The Marshall Review,* 4 (2): 33–40. 1941.

——, Representatives of the genus *Gyrinophilus* in West Virginia. *Proc. W. Va. Acad. Sci.*, 15: 179–183. 1942.

——, and N. D. Richmond, Two amphibians new to the herpetofauna of West Virginia. *Copeia,* no. 2, p. 127. 1940.

Llewellyn, L. M., The amphibians and reptiles of Mineral County, West Virginia. *Proc. W. Va. Acad. Sci.*, 14: 148–150. 1940.

Mohr, C. E., *see* Pennsylvania. 1937.

Netting, M. G., The amphibians of West Virginia. *W. Va. Wild Life,* 11 (3–4): 5–6, 15. 1933.

——, Wehrle's salamander, *Plethodon wehrlei* Fowler and Dunn, in West Virginia. *Proc. W. Va. Acad. Sci.*, 10: 89–93. 1936.

——, *see* Pennsylvania. 1939.

——, and M. B. Mittleman, Description of *Plethodon richmondi,* a new salamander from West Virginia and Ohio. *Ann. Carnegie Mus.*, 27: 287–293, pl. 30. 1938.

——, and Neil Richmond, The green salamander, *Aneides æneus,* in northern West Virginia. *Copeia,* no. 2, pp. 101–102. 1932.

Noble, G. K., and Gertrude Evans, Observations and experiments on the life

history of the salamander, *Desmognathus fuscus fuscus* (Rafinesque). *Amer. Mus. Nov.*, no. 533, pp. 1–16. 1932.
Porter, H. C., A case of nearly perfect cauda bifida in *Triturus viridescens.—Proc. W. Va. Acad. Sci.*, 7: 33–34. 1933.
Reese, A. M., Variations in the vermilion-spotted newt, *D. viridescens.—Amer. Nat.*, 50: 316–319, figs. 1–24. 1916.
——, The fauna of West Virginia caves. *Proc. W. Va. Acad. Sci.*, 7: 39–53. 1933.
Richmond, N. D., and G. S. Boggess, Amphibians of Marion County, West Virginia. *Proc. W. Va. Acad. Sci.*, 12: 57–60. 1939.
Strader, L. D., Herpetology of the eastern panhandle of West Virginia. *Proc. W. Va. Acad. Sci.*, 9: 32–35. 1936.
Wright, A. H., and Harold Trapido, *Pseudotriton montanus montanus* in West Virginia. *Copeia*, no. 2, p. 133. 1940.

## WISCONSIN

Bishop, S. C., *see above*, General works. 1941.
Burt, C. E., *see* New York. 1932.
Cahn, A. R., The herpetology of Waukesha County, Wisconsin. *Copeia*, no. 170, pp. 4–8. 1929.
——, and Waldo Shumway, Color variations in larvae of *Necturus maculosus.—Copeia*, no. 151, pp. 106–107. 1926.
Higley, W. K., Reptilia and Batrachia of Wisconsin. *Trans. Wis. Acad. Sci., Arts and Letters*, 7: 155–176. 1889.
Hoy, P. R., On the *Amblystoma luridum*, a salamander inhabiting Wisconsin. *9th Ann. Rept. Smiths. Inst.*, p. 295. 1855.
——, Catalogue of the cold-blooded vertebrates of Wisconsin. *Geol. Surv. Wis., 1873–1879*, 1: 422–435. 1883.
Lapham, I. A., Fauna and flora of Wisconsin. *Trans. Wis. State. Agr. Soc. for 1852*, 2: 337–445. 1853.
Necker, W. L., *see* Illinois. 1939.
Pearse, A. S., Habits of the mud-puppy *Necturus*, an enemy of food fishes. *Bureau of Fisheries, Economic Circular*, no. 49, pp. 1–8. 1921.
Pope, T. E. B., Wisconsin herpetological notes. *Year Book, Public Mus. City of Milwaukee*, 8: 177–184. 1928.
——, Wisconsin herpetological notes. *Trans. Wis. Acad. Sci., Arts and Letters*, 25: 273–284. 1930.
——, Wisconsin herpetological notes. *Trans. Wis. Acad. Sci., Arts and Letters*, 26: 321–329. 1931.
——, and W. E. Dickinson, The amphibians and reptiles of Wisconsin. *Bul. Public Mus. City of Milwaukee*, 8 (1): 1–138, pls. 1–21, figs. 1–28. 1928.
Schmidt, F. J. W., List of the amphibians and reptiles of Worden Township, Clark County, Wisconsin. *Copeia*, no. 154, pp. 131–132. 1926.
Schmidt, K. P., and W. L. Necker, *see* Illinois. 1935.

Smith, B. G., The nests and larvae of *Necturus.—Biol. Bul.,* 20 (4): 191–200, figs. 1–7. 1911.
——, Notes on the natural history of *Ambystoma jeffersonianum, A. punctatum* and *A. tigrinum.—Bul. Wis. Nat. Hist. Soc.,* 9 (1–2): 14–27, pls. 1–3. 1911 a.

#### WYOMING

Carlin, W. E., Observations on *Siredon lichenoides.—Proc. U. S. Nat. Mus.,* 4: 120–121. See also review: Habits of the Rocky Mountain axolotl. *Amer. Nat.,* 15 (10): 810. 1881.
Cary, Merritt, Life zone investigations in Wyoming. *N. Amer. Fauna,* no. 42, pp. 1–95, pls. 1–15, figs. 1–17. 1917.
Marsh, O. C., Observations on the metamorphosis of *Siredon* into *Amblystoma.—Amer. Jour. Sci. and Arts.,* 2nd ser., 46 (138): 364-374, 1 pl. 1868.
——, Siredon, a larval salamander. *Amer. Nat.,* 2 (9): 493. 1868 a.
Rahn, Hermann, The axolotl, or water dog. *Wyoming Wild Life,* 6 (2): 12–16, 3 pls., 1 fig. 1941.

## Dominion of Canada

#### THE DOMINION

Logier, E. B. S., and G. C. Toner, Amphibians and reptiles of Canada. *Can. Field-Nat.,* 56 (2): 15–16. 1942.
Preble, E. A., A biological survey of the Hudson Bay region. Batrachians of Keewatin. *North Amer. Fauna,* no. 22, pp. 133–134. 1902.
Provancher, L'Abbé, Fauna canadienne: Les reptiles. *Le naturaliste canadien,* 7 (3): 65–73. 1875.
Anonymous, Biography of Robert Kennicott. *Trans. Chi. Acad. Sci.,* 1 (6): 133–226. 1869.
——, Les salamandres en Canada. *Le naturaliste canadien,* 2 (4): 119. 1870.

#### ALBERTA

Fowler, R. L., Some amphibians and reptiles of the district around High River, Alberta, 1933. *Can. Field-Nat.,* 48 (9): 139–140. 1934.
——, A note on the migration of the tiger salamander, *Ambystoma tigrinum.—Can. Field-Nat.,* 49 (3): 59–60. 1935.
Patch, C. L., Some amphibians of western North America. *Can. Field-Nat.,* 43 (6): 137–138. 1929.

#### BRITISH COLUMBIA

Bennett, W. H., *see* Ontario. 1937.
Carl, G. C., The red salamander (*Ensatina eschscholtzii* Gray) on Vancouver Island. *Copeia,* no. 2, p. 129. 1940.

——, The long-toed salamander on Vancouver Island. *Copeia,* no. 1, p. 56. 1942.
——, *see above,* General works. 1886.
——, A contribution to the herpetology of British Columbia. *Proc. Acad. Nat. Sci. Phila. for 1893,* pp. 181–184. 1894.
Cowan, I. M., A review of the reptiles and amphibians of British Columbia. *Rept. Prov. Mus. Nat. Hist. for 1936,* pp. K 16–25. 1937.
Dunn, E. R., *see* Alaska. 1930.
Fanning, J., A preliminary catalogue of the collections of natural history and ethnology in the provincial museum, Victoria, British Columbia, pp. 1–196. 1898.
Hardy, G. A., Amphibia of British Columbia. *Rept. Prov. Mus. Nat. Hist. for 1925,* pp. C 21–24. 1926.
——, Amphibia of British Columbia, additional notes and corrections. *Rept. Prov. Mus. Nat. Hist. for 1926,* pp. C 37–38. 1927.
——, Report on a collecting trip to Garibaldi Park, British Columbia. *Rept. Prov. Mus. Nat. Hist. for 1926,* pp. C 15–26. 1927.
——, Amphibia of British Columbia, additional notes and corrections. *Rept. Prov. Mus. Nat. Hist. for 1927,* p. E 17. 1928.
Hollister, N., List of reptiles and batrachians of the Alpine Club expedition to the Mount Robson region. *Can. Alpine Jour.,* special no., pp. 45–46. 1912.
Kermode, Francis, "Accessions." *Rept. Prov. Mus. Nat. Hist. for 1925,* p. C 35. 1926.
——, "Accessions." *Rept. Prov. Mus. Nat. Hist. for 1926,* p. C 39. 1927.
——, "Accessions." *Rept. Prov. Mus. Nat. Hist. for 1927,* pp. E 17–22. 1928.
——, "Accessions." *Rept. Prov. Mus. Nat. Hist. for 1930,* p. C 19. 1931.
Logier, E. B. S., Some account of the amphibians and reptiles of British Columbia. *Trans. Royal Can. Inst.,* 18 (2): 311–336. 1932.
Lord, J. K., The naturalist in Vancouver Island and British Columbia, Appendix, 2: 308–309. 1866.
Myers, G. S., *see* California. 1930.
——, *see* Oregon. 1942.
Newcombe, W. A., Accession Notes. *Rept. Prov. Mus. Nat. Hist. for 1931,* pp. B 8–16. 1932.
Patch, C. L., Some amphibians and reptiles from British Columbia. *Copeia,* no. 111, pp. 74–79. 1922.
——, *see* Alberta. 1929.
Slater, J. R., Salamander records from British Columbia. *Occ. Papers Dept. Zool. College Puget Sound,* no. 9, pp. 43–44. 1940.
Smith, G. M., The detailed anatomy of *Triturus torosus.—Trans. Royal Soc. Can.,* 21 (2): 452. 1927.
Van Denburgh, John, *see* California. 1916.
Watney, G. M. S., A new record of *Plethodon vehiculus* (Cooper) from Vancouver, British Columbia. *Copeia,* no. 2, p. 89. 1938.
——, Notes on the life history of *Ambystoma gracile* Baird. *Copeia,* no. 1, pp. 14–17. 1941.

### LABRADOR—NEWFOUNDLAND

Packard, A. S., Jr., List of vertebrates observed at Okak, Labrador, by Rev. Samuel Weiz, with annotations. *Proc. Bost. Soc. Nat. Hist.,* 10: 264–277. 1866.

### MANITOBA

Bennett, W. H., *see* Ontario. 1937.
Bird, R. D., The great horned owl in Manitoba. *Can. Field-Nat.,* 43 (4): 79–83. 1929.
Cox, Philip, Lizards and salamanders of Canada. *Proc. Miramichi Nat. Hist. Assn.,* no. 5, pp. 46–55. 1907.
Criddle, Norman, Additional notes on Manitoba turtles, snakes, and batrachians. *Ottawa Nat.,* 32 (7): 135. 1919.
Jackson, V. W., *A manual of vertebrates of Manitoba.* Pp. 1–48, 178 ills., 7 maps. Winnipeg, Canada, 1934.
Patch, C. L., and D. A. Stewart, The tiger salamander at Ninette, Manitoba. *Can. Field-Nat.,* 38 (5): 81–82, 1 fig. 1924.
Seton, E. T., A list of the turtles, snakes, and batrachians of Manitoba. *Ottawa Nat.,* 32 (5): 79–83. 1918.

### NEW BRUNSWICK

Cox, Philip, Batrachia of New Brunswick. *Bul. Nat. Hist. Soc. New Brunswick,* 4 (1): 64–66. 1898.
——, *see* Quebec. 1899.
——, *see* Quebec. 1899 a.
——, *see* Manitoba. 1907.

### NOVA SCOTIA

Cox, Philip, *see* Quebec 1899.
——, *see* Manitoba. 1907.
Jones, J. M., Contributions to the natural history of Nova Scotia. Reptilia. *Proc. and Trans. Nova Scotian Inst. Nat. Sci.,* 1 (3): 128–144, 1865.
MacKay, A. H., Batrachia and Reptilia of Nova Scotia. *Proc. and Trans. Nova Scotian Inst. Sci.,* 9 (2): xli–xliii. 1896.
Piers, Harry, Notes on Nova Scotian zoology, no. 2. *Proc. and Trans. Nova Scotian Inst. Sci.,* 2nd ser., 1 (2): 175–184. 1892.
——, Notes on Nova Scotian zoology, no. 4. *Proc. and Trans. Nova Scotian Inst. Sci.,* 9 (3): 255–267. 1897.
Storer, H. R., *see* Maine. 1852.

### ONTARIO

Allin, A. E., The vertebrate fauna of Darlington Township, Durham County, Ontario. *Trans. Roy. Can. Inst.,* 23 (49): 83–118. 1940.

Bennett, W. H., Notes on the care and habits of some interesting urodeles. *Can. Field-Nat.,* 51 (2): 17–20. 1937.
Brown, J. R., The herpetology of Hamilton, Ontario, and district. *Can. Field-Nat.,* 42 (5): 125–127. 1928.
Bull, W. P., From amphibians to reptiles. *Perkins Bul. Foundation,* pp. 1–89, 11 pls., 17 figs., 1 map, 1 chart. Toronto, 1938.
Coventry, A. F., Amphibia, Reptilia and Mammalia of the Temagami district, Ontario. *Can. Field-Nat.,* 45: 109–113. 1931.
——, Further notes on the Amphibia and Mammalia of the Temagami district, Ontario. *Can. Field-Nat.,* 46 (7): 147–149. 1932.
Cox, Philip, *see* Manitoba. 1907.
[Elliott, Robert], Extracts from the diary of the late Robert Elliott. *Ottawa Nat.,* 19 (9): 173–178. 1905.
——, Extracts from the diary of the late Robert Elliott. *Ottawa Nat.,* 20 (6): 120–126. 1906.
Garnier, J. H., List of Reptilia of Ontario. *Can. Sportsman and Nat.,* 1 (5): 37–39. 1881.
——, [Dr. Garnier on a new species of *Menobranchus.*]—*Proc. Can. Inst.,* ser. 3, 5: 218–219. 1888.
Groh, H., Salamanders lost, strayed or? *Can. Field-Nat.,* 38: 159. 1924.
G[roh], H., Excursions. *Ottawa Nat.,* 24 (4): 76. 1910.
Harrington, W. H., Mud puppies. *Ottawa Nat.,* 16 (11): 223–225. 1903.
Hodgins, J. G., Remarks on a Canadian specimen of the Proteus of the Lakes. *Can. Jour.,* n.s., 1 (1): 19–23. 1856.
Latchford, F. R., Notes on Ottawa salamanders. *Ottawa Nat.,* 1 (8): 105–107. 1887.
Logier, E. B. S., The amphibians and reptiles of the Lake Nipigon region. *Trans. Royal Can. Inst.,* 16 (2): 279–291. 1928.
——, A faunal investigation of King Township, York County, Ontario. IV. The amphibians and reptiles of King Township. *Trans. Royal Can. Inst.,* 17 (2): 203–208, pls. 2–4. 1930.
——, A faunal investigation of Long Point and vicinity, Norfolk County, Ontario. IV. The amphibians and reptiles of Long Point. *Trans. Royal Can. Inst.,* 18 (1): 229–236. 1931.
——, The salamanders of Ontario. *The School,* April, May, June. Ontario, Canada, 1937.
——, The amphibians of Ontario. *Handbook No. 3, Royal Ont. Mus. Zool.,* pp. 1–16, figs. 1–20. 1937 a.
——, The amphibians and reptiles of Prince Edward County, Ontario. *Univ. Toronto Studies,* biol. ser. 48, pp. 93–106. 1941.
Macallum, A. B., Studies on the blood of Amphibia. *Trans. Can. Inst.,* 2 (2): 221–260, 1 pl. 1892.
Meek, S. E., and D. G. Elliot, Notes on a collection of fishes and amphibians from Muskoka and Gull Lakes. *Field Columbian Mus., zool. ser.,* 1: 305–311. 1899.
——, and H. W. Clark, Notes on a collection of cold-blooded vertebrates from Ontario. *Field Columbian Museum, zool. ser.,* 3 (7): 131–140. 1902.

Montgomery, Henry, Some observations on the *Menobranchus maculatus.—Can. Nat.*, n.s., 9 (3): 160–164. 1879.

Müller, F., Sechster Nachtrag zum Katalog der herpetologischen Sammlung des Basler Museums. *Verhandlungen der Naturforschende Gesellschaft in Basel,* 8: 685–705, pl. 10. 1890.

Nash, C. W., Check list of the vertebrates of Ontario, and catalogue of the specimens of the biological section of the provincial museum. Batrachians, reptiles, and mammals, pp. 1–32. Toronto: Department of Education, 1905.

——, Batrachians and reptiles of Ontario. In *Vertebrates of Ontario,* pp. 5–18. Toronto: Department of Education, 1908.

Nicholson, H. A., Contributions to a Fauna Canadensis; being an account of the animals dredged in Lake Ontario in 1872. *Can. Jour.*, n.s., 13 (78): 490–506. 1873.

Odell, W. S., The two-lined salamander, *Spelerpes bilineatus* (Green). *Ottawa Nat.*, 14 (3): 53–55. 1900.

Patch, C. L., A list of amphibians and reptiles of the Ottawa, Ontario, district. *Ottawa Nat.*, 32 (3): 53. 1918.

Piersol, W. H., The habits and larval state of *Plethodon erythronotus.—Trans. Can. Inst.*, 8 (4): 469–493, figs. 1–14. 1910.

——, Spawn and larva of *Ambystoma jeffersonianum.—Amer. Nat.*, 44 (582): 732–738, figs. 1–4. 1910 a.

——, Amphibia. Chap. 18 in *The natural history of the Toronto region, Ontario, Canada.* Pp. 1–419, 7 pls., 5 maps. Toronto: Canadian Institute, 1913.

Small, H. B., and W. P. Lett, Report of the zoological branch. *Ottawa Field-Nat. Club Trans.*, 5, 2 (1): 148–151. 1884.

Toner, G. C., and W. E. Edwards, Cold-blooded vertebrates of Grippen Lake, Leeds County, Ontario. *Can. Field-Nat.*, 52 (3): 40–43. 1938.

——, and N. de St. Remy, Amphibians of eastern Ontario. *Copeia,* no. 1, pp. 10–13. 1941.

Wilson, A. W. G., Geology of the Nipigon basin, Ontario. *Memoir 1, Dept. Mines, Geol. Surv. Br. Canada,* pp. 1–152, pls. 1–16, figs. 1–4, map of Lake Nipigon. 1910.

Wright, A. H., and S. E. R. Simpson, The vertebrates of the Otter Lake region, Dorset, Ontario. III. The bactrachians and reptiles. *Can. Field-Nat.*, 34 (8): 141–145. 1920.

Anonymous, The hand-book of Toronto. Pp. i–viii, 9–272, frontispiece. Toronto: Lovell and Gibson, 1858.

#### PRINCE EDWARD ISLAND

Cox, Philip, *see* Quebec. 1899 a.

——, *see* Manitoba. 1907.

**QUEBEC**

Ball, S. C., Amphibians of Gaspé County, Quebec. *Copeia,* no. 4, p. 230. 1937.

Cox, Philip, Freshwater fishes and batrachians of the peninsula of Gaspé, P. Q., and their distribution in the Maritime Provinces. *Trans. Royal Soc. Canada,* 2nd ser., 5 (4): 141–154. 1899.

——, Preliminary list of the Batrachia of the Gaspé Peninsula and the Maritime Provinces. *Ottawa Nat.,* 13 (8): 194–195. 1899 a.

——, *see* Manitoba. 1907.

Huard, A., Une salamandre (*Amblystoma*) nouvelle dans la province de Quebec. *Le naturaliste canadien,* 29: 33–35. 1902.

Small, H. B., and W. P. Lett, *Report of the zoological branch, Ottawa Field-Nat. Club.* Trans. no. 5, 2 (1): 148–151. 1884.

——, and W. P. Lett, *Report of the zoological branch.* Trans. no. 6, 2 (2): 280–283. 1885.

Sternberg, C. M., Notes on the feeding habits of two salamanders in captivity. *Ottawa Nat.,* 30: 129–130. 1917.

Trapido, Harold, and R. T. Clausen, Amphibians and reptiles of eastern Quebec. *Copeia,* no. 3, pp. 117–125. 1938.

——, and R. T. Clausen, The larvae of *Eurycea bislineata major.—Copeia,* no. 4, pp. 244–246, fig. 1. 1940.

Anonymous, La salamandre saumonée. *Le Nat. Canadien,* 10 (7): 221–222. 1878.

**SASKATCHEWAN**

Dunn, E. R., *see above,* General works. 1940.

# Baja California

Dunn, E. R., *see* California. 1922.

Gadow, Hans, The distribution of Mexican amphibians and reptiles. *Proc. Zool. Soc. London,* 2: 191–244, figs. 29–32, pls. 6–7. 1905.

Myers, G. S., *see* California. 1942.

Nelson, E. W., Lower California and its natural resources. *Mem. Nat. Acad. Sci.,* 16: 1–194, pls. 1–34, maps. Amphibians, p. 113. 1922.

Schmidt, K. P., The amphibians and reptiles of Lower California and the neighboring islands. *Bul. Amer. Mus. Nat. Hist.,* 46: 607–707, pls. 47–57, figs. 1–13. 1922.

Van Denburgh, John, A review of the herpetology of Lower California. Part 2, Batrachians. *Proc. Cal. Acad. Sci.,* 2nd ser., 5: 556–561. 1895.

——, and J. R. Slevin, A list of the amphibians and reptiles of the peninsula of Lower California, with notes on the species in the collection of the Academy. *Proc. Cal. Acad. Sci.,* 4th ser., 11 (4): 49–72. 1921.

# INDEX

[*Please Note.*—The numbers all refer to pages. The boldface numbers refer to the Accounts of Species. Within each specific account, in a regular order, will be found the common name and the scientific name, reference to the appropriate figure and map, and the details of type locality, range, habitat, size, description, color, breeding habits, etc.]

Despite their abundance in many parts of North America, salamanders have generally been neglected by all but a few specialists. In this book Sherman C. Bishop discusses in a lively yet authoritative manner the 126 species and subspecies of salamanders that are known to exist in the United States, Canada, and Lower California.

Group by group, salamanders are described in accounts which give the common name, technical name, type of locality, range, habitat, size, anatomical characteristics, color, breeding habits, and relationships—all in a uniform plan of arrangement which makes the Handbook especially convenient for use in connection with the study of living animals or of laboratory specimens. Numerous keys for identification are provided which are designed to aid the amateur naturalist as well as the herpetologist.

The descriptions of salamanders